Student Solutions Manual

Laurel Technical Services

Introductory Algebra

K. Elayn Martin-Gay

PRENTICE HALL, Upper Saddle River, NJ 07458

Senior Editor: Karin Wagner
Supplement Editor: Kate Marks
Special Projects Manager: Barbara A. Murray
Production Editor: Michele Wells
Supplement Cover Manager: Paul Gourhan
Supplement Cover Designer: Liz Nemeth
Manufacturing Buyer: Alan Fischer

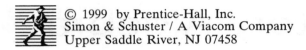

© 1999 by Prentice-Hall, Inc.
Simon & Schuster / A Viacom Company
Upper Saddle River, NJ 07458

Printed in the United States of America

10 9 8 7 6 5 4 3 2

ISBN   0-13-862525-5

Prentice-Hall International (UK) Limited, *London*
Prentice-Hall of Australia Pty. Limited, *Sydney*
Prentice-Hall Canada, Inc., *London*
Prentice-Hall Hispanoamericana, S.A., *Mexico*
Prentice-Hall of India Private Limited, *New Delhi*
Prentice-Hall of Japan, Inc., *Tokyo*
Simon & Schuster Asia Pte. Ltd., *Singapore*
Editora Prentice-Hall do Brazil, Ltda., *Rio de Janeiro*

Table of Contents

Chapter R

Pretest

1. $12 = 1 \cdot 12,\ 12 = 2 \cdot 6,\ 12 = 3 \cdot 4$
The factors of 12 are 1, 2, 3 4, 6, and 12.

2. $150 = 2 \cdot 3 \cdot 5 \cdot 5$

3. $8 = 2 \cdot 2 \cdot 2$
$14 = 2 \cdot 7$
$20 = 2 \cdot 2 \cdot 5$
$\text{LCM} = 2 \cdot 2 \cdot 2 \cdot 5 \cdot 7 = 280$

4. $\dfrac{7}{8} = \dfrac{7 \cdot 5}{8 \cdot 5} = \dfrac{35}{40}$

5. $\dfrac{24}{40} = \dfrac{2 \cdot 2 \cdot 2 \cdot 3}{2 \cdot 2 \cdot 2 \cdot 5} = \dfrac{3}{5}$

6. $\dfrac{120}{250} = \dfrac{2 \cdot 2 \cdot 2 \cdot 3 \cdot 5}{2 \cdot 5 \cdot 5 \cdot 5} = \dfrac{2 \cdot 2 \cdot 3}{5 \cdot 5} = \dfrac{12}{25}$

7. $\dfrac{2}{9} \cdot \dfrac{3}{8} = \dfrac{2 \cdot 3}{3 \cdot 3 \cdot 2 \cdot 2 \cdot 2} = \dfrac{1}{3 \cdot 2 \cdot 2} = \dfrac{1}{12}$

8. $\dfrac{1}{4} + \dfrac{5}{6} = \dfrac{1 \cdot 3}{4 \cdot 3} + \dfrac{5 \cdot 2}{6 \cdot 2} = \dfrac{3}{12} + \dfrac{10}{12} = \dfrac{3 + 10}{12} = \dfrac{13}{12}$

9. $\dfrac{3}{7} \div \dfrac{7}{10} = \dfrac{3}{7} \cdot \dfrac{10}{7} = \dfrac{3 \cdot 10}{7 \cdot 7} = \dfrac{30}{49}$

10. $\dfrac{2}{3} - \dfrac{5}{9} = \dfrac{2 \cdot 3}{3 \cdot 3} - \dfrac{5}{9} = \dfrac{6}{9} - \dfrac{5}{9}$
$= \dfrac{6 - 5}{9} = \dfrac{1}{9}$

11.
```
  76
  0.5
+ 2.03
─────
78.53
```

12.
```
 18.00
-12.67
──────
 5.33
```

13.
```
  12.8
× 0.19
──────
 1152
  128
──────
2.432
```

14.
```
        34.9
75) 2617.5
    -225
    ────
    367
    -300
    ────
     675
    -675
    ────
      0
```

15. $7.16 = \dfrac{716}{100}$

16.
```
       0.1875
16) 3.0000
   -1 6
   ────
    1 40
   -1 28
   ─────
     120
    -112
    ────
      80
     -80
     ───
       0
```

$\dfrac{3}{16} = 0.1875$

17.
```
      0.8333
6) 5.0000
  -48
  ───
   20
  -18
  ───
   20
  -18
  ───
   20
  -18
  ───
    2
```

This pattern will continue so that
$\dfrac{5}{6} = 0.8333....$
$\dfrac{5}{6} = 0.8\overline{3}$

18. 78.6

19. 78.62

20. $80.6\% = 0.806$

21. $0.3 = 30\%$

1

Section R.1

Practice Problems

1. $4 = 1 \cdot 4, 4 = 2 \cdot 2$
   The factors of 4 are 1, 2, and 4.

2. $18 = 1 \cdot 18, 18 = 2 \cdot 9, 18 = 3 \cdot 6$
   The factors of 18 are 1, 2, 3, 6, 9, and 18.

3. 5 and 11 are prime numbers. The factors of 5 are 1 and 5 only. The factors of 11 are 1 and 11 only.
   6 and 18 are composite numbers. The factors of 6 are 1, 2, 3, and 6. The factors of 18 are 1, 2, 3, 6, and 18.

4. $28 = 2 \cdot 2 \cdot 7$

5. $60 = 2 \cdot 2 \cdot 3 \cdot 5$

6. $297 = 3 \cdot 3 \cdot 3 \cdot 11$

7. $14 = 2 \cdot 7$
   $35 = 5 \cdot 7$
   $\text{LCM} = 2 \cdot 5 \cdot 7 = 70$

8. $5 = 5$
   $9 = 3 \cdot 3$
   $\text{LCM} = 3 \cdot 3 \cdot 5 = 45$

9. $4 = 2 \cdot 2$
   $15 = 3 \cdot 5$
   $10 = 2 \cdot 5$
   $\text{LCM} = 2 \cdot 2 \cdot 3 \cdot 5 = 60$

Exercise Set R.1

1. $9 = 1 \cdot 9, 9 = 3 \cdot 3$
   The factors of 9 are 1, 3, and 9.

3. $24 = 1 \cdot 24, 24 = 2 \cdot 12, 24 = 3 \cdot 8, 24 = 4 \cdot 6$
   The factors of 24 are 1, 2, 3, 4, 6, 8, 12, and 24.

5. $42 = 1 \cdot 42, 42 = 21 \cdot 2, 42 = 3 \cdot 14,$
   $42 = 6 \cdot 7$
   The factors of 42 are 1, 2, 3, 6, 7, 14, 21, and 42.

7. $80 = 1 \cdot 80, 80 = 2 \cdot 40, 80 = 4 \cdot 20,$
   $80 = 5 \cdot 16, 80 = 8 \cdot 10$
   The factors of 80 are 1, 2, 4, 5, 8, 10, 16, 20, 40, and 80.

9. 13 is a prime number. Its factors are 1 and 13 only.

11. 21 is a composite number. Its factors are 1, 3, 7, and 21.

13. 37 is a prime number. Its factors are 1 and 37 only.

15. 2065 is a composite number. Its factors are 1, 5, 413, and 2065.

17. $18 = 2 \cdot 3 \cdot 3$

19. $20 = 2 \cdot 2 \cdot 5$

21. $56 = 2 \cdot 2 \cdot 2 \cdot 7$

23. $300 = 2 \cdot 2 \cdot 3 \cdot 5 \cdot 5$

25. $81 = 3 \cdot 3 \cdot 3 \cdot 3$

27. $588 = 2 \cdot 2 \cdot 3 \cdot 7 \cdot 7$

29. $6 = 2 \cdot 3$
    $14 = 2 \cdot 7$
    $\text{LCM} = 2 \cdot 3 \cdot 7 = 42$

31. $3 = 3$
    $4 = 2 \cdot 2$
    $\text{LCM} = 2 \cdot 2 \cdot 3 = 12$

33. $20 = 2 \cdot 2 \cdot 5$
    $30 = 2 \cdot 3 \cdot 5$
    $\text{LCM} = 2 \cdot 2 \cdot 3 \cdot 5 = 60$

35. $5 = 5$
    $7 = 7$
    $\text{LCM} = 5 \cdot 7 = 35$

37. $6 = 2 \cdot 3$
    $12 = 2 \cdot 2 \cdot 3$
    $\text{LCM} = 2 \cdot 2 \cdot 3 = 12$

39. $12 = 2 \cdot 2 \cdot 3$
    $20 = 2 \cdot 2 \cdot 5$
    $\text{LCM} = 2 \cdot 2 \cdot 3 \cdot 5 = 60$

41. $50 = 2 \cdot 5 \cdot 5$
$70 = 2 \cdot 5 \cdot 7$
LCM $= 2 \cdot 5 \cdot 5 \cdot 7 = 350$

43. $24 = 2 \cdot 2 \cdot 2 \cdot 3$
$36 = 2 \cdot 2 \cdot 3 \cdot 3$
LCM $= 2 \cdot 2 \cdot 2 \cdot 3 \cdot 3 = 72$

45. $5 = 5$
$10 = 2 \cdot 5$
$12 = 2 \cdot 2 \cdot 3$
LCM $= 2 \cdot 2 \cdot 3 \cdot 5 = 60$

47. $2 = 2$
$3 = 3$
$5 = 5$
LCM $= 2 \cdot 3 \cdot 5 = 30$

49. $8 = 2 \cdot 2 \cdot 2$
$18 = 2 \cdot 3 \cdot 3$
$30 = 2 \cdot 3 \cdot 5$
LCM $= 2 \cdot 2 \cdot 2 \cdot 3 \cdot 3 \cdot 5 = 360$

51. $4 = 2 \cdot 2$
$8 = 2 \cdot 2 \cdot 2$
$24 = 2 \cdot 2 \cdot 2 \cdot 3$
LCM $= 2 \cdot 2 \cdot 2 \cdot 3 = 24$

53. $315 = 3 \cdot 3 \cdot 5 \cdot 7$
$504 = 2 \cdot 2 \cdot 2 \cdot 3 \cdot 3 \cdot 7$
LCM $= 2 \cdot 2 \cdot 2 \cdot 3 \cdot 3 \cdot 5 \cdot 7 = 2520$

55. every $5 \cdot 7 = 35$ days

Section R.2

Practice Problems

1. $\dfrac{1}{4} = \dfrac{1 \cdot 5}{4 \cdot 5} = \dfrac{5}{20}$

2. $\dfrac{20}{35} = \dfrac{2 \cdot 2 \cdot 5}{5 \cdot 7} = \dfrac{2 \cdot 2}{7} = \dfrac{4}{7}$

3. $\dfrac{7}{20}$ is already in lowest terms.

4. $\dfrac{12}{40} = \dfrac{2 \cdot 2 \cdot 3}{2 \cdot 2 \cdot 2 \cdot 5} = \dfrac{3}{2 \cdot 5} = \dfrac{3}{10}$

5. $\dfrac{4}{4} = 4 \div 4 = 1$

6. $\dfrac{9}{3} = 9 \div 3 = 3$

7. $\dfrac{10}{10} = 10 \div 10 = 1$

8. $\dfrac{5}{1} = 5 \div 1 = 5$

9. $\dfrac{3}{7} \cdot \dfrac{3}{5} = \dfrac{3 \cdot 3}{7 \cdot 5} = \dfrac{9}{35}$

10. $\dfrac{2}{9} \div \dfrac{3}{4} = \dfrac{2}{9} \cdot \dfrac{4}{3} = \dfrac{2 \cdot 4}{9 \cdot 3} = \dfrac{8}{27}$

11. $\dfrac{8}{11} \div 24 = \dfrac{8}{11} \div \dfrac{24}{1} = \dfrac{8}{11} \cdot \dfrac{1}{24} = \dfrac{8 \cdot 1}{11 \cdot 3 \cdot 8}$
$= \dfrac{1}{11 \cdot 3} = \dfrac{1}{33}$

12. $\dfrac{5}{4} \div \dfrac{5}{8} = \dfrac{5}{4} \cdot \dfrac{8}{5} = \dfrac{5 \cdot 2 \cdot 4}{4 \cdot 5} = \dfrac{2}{1} = 2$

13. $\dfrac{2}{11} + \dfrac{5}{11} = \dfrac{2 + 5}{11} = \dfrac{7}{11}$

14. $\dfrac{1}{8} + \dfrac{3}{8} = \dfrac{1 + 3}{8} = \dfrac{4}{8} = \dfrac{4}{2 \cdot 4} = \dfrac{1}{2}$

15. $\dfrac{13}{10} - \dfrac{3}{10} = \dfrac{13 - 3}{10} = \dfrac{10}{10} = 1$

16. $\dfrac{7}{6} - \dfrac{2}{6} = \dfrac{7 - 2}{6} = \dfrac{5}{6}$

17. $\dfrac{3}{8} + \dfrac{1}{20} = \dfrac{3 \cdot 5}{8 \cdot 5} + \dfrac{1 \cdot 2}{20 \cdot 2} = \dfrac{15}{40} + \dfrac{2}{40}$
$= \dfrac{17}{40}$

18. $\dfrac{8}{15} - \dfrac{1}{3} = \dfrac{8}{15} - \dfrac{1 \cdot 5}{3 \cdot 5} = \dfrac{8}{15} - \dfrac{5}{15} = \dfrac{8 - 5}{15} = \dfrac{3}{15}$
$= \dfrac{3}{3 \cdot 5} = \dfrac{1}{5}$

Exercise Set R.2

1. $\dfrac{7}{10} = \dfrac{7 \cdot 3}{10 \cdot 3} = \dfrac{21}{30}$

3. $\dfrac{2}{9} = \dfrac{2 \cdot 2}{9 \cdot 2} = \dfrac{4}{18}$

5. $\dfrac{4}{5} = \dfrac{4 \cdot 4}{5 \cdot 4} = \dfrac{16}{20}$

7. $\dfrac{2}{2 \cdot 2} = \dfrac{1}{2}$

9. $\dfrac{10}{15} = \dfrac{2 \cdot 5}{3 \cdot 5} = \dfrac{2}{3}$

11. $\dfrac{3}{7} = \dfrac{3}{7}$

13. $\dfrac{20}{20} = \dfrac{2 \cdot 2 \cdot 5}{2 \cdot 2 \cdot 5} = 1$

15. $\dfrac{35}{7} = \dfrac{5 \cdot 7}{7} = 5$

17. $\dfrac{18}{30} = \dfrac{2 \cdot 3 \cdot 3}{2 \cdot 3 \cdot 5} = \dfrac{3}{5}$

19. $\dfrac{16}{20} = \dfrac{2 \cdot 2 \cdot 2 \cdot 2}{2 \cdot 2 \cdot 5} = \dfrac{2 \cdot 2}{5} = \dfrac{4}{5}$

21. $\dfrac{66}{48} = \dfrac{2 \cdot 3 \cdot 11}{2 \cdot 2 \cdot 2 \cdot 2 \cdot 3} = \dfrac{11}{2 \cdot 2 \cdot 2} = \dfrac{11}{8}$

23. $\dfrac{120}{244} = \dfrac{2 \cdot 2 \cdot 2 \cdot 3 \cdot 5}{2 \cdot 2 \cdot 61} = \dfrac{2 \cdot 3 \cdot 5}{61} = \dfrac{30}{61}$

25. $\dfrac{1}{2} \cdot \dfrac{3}{4} = \dfrac{1 \cdot 3}{2 \cdot 4} = \dfrac{3}{8}$

27. $\dfrac{2}{3} \cdot \dfrac{3}{4} = \dfrac{2 \cdot 3}{3 \cdot 2 \cdot 2} = \dfrac{1}{2}$

29. $\dfrac{1}{2} \div \dfrac{7}{12} = \dfrac{1}{2} \cdot \dfrac{12}{7} = \dfrac{1 \cdot 2 \cdot 2 \cdot 3}{2 \cdot 7} = \dfrac{6}{7}$

31. $\dfrac{3}{4} \div \dfrac{1}{20} = \dfrac{3}{4} \cdot \dfrac{20}{1} = \dfrac{3 \cdot 2 \cdot 2 \cdot 5}{2 \cdot 2 \cdot 1} = 15$

33. $\dfrac{7}{10} \cdot \dfrac{5}{21} = \dfrac{7 \cdot 5}{2 \cdot 5 \cdot 3 \cdot 7} = \dfrac{1}{6}$

35. $\dfrac{9}{20} \div 12 = \dfrac{9}{20} \cdot \dfrac{1}{12} = \dfrac{3 \cdot 3 \cdot 1}{2 \cdot 2 \cdot 5 \cdot 2 \cdot 2 \cdot 3} = \dfrac{3}{80}$

37. $\dfrac{4}{5} + \dfrac{1}{5} = \dfrac{4 + 1}{5} = \dfrac{5}{5} = 1$

39. $\dfrac{4}{5} - \dfrac{1}{5} = \dfrac{4 - 1}{5} = \dfrac{3}{5}$

41. $\dfrac{23}{105} + \dfrac{4}{105} = \dfrac{23 + 4}{105} = \dfrac{27}{105} = \dfrac{3 \cdot 9}{3 \cdot 35} = \dfrac{9}{35}$

43. $\dfrac{17}{21} - \dfrac{10}{21} = \dfrac{17 - 10}{21} = \dfrac{7}{21} = \dfrac{7}{3 \cdot 7} = \dfrac{1}{3}$

45. $\dfrac{2}{3} + \dfrac{3}{7} = \dfrac{2 \cdot 7}{3 \cdot 7} + \dfrac{3 \cdot 3}{7 \cdot 3}$
$= \dfrac{14}{21} + \dfrac{9}{21} = \dfrac{14 + 9}{21}$
$= \dfrac{23}{21}$

47. $\dfrac{10}{3} - \dfrac{5}{21} = \dfrac{10 \cdot 7}{3 \cdot 7} - \dfrac{5}{21}$
$= \dfrac{70}{21} - \dfrac{5}{21} = \dfrac{65}{21}$

49. $\dfrac{10}{21} + \dfrac{5}{21} = \dfrac{10 + 5}{21}$
$= \dfrac{15}{21} = \dfrac{3 \cdot 5}{3 \cdot 7} = \dfrac{5}{7}$

51. $\dfrac{5}{22} - \dfrac{5}{33} = \dfrac{5 \cdot 3}{22 \cdot 3} - \dfrac{5 \cdot 2}{33 \cdot 2}$
$= \dfrac{15}{66} - \dfrac{10}{66} = \dfrac{15 - 10}{66}$
$= \dfrac{5}{66}$

53. $\dfrac{12}{5} - 1 = \dfrac{12}{5} - \dfrac{1 \cdot 5}{1 \cdot 5}$
$= \dfrac{12}{5} - \dfrac{5}{5}$
$= \dfrac{12 - 5}{5} = \dfrac{7}{5}$

55. $\dfrac{2}{3} - \dfrac{5}{9} + \dfrac{5}{6} = \dfrac{2 \cdot 6}{3 \cdot 6} - \dfrac{5 \cdot 2}{9 \cdot 2} + \dfrac{5 \cdot 3}{6 \cdot 3}$
$= \dfrac{12}{18} - \dfrac{10}{18} + \dfrac{15}{18}$
$= \dfrac{12 - 10 + 15}{18}$
$= \dfrac{17}{18}$

57. $1 - \dfrac{3}{10} - \dfrac{5}{10} = \dfrac{1 \cdot 10}{1 \cdot 10} - \dfrac{3}{10} - \dfrac{5}{10}$

$\qquad = \dfrac{10}{10} - \dfrac{3}{10} - \dfrac{5}{10}$

$\qquad = \dfrac{10 - 3 - 5}{10}$

$\qquad = \dfrac{2}{10} = \dfrac{1}{5}$

The unknown part is $\dfrac{1}{5}$.

59. $1 - \dfrac{1}{4} - \dfrac{3}{8} = \dfrac{1 \cdot 8}{1 \cdot 8} - \dfrac{1 \cdot 2}{4 \cdot 2} - \dfrac{3}{8}$

$\qquad = \dfrac{8}{8} - \dfrac{2}{8} - \dfrac{3}{8}$

$\qquad = \dfrac{8 - 2 - 3}{8} = \dfrac{3}{8}$

The unknown part is $\dfrac{3}{8}$.

61. $9\dfrac{1}{4} - 8\dfrac{1}{2} = \dfrac{37}{4} - \dfrac{17}{2}$

$\qquad = \dfrac{37}{4} - \dfrac{17 \cdot 2}{2 \cdot 2}$

$\qquad = \dfrac{37}{4} - \dfrac{34}{4}$

$\qquad = \dfrac{37 - 34}{4} = \dfrac{3}{4}$

There was $\dfrac{3}{4}$ of a year more work credit needed in 1998 than in 1985.

63. a. $\dfrac{1}{5}$

    b. $\dfrac{13}{50}$

    c. $1 - \dfrac{1}{5} - \dfrac{13}{50} - \dfrac{29}{100} = \dfrac{1 \cdot 100}{1 \cdot 100} - \dfrac{1 \cdot 20}{5 \cdot 20} - \dfrac{13 \cdot 2}{50 \cdot 2} - \dfrac{29}{100}$

$\qquad\qquad = \dfrac{100}{100} - \dfrac{20}{100} - \dfrac{26}{100} - \dfrac{29}{100} = \dfrac{100 - 20 - 26 - 29}{100}$

$\qquad\qquad = \dfrac{25}{100} = \dfrac{5 \cdot 5}{4 \cdot 5 \cdot 5} = \dfrac{1}{4}$

The fraction of land that falls into the other category is $\dfrac{1}{4}$.

d. $\dfrac{29}{100}+\dfrac{13}{50}=\dfrac{29}{100}+\dfrac{13\cdot2}{50\cdot2}$

$\quad=\dfrac{29}{100}+\dfrac{26}{100}=\dfrac{55}{100}$

$\quad=\dfrac{5\cdot11}{5\cdot20}=\dfrac{11}{20}$

The fraction of land that is either meadows and pastures or forest and woodland is $\dfrac{11}{20}$.

65. Answers may vary.

Section R.3

Practice Problems

1. $0.27=\dfrac{27}{100}$

2. $5.1=\dfrac{51}{10}$

3. $7.685=\dfrac{7685}{1000}$

4. a.
$$\begin{array}{r}7.19\\19.782\\+\ 1.006\\\hline27.978\end{array}$$

b.
$$\begin{array}{r}12\\0.79\\+\ 0.03\\\hline12.82\end{array}$$

5. a.
$$\begin{array}{r}84.230\\-\ 26.982\\\hline57.248\end{array}$$

b.
$$\begin{array}{r}90.00\\-\ 0.19\\\hline89.81\end{array}$$

6. a.
$$\begin{array}{r}0.31\\\times\ 4.6\\\hline186\\1\ 24\\\hline1.426\end{array}$$

b.
$$\begin{array}{r}1.26\\\times\ 0.03\\\hline0.0378\end{array}$$

7. a.
$$\begin{array}{r}43.5\\5\overline{)217.5}\\20\\\hline17\\15\\\hline2\ 5\\2\ 5\\\hline0\end{array}$$

Thus, $21.75\div0.5=43.5$.

b.
$$\begin{array}{r}2600\\6\overline{)15600}\\12\\\hline36\\36\\\hline0\end{array}$$

8. 12.9187 rounded to the nearest hundredth is 12.92.

9. 245.348 rounded to the nearest tenth is 245.3.

10.
$$\begin{array}{r}0.4\\5\overline{)2.0}\\2.0\\\hline0\end{array}$$

$\dfrac{2}{5}=0.4$

11.
$$\begin{array}{r}0.833\\6\overline{)5.000}\\48\\\hline20\\18\\\hline20\\18\\\hline2\end{array}$$

This pattern will continue so that

$\dfrac{5}{6}=0.833\ldots$

$\dfrac{5}{6}=0.8\overline{3}$

12.

$$9\overline{)1.0000} \quad \begin{array}{r} 0.1111 \\ \end{array}$$

$$
\begin{array}{r}
0.1111 \\
9\overline{)1.0000} \\
\underline{9} \\
10 \\
\underline{9} \\
10 \\
\underline{9} \\
10 \\
\underline{9} \\
1 \\
\end{array}
$$

This pattern will continue so that

$\dfrac{1}{9} = 0.1111\ldots$

$\dfrac{1}{9} = 0.\overline{1} \approx 0.111$

13. a. $20\% = 0.20$

   b. $1.2\% = 0.012$

   c. $465\% = 4.65$

14. a. $0.42 = 42\%$

   b. $0.003 = 0.3\%$

   c. $2.36 = 236\%$

   d. $0.7 = 70\%$

Exercise Set R.3

1. $0.6 = \dfrac{6}{10}$

3. $1.86 = \dfrac{186}{100}$

5. $0.144 = \dfrac{114}{1000}$

7. $123.1 = \dfrac{1231}{10}$

9.
$$
\begin{array}{r}
5.7 \\
+\ 1.13 \\
\hline
6.83 \\
\end{array}
$$

11.
$$
\begin{array}{r}
24.6 \\
2.39 \\
+\ 0.0678 \\
\hline
27.0578 \\
\end{array}
$$

13.
$$
\begin{array}{r}
8.8 \\
-\ 2.3 \\
\hline
6.5 \\
\end{array}
$$

15.
$$
\begin{array}{r}
18.00 \\
-\ 2.78 \\
\hline
15.22 \\
\end{array}
$$

17.
$$
\begin{array}{r}
45.02 \\
3.006 \\
+\ 8.405 \\
\hline
56.431 \\
\end{array}
$$

19.
$$
\begin{array}{r}
654.90 \\
-\ 56.67 \\
\hline
598.23 \\
\end{array}
$$

21.
$$
\begin{array}{r}
0.2 \\
\times\ 0.6 \\
\hline
0.12 \\
\end{array}
$$

23.
$$
\begin{array}{r}
6.75 \\
\times\ 10 \\
\hline
000 \\
675 \\
\hline
67.50 \\
\end{array}
$$

25.
$$
\begin{array}{r}
5.62 \\
\times\ 7.7 \\
\hline
3934 \\
3934 \\
\hline
43.274 \\
\end{array}
$$

27.
$$
\begin{array}{r}
16.003 \\
\times\ 5.31 \\
\hline
16003 \\
48009 \\
80015 \\
\hline
84.97593 \\
\end{array}
$$

29.

$$
\begin{array}{r}
0.094 \\
5\overline{)\ 0.470} \\
\underline{-0} \\
47 \\
\underline{-45} \\
20 \\
\underline{-20}
\end{array}
$$

31.

$$
\begin{array}{r}
70 \\
6\overline{)\ 420} \\
\underline{-42} \\
0 \\
\underline{-0}
\end{array}
$$

33.

$$
\begin{array}{r}
5.8 \\
82\overline{)\ 475.6} \\
\underline{-410} \\
656 \\
\underline{-656}
\end{array}
$$

35.

$$
\begin{array}{r}
840 \\
63\overline{)\ 52920} \\
\underline{-504} \\
252 \\
\underline{-252} \\
0 \\
\underline{-0}
\end{array}
$$

37. 0.6

39. 0.23

41. 0.594

43. 98,207.2

45. 12.3

47.

$$
\begin{array}{r}
0.75 \\
4\overline{)\ 3.00} \\
\underline{-28} \\
20 \\
\underline{-20} \\
0
\end{array}
$$

$\dfrac{3}{4} = 0.75$

49.

$$
\begin{array}{r}
0.333 \\
3\overline{)\ 1.00} \\
\underline{-9} \\
10 \\
\underline{-9} \\
10 \\
\underline{-9} \\
1
\end{array}
$$

This pattern will continue so that

$\dfrac{1}{3} = 0.3333....$

$\dfrac{1}{3} = 0.\overline{3} \approx 0.33$

51.

$$
\begin{array}{r}
0.4375 \\
16\overline{)\ 7.0000} \\
\underline{-64} \\
60 \\
\underline{-48} \\
120 \\
\underline{-112} \\
80 \\
\underline{-80} \\
0
\end{array}
$$

$\dfrac{7}{16} = 0.4375$

53.

$$
\begin{array}{r}
0.5454 \\
11\overline{)\ 6.0000} \\
\underline{-55} \\
50 \\
\underline{-44} \\
60 \\
\underline{-55} \\
50 \\
\underline{-44} \\
6
\end{array}
$$

This pattern will continue so that

$\dfrac{6}{11} = 0.5454....$

$\dfrac{6}{11} = 0.\overline{54} \approx 0.55$

55. $28\% = 0.28$

57. $3.1\% = 0.031$

59. $135\% = 1.35$

61. $96.55\ \% = 0.9655$

63. $61\% = 0.61$

65. $0.68 = 68\%$

67. $0.876 = 87.6\%$

69. $1 = 100\%$

71. $0.5 = 50\%$

73.
$$\begin{array}{r} 82.65 \\ -75.67 \\ \hline 6.98 \end{array}$$

Females are expected to live 6.98 years longer than males.

75.
$$\begin{array}{r} 0.64 \\ 25\overline{)16.00} \\ -150 \\ \hline 100 \\ -100 \\ \hline \end{array}$$

$\frac{16}{25} = 0.64 = 64\%$

64% of Americans own credit cards.

Chapter R Review

1. $42 = 2 \cdot 3 \cdot 7$

2. $800 = 2 \cdot 2 \cdot 2 \cdot 2 \cdot 2 \cdot 5 \cdot 5$

3. $12 = 2 \cdot 2 \cdot 3$
$30 = 2 \cdot 3 \cdot 5$
$LCM = 2 \cdot 2 \cdot 3 \cdot 5 = 60$

4. $7 = 7$
$42 = 2 \cdot 3 \cdot 7$
$LCM = 2 \cdot 3 \cdot 7 = 42$

5. $4 = 2 \cdot 2$
$6 = 2 \cdot 3$
$10 = 2 \cdot 5$
$LCM = 2 \cdot 2 \cdot 3 \cdot 5 = 60$

6. $2 = 2$
$5 = 5$
$7 = 7$
$LCM = 2 \cdot 5 \cdot 7 = 70$

7. $\frac{5}{8} = \frac{5 \cdot 3}{8 \cdot 3} = \frac{15}{24}$

8. $\frac{2}{3} = \frac{2 \cdot 20}{3 \cdot 20} = \frac{40}{60}$

9. $\frac{8}{20} = \frac{2 \cdot 2 \cdot 2}{2 \cdot 2 \cdot 5} = \frac{2}{5}$

10. $\frac{15}{100} = \frac{3 \cdot 5}{2 \cdot 2 \cdot 5 \cdot 5} = \frac{3}{2 \cdot 2 \cdot 5} = \frac{3}{20}$

11. $\frac{12}{6} = \frac{2 \cdot 2 \cdot 3}{2 \cdot 3} = 2$

12. $\frac{8}{8} = \frac{2 \cdot 2 \cdot 2}{2 \cdot 2 \cdot 2} = 1$

13. $\frac{1}{7} \cdot \frac{8}{11} = \frac{1 \cdot 8}{7 \cdot 11} = \frac{8}{77}$

14. $\frac{5}{12} + \frac{2}{15}$
$= \frac{5 \cdot 5}{12 \cdot 5} + \frac{2 \cdot 4}{15 \cdot 4}$
$= \frac{25}{60} + \frac{8}{60}$
$= \frac{25 + 8}{60}$
$= \frac{33}{60}$
$= \frac{3 \cdot 11}{3 \cdot 20}$
$= \frac{11}{20}$

15. $\frac{3}{10} \div 6$
$= \frac{3}{10} \cdot \frac{1}{6}$
$= \frac{3 \cdot 1}{2 \cdot 5 \cdot 2 \cdot 3}$
$= \frac{1}{20}$

16. $\frac{7}{9} - \frac{1}{6} = \frac{7 \cdot 2}{9 \cdot 2} - \frac{1 \cdot 3}{6 \cdot 3} = \frac{14}{18} - \frac{3}{18} = \frac{14 - 3}{18}$
$= \frac{11}{18}$

17. $A = \ell w$
$A = \frac{11}{12} \cdot \frac{3}{5} = \frac{11 \cdot 3}{2 \cdot 2 \cdot 3 \cdot 5} = \frac{11}{20}$
The area of the rectangle is $\frac{11}{20}$ sq. mile.

18. $A = \frac{1}{2}bh$

$A = \frac{1}{2} \cdot \frac{5}{4} \cdot \frac{1}{2} = \frac{1 \cdot 5 \cdot 1}{2 \cdot 4 \cdot 2} = \frac{5}{16}$

The area of the triangle is $\frac{5}{16}$ sq. meter.

19. $1.81 = \frac{181}{100}$

20. $0.035 = \frac{35}{1000}$

21.
$$\begin{array}{r} 76.358 \\ + 18.76 \\ \hline 95.118 \end{array}$$

22.
$$\begin{array}{r} 35 \\ 0.02 \\ + 1.765 \\ \hline 36.785 \end{array}$$

23.
$$\begin{array}{r} 18.00 \\ - 4.62 \\ \hline 13.38 \end{array}$$

24.
$$\begin{array}{r} 804.062 \\ - 112.489 \\ \hline 691.573 \end{array}$$

25.
$$\begin{array}{r} 7.6 \\ \times 12 \\ \hline 152 \\ 76 \\ \hline 91.2 \end{array}$$

26.
$$\begin{array}{r} 14.63 \\ \times 3.2 \\ \hline 2926 \\ 4389 \\ \hline 46.816 \end{array}$$

27.
$$\begin{array}{r} 28.6 \\ 27 \overline{)\ 772.2} \\ -54 \\ \hline 232 \\ -216 \\ \hline 162 \end{array}$$

28.
$$\begin{array}{r} 230 \\ 6 \overline{)\ 1380} \\ -12 \\ \hline 18 \\ -18 \\ \hline 0 \\ -0 \\ \hline \end{array}$$

29. 0.77

30. 25.6

31.
$$\begin{array}{r} 0.5 \\ 2 \overline{)\ 1.0} \\ -10 \\ \hline 0 \end{array}$$

$\frac{1}{2} = 0.5$

32.
$$\begin{array}{r} 0.375 \\ 8 \overline{)\ 3.000} \\ -24 \\ \hline 60 \\ -56 \\ \hline 40 \\ -40 \\ \hline 0 \end{array}$$

$\frac{3}{8} = 0.375$

33.
$$\begin{array}{r} 0.3636 \\ 11 \overline{)\ 4.0000} \\ -33 \\ \hline 70 \\ -66 \\ \hline 40 \\ -33 \\ \hline 70 \\ -66 \\ \hline 4 \end{array}$$

This pattern will continue so that
$\frac{4}{11} = 0.3636....$

$\frac{4}{11} = 0.\overline{36} \approx 0.364$

34.

$$
\begin{array}{r}
0.8333 \\
6\overline{)\,5.0000} \\
-48 \\
\hline
20 \\
-18 \\
\hline
20 \\
-18 \\
\hline
20 \\
-18 \\
\hline
2
\end{array}
$$

This pattern will continue so that

$\dfrac{5}{6} = 0.8333....$

$\dfrac{5}{6} = 0.8\overline{3} \approx 0.833$

35. $29\% = 0.29$

36. $1.4\% = 0.014$

37. $0.39 = 39\%$

38. $1.2 = 120\%$

39. $70.8\% = 0.708$

40. b; $5 = 500\%$

Chapter R Test

1. $72 = 2 \cdot 2 \cdot 2 \cdot 3 \cdot 3$

2. $15 = 3 \cdot 5$
$18 = 2 \cdot 3 \cdot 3$
$20 = 2 \cdot 2 \cdot 5$
$\text{LCM} = 2 \cdot 2 \cdot 3 \cdot 3 \cdot 5 = 180$

3. $\dfrac{5}{12} = \dfrac{5 \cdot 5}{12 \cdot 5} = \dfrac{25}{60}$

4. $\dfrac{15}{20} = \dfrac{3 \cdot 5}{2 \cdot 2 \cdot 5} = \dfrac{3}{2 \cdot 2} = \dfrac{3}{4}$

5. $\dfrac{48}{100} = \dfrac{2 \cdot 2 \cdot 2 \cdot 2 \cdot 3}{2 \cdot 2 \cdot 5 \cdot 5} = \dfrac{2 \cdot 2 \cdot 3}{5 \cdot 5} = \dfrac{12}{25}$

6. $1.3 = \dfrac{13}{10}$

7. $\dfrac{5}{8} + \dfrac{7}{10} = \dfrac{5 \cdot 5}{8 \cdot 5} + \dfrac{7 \cdot 4}{10 \cdot 4} = \dfrac{25}{40} + \dfrac{28}{40}$
$= \dfrac{25 + 28}{40} = \dfrac{53}{40}$

8. $\dfrac{2}{3} \cdot \dfrac{27}{49} = \dfrac{2 \cdot 3 \cdot 3 \cdot 3}{3 \cdot 7 \cdot 7} = \dfrac{2 \cdot 3 \cdot 3}{7 \cdot 7} = \dfrac{18}{49}$

9. $\dfrac{9}{10} \div 18 = \dfrac{9}{10} \cdot \dfrac{1}{18} = \dfrac{3 \cdot 3 \cdot 1}{2 \cdot 5 \cdot 2 \cdot 3 \cdot 3}$
$= \dfrac{1}{2 \cdot 5 \cdot 2} = \dfrac{1}{20}$

10. $\dfrac{8}{9} - \dfrac{1}{12}$
$= \dfrac{8 \cdot 4}{9 \cdot 4} - \dfrac{1 \cdot 3}{12 \cdot 3}$
$= \dfrac{32}{36} - \dfrac{3}{36}$
$= \dfrac{32 - 3}{36}$
$= \dfrac{29}{36}$

11.
$$
\begin{array}{r}
43 \\
0.21 \\
+\,1.9 \\
\hline
45.11
\end{array}
$$

12.
$$
\begin{array}{r}
123.60 \\
-\,57.72 \\
\hline
65.88
\end{array}
$$

13.
$$
\begin{array}{r}
7.93 \\
\times\,1.6 \\
\hline
4758 \\
793 \\
\hline
12.688
\end{array}
$$

14.
$$
\begin{array}{r}
320 \\
25\overline{)\,8000} \\
-75 \\
\hline
50 \\
-50 \\
\hline
0 \\
-0 \\
\hline
0
\end{array}
$$

15. 23.73

16.

$$\begin{array}{r} 0.875 \\ 8\overline{)\,7.000} \\ \underline{-64} \\ 60 \\ \underline{-56} \\ 40 \\ \underline{-40} \\ 0 \end{array}$$

$$\frac{7}{8} = 0.875$$

17.

$$\begin{array}{r} 0.1666 \\ 6\overline{)\,1.0000} \\ \underline{-6} \\ 40 \\ \underline{-36} \\ 40 \\ \underline{-36} \\ 40 \\ \underline{-36} \\ 4 \end{array}$$

This pattern will continue so that

$$\frac{1}{6} = 0.1666\ldots$$

$$\frac{1}{6} = 0.1\overline{6} \approx 0.167$$

18. $58.1\% = 0.581$

19. $0.07 = 7\%$

20.

$$\begin{array}{r} 0.75 \\ 4\overline{)\,3.00} \\ \underline{-28} \\ 20 \\ \underline{-20} \\ 0 \end{array}$$

$$\frac{3}{4} = 0.75 = 75\%$$

21. From the graph, we can see that $\frac{3}{4}$ of the fresh water is icecaps and glaciers.

22. From the graph, we can see that $\frac{1}{200}$ of the fresh water is active water.

23. From the graph, we can see that

$$1 - \frac{3}{4} - \frac{1}{200} = \frac{200}{200} - \frac{150}{200} - \frac{1}{200}$$

$$= \frac{49}{200} \text{ of the fresh water is groundwater.}$$

24. $\dfrac{3}{4} + \dfrac{49}{200} = \dfrac{3 \cdot 50}{4 \cdot 50} + \dfrac{49}{200}$

$$= \frac{150}{200} + \frac{49}{200}$$

$$= \frac{150 + 49}{200} = \frac{199}{200}$$

$\dfrac{199}{200}$ of the fresh water is groundwater or icecaps and glaciers.

Chapter 1

1. $0 > -3$

2. $-10 < -8$

3. $1.7 > 1.07$

4. $|5| = 5$

5. $|-1.2| = 1.2$

6. $|0| = 0$

7. Replace x with 2 and y with 5.
   $xy - x^2 = 2 \cdot 5 - 2^2 = 2 \cdot 5 - 4 = 10 - 4$
   $= 6$

8. $2x - 10$

9. $4^3 = 4 \cdot 4 \cdot 4 = 64$

10. $-3^2 = -(3)(3) = -9$

11. The opposite of $-\dfrac{3}{5}$ is $\dfrac{3}{5}$.

12. The reciprocal of $\dfrac{1}{8}$ is 8, since $\dfrac{1}{8} \cdot 8 = 1$.

13. $3 + 2 \cdot 5^2 = 3 + 2 \cdot 25 = 3 + 50 = 53$

14. $-10 + 13 = 3$

15. $-6 - 21 = -6 + (-21) = -27$

16. $(-7)(-8) = 56$

17. $-2.8 \div 0.04 = -2.8 \cdot \dfrac{1}{0.04} = -70$

18. $\dfrac{-4 - 6^2}{5(-2)} = \dfrac{-4 - 36}{5(-2)} = \dfrac{-4 - 36}{-10} = \dfrac{-4 + (-36)}{-10}$
    $= \dfrac{-40}{-10} = 4$

19. Replace x with -4 and y with -7.
    $x^2 - 2xy = (-4)^2 - 2(-4)(-7)$
    $= 16 - 2(-4)(-7) = 16 - (-8)(-7)$
    $= 16 - 56 = -40$

20. Replace x with 7 and see if a true statement results.
    $x - 12 = 5$
    $7 - 12 \stackrel{?}{=} 5$
    $-5 = 5$
    Since $-5 = 5$ is a false statement, 7 is not a solution of the equation.

21. Replace x with 40 and see if a true statement results.
    $\dfrac{x}{8} + 2 = 7$
    $\dfrac{40}{8} + 2 \stackrel{?}{=} 7$
    $5 + 2 \stackrel{?}{=} 7$
    $7 = 7$
    Since $7 = 7$ is a true statement, 40 is a solution of the equation.

22. The temperature rise is the difference of the high and low temperatures.
    $22 - (-14) = 22 + 14 = 36$
    The temperature rose $36°F$.

23. $2y + 5 = 5 + 2y$

24. $4(3 + 2t) = 4(3) + 4(2t) = 12 + 8t$

25. Additive inverse property

Section 1.1

Practice Problems

1. False, since 8 is to the right of 6 on the number line.

2. True, since 100 is to the right of 10 on the number line.

3. True, since $21 = 21$ is true.

4. True, since $21 = 21$ is true.

5. True, since 0 is to the left of 5 on the number line.

6. True, since 25 is to the right of 22 on the number line.

7. a. $14 \geq 14$

   b. $0 < 5$

   c. $9 \neq 10$

8. The integer -282 represents 282 feet below sea level.

9.

10. a. The natural numbers are 6 and 913.

   b. The whole numbers are 0, 6, and 913.

   c. The integers are -100, 0, 6, and 913.

   d. The rational numbers are -100, $-\frac{2}{5}$, 0, 6, and 913.

   e. The irrational number is π.

   f. The real numbers are all numbers in the given set.

11. a. $|7| = 7$ since 7 is 7 units from 0 on the number line.

   b. $|-8| = 8$ since -8 is 8 units from 0 on the number line.

   c. $\left|-\frac{2}{3}\right| = \frac{2}{3}$ since $-\frac{2}{3}$ is $\frac{2}{3}$ of a unit from 0 on the number line.

12. a. $|-4| = 4$

   b. $-3 < |0|$ since $-3 < 0$.

   c. $|-2.7| > |-2|$ since $2.7 > 2$.

d. $|6| < |16|$ since $6 < 16$.

e. $|-6| < |-16|$ since $6 < 16$.

Exercise Set 1.1

1. $4 < 10$

3. $7 > 3$

5. $6.26 = 6.26$

7. $0 < 7$

9. $32 < 212$

11. True, since $11 = 11$ is true.

13. False, since 10 is to the left of 11 on the number line.

15. False, since 11 is to the left of 24 on the number line.

17. True, since 7 is to the right of 0 on the number line.

19. $30 \leq 45$

21. $20 \leq 25$

23. $6 > 0$

25. $-12 < -10$

27. $8 < 12$

29. $5 \geq 4$

31. $15 \neq -2$

33. The integer 535 represents 535 feet. The integer -8 represents 8 feet below sea level.

35. The integer $-398,000$ represents a decrease in attendance of 398,000.

37. The integer 350 represents a deposit of $350. The integer -126 represents a withdrawal of $126.

39.

41.

43. To help graph the improper fractions, first write them as mixed numbers.

$$\frac{7}{4} = 1\frac{3}{4}, \ -\frac{3}{2} = -1\frac{1}{2}$$

45. The number 0 belongs to the sets of: whole numbers, integers, rational numbers, and real numbers.

47. The number –2 belongs to the sets of: integers, rational numbers, and real numbers.

49. The number 6 belongs to the sets of: natural numbers, whole numbers, integers, rational numbers, and real numbers.

51. The number $\frac{2}{3}$ belongs to the sets of: rational numbers and real numbers.

53. False; rational numbers can be either nonintegers, such as $\frac{1}{2}$, or integers, such as 2.

55. True; the set of natural numbers is the set of all positive integers.

57. True; 0 corresponds to a point on the number line. Therefore, 0 is a real number.

59. True; all whole numbers belong to the set of integers.

61. $|-5| > -4$ since $|-5| = 5$ and $5 > -4$.

63. $|-1| = |1|$ since $|-1| = 1$ and $|1| = 1$.

65. $|-2| < |-3|$ since $|-2| = 2$ and $|-3| = 3$ and $2 < 3$.

67. $|0| < |-8|$ since $|0| = 0$ and $|-8| = 8$ and $0 < 8$.

69. False, since $\frac{1}{2}$ is to the right of $\frac{1}{3}$ on the number line.

71. True, since $|-5.3| = |5.3|$ is true.

73. False, since –9.6 is to the left of –9.1 on the number line.

75. True, since $-\frac{2}{3} < -\frac{1}{5}$ is true.

77. Bill's highest quiz score is 90.

79. The score for Quiz 2 = 70.
The score for Quiz 3 = 90.
$70 \le 90$

81. $-0.04 > -26.7$

83. The sun is brighter, since $-26.7 < -0.04$.

85. The sun; since on the number line –26.7 is to the left of all other numbers listed, and therefore, –26.7 is smaller than all other numbers listed.

87. Answers may vary.

Section 1.2

Practice Problems

1. a. $4^2 = 4 \cdot 4 = 16$

   b. $2^2 = 2 \cdot 2 = 4$

   c. $3^4 = 3 \cdot 3 \cdot 3 \cdot 3 = 81$

   d. $9^1 = 9$

   e. $\left(\frac{2}{5}\right)^2 = \frac{2}{5} \cdot \frac{2}{5} = \frac{2 \cdot 2}{5 \cdot 5} = \frac{4}{25}$

2. $3 + 2 \cdot 4^2 = 3 + 2 \cdot 16 = 3 + 32 = 35$

3. $\frac{9}{5} \cdot \frac{1}{3} - \frac{1}{3} = \frac{9}{15} - \frac{1}{3} = \frac{9}{15} - \frac{5}{15} = \frac{4}{15}$

4. $8[2(6 + 3) - 9] = 8[2(9) - 9] = 8[18 - 9]$
$= 8[9] = 72$

5. $\dfrac{1 + |7 - 4| + 3^2}{8 - 5} = \dfrac{1 + |3| + 3^2}{8 - 5}$

$= \dfrac{1 + 3 + 3^2}{3} = \dfrac{1 + 3 + 9}{3} = \dfrac{13}{3}$

6. Replace x with 1 and y with 4. Then simplify.

   a. $2y - x = 2(4) - 1 = 8 - 1 = 7$

   b. $\dfrac{8x}{3y} = \dfrac{8 \cdot 1}{3 \cdot 4} = \dfrac{2 \cdot 4 \cdot 1}{3 \cdot 4} = \dfrac{2 \cdot 1}{3} = \dfrac{2}{3}$

   c. $\dfrac{x}{y} + \dfrac{5}{y} = \dfrac{1}{4} + \dfrac{5}{4} = \dfrac{6}{4} = \dfrac{2 \cdot 3}{2 \cdot 2} = \dfrac{3}{2}$

   d. $y^2 - x^2 = 4^2 - 1^2 = 16 - 1 = 15$

7. Replace x with 3 and see if a true statement results.
$5x - 10 = x + 2$
$5(3) - 10 \overset{?}{=} 3 + 2$
$15 - 10 \overset{?}{=} 3 + 2$
$5 = 5$
Since we arrived at a true statement, 3 is a solution of the system.

8. a. $5x$

   b. $x + 7$

   c. $3x$

   d. $8 - x$

   e. $2x + 1$

9. a. $6x = 24$

   b. $10 - x = 18$

   c. $2x - 1 = 99$

Exercise Set 1.2

1. $3^5 = 3 \cdot 3 \cdot 3 \cdot 3 \cdot 3 = 243$

3. $3^3 = 3 \cdot 3 \cdot 3 = 27$

5. $1^5 = 1 \cdot 1 \cdot 1 \cdot 1 \cdot 1 = 1$

7. $5^1 = 5$

9. $\left(\dfrac{1}{5}\right)^3 = \left(\dfrac{1}{5}\right)\left(\dfrac{1}{5}\right)\left(\dfrac{1}{5}\right) = \dfrac{1 \cdot 1 \cdot 1}{5 \cdot 5 \cdot 5} = \dfrac{1}{125}$

11. $\left(\dfrac{2}{3}\right)^4 = \left(\dfrac{2}{3}\right)\left(\dfrac{2}{3}\right)\left(\dfrac{2}{3}\right)\left(\dfrac{2}{3}\right) = \dfrac{2 \cdot 2 \cdot 2 \cdot 2}{3 \cdot 3 \cdot 3 \cdot 3} = \dfrac{16}{81}$

13. $7^2 = 7 \cdot 7 = 49$

15. $(1.2)^2 = 1.2 \cdot 1.2 = 1.44$

17. $(5 \cdot 5)$ square meters $= 5^2$ square meters

19. $5 + 6 \cdot 2 = 5 + 12 = 17$

21. $4 \cdot 8 - 6 \cdot 2 = 32 - 12 = 20$

23. $2(8 - 3) = 2(5) = 10$

25. $2 + (5 - 2) + 4^2 = 2 + 3 + 4^2$
$= 2 + 3 + 16 = 5 + 16$
$= 21$

27. $5 \cdot 3^2 = 5 \cdot 9 = 45$

29. $\dfrac{1}{4} \cdot \dfrac{2}{3} - \dfrac{1}{6} = \dfrac{2}{12} - \dfrac{1}{6} = \dfrac{1}{6} - \dfrac{1}{6} = 0$

31. $\dfrac{6 - 4}{9 - 2} = \dfrac{2}{7}$

33. $2[5 + 2(8 - 3)] = 2[5 + 2(5)]$
$= 2[5 + 10] = 2[15] = 30$

35. $\dfrac{19 - 3 \cdot 5}{6 - 4} = \dfrac{19 - 15}{2} = \dfrac{4}{2} = 2$

37. $\dfrac{|6 - 2| + 3}{8 + 2 \cdot 5} = \dfrac{4 + 3}{8 + 10} = \dfrac{7}{18}$

39. $\dfrac{3 + 3(5 + 3)}{3^2 + 1} = \dfrac{3 + 3(8)}{9 + 1} = \dfrac{3 + 24}{10}$
$= \dfrac{27}{10}$

41. $\dfrac{6+|8-2|+3^2}{18-3} = \dfrac{6+|6|+3^2}{15}$

$= \dfrac{6+6+9}{15} = \dfrac{21}{15} = \dfrac{3\cdot 7}{3\cdot 5} = \dfrac{7}{5}$

43. No; since in the absence of grouping symbols we always perform multiplications or divisions before additions or subtractions in any expression. Thus, in the expression $2 + 3 \cdot 5$, we should first find the product $3 \cdot 5$ and then add 2.
$2 + 3 \cdot 5 = 2 + 15 = 17$

45. Replace y with 3.
$3y = 3(3) = 9$

47. Replace x with 1 and z with 5.
$\dfrac{z}{5x} = \dfrac{5}{5(1)} = \dfrac{5}{5} = 1$

49. Replace x with 1.
$3x - 2 = 3(1) - 2 = 3 - 2 = 1$

51. Replace x with 1 and y with 3.
$|2x + 3y| = |2(1) + 3(3)|$
$= |2 + 9| = |11| = 11$

53. Replace x with 1, y with 3, and z with 5.
$xy + z = (1)(3) + 5 = 3 + 5 = 8$

55. Replace y with 3.
$5y^2 = 5 \cdot 3^2 = 5 \cdot 9 = 45$

57. Replace z with 3.
$5z = 5(3) = 15$

59. Replace x with 2 and y with 6.
$\dfrac{y}{x} = \dfrac{6}{2} = 3$

61. Replace x with 2 and y with 6.
$\dfrac{y}{x} + \dfrac{y}{x} = \dfrac{6}{2} + \dfrac{6}{2} = 3 + 3 = 6$

63. We evaluate the expression $16t^2$ for each value of t.
When $t = 1$, we have
$16t^2 = 16 \cdot 1^2 = 16 \cdot 1 = 16.$

When $t = 2$ we have
$16t^2 = 16 \cdot 2^2 = 16 \cdot 4 = 64.$
When $t = 3$ we have
$16t^2 = 16 \cdot 3^2 = 16 \cdot 9 = 144.$
When $t = 4$, we have
$16t^2 = 16 \cdot 4^2 = 16 \cdot 16 = 256.$

Time t (in seconds)	Distance $16t^2$ (in feet)
1	16
2	64
3	144
4	256

65. Replace x with 5 and see if a true statement results.
$3x - 6 = 9$
$3(5) - 6 \overset{?}{=} 9$
$15 - 6 \overset{?}{=} 9$
$9 = 9$
Since we arrived at a true statement, 5 is a solution of the equation $3x - 6 = 9$.

67. Replace x with 0 and see if a true statement results.
$2x + 6 = 5x - 1$
$2(0) + 6 \overset{?}{=} 5(0) - 1$
$6 \neq -1$
Because $2(0) + 6 = 5(0) - 1$ is not a true statement, 0 is not a solution of $2x + 6 = 5x - 1$.

69. Replace x with 8 and see if a true statement results.
$2x - 5 = 5$
$2(8) - 5 \overset{?}{=} 5$
$16 - 5 \overset{?}{=} 5$
$11 \neq 5$
Because $2(8) - 5 = 5$ is not a true statement, 8 is not a solution of $2x - 5 = 5$.

71. Replace x with 2 and see if a true statement results.

$x + 6 = x + 6$

$2 + 6 \overset{?}{=} 2 + 6$

$8 = 8$

Since we arrived at a true statement, 2 is a solution of the equation $x + 6 = x + 6$.

73. Replace x with 0 and see if a true statement results.

$x = 5x + 15$

$0 \overset{?}{=} 5(0) + 15$

$0 \overset{?}{=} 0 + 15$

$0 \neq 15$

Because $0 = 5(0) + 15$ is not a true statement, 0 is not a solution of $x = 5x + 15$.

75. $x + 15$

77. $x - 5$

79. $3x + 22$

81. $1 + 2 = 9 \div 3$

83. $3 \neq 4 \div 2$

85. $5 + x = 20$

87. $13 - 3x = 13$

89. $\dfrac{12}{x} = \dfrac{1}{2}$

91. The expression $20 - 4 \cdot 4 \div 2$ simplifies to 32 if we subtract 4 from 20 first. Therefore, we group $20 - 4$ with parentheses and simplify as follows.
$(20 - 4) \cdot 4 \div 2 = 16 \cdot 4 \div 2 = 64 \div 2 = 32$

93. Answers may vary.

95. Answers may vary.

Section 1.3

Practice Problems

1.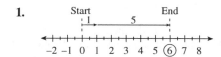

$1 + 5 = 6$

2.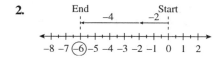

$-2 + (-4) = -6$

3.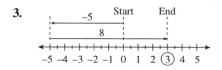

$-5 + 8 = 3$

4.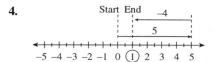

$5 + (-4) = 1$

5. $(-8) + (-5) = -13$

6. $(-14) + 6 = -8$

7. $(-17) + (-10) = -27$

8. $(-4) + 12 = 8$

9. $1.5 + (-3.2) = -1.7$

10. $-\dfrac{6}{11} + \left(-\dfrac{3}{11}\right) = -\dfrac{9}{11}$

11. $12.8 + (-3.6) = 9.2$

12. $-\dfrac{4}{5} + \dfrac{2}{3} = -\dfrac{4 \cdot 3}{5 \cdot 3} + \dfrac{2 \cdot 5}{3 \cdot 5} = -\dfrac{12}{15} + \dfrac{10}{15} = -\dfrac{2}{15}$

13. a. $16 + (-9) + (-9) = 7 + (-9) = -2$

  b. $[3 + (-13)] + [-4 + (-7)]$
       $= [-10] + [-11]$
       $= -21$

14. The overall gain or loss is the sum of the gains and losses. Gains can be represented by positive numbers. Losses can be represented by negative numbers.
$-2 + (-1) + 3 + 3 = 3$
The overall gain was $3.

15. The opposite of -35 is 35.

16. The opposite of 12 is -12.

17. The opposite of $-\dfrac{3}{11}$ is $\dfrac{3}{11}$.

18. The opposite of 1.9 is -1.9.

19. a. $-(-22) = 22$

    b. $-\left(-\dfrac{2}{7}\right) = \dfrac{2}{7}$

    c. $-(-x) = x$

    d. $-|-14| = -14$

Exercise Set 1.3

1. $6 + 3 = 9$

3. $-6 + (-8) = -14$

5. $8 + (-7) = 1$

7. $-14 + 2 = -12$

9. $-2 + (-3) = -5$

11. $-9 + (-3) = -12$

13. $-7 + 3 = -4$

15. $10 + (-3) = 7$

17. $5 + (-7) = -2$

19. $-16 + 16 = 0$

21. $27 + (-46) = -19$

23. $-18 + 49 = 31$

25. $-33 + (-14) = -47$

27. $6.3 + (-8.4) = -2.1$

29. $|-8| + (-16) = 8 + (-16) = -8$

31. $117 + (-79) = 38$

33. $-9.6 + (-3.5) = -13.1$

35. $-\dfrac{3}{8} + \dfrac{5}{8} = \dfrac{2}{8} = \dfrac{1}{4}$

37. $-\dfrac{7}{16} + \dfrac{1}{4} = -\dfrac{7}{16} + \dfrac{4 \cdot 1}{4 \cdot 4}$
       $= -\dfrac{7}{16} + \dfrac{4}{16} = -\dfrac{3}{16}$

39. $-\dfrac{7}{10} + \left(-\dfrac{3}{5}\right) = -\dfrac{7}{10} + \left(-\dfrac{2 \cdot 3}{2 \cdot 5}\right)$
       $= -\dfrac{7}{10} + \left(-\dfrac{6}{10}\right) = -\dfrac{13}{10}$

41. $-15 + 9 + (-2) = -6 + (-2) = -8$

43. $-21 + (-16) + (-22) = -37 + (-22) = -59$

45. $-23 + 16 + (-2) = -7 + (-2) = -9$

47. $|5 + (-10)| = |-5| = 5$

49. $6 + (-4) + 9 = 2 + 9 = 11$

51. $[-17 + (-4)] + [-12 + 15] = [-21] + [3]$
       $= -18$

53. $|9 + (-12)| + |-16| = |-3| + |-16| = 3 + 16 = 19$

55. $-13 + [5 + (-3) + 4] = -13 + [2 + 4]$
       $= -13 + [6] = -7$

57. Answers may vary.

59. $-15 + 9 = -6$
The high temperature for the day was $-6°$.

61. $-1296 + 658 = -638$
You are at an elevation of -638 feet (638 feet below sea level).

63. The overall change is the sum of the changes.

$$-1\frac{5}{8}+\left(-2\frac{1}{2}\right)=-\frac{13}{8}+\left(-\frac{5}{2}\right)$$
$$=-\frac{13}{8}+\left(-\frac{4\cdot5}{4\cdot2}\right)=-\frac{13}{8}+\left(-\frac{20}{8}\right)$$
$$=-\frac{33}{8}=-4\frac{1}{8}$$

The combined change is $-4\frac{1}{8}$ points.

65. His total overall score is the sum of the scores for all four rounds of play.
$0+(-4)+(-2)+(-4)$
$=-4+(-2)+(-4)$
$=-6+(-4)$
$=-10$
His total overall score was -10 (10 below par).

67. The total net profit is the sum of the net profits for the three years.
$-3.0+(-22.0)+(-8.2)$
$=-25+(-8.2)$
$=-33.2$
The total net profit for these three years was $-\$33.2$ million.

69. The opposite of 6 is -6.

71. The opposite of -2 is 2.

73. The opposite of 0 is 0.

75. Since $|-6|=6$, the opposite of $|-6|$ is -6.

77. Answers may vary.

79. Since $|-2|=2$, $-|-2|=-2$.

81. $-|0|=-0=0$

83. Since $\left|-\frac{2}{3}\right|=\frac{2}{3}$, $-\left|-\frac{2}{3}\right|=-\frac{2}{3}$.

85. The highest temperature is represented by the bar that is farthest above 0 degrees. From the graph, we can see that the highest temperature occurred on Tuesday.

87. The highest temperature shown is $7°$.

89. To find the average of the five temperatures, we first find the sum and then divide by 5.
$$\frac{-4+3+7+(-2)+1}{5}=\frac{-1+7+(-2)+1}{5}$$
$$=\frac{6+(-2)+1}{5}=\frac{4+1}{5}=\frac{5}{5}$$
$$=\frac{5}{5}=1$$
The average daily low temperature for Sunday through Thursday was $1°$.

91. $-a$ is a negative number.

93. $a+a$ is a positive number.

Section 1.4

Practice Problems

1. a. $-20-6=-20+(-6)=-26$

   b. $3-(-5)=3+5=8$

   c. $7-17=7+(-17)=-10$

   d. $-4-(-9)=-4+9=5$

2. $9.6-(-5.7)=9.6+5.7=15.3$

3. $-\frac{4}{9}-\frac{2}{9}=-\frac{4}{9}+\left(-\frac{2}{9}\right)=-\frac{6}{9}=-\frac{2\cdot3}{3\cdot3}$
$$=-\frac{2}{3}$$

4. $-\frac{1}{4}-\left(-\frac{2}{5}\right)=-\frac{1}{4}+\frac{2}{5}=-\frac{1\cdot5}{4\cdot5}+\frac{2\cdot4}{5\cdot4}$
$$=-\frac{5}{20}+\frac{8}{20}=\frac{3}{20}$$

5. $-11-7=-11+(-7)=-18$

6. a. $-20-5+12-(-3)$
$=-20+(-5)+12+3=-25+12+3$
$=-13+3=-10$

   b. $5.2-(-4.4)+(-8.8)$
$=5.2+4.4+(-8.8)=9.6+(-8.8)$
$=0.8$

7. **a.** $-9 + [(-4 - 1) - 10]$
      $= -9 + [(-4 + (-1)) - 10]$
      $= -9 + [(-5) - 10] = -9 + [-5 + (-10)]$
      $= -9 + [-15] = -24$

  b. $5^2 - 20 + [-11 - (-3)]$
      $= 5^2 - 20 + [-11 + 3]$
      $= 5^2 - 20 + [-8] = 25 - 20 + (-8)$
      $= 25 + (-20) + (-8) = 5 + (-8) = -3$

8. **a.** Replace x with 1 and y with -4.
      $\dfrac{x - y}{14 + x} = \dfrac{1 - (-4)}{14 + 1} = \dfrac{1 + 4}{14 + 1} = \dfrac{5}{15} = \dfrac{1}{3}$

  b. Replace x with 1 and y with -4.
      $x^2 - y = 1^2 - (-4) = 1^2 + 4 = 1 + 4 = 5$

9. Replace x with -2 and see if a true statement results.
   $-1 + x = 1$
   $-1 + (-2) \overset{?}{=} 1$
   $-3 \neq 1$
   Since $-3 = 1$ is a false statement, -2 is not a solution of the equation $-1 + x = 1$.

10. The overall change is the difference of the temperatures.
   $-23 - 14 = -23 + (-14) = -37$
   The overall change was $-37°$.

11. **a.** $x = 180° - 78° = 102°$

  b. $y = 90° - 81° = 9°$

Exercise Set 1.4

1. $-6 - 4 = -6 + (-4) = -10$

3. $4 - 9 = 4 + (-9) = -5$

5. $16 - (-3) = 16 + (3) = 19$

7. $\dfrac{1}{2} - \dfrac{1}{3} = \dfrac{1}{2} + \left(-\dfrac{1}{3}\right) = \dfrac{3 \cdot 1}{3 \cdot 2} + \left(-\dfrac{2 \cdot 1}{2 \cdot 3}\right)$
   $= \dfrac{3}{6} + \left(-\dfrac{2}{6}\right) = \dfrac{1}{6}$

9. $-16 - (-18) = -16 + (18) = 2$

11. $-6 - 5 = -6 + (-5) = -11$

13. $7 - (-4) = 7 + (4) = 11$

15. $-6 - (-11) = -6 + (11) = 5$

17. $16 - (-21) = 16 + (21) = 37$

19. $9.7 - 16.1 = 9.7 + (-16.1) = -6.4$

21. $-44 - 27 = -44 + (-27) = -71$

23. $-21 - (-21) = -21 + (21) = 0$

25. $-2.6 - (-6.7) = -2.6 + (6.7) = 4.1$

27. $-\dfrac{3}{11} - \left(-\dfrac{5}{11}\right) = -\dfrac{3}{11} + \left(\dfrac{5}{11}\right) = \dfrac{2}{11}$

29. $-\dfrac{1}{6} - \dfrac{3}{4} = -\dfrac{1}{6} + \left(-\dfrac{3}{4}\right)$
   $= -\dfrac{4 \cdot 1}{4 \cdot 6} + \left(-\dfrac{6 \cdot 3}{6 \cdot 4}\right) = -\dfrac{4}{24} + \left(-\dfrac{18}{24}\right)$
   $= -\dfrac{22}{24} = -\dfrac{2 \cdot 11}{2 \cdot 12} = -\dfrac{11}{12}$

31. $8.3 - (-0.62) = 8.3 + (0.62) = 8.92$

33. Sometimes positive and sometimes negative. If a and b are positive numbers and $a \geq b$, then $a - b \geq 0$. If a and b are positive numbers and $a \leq b$, then $a - b \leq 0$.

35. $8 - (-5) = 8 + (5) = 13$

37. $-6 - (-1) = -6 + (1) = -5$

39. $7 - 8 = 7 + (-8) = -1$

41. $-8 - 15 = -8 + (-15) = -23$

43. $-10 - (-8) + (-4) - 20$
   $= -10 + 8 + (-4) + (-20)$
   $= -2 + (-4) + (-20) = -6 + (-20) = -26$

45. $5 - 9 + (-4) - 8 - 8$
   $= 5 + (-9) + (-4) + (-8) + (-8)$
   $= -4 + (-4) + (-8) + (-8)$
   $= -8 + (-8) + (-8)$
   $= -16 + (-8)$
   $= -24$

47. $-6 - (2 - 11) = -6 - (-9) = -6 + 9 = 3$

49. $3^3 - 8 \cdot 9 = 27 - 8 \cdot 9 = 27 - 72$
   $= 27 + (-72) = -45$

51. $2 - 3(8 - 6) = 2 - 3(8 + (-6))$
$= 2 - 3(2) = 2 - 6 = 2 + (-6)$
$= -4$

53. $(3 - 6) + 4^2 = (3 + (-6)) + 4^2 = (-3) + 4^2$
$= -3 + 16 = 13$

55. $-2 + [(8 - 11) - (-2 - 9)]$
$= -2 + [(8 + (-11)) - (-2 + (-9))]$
$= -2 + [(-3) - (-11)]$
$= -2 + [-3 + 11]$
$= -2 + [8]$
$= 6$

57. $|-3| + 2^2 + [-4 - (-6)] = 3 + 2^2 + [-4 + 6]$
$= 3 + 2^2 + [2] = 3 + 4 + 2$
$= 7 + 2 = 9$

59. Replace x with -5 and y with 4.
$x - y = -5 - 4 = -5 + (-4) = -9$

61. Replace x with -5, y with 4, and t with 10.
$|x| + 2t - 8y = |-5| + 2(10) - 8(4)$
$= 5 + 20 - 32 = 5 + 20 + (-32)$
$= 25 + (-32) = -7$

63. Replace x with -5 and y with 4.
$\dfrac{9 - x}{y + 6} = \dfrac{9 - (-5)}{4 + 6} = \dfrac{9 + 5}{4 + 6} = \dfrac{14}{10} = \dfrac{2 \cdot 7}{2 \cdot 5} = \dfrac{7}{5}$

65. Replace x with -5 and y with 4.
$y^2 - x = 4^2 - (-5) = 16 - (-5)$
$= 16 + 5 = 21$

67. Replace x with -5 and t with 10.
$\dfrac{|x - (-10)|}{2t} = \dfrac{|-5 - (-10)|}{2(10)} = \dfrac{|-5 + 10|}{20}$
$= \dfrac{|5|}{20} = \dfrac{5}{20} = \dfrac{5}{4 \cdot 5} = \dfrac{1}{4}$

69. Replace x with -4 and see if a true statement results.
$x - 9 = 5$
$-4 - 9 \stackrel{?}{=} 5$
$-4 + (-9) \stackrel{?}{=} 5$
$-13 \neq 5$
-4 is not a solution of $x - 9 = 5$.

71. Replace x with -2 and see if a true statement results.
$-x + 6 = -x - 1$
$-(-2) + 6 \stackrel{?}{=} -(-2) - 1$
$2 + 6 \stackrel{?}{=} 2 - 1$
$8 \stackrel{?}{=} 2 + (-1)$
$8 \neq 1$
-2 is not a solution of $-x + 6 = -x - 1$.

73. Replace x with 2 and see if a true statement results.
$-x - 13 = -15$
$-2 - 13 \stackrel{?}{=} -15$
$-2 + (-13) \stackrel{?}{=} -15$
$-15 = -15$
2 is a solution of $-x - 13 = -15$.

75. To find the drop in temperature, find the difference between the high temperature and low temperature.
$44 - (-56) = 44 + 56 = 100$
The temperature dropped $100°$.

77. The total gain or loss of yardage is the sum of the gains and losses. Gains are represented by positive numbers and losses are represented by negative numbers.
$2 + (-5) + (-20) = -3 + (-20) = -23$
The 49ers lost a total of 23 yards.

79. If Aristotle died in the year -322 and he was 62 years old when he died, then the year of his birth is found by subtracting 62 from -322.
$-322 - 62 = -322 + (-62) = -384$
Aristotle was born in 384 B.C.

81. The overall vertical change is the sum of the gains and losses in altitude. Gains are represented by positive numbers and losses are represented by negative numbers.
$-250 + 120 + (-178) = -130 + (-178)$
$= -308$
The overall vertical change is -308 feet.

83. To find how much higher Mount Aconcagua is than the Valdes Peninsula, find the difference between the elevations.
$22,834 - (-131) = 22,834 + 131 = 22,965$
Mount Aconcagua is 22,965 feet higher than the Valdes Peninsula.

85. These angles are supplementary, so their sum is 180°.
$y = 180° - 50° = 130°$

87. These angles are complementary, so their sum is 90°.
$x = 90° - 60° = 30°$

89. The daily change in temperature is found by subtracting yesterday's temperature from today's temperature. A temperature increase is represented by a positive number. A temperature decrease is represented by a negative number. On Monday, the temperature increased from –4° to 3°.
$3 - (-4) = 3 + 4 = 7$
On Monday, the temperature increased 7°.

On Tuesday, the temperature increased from 3° to 7°.
$7 - 3 = 7 + (-3) = 4$
On Tuesday, the temperature increased 4°.

On Wednesday, the temperature decreased from 7° to –2°.
$-2 - 7 = -2 + (-7) = -9$
On Wednesday, the temperature decreased 9°.

On Thursday, the temperature increased from –2° to 1°.
$1 - (-2) = 1 + 2 = 3$
On Thursday, the temperature increased 3°.

On Friday, the temperature decreased from 1° to –5°.
$-5 - 1 = -5 + (-1) = -6$
On Friday, the temperature decreased 6°.

Finally, on Saturday, the temperature increased from –5° to 1°.
$1 - (-5) = 1 + 5 = 6$
On Saturday, the temperature increased 6°.

Day	Daily Increase or Decrease
Monday	7°
Tuesday	4°
Wednesday	–9°
Thursday	3°
Friday	–6°
Saturday	6°

91. Wednesday

93. True

95. True

97. $4.362 + (-7.0086)$
Negative since $7.0086 > 4.362$
$4.362 - 7.0086 = -2.6466$

Section 1.5

Practice Problems

1. $-8(3) = -24$

2. $5(-30) = -150$

3. $-4(-12) = 48$

4. $-\dfrac{5}{6} \cdot \dfrac{1}{4} = -\dfrac{5 \cdot 1}{6 \cdot 4} = -\dfrac{5}{24}$

5. $6(-2.3) = -13.8$

6. $-15(-2) = 30$

7. a. $5(0)(-3) = 0(-3) = 0$

   b. $(-1)(-6)(-7) = (6)(-7) = -42$

   c. $(-2)(4)(-8) = (-8)(-8) = 64$

   d. $(-2)^2 = (-2)(-2) = 4$

   e. $-3(-9) - 4(-4) = 27 - (-16) = 27 + 16$
      $= 43$

8. Replace x with -1 and y with -5.

    a. $\quad 3x - y = 3(-1) - (-5) = -3 - (-5)$
          $= -3 + 5 = 2$

    b. $\quad x^2 - y^3 = (-1)^2 - (-5)^3 = 1 - (-125)$
          $= 1 + 125 = 126$

9. Replace x with -10 and see if a true
statement results.
$3x + 4 = -26$
$3(-10) + 4 \overset{?}{=} -26$
$-30 + 4 \overset{?}{=} -26$
$-26 = -26$
Since $-26 = -26$ is a true statement, -10 is a
solution of the equation.

Exercise Set 1.5

1. $-6(4) = -24$

3. $2(-1) = -2$

5. $-5(-10) = 50$

7. $-3 \cdot 4 = -12$

9. $-6(-7) = 42$

11. $2(-9) = -18$

13. $-\dfrac{1}{2}\left(-\dfrac{3}{5}\right) = -\left(-\dfrac{1 \cdot 3}{2 \cdot 5}\right) = -\left(-\dfrac{3}{10}\right) = \dfrac{3}{10}$

15. $-\dfrac{3}{4}\left(-\dfrac{8}{9}\right) = -\left(-\dfrac{3 \cdot 8}{4 \cdot 9}\right) = -\left(-\dfrac{24}{36}\right)$
    $= \dfrac{24}{36} = \dfrac{2 \cdot 12}{3 \cdot 12} = \dfrac{2}{3}$

17. $5(-1.4) = -7.0$

19. $-0.2(-0.7) = 0.14$

21. $-10(80) = -800$

23. $4(-7) = -28$

25. $(-5)(-5) = 25$

27. $\dfrac{2}{3}\left(-\dfrac{4}{9}\right) = -\dfrac{2 \cdot 4}{3 \cdot 9} = -\dfrac{8}{27}$

29. $-11(11) = -121$

31. $-\dfrac{20}{25}\left(\dfrac{5}{16}\right) = -\dfrac{20 \cdot 5}{25 \cdot 16} = -\dfrac{100}{400}$
    $= -\dfrac{4 \cdot 5 \cdot 5}{5 \cdot 5 \cdot 4 \cdot 4} = -\dfrac{1}{4}$

33. $-2.1(-0.4) = 0.84$

35. $(-1)(2)(-3)(-5) = (-2)(-3)(-5)$
    $= (6)(-5) = -30$

37. $(2)(-1)(-3)(5)(3) = (-2)(-3)(5)(3)$
    $= (6)(5)(3) = (30)(3) = 90$

39. $(-4)^2 = (-4)(-4) = 16$

41. $(-6)(3)(-2)(-1) = (-18)(-2)(-1)$
    $= (36)(-1) = -36$

43. $(-5)^3 = (-5)(-5)(-5) = (25)(-5) = -125$

45. $-4^2 = -(4)(4) = -16$

47. $-3(2 - 8) = -3(2 + (-8)) = -3(-6) = 18$

49. $6(3 - 8) = 6(3 + (-8)) = 6(-5) = -30$

51. $-3[(2 - 8) - (-6 - 8)]$
    $= -3[(2 + (-8)) - (-6 + (-8))]$
    $= -3[-6 - (-14)]$
    $= -3[-6 + 14]$
    $= -3[8] = -24$

53. $\left(-\dfrac{3}{4}\right)^2 = \left(-\dfrac{3}{4}\right)\left(-\dfrac{3}{4}\right) = -\left(-\dfrac{3 \cdot 3}{4 \cdot 4}\right)$
    $= -\left(-\dfrac{9}{16}\right) = \dfrac{9}{16}$

55. True

57. False

59. Replace x with -5 and y with -3 and
simplify.
$3x + 2y = 3(-5) + 2(-3) = -15 - 6$
$= -15 + (-6) = -21$

61. Replace x with -5 and y with -3 and
simplify.
$2x^2 - y^2 = 2(-5)^2 - (-3)^2 = 2(25) - (9)$
$= 50 - 9 = 50 + (-9) = 41$

63. Replace x with -5 and y with -3 and simplify.
$$x^3 + 3y = (-5)^3 + 3(-3) = -125 - 9$$
$$= -125 + (-9) = -134$$

65. Replace x with 7 and see if a true statement results.
$$-5x = -35$$
$$-5(7) \overset{?}{=} -35$$
$$-35 = -35$$
Since $-35 = -35$ is a true statement, 7 is a solution of the equation.

67. Replace x with 5 and see if a true statement results.
$$-3x - 5 = -20$$
$$-3(5) - 5 \overset{?}{=} -20$$
$$-15 - 5 \overset{?}{=} -20$$
$$-15 + (-5) \overset{?}{=} -20$$
$$-20 = -20$$
Since $-20 = -20$ is a true statement, 5 is a solution of the equation.

69. Replace x with -1 and see if a true statement results.
$$9x + 1 = 14$$
$$9(-1) + 1 \overset{?}{=} 14$$
$$-9 + 1 \overset{?}{=} 14$$
$$-8 \neq 14$$
Since $-8 = 14$ is not a true statement, -1 is not a solution of the equation.

71. Replace x with 5 and see if a true statement results.
$$3x - 20 = -5$$
$$3(5) - 20 \overset{?}{=} -5$$
$$15 - 20 \overset{?}{=} -5$$
$$15 + (-20) \overset{?}{=} -5$$
$$-5 = -5$$
Since $-5 = -5$ is a true statement, 5 is a solution of the equation.

73. Replace x with -2 and see if a true statement results.
$$17 - 4x = x + 27$$
$$17 - 4(-2) \overset{?}{=} -2 + 27$$
$$17 - (-8) \overset{?}{=} -2 + 27$$
$$17 + 8 \overset{?}{=} 25$$
$$25 = 25$$

Since $25 = 25$ is a true statement, -2 is a solution of the equation.

75. Positive; since the product $q \cdot r$ is positive, the product $(q \cdot r) \cdot t$ is positive.

77. Can't determine.

79. Negative; since $q + r$ is negative, the product $t(q \cdot r)$ is negative.

81. Since the share price decreases in value by $1.50 for each day that goes by, the stock decreases by $1 \cdot \$1.50 = \$1.50$ on the first day, $2 \cdot \$1.50 = \$3.00$ after 2 days, $3 \cdot \$1.50 = \$4.50$ after 3 days, and so on. We can express this trend in a more general way by letting x represent the number of days. Then $x \cdot \$1.50$ represents the decrease in value of the stock after x days. To find the final value of the stock we need to subtract the loss, $x \cdot \$1.50$, from the initial price, $38. Thus, the expression $38 - 1.50x$ can be used to find the price of the stock after x days. If we replace x with 4 we obtain
$$38 - 1.50x = 38 - 1.50(4) = 38 - 6$$
$$= 38 + (-6) = 32$$
Thus, 4 days after September 20, the value of the stock will be $32. To find when the stock price will reach $20 per share, we continue evaluating the expression $38 - 1.50x$ with increasing values of x until we find the value of x that makes $38 - 1.50x = 20$ a true statement. If we do this, we find that when $x = 12$, $38 - 1.50x = 20$ is a true statement.
$$38 - 1.50x = 20$$
$$38 - 1.50(12) \overset{?}{=} 20$$
$$38 - 18 \overset{?}{=} 20$$
$$38 + (-18) \overset{?}{=} 20$$
$$20 = 20$$
Therefore, 12 days after September 20, on Saturday, October 2, the stock will be worth $20 per share.

83. No; answers may vary.

Section 1.6

Practice Problems

1. The reciprocal of 13 is $\frac{1}{13}$ since $13 \cdot \frac{1}{13} = 1$.

2. The reciprocal of $\frac{7}{15}$ is $\frac{15}{7}$ since
   $\frac{7}{15} \cdot \frac{15}{7} = 1$.

3. The reciprocal of -5 is $-\frac{1}{5}$ since
   $-5 \cdot -\frac{1}{5} = 1$.

4. The reciprocal of $-\frac{8}{11}$ is $-\frac{11}{8}$ since
   $-\frac{8}{11} \cdot -\frac{11}{8} = 1$.

5. The reciprocal of 7.9 is $\frac{1}{7.9}$ since
   $7.9 \cdot \frac{1}{7.9} = 1$.

6. a. $-12 \div 4 = -12 \cdot \frac{1}{4} = -3$

   b. $\frac{-20}{-10} = -20 \cdot -\frac{1}{10} = 2$

   c. $\frac{36}{-4} = 36 \cdot -\frac{1}{4} = -9$

7. $\frac{-25}{5} = -5$

8. $\frac{-48}{-6} = 8$

9. $\frac{50}{-2} = -25$

10. $\frac{-72}{0.2} = -360$

11. $-\frac{5}{9} \div \frac{2}{3} = -\frac{5}{9} \cdot \frac{3}{2} = -\frac{5 \cdot 3}{9 \cdot 2} = -\frac{15}{18} = -\frac{5}{6}$

12. $-\frac{2}{7} \div \left(-\frac{1}{5}\right) = -\frac{2}{7} \cdot \left(-\frac{5}{1}\right) = \frac{10}{7}$

13. $\frac{-7}{0}$ is undefined.

14. $\frac{0}{-2} = 0$

15. $\frac{0(-5)}{3} = \frac{0}{3} = 0$

16. a. $\frac{-7(-4)+2}{-10-(-5)} = \frac{28+2}{-10+5} = \frac{30}{-5} = -6$

    b. $\frac{5(-2)^3 + 52}{-4+1} = \frac{5(-8)+52}{-4+1} = \frac{-40+52}{-4+1}$
       $= \frac{12}{-3} = -4$

17. Replace x with -1 and y with -5 and simplify.
    $\frac{x+y}{3x} = \frac{-1+(-5)}{3(-1)} = \frac{-1+(-5)}{-3} = \frac{-6}{-3} = 2$

18. $\frac{x}{4} - 3 = x + 3$
    $\frac{-8}{4} - 3 \overset{?}{=} -8 + 3$
    $-2 - 3 \overset{?}{=} -8 + 3$
    $-5 = -5$
    Since $-5 = -5$ is a true statement, -8 is a solution of the equation.

Exercise Set 1.6

1. The reciprocal of 9 is $\frac{1}{9}$ since $9 \cdot \frac{1}{9} = 1$.

3. The reciprocal of $\frac{2}{3}$ is $\frac{3}{2}$ since $\frac{2}{3} \cdot \frac{3}{2} = 1$.

5. The reciprocal of -14 is $-\frac{1}{14}$ since
   $(-14)\left(-\frac{1}{14}\right) = 1$.

7. The reciprocal of $-\frac{3}{11}$ is $-\frac{11}{3}$ since
   $\left(-\frac{3}{11}\right)\left(-\frac{11}{3}\right) = 1$.

9. The reciprocal of 0.2 is $\frac{1}{0.2}$ since $0.2 \cdot \frac{1}{0.2} = 1$.

11. The reciprocal of $\frac{1}{-6.3}$ is -6.3 since $\left(\frac{1}{-6.3}\right)(-6.3) = 1$.

13. $1, -1$

15. $\frac{18}{-2} = 18 \cdot -\frac{1}{2} = -9$

17. $\frac{-16}{-4} = -16 \cdot -\frac{1}{4} = 4$

19. $\frac{-48}{12} = -48 \cdot \frac{1}{12} = -4$

21. $\frac{0}{-4} = 0 \cdot -\frac{1}{4} = 0$

23. $-\frac{15}{3} = -15 \cdot \frac{1}{3} = -5$

25. $\frac{5}{0}$ is undefined

27. $\frac{-12}{-4} = -12 \cdot -\frac{1}{4} = 3$

29. $\frac{30}{-2} = 30 \cdot -\frac{1}{2} = -15$

31. $\frac{6}{7} \div \left(-\frac{1}{3}\right) = \frac{6}{7} \cdot \left(-\frac{3}{1}\right) = -\frac{18}{7}$

33. $-\frac{5}{9} \div \left(-\frac{3}{4}\right) = -\frac{5}{9} \cdot \left(-\frac{4}{3}\right) = \frac{20}{27}$

35. $-\frac{4}{9} \div \frac{4}{9} = -\frac{4}{9} \cdot \frac{9}{4} = -1$

37. $-\frac{5}{8} \div \frac{3}{4} = -\frac{5}{8} \cdot \frac{4}{3} = -\frac{20}{24} = -\frac{5}{6}$

39. $-48 \div 1.2 = -48 \cdot \frac{1}{1.2} = -40$

41. $-3.2 \div -0.02 = -3.2 \cdot -\frac{1}{0.02} = 160$

43. $\frac{-9(-3)}{-6} = \frac{27}{-6} = -\frac{9}{2}$

45. $\frac{12}{9-12} = \frac{12}{-3} = -4$

47. $\frac{-6^2+4}{-2} = \frac{-36+4}{-2} = \frac{-32}{-2} = 16$

49. $\frac{8+(-4)^2}{4-12} = \frac{8+16}{4-12} = \frac{24}{-8} = -3$

51. $\frac{22+(3)(-2)}{-5-2} = \frac{22-6}{-5-2} = \frac{16}{-7} = -\frac{16}{7}$

53. $\frac{-3-5^2}{2(-7)} = \frac{-3-25}{-14} = \frac{-28}{-14} = 2$

55. $\frac{6-2(-3)}{4-3(-2)} = \frac{6+6}{4+6} = \frac{12}{10} = \frac{6}{5}$

57. $\frac{-3-2(-9)}{-15-3(-4)} = \frac{-3+18}{-15+12} = \frac{15}{-3} = -5$

59. $\frac{|5-9|+|10-15|}{|2(-3)|} = \frac{|-4|+|-5|}{|-6|} = \frac{4+5}{6}$
$= \frac{9}{6} = \frac{3}{2}$

61. Replace x with -5 and y with -3 and simplify.
$\frac{2x-5}{y-2} = \frac{2(-5)-5}{-3-2} = \frac{-10-5}{-5}$
$= \frac{-15}{-5} = 3$

63. Replace x with -5 and y with -3 and simplify.
$\frac{6-y}{x-4} = \frac{6-(-3)}{-5-4} = \frac{6+3}{-5-4}$
$= \frac{9}{-9} = -1$

65. Replace x with -5 and y with -3 and simplify.
$\frac{x+y}{3y} = \frac{-5+(-3)}{3(-3)} = \frac{-8}{-9} = \frac{8}{9}$

67. Replace x with 2.

$$\frac{-10}{x} = -5$$

$$\frac{-10}{2} \stackrel{?}{=} -5$$

$$-5 = -5$$

Since $-5 = -5$ is a true statement, 2 is a solution of the equation.

69. Replace x with 15.

$$\frac{x}{5} + 2 = -1$$

$$\frac{15}{5} + 2 \stackrel{?}{=} -1$$

$$3 + 2 \stackrel{?}{=} -1$$

$$5 \neq -1$$

Since $5 = -1$ is a false statement, 15 is not a solution of the equation.

71. Replace x with -30.

$$\frac{x+4}{5} = -6$$

$$\frac{-30+4}{5} \stackrel{?}{=} -6$$

$$\frac{-26}{5} = -6$$

Since $-\frac{26}{5} = -6$ is a false statement, -30 is not a solution of the equation.

73. $\dfrac{0}{5} - 7 = 0 - 7 = -7$

75. $(-8)(-5) + (-1) = 40 - 1 = 39$

77. $\dfrac{-8}{-20} = \dfrac{2}{5}$

79. The quotient of two numbers with different signs is a negative number. Thus, if a is a positive number and b is a negative number, then $\frac{a}{b}$ is a negative number.

81. The quotient of two numbers with different signs is a negative number. If a is a positive number and b is a negative number, then $b + b$ is a negative number and $a + a$ is a positive number. Therefore $\frac{b+b}{a+a}$ is a negative number.

83. Answers may vary.

Integrated Review

1. $5(-7) = -35$

2. $-3(-10) = 30$

3. $\dfrac{-20}{-4} = 5$

4. $\dfrac{30}{-6} = -5$

5. $7 - (-3) = 7 + 3 = 10$

6. $-8 - 10 = -8 + (-10) = -18$

7. $-14 - (-12) = -14 + 12 = -2$

8. $-3 - (-1) = -3 + 1 = -2$

9. $-\dfrac{1}{2}\left(-\dfrac{3}{4}\right) = -\left(-\dfrac{1\cdot 3}{2\cdot 4}\right) = -\left(-\dfrac{3}{8}\right) = \dfrac{3}{8}$

10. $-\dfrac{2}{7}\left(\dfrac{11}{12}\right) = -\dfrac{2\cdot 11}{7\cdot 12} = -\dfrac{22}{84} = -\dfrac{11}{42}$

11. $\dfrac{-12}{0.2} = -60$

12. $\dfrac{-3.8}{-2} = 1.9$

13. $-19 + (-23) = -42$

14. $18 + (-25) = -7$

15. $-15 + 17 = 2$

16. $-2 + (-37) = -39$

17. $(-8)^2 = (-8)(-8) = 64$

18. $-9^2 = -(9)(9) = -81$

19. $-3^3 = -(3)(3)(3) = -27$

20. $(-2)^4 = (-2)(-2)(-2)(-2) = 16$

21. $(2)(-8)(-3) = -16(-3) = 48$

22. $3(-2)(5) = -6(5) = -30$

23. $-6(2) - 5(2) - 4 = -12 - 10 - 4$
$= -12 + (-10) + (-4) = -22 + (-4) = -26$

24. $(7 - 10)(4 - 6) = (7 + (-10))(4 + (-6))$
$= (-3)(-2) = 6$

25. $2(19 - 17)^3 - 3(7 - 9)^2$
$= 2(19(-17))^3 - 3(7 + (-9))^2$
$= 2(2)^3 - 3(-2)^2$
$= 2(8) - 3(4) = 16 - 12$
$= 16 + (-12) = 4$

26. $3(10 - 9)^2 - 6(20 - 19)^3$
$= 3(10 + (-9))^2 - 6(20 + (-19))^3$
$= 3(1)^2 - 6(1)^3$
$= 3(1) - 6(1) = 3 - 6$
$= -3$

27. $\dfrac{19 - 25}{3(-1)} = \dfrac{19 + (-25)}{3(-1)} = \dfrac{-6}{-3} = 2$

28. $\dfrac{8(-4)}{-2} = \dfrac{-32}{-2} = 16$

29. $\dfrac{-2(3 - 6) - 6(10 - 9)}{-6 - (-5)}$
$= \dfrac{-2(3 + (-6)) - 6(10 + (-9))}{-6 + 5}$
$= \dfrac{-2(-3) - 6(1)}{-1} = \dfrac{6 - 6}{-1}$
$= \dfrac{0}{-1} = 0$

30. $\dfrac{-4(8 - 10)^3}{-2 - 1 - 12} = \dfrac{-4(8 + (-10))^3}{-2 + (-1) + (-12)}$
$= \dfrac{-4(-2)^3}{-2 + (-1) + (-12)} = \dfrac{-4(-8)}{-2 + (-1) + (-12)}$
$= \dfrac{32}{-2 + (-1) + (-12)} = \dfrac{32}{-3 + (-12)}$
$= -\dfrac{32}{15}$

Section 1.7

Practice Problems

1. a. $7 \cdot y = y \cdot 7$

b. $4 + x = x + 4$

2. a. $5 \cdot (-3 \cdot 6) = (5 \cdot -3) \cdot 6$

b. $(-2 + 7) + 3 = -2 + (7 + 3)$

3. Commutative property, since the order was changed.

4. Associative property, since the grouping was changed.

5. $(-3 + x) + 17 = (x + (-3)) + 17$
$= x + (-3 + 17) = x + 14$

6. $4(5x) = (4 \cdot 5)x = 20x$

7. $5(x + y) = 5(x) + 5(y) = 5x + 5y$

8. $-3(2 + 7x) = -3(2) + (-3)(7x) = -6 - 21x$

9. $4(x + 6y - 2z) = 4(x) + 4(6y) - 4(2z)$
$= 4x + 24y - 8z$

10. $-1(3 - a) = -1(3) - (-1)(a) = -3 + a$

11. $-(8 + a - b) = -1(8 + a - b)$
$= -1(8) + (-1)(a) - (-1)(b) = -8 - a + b$

12. $9(2x + 4) + 9 = 9(2x) + 9(4) + 9$
$= 18x + 36 + 9 = 18x + 45$

13. $9 \cdot 3 + 9 \cdot y = 9(3 + y)$

14. $4x + 4y = 4(x + y)$

15. Additive inverse property

16. Commutative property of addition

17. Associative property of multiplication

18. Commutative peroperty of addition

19. Multiplicative inverse property

20. Identity element for addition

21. Commutative and associative properties of multiplication

Exercise Set 1.7

1. $x + 16 = 16 + x$

3. $-4 \cdot y = y \cdot (-4)$

5. $xy = yx$

7. $2x + 13 = 13 + 2x$

9. $(xy) \cdot z = x \cdot (yz)$

11. $2 + (a + b) = (2 + a) + b$

13. $4 \cdot (ab) = 4a \cdot (b)$

15. $(a + b) + c = a + (b + c)$

17. $8 + (9 + b) = (8 + 9) + b = 17 + b$

19. $4(6y) = (4 \cdot 6) \cdot y = 24 \cdot y = 24y$

21. $\frac{1}{5}(5y) = \left(\frac{1}{5} \cdot 5\right) \cdot y = 1 \cdot y = y$

23. $(13 + a) + 13 = (a + 13) + 13 = a + (13 + 13)$
     $= a + 26$

25. $-9(8x) = (-9 \cdot 8) \cdot x = -72 \cdot x = -72x$

27. $\frac{3}{4}\left(\frac{4}{3}s\right) = \left(\frac{3}{4} \cdot \frac{4}{3}\right) \cdot s = 1 \cdot s = s$

29. Answers may vary.

31. $4(x + y) = 4(x) + 4(y) = 4x + 4y$

33. $9(x - 6) = 9(x) - 9(6) = 9x - 54$

35. $2(3x + 5) = 2(3x) + 2(5) = 6x + 10$

37. $7(4x - 3) = 7(4x) - 7(3) = 28x - 21$

39. $3(6 + x) = 3(6) + 3(x) = 18 + 3x$

41. $-2(y - z) = -2(y) - (-2)(z) = -2y + 2z$

43. $-7(3y + 5) = -7(3y) + (-7)(5) = -21y - 35$

45. $5(x + 4m + 2) = 5(x) + 5(4m) + 5(2)$
     $= 5x + 20m + 10$

47. $-4(1 - 2m + n) = -4(1) - (-4)(2m) + (-4)(n)$
     $= -4 + 8m - 4n$

49. $-(5x + 2) = -1(5x + 2) = (-1)(5x) + (-1)(2)$
     $= -5x - 2$

51. $-(r - 3 - 7p) = -1(r - 3 - 7p)$
     $= (-1)(r) - (-1)(3) - (-1)(7p)$
     $= -r + 3 + 7p$

53. $\frac{1}{2}(6x + 8) = \frac{1}{2}(6x) + \frac{1}{2}(8)$
     $= \left(\frac{1}{2} \cdot 6\right)x + \left(\frac{1}{2} \cdot 8\right) = 3x + 4$

55. $-\frac{1}{3}(3x - 9y) = -\frac{1}{3}(3x) - \left(-\frac{1}{3}\right)(9y)$
     $= \left(-\frac{1}{3} \cdot 3\right)x - \left(-\frac{1}{3} \cdot 9\right)y = (-1)x - (-3)y$
     $= -x + 3y$

57. $3(2r + 5) - 7 = 3(2r) + 3(5) - 7$
     $= 6r + 15 - 7 = 6r + 8$

59. $-9(4x + 8) + 2 = -9(4x) + (-9)(8) + 2$
     $= -36x - 72 + 2 = -36x - 70$

61. $-4(4x + 5) - 5 = -4(4x) + (-4)(5) - 5$
     $= -16x - 20 - 5 = -16x - 25$

63. $4 \cdot 1 + 4 \cdot y = 4(1 + y)$

65. $11x + 11y = 11(x + y)$

67. $(-1) \cdot 5 + (-1) \cdot x = -1(5 + x) = -(5 + x)$

69. $30a + 30b = 30(a + b)$

71. The additive inverse of 16 is -16 since
     $16 + (-16) = 0$.

73. The additive inverse of -8 is 8 since
     $(-8) + 8 = 0$.

75. The additive inverse of $-(-1.2)$ is -1.2 since
     $-(-1.2) + (-1.2) = 1.2 + (-1.2) = 0$

77. The additive inverse of $-|-2|$ is 2 since
     $-|-2| + 2 = -2 + 2 = 0$

79. The multiplicative inverse of $\frac{2}{3}$ is $\frac{3}{2}$ since

$\frac{2}{3} \cdot \frac{3}{2} = 1$.

81. The multiplicative inverse of $-\frac{5}{6}$ is $-\frac{6}{5}$

since $-\frac{5}{6} \cdot \left(-\frac{6}{5}\right) = 1$.

83. The multiplicative inverse of $3\frac{5}{6}$ is $\frac{6}{23}$

since

$3\frac{5}{6} \cdot \frac{6}{23} = \frac{23}{6} \cdot \frac{6}{23} = 1$

85. The multiplicative inverse of -2 is $-\frac{1}{2}$ since

$-2 \cdot \left(-\frac{1}{2}\right) = 1$.

87. The commutative property of multiplication

89. The associative property of addition

91. The distributive property

93. The associative property of multiplication

95. The identity property of addition

97. The distributive property

99. The associative and commutative properties of multiplication

101.

Expression	Opposite	Reciprocal
8	−8	$\frac{1}{8}$

103.

Expression	Opposite	Reciprocal
x	$-x$	$\frac{1}{x}$

105.

Expression	Opposite	Reciprocal
$2x$	$-2x$	$\frac{1}{2x}$

107. a. $\triangle + (\square + \bigcirc) = (\square + \bigcirc) + \triangle$ by the commutative property of addition

    b. $(\square + \bigcirc) + \triangle = (\bigcirc + \square) + \triangle$ by the commutative property of addition

    c. $(\bigcirc + \square) + \triangle = \bigcirc + (\square + \triangle)$ by the associative property of addition

109. Answers may vary.

Section 1.8

Practice Problems

1. a. East Mississippi Electric charges approximately 8¢ per kilowatt-hour.

    b. Southern California Edison charges approximately 12¢ per kilowatt-hour.

    c. The difference in rates for East Mississippi Electric and Southern California Edison is approximately 12¢ − 8¢ or 4¢.

2. a. *Snow White and the Seven Dwarfs* generated approximately 175 million dollars.

    b. *The Jungle Book* generated approximately 130 million dollars. *Bambi* generated approximately 110 million dollars.
130 − 110 = 20
The Jungle Book generated approximately 20 million dollars more than *Bambi*.

3. a. The total cost of renting the truck if 50 miles were driven is approximately $50.

    b. The truck is driven approximately 180 miles.

4. a. The pulse rate is 70 beats per minute 40 minutes after lighting a cigarette.

    b. The pulse rate is 60 beats per minute when the cigarette is being lit.

c. By locating the highest point of the line graph, we see that the pulse rate is the highest 5 minutes after lighting a cigarette.

Exercise Set 1.8

1. The number of teenagers expected to use the internet in 1999 is about 7.8 million.

3. From the graph, the greatest increase in the heights of successive bars is from 1999 to 2002. The greatest increase is in 2002.

5. Look for the shortest bar, which is the bar representing the PGA/LPGA tours. The PGA/LPGA tours spent the least amount of money on advertising.

7. Major League Baseball and the NBA each spent over $10,000,000 on advertising.

9. The NBA spent about $15 million on advertising.

11. The tallest bar corresponds to the sports team that has gone the longest time without being in a playoff. The Cleveland Indians baseball team has gone the longest time without being in a playoff.

13. Each team has gone an equal number of years without being in a playoff.

15. The longest time a football team has gone without being in a playoff is about 12 years. The longest time a basketball team has gone without being in a playoff is about 9 years.
$12 - 9 = 3$
In football, a team has gone approximately 3 years longer without being in a playoff than in basketball.

17. *Pocahontas* generated approximately 142 million dollars, or $142,000,000.

19. *Snow White and the Seven Dwarfs*

21. Answers may vary.

23. Find the highest point of the line graph. The highest percent of arson fires started by juveniles occurred in 1994.

25. 1986 and 1987 or 1989 and 1990

27. The percent of arson fires started by juveniles in 1995 was approximately 52%.

29. The greatest increase in the percent of arson fires started by juveniles occurred in 1994. Notice that the line graph is steepest between the years 1993 and 1994.

31. The pulse rate was approximately 59 beats per minute 5 minutes before lighting a cigarette.

33. 5 minutes before lighting a cigarette, the pulse rate was approximately 59 beats per minute. 10 minutes after lighting a cigarette, the pulse rate had increased to about 85 beats per minute.
$85 - 59 = 26$
The pulse rate increased by 26 beats per minute between 5 minutes before and 10 minutes after lighting a cigarette.

35. In 1992 the cost of newsprint was less than $500 per metric ton.

37. The cost of newsprint was lowest in 1992, when it cost approximately $450 per metric ton.

39. New Orleans, Louisiana is located at latitude 30° north and longitude 90° west.

41. Answers may vary.

Chapter 1 Review

1. $8 < 10$

2. $7 > 2$

3. $-4 > -5$

4. $\frac{12}{2} > -8$

5. $|-7| < |-8|$

6. $|-9| > -9$

7. $-|-1| = -1$

8. $|-14| = -(-14)$

9. $1.2 > 1.02$

10. $-\dfrac{3}{2} < -\dfrac{3}{4}$

11. $4 \geq -3$

12. $6 \neq 5$

13. $0.03 < 0.3$

14. $50 > 40$

15. **a.** The natural numbers are 1 and 3.

     b. The whole numbers are 0, 1, and 3.

     c. The integers are –6, 0, 1, and 3.

     d. The rational numbers are –6, 0, 1, $1\frac{1}{2}$, 3, and 9.62.

     e. The irrational number is π.

     f. The real numbers are all numbers in the given set.

16. **a.** The natural numbers are 2 and 5.

     b. The whole numbers are 2 and 5.

     c. The integers are –3, 2, and 5.

     d. The rational numbers are –3, –1.6, 2, 5, $\frac{11}{2}$, and 15.1.

     e. The irrational numbers are $\sqrt{5}$ and 2π.

     f. The real numbers are all numbers in the given set.

17. Losses are represented by negative numbers. The greatest loss occurred on Friday when the stock lost 4 dollars.

18. Gains are represented by positive numbers. The greatest gain occurred on Wednesday when the stock gained 5 dollars.

19. c
$$6 \cdot 3^2 + 2 \cdot 8 = 6 \cdot 9 + 2 \cdot 8$$
$$= 54 + 16$$
$$= 70$$

20. b
$$68 - 5 \cdot 2^3 = 68 - 5 \cdot 8 = 68 - 40$$
$$= 68 + (-40) = 28$$

21. $3(1 + 2 \cdot 5) + 4 = 3(1 + 10) + 4$
$$= 3(11) + 4 = 33 + 4 = 37$$

22. $8 + 3(2 \cdot 6 - 1) = 8 + 3(12 - 1)$
$$= 8 + 3(12 + (-1)) = 8 + 3(11)$$
$$= 8 + 33 = 41$$

23. $\dfrac{4 + |6 - 2| + 8^2}{4 + 6 \cdot 4} = \dfrac{4 + |4| + 64}{4 + 24}$
$$= \dfrac{4 + 4 + 64}{4 + 24} = \dfrac{72}{28} = \dfrac{4 \cdot 18}{4 \cdot 7} = \dfrac{18}{7}$$

24. $5[3(2 + 5) - 5] = 5[3(7) - 5)] = 5[21 - 5]$
$$= 5[21 + (-5)] = 5[16] = 80$$

25. $20 - 12 = 2 \cdot 4$

26. $\dfrac{9}{2} > -5$

27. Replace x with 6 and y with 2 then simplify.
$$2x + 3y = 2(6) + 3(2) = 12 + 6 = 18$$

28. Replace x with 6, y with 2, and z with 8 then simplify.
$$x(y + 2z) = 6(2 + 2(8)) = 6(2 + 16)$$
$$= 6(18) = 108$$

29. Replace x with 6, y with 2, and z with 8 then simplify.
$$\frac{x}{y} + \frac{z}{2y} = \frac{6}{2} + \frac{8}{2(2)} = \frac{6}{2} + \frac{8}{4} = 3 + 2 = 5$$

30. Replace x with 6 and y with 2 then simplify.
$$x^2 - 3y^2 = 6^2 - 3\left(2^2\right) = 36 - 3(4)$$
$$= 36 - 12 = 36 + (-12) = 24$$

31. $180 - a - b = 180 - 37 - 80$
$= 180 + (-37) + (-80)$
$= 143 + (-80) = 63$
The measure of the unknown angle is 63°.

32.　　$7x - 3 = 18$
$7(3) - 3 \stackrel{?}{=} 18$
$21 - 3 \stackrel{?}{=} 18$
$18 = 18$
Since $18 = 18$ is a true statement, 3 is a solution of the equation.

33.　　$3x^2 + 4 = x - 1$
$3(1^2) + 4 \stackrel{?}{=} 1 - 1$
$3(1) + 4 \stackrel{?}{=} 0$
$3 + 4 \stackrel{?}{=} 0$
$7 \neq 0$

Since $7 = 0$ is a false statement, 1 is not a solution of the equation.

34. The additive inverse of –9 is 9.

35. The additive inverse of $\frac{2}{3}$ is $-\frac{2}{3}$.

36. The additive inverse of $|-2|$ is –2, since $|-2| = 2$ and the additive inverse of 2 is –2.

37. The additive inverse of $-|-7|$ is 7 since $-|-7| = -7$ and the additive inverse of –7 is 7.

38. $-15 + 4 = -11$

39. $-6 + (-11) = -17$

40. $\frac{1}{16} + \left(-\frac{1}{4}\right) = \frac{1}{16} + \left(-\frac{4 \cdot 1}{4 \cdot 4}\right) = \frac{1}{16} + \left(-\frac{4}{16}\right)$
$= -\frac{3}{16}$

41. $-8 + |-3| = -8 + 3 = -5$

42. $-4.6 + (-9.3) = -13.9$

43. $-2.8 + 6.7 = 3.9$

44. $6 - 20 = 6 + (-20) = -14$

45. $-3.1 - 8.4 = -3.1 + (-8.4) = -11.5$

46. $-6 - (-11) = -6 + 11 = 5$

47. $4 - 15 = 4 + (-15) = -11$

48. $-21 - 16 + 3(8 - 2)$
$= -21 + (-16) + 3(8 + (-2))$
$= -21 + (-16) + 3(6)$
$= -21 + (-16) + 18$
$= -37 + 18 = -19$

49. $\frac{11 - (-9) + 6(8 - 2)}{2 + 3 \cdot 4} = \frac{11 + 9 + 6(8 + (-2))}{2 + 3 \cdot 4}$
$= \frac{11 + 9 + 6(6)}{2 + 3 \cdot 4} = \frac{11 + 9 + 36}{2 + 3 \cdot 4} = \frac{11 + 9 + 36}{2 + 12}$
$= \frac{20 + 36}{14} = \frac{56}{14} = \frac{2 \cdot 4 \cdot 7}{2 \cdot 7} = 4$

50. a
Replace x with 3, y with –6, and z with –9, then simplify.
$2x^2 - y + z = 2(3^2) - (-6) + (-9)$
$= 2(9) + 6 + (-9) = 18 + 6 + (-9)$
$= 24 + (-9) = 15$

51. d
Replace x with 3 and y with –6, then simplify.
$\frac{y - 4x}{2x} = \frac{-6 - 4(3)}{2(3)} = \frac{-6 - 12}{6}$
$= \frac{-6 + (-12)}{6} = \frac{-18}{6} = \frac{-2 \cdot 3 \cdot 3}{2 \cdot 3} = -3$

52. We must first find the overall gain or loss for the stock for the five days. The overall gain or loss is the sum of the daily gains and losses.
$1 + (-2) + 5 + 1 + (-4) = -1 + 5 + 1 + (-4)$
$= 4 + 1 + (-4) = 5 + (-4) = 1$
Thus, the stock gained \$1. To find the share price at week's end we add the gain to the beginning stock price.
$50 + 1 = 51$
Therefore, the share price at the end of the week was \$51.

53. The multiplicative inverse of –6 is $-\frac{1}{6}$ since
$-6 \cdot -\frac{1}{6} = 1$.

54. The multiplicative inverse of $\frac{3}{5}$ is $\frac{5}{3}$ since $\frac{3}{5} \cdot \frac{5}{3} = 1$.

55. $6(-8) = -48$

56. $(-2)(-14) = 28$

57. $\frac{-18}{-6} = 3$

58. $\frac{42}{-3} = -14$

59. $-3(-6)(-2) = 18(-2) = -36$

60. $(-4)(-3)(0)(-6) = 12(0)(-6) = 0(-6) = 0$

61. $\dfrac{4 \cdot (-3) + (-8)}{2 + (-2)} = \dfrac{-12 + (-8)}{2 + (-2)} = \dfrac{-20}{0}$
The expression is undefined.

62. $\dfrac{3(-2)^2 - 5}{-14} = \dfrac{3(4) - 5}{-14} = \dfrac{12 - 5}{-14}$
$= \dfrac{7}{-14} = -\dfrac{1}{2}$

63. The commutative property of addition

64. The multiplicative identity property

65. The distributive property

66. The additive inverse property

67. The associate property of addition

68. The commutative property of multiplication

69. The distributive property

70. The associative property of multiplication

71. The multiplicative inverse property

72. The additive identity property

73. The commutative property of addition

74. The height of the bar corresponding to 1994 is approximately $1,800 million. Therefore, in 1994 Disney's consumer products revenue was $1800 million.

75. The consumer products revenue was approximately $700 million in 1991 and $1000 million in 1992. Thus, the difference in revenues is given by $1000 - 700 = 400$. We conclude that consumer products revenue increased by $400 million in 1992.

76. Since these bars are arranged vertically, we look for the tallest bar, which is the bar representing the year 1994.

77. The revenue from consumer products is increasing.

Chapter 1 Test

1. $|-7| > 5$

2. $(9 + 5) \geq 4$

3. $-13 + 8 = -5$

4. $-13 - (-2) = -13 + 2 = -11$

5. $6 \cdot 3 - 8 \cdot 4 = 18 - 32 = 18 + (-32) = -14$

6. $(13)(-3) = -39$

7. $(-6)(-2) = 12$

8. $\dfrac{|-16|}{-8} = \dfrac{16}{-8} = -2$

9. $\dfrac{-8}{0}$ is undefined.

10. $\dfrac{|-6| + 2}{5 - 6} = \dfrac{6 + 2}{5 + (-6)} = \dfrac{8}{-1} = -8$

11. $\dfrac{1}{2} - \dfrac{5}{6} = \dfrac{3 \cdot 1}{3 \cdot 2} - \dfrac{5}{6} = \dfrac{3}{6} - \dfrac{5}{6} = -\dfrac{2}{6} = -\dfrac{1}{3}$

12. $-1\dfrac{1}{8} + 5\dfrac{3}{4} = -\dfrac{9}{8} + \dfrac{23}{4} = -\dfrac{9}{8} + \dfrac{2 \cdot 23}{2 \cdot 4}$
$= -\dfrac{9}{8} + \dfrac{46}{8} = \dfrac{37}{8} = 4\dfrac{5}{8}$

13. $-\dfrac{3}{5}+\dfrac{15}{8}=-\dfrac{8\cdot 3}{8\cdot 5}+\dfrac{5\cdot 15}{5\cdot 8}=-\dfrac{24}{40}+\dfrac{75}{40}=\dfrac{51}{40}$

14. $3(-4)^2-80=3(16)-80=48-80$
    $=48+(-80)=-32$

15. $6[5+2(3-8)-3]$
    $=6[5+2(3+(-8))+(-3)]$
    $=6[5+2(-5)+(-3)]$
    $=6[5-10+(-3)]$
    $=6[5+(-10)+(-3)]$
    $=6[-5+(-3)]=6[-8]=-48$

16. $\dfrac{-12+3\cdot 8}{4}=\dfrac{-12+24}{4}=\dfrac{12}{4}=\dfrac{3\cdot 4}{4}=3$

17. $\dfrac{(-2)(0)(-3)}{-6}=\dfrac{0(-3)}{-6}=\dfrac{0}{-6}=0$

18. $-3>-7$

19. $4>-8$

20. $|-3|>2$

21. $|-2|=-1-(-3)$

22. **a.** The natural numbers are 1 and 7.

    b. The whole numbers are 0, 1, and 7.

    c. The integers are $-5, -1, 0, 1,$ and 7.

    d. The rational numbers are $-5, -1, \dfrac{1}{4}, 0,$
    $1, 7, 11.6.$

    e. The irrational numbers are $\sqrt{7}$ and 3π.

    f. The real numbers are all the numbers in
    the given set.

23. Replace x with 6 and y with -2, then
    simplify.
    $x^2+y^2=6^2+(-2)^2=36+4=40$

24. Replace x with 6, y with -2, and z with -3,
    then simplify.
    $x+yz=6+(-2)(-3)=6+6=12$

25. Replace x with 6 and y with -2, then
    simplify.
    $2+3x-y=2+3(6)-(-2)=2+18+2$
    $=20+2=22$

26. Replace x with 6, y with -2, and z with -3,
    then simplify.
    $\dfrac{y+z-1}{x}=\dfrac{-2+(-3)-1}{6}=\dfrac{-2+(-3)+(-1)}{6}$
    $=\dfrac{-5+(-1)}{6}=\dfrac{-6}{6}=-1$

27. The associative property of addition

28. The commutative property of multiplication

29. The distributive property

30. The multiplicative inverse property

31. The opposite of -9 is 9.

32. The reciprocal of $-\dfrac{1}{3}$ is -3.

33. Losses are represented by negative numbers.
    The greatest loss occurred on second down
    when the Saints lost 10 yards.

34. Yes, to score a touchdown the Saints needed
    to gain 22 yards. The overall gain or loss in
    yardage is the sum of the losses and gains
    for the four downs. Since losses are
    represented by negative numbers and gains
    are represented by positive numbers we can
    form the following mathematical statement.
    $5+(-10)+(-2)+29=-5+(-2)+29$
    $=-7+29=22$
    The Saints gained 22 yards on four downs,
    so they did score a touchdown.

35. Since $-14+31=17$, the termperature at
    noon was $17°$.

36. If each share of stock decreased in value by
    \$1.50 and she sold 280 shares, then her total
    loss is given by product: $280\cdot 1.50=\$420.$

37. Intel's revenue in 1993 was \$8 billion.

38. Intel's revenue in 1989 was approximately
    \$3 billion.

39. From the graph, we can see that Intel's
revenue in 1993 was $8 billion. Intel's
revenue in 1995 was $13.5 billion. Thus,
Intel's revenue increased by
$13.5 - 8 = \$5.5$ billion between the years
1993 and 1995.

40. The greatest increase in revenue occurred in
1994. Notice that the line graph is steepest
between the years 1993 and 1994.

Chapter 2

1. $3c - 4 + 6c - 9 = 3c + 6c + (-4 - 9)$
$= (3 + 6)c + (-4 - 9) = 9c - 13$

2. $-5(2y - 3) - 7y + 1 = -10y + 15 - 7y + 1$
$= -10y - 7y + (15 + 1)$
$= (-10 - 7)y + (15 + 1)$
$= -17y + 16$

3. $3 - x = -12$
$3 - x - 3 = -12 - 3$
$-x = -15$
$\dfrac{-x}{-1} = \dfrac{-15}{-1}$
$x = 15$

4. $12 - (5 - 4b) = 9 + 3b$
$12 - 5 + 4b = 9 + 3b$
$7 + 4b = 9 + 3b$
$7 + 4b - 3b = 9 + 3b - 3b$
$7 + b = 9$
$7 + b - 7 = 9 - 7$
$b = 2$

5. $\dfrac{2}{3}m = -8$
$\dfrac{3}{2}\left(\dfrac{2}{3}m\right) = \dfrac{3}{2}(-8)$
$m = -12$

6. $-7 - 3y = 17 + 5y$
$-7 - 3y + 3y = 17 + 5y + 3y$
$-7 = 17 + 8y$
$-7 - 17 = 17 + 8y - 17$
$-24 = 8y$
$\dfrac{-24}{8} = \dfrac{8y}{8}$
$-3 = y$

7. $3(1 - 4x) + 2(5x) = 9$
$3 - 12x + 10x = 9$
$3 - 2x = 9$
$3 - 2x - 3 = 9 - 3$
$-2x = 6$
$\dfrac{-2x}{-2} = \dfrac{6}{-2}$
$x = -3$

8. $0.20x + 0.15(60) = 0.75(18)$
$100[0.20x + 0.15(60)] = 100[0.75(18)]$
$20x + 15(60) = 75(18)$
$20x + 900 = 1350$
$20x + 900 - 900 = 1350 - 900$
$20x = 450$
$\dfrac{20x}{20} = \dfrac{450}{20}$
$x = 22.5$

9. $2(x - 1) = 2x + 5$
$2x - 2 = 2x + 5$
$2x - 2 - 2x = 2x + 5 - 2x$
$-2 = 5$
Since there is no value for x that makes $-2 = 5$ a true equation, there is no solution to the equation $2(x - 1) = 2x + 5$.

10. Let $x =$ the unknown number
$3(x + (-2)) = x + 2$
$3(x - 2) = x + 2$
$3x - 6 = x + 2$
$3x - 6 + 6 = x + 2 + 6$
$3x = x + 8$
$3x - x = x + 8 - x$
$2x = 8$
$\dfrac{2x}{2} = \dfrac{8}{2}$
$x = 4$
The number is 4.

11. Let $x =$ the smaller even integer. Then $x + 2 =$ the next consecutive even integer.
$3x = 2(x + 2) + 16$
$3x = 2x + 4 + 16$
$3x = 2x + 20$
$3x - 2x = 2x + 20 - 2x$
$x = 20$
The integers are 20 and 22.

12. $V = \dfrac{1}{3}Ah$

$60 = \dfrac{1}{3}A(4)$

$60 = \dfrac{4}{3}A$

$\dfrac{3}{4}(60) = \dfrac{3}{4}\left(\dfrac{4}{3}A\right)$

$45 = A$

13. The area of a triangle is given by the formula $A = \dfrac{1}{2}bh$.

$A = \dfrac{1}{2}bh$

$18 = \dfrac{1}{2}b(4)$

$18 = 2b$

$\dfrac{18}{2} = \dfrac{2b}{2}$

$9 = b$

The base of the sign is 9 feet.

14. $2x + y = 8$

$2x + y - 2x = 8 - 2x$

$y = 8 - 2x$

15. Let x = the unknown number.

$x = (0.22)(90)$

$x = 19.8$

16. 5 gallons $= 5 \cdot 4$ quarts $= 20$ quarts

The ratio 4 quarts to 5 gallons is $\dfrac{4}{20}$ or $\dfrac{1}{5}$ in lowest terms.

17. $\dfrac{3x}{8} = \dfrac{9}{7}$

$(3x)(7) = (8)(9)$

$21x = 72$

$\dfrac{21x}{21} = \dfrac{72}{21}$

$x = \dfrac{24}{7}$

18. $-4 + x \le 2$

$-4 + x + 4 \le 2 + 4$

$x \le 6$

19. $-\dfrac{3}{2}y > 6$

$-\dfrac{2}{3}\left(-\dfrac{3}{2}y\right) < -\dfrac{2}{3}(6)$

$y < -4$

20. $-5x + 3 \le 4(x - 6)$

$-5x + 3 \le 4x - 24$

$-5x + 3 - 3 \le 4x - 24 - 3$

$-5x \le 4x - 27$

$-5x - 4x \le 4x - 27 - 4x$

$-9x \le -27$

$\dfrac{-9x}{-9} \ge \dfrac{-27}{-9}$

$x \ge 3$

Section 2.1

Practice Problems

1. a. The numerical coefficient of $-4x$ is -4.

   b. The numerical coefficient of $15y^3$ is 15.

   c. The numerical coefficient of x is 1.

   d. The numerical coefficient of $-y$ is -1.

   e. The numerical coefficient of $\dfrac{z}{4}$ is $\dfrac{1}{4}$.

2. a. Like terms, since the variable and its exponent match.

   b. Like terms, since each variable and its exponent match.

   c. Like terms, since each variable and its exponent match.

3. a. $9y - 4y = (9 - 4)y = 5y$

b. $11x^2 + x^2 = (11 + 1)x^2 = 12x^2$

c. $5y - 3x + 6x = 5y + (-3 + 6)x = 5y + 3x$

4. $7y + 2y + 6 + 10 = (7 + 2)y + (6 + 10)$
$= 9y + 16$

5. $-2x + 4 + x - 11 = -2x + x + 4 - 11$
$= (-2 + 1)x + (4 - 11) = -x - 7$

6. $3z - 3z^2$
These terms cannot be combined because they are unlike terms.

7. $8.9y + 4.2y - 3 = (8.9 + 4.2)y - 3 = 13.1y - 3$

8. $3(y + 6) = 3(y) + 3(6) = 3y + 18$

9. $-4(x + 0.2y - 3)$
$= -4(x) + (-4)(0.2y) + (-4)(-3)$
$= -4x - 0.8y + 12$

10. $-(3x + 2y + z - 1) = -1(3x + 2y + z - 1)$
$= -1(3x) + (-1)(2y) + (-1)(z) + (-1)(-1)$
$= -3x - 2y - z + 1$

11. $4(x - 6) + 20 = 4x - 24 + 20 = 4x - 4$

12. $5 - (3x + 9) = 5 - 3x - 9 = 5 - 9 - 3x$
$= -4 - 3x = -3x - 4$

13. $-3(7x + 1) - (4x - 2) = -21x - 3 - 4x + 2$
$= -21x - 4x - 3 + 2 = -25x - 1$

14. $(4x - 3) - (9x - 10) = 4x - 3 - 9x + 10$
$= 4x - 9x - 3 + 10 = -5x + 7$

15. $10 - 3x$

16. $\dfrac{x + 2}{5}$

17. $3x + (2x + 6) = 3x + 2x + 6 = 5x + 6$

Mental Math

1. The numerical coefficient of $-7y$ is –7.

2. The numerical coefficient of $3x$ is 3.

3. The numerical coefficient of x is 1 since x is $1 \cdot x$.

4. The numerical coefficient of $-y$ is –1 since $-y$ is $-1 \cdot y$.

5. The numerical coefficient of $17x^2y$ is 17.

6. The numerical coefficient of $1.2xyz$ is 1.2.

7. Like terms, since the variable and its exponent match.

8. Unlike terms, since the exponents on x are not the same.

9. Unlike terms, since the exponents on z are not the same.

10. Like terms, since each variable and its exponent match.

11. Like terms, since $wz = zw$ by the commutative property.

12. Unlike terms, since the variables do not match.

Exercise Set 2.1

1. $7y + 8y = (7 + 8)y = 15y$

3. $8w - w + 6w = (8 - 1 + 6)w = 13w$

5. $3b - 5 - 10b - 4 = 3b - 10b + (-5 - 4)$
$= (3 - 10)b + (-5 - 4) = -7b - 9$

7. $m - 4m + 2m - 6 = (1 - 4 + 2)m - 6$
$= (-3 + 2)m - 6 = -m - 6$

9. $5g - 3 - 5 - 5g = 5g - 5g + (-3 - 5)$
$= (5 - 5)g + (-3 - 5) = 0 \cdot g + (-3 - 5)$
$= -8$

11. $6.2x - 4 + x - 1.2 = 6.2x + x + (-4 - 1.2)$
$= (6.2 + 1)x + (-4 - 1.2) = 7.2x - 5.2$

13. $2k - k - 6 = (2 - 1)k - 6 = k - 6$

15. $-9x + 4x + 18 - 10x = -9x + 4x - 10x + 18$
$= (-9 + 4 - 10)x + 18 = -15x + 18$

17. $6x - 5x + x - 3 + 2x = 6x - 5x + x + 2x - 3$
$= (6 - 5 + 1 + 2)x - 3 = 4x - 3$

19. $7x^2 + 8x^2 - 10x^2 = (7 + 8 - 10)x^2 = 5x^2$

21. $3.4m - 4 - 3.4m - 7$
$= 3.4m - 3.4m + (-4 - 7)$
$= (3.4 - 3.4)m + (-4 - 7)$
$= 0 \cdot m - 11 = -11$

23. $6x + 0.5 - 4.3x - 0.4x + 3$
$= 6x - 4.3x - 0.4x + (0.5 + 3)$
$= (6 - 4.3 - 0.4)x + (0.5 + 3)$
$= 1.3x + 3.5$

25. $5(y + 4) = 5(y) + 5(4) = 5y + 20$

27. $-2(x + 2) = -2(x) + (-2)(2) = -2x - 4$

29. $-5(2x - 3y + 6)$
$= -5(2x) + (-5)(-3y) + (-5)(6)$
$= -10x + 15y - 30$

31. $-(3x - 2y + 1) = -1(3x - 2y + 1)$
$= -1(3x) + (-1)(-2y) + (-1)(1)$
$= -3x + 2y - 1$

33. $7(d - 3) + 10 = 7d - 21 + 10 = 7d - 11$

35. $-4(3y - 4) + 12y = -12y + 16 + 12y$
$= -12y + 12y + 16 = 16$

37. $3(2x - 5) - 5(x - 4) = 6x - 15 - 5x + 20$
$= 6x - 5x - 15 + 20 = x + 5$

39. $-2(3x - 4) + 7x - 6 = -6x + 8 + 7x - 6$
$= -6x + 7x + 8 - 6 = x + 2$

41. $5k - (3k - 10) = 5k - 3k + 10 = 2k + 10$

43. $(3x + 4) - (6x - 1) = 3x + 4 - 6x + 1$
$= 3x - 6x + 4 + 1 = -3x + 5$

45. $5(x + 2) - (3x - 4) = 5x + 10 - 3x + 4$
$= 5x - 3x + 10 + 4 = 2x + 14$

47. $0.5(m + 2) + 0.4m = 0.5m + 1 + 0.4m$
$= 0.5m + 0.4m + 1 = 0.9m + 1$

49. Answers may vary.

51. $(6x + 7) + (4x - 10) = 6x + 7 + 4x - 10$
$= 6x + 4x + 7 - 10 = 10x - 3$

53. $(3x - 8) - (7x + 1) = 3x - 8 - 7x - 1$
$= 3x - 7x - 8 - 1 = -4x - 9$

55. $(m - 9) - (5m - 6) = m - 9 - 5m + 6$
$= m - 5m - 9 + 6 = -4m - 3$

57. $2x - 4$

59. $\frac{3}{4}x + 12$

61. $(5x - 2) + 7x = 5x + 7x - 2 = 12x - 2$

63. $8(x + 6) = 8x + 48$

65. $2x - (x + 10) = 2x - x - 10 = x - 10$

67. $y - x^2 = 3 - (-1)^2 = 3 - 1 = 3 + (-1) = 2$

69. $a - b^2 = 2 - (-5)^2 = 2 - 25 = 2 + (-25)$
$= -23$

71. $yz - y^2 = (-5)(0) - (-5)^2 = 0 - 25 = -25$

73. To determine if the scale is balanced, find the number of cubes on each side of the scale and see if they are equal.

Left side	Right side
1 cone = 1 cube 1 cylinder = 2 cubes	3 cubes
Total = 3 cubes	Total = 3 cubes

The scale is balanced.

75. To determine if the scale is balanced, find the number of cubes on each side of the scale and see if they are equal.

Left side	Right side
2 cylinders = 4 cubes 1 cube = 1 cube	3 cones = 3 cubes 2 cubes = 2 cubes
Total = 5 cubes	Total = 5 cubes

The scale is balanced.

77. $5x + (4x - 1) + 5x + (4x - 1)$
$= 5x + 4x - 1 + 5x + 4x - 1$
$= 5x + 4x + 5x + 4x + (-1 - 1)$
$= (5 + 4 + 5 + 4)x + (-1 - 1)$
$= 18x - 2$
The perimeter of the rectangle is
$(18x - 2)$ feet.

79. The length of the first board is
$(x + 2)$ feet = $12(x + 2)$ inches. The length of
the second board is $(3x - 1)$ inches.
$12(x + 2) + (3x - 1) = 12x + 24 + 3x - 1$
$= 12x + 3x + (24 - 1) = (12 + 3)x + (24 - 1)$
$= 15x + 23$
The total length of the boards is
$(15x + 23)$ inches.

Section 2.2

Practice Problems

1. $x - 5 = 8$
$x - 5 + 5 = 8 + 5$
$x = 13$

2. $y + 1.7 = 0.3$
$y + 1.7 - 1.7 = 0.3 - 1.7$
$y = -1.4$

3. $\frac{7}{8} = y - \frac{1}{3}$
$\frac{7}{8} + \frac{1}{3} = y - \frac{1}{3} + \frac{1}{3}$
$\frac{7}{8} + \frac{1}{3} = y$
$\frac{21}{24} + \frac{8}{24} = y$
$\frac{29}{24} = y$

4. $3x + 10 = 4x$
$3x + 10 - 3x = 4x - 3x$
$10 = x$

5. $10w + 3 - 4w + 4 = -2w + 3 + 7w$
$6w + 7 = 5w + 3$
$6w + 7 - 5w = 5w + 3 - 5w$
$w + 7 = 3$
$w + 7 - 7 = 3 - 7$
$w = -4$

6. $3(2w - 5) - (5w + 1) = -3$
$6w - 15 - 5w - 1 = -3$
$w - 16 = -3$
$w - 16 + 16 = -3 + 16$
$w = 13$

7. $12 - y = 9$
$12 - y - 12 = 9 - 12$
$-y = -3$
If $-y = -3$ then $y = 3$.

8. If x is one number, then $11 - x$ is the other
number since $x + (11 - x) = 11$.

Mental Math

1. $x + 4 = 6$
$x = 2$

2. $x + 7 = 10$
$x = 3$

3. $n + 18 = 30$
$n = 12$

4. $z + 22 = 40$
$z = 18$

5. $b - 11 = 6$
$b = 17$

6. $d - 16 = 5$
$d = 21$

Exercise Set 2.2

1. $x + 7 = 10$
$x + 7 - 7 = 10 - 7$
$x = 3$

Check: $x + 7 = 10$
$3 + 7 \overset{?}{=} 10$
$10 = 10$

The solution is 3.

3. $x - 2 = -4$
$x - 2 + 2 = -4 + 2$
$x = -2$
Check: $x - 2 = -4$
$-2 - 2 \overset{?}{=} -4$
$-4 = -4$

The solution is -2.

42

5. $3 + x = -11$

$3 + x - 3 = -11 - 3$

$x = -14$

Check: $3 + x = -11$

$3 + (-14) \overset{?}{=} -11$

$11 = -11$

The solution is -14.

7. $r - 8.6 = -8.1$

$r - 8.6 + 8.6 = -8.1 + 8.6$

$r = 0.5$

Check: $r - 8.6 = -8.1$

$0.5 - 8.6 \overset{?}{=} -8.1$

$-8.1 = -8.1$

The solution is 0.5.

9. $\frac{1}{3} + f = \frac{3}{4}$

$\frac{1}{3} + f - \frac{1}{3} = \frac{3}{4} - \frac{1}{3}$

$f = \frac{9}{12} - \frac{4}{12}$

$f = \frac{5}{12}$

Check: $\frac{1}{3} + f = \frac{3}{4}$

$\frac{1}{3} + \frac{5}{12} \overset{?}{=} \frac{3}{4}$

$\frac{4}{12} + \frac{5}{12} \overset{?}{=} \frac{3}{4}$

$\frac{9}{12} \overset{?}{=} \frac{3}{4}$

$\frac{3}{4} = \frac{3}{4}$

The solution is $\frac{5}{12}$.

11. $5b - 0.7 = 6b$

$5b - 0.7 - 5b = 6b - 5b$

$-0.7 = b$

Check: $5b - 07 = 6b$

$5(-0.7) - 0.7 \overset{?}{=} 6(-0.7)$

$-3.5 - 0.7 \overset{?}{=} -4.2$

$-4.2 = -4.2$

The solution is -0.7.

13. $7x - 3 = 6x$

$7x - 3 - 6x = 6x - 6x$

$x - 3 = 0$

$x - 3 + 3 = 0 + 3$

$x = 3$

Check: $7x - 3 = 6x$

$7(3) - 3 \overset{?}{=} 6(3)$

$21 - 3 \overset{?}{=} 18$

$18 = 18$

The solution is 3.

15. Answers may vary.

17. $7x + 2x = 8x - 3$

$9x = 8x - 3$

$9x - 8x = 8x - 3 - 8x$

$x = -3$

Check: $7x + 2x = 8x - 3$

$7(-3) + 2(-3) \overset{?}{=} 8(-3) - 3$

$-21 - 6 \overset{?}{=} -24 - 3$

$-27 = -27$

The solution is -3.

19. $2y + 10 = 5y - 4y$

$2y + 10 = y$

$2y + 10 - y = y - y$

$y + 10 = 0$

$y + 10 - 10 = 0 - 10$

$y = -10$

Check: $2y + 10 = 5y - 4y$

$2(-10) + 10 \overset{?}{=} 5(-10) - 4(-10)$

$-20 + 10 \overset{?}{=} -50 + 40$

$-10 = -10$

The solution is -10.

21. $3x - 6 = 2x + 5$

$3x - 6 - 2x = 2x + 5 - 2x$

$x - 6 = 5$

$x - 6 + 6 = 5 + 6$

$x = 11$

Check: $3x - 6 = 2x + 5$

$3(11) - 6 \overset{?}{=} 2(11) + 5$

$33 - 6 \overset{?}{=} 22 + 5$

$27 = 27$

The solution is 11.

23. $5x - 6 = 6x - 5$
$5x - 6 - 5x = 6x - 5 - 5x$
$-6 = x - 5$
$-6 + 5 = x - 5 + 5$
$-1 = x$

Check: $5x - 6 = 6x - 5$
$5(-1) - 6 \overset{?}{=} 6(-1) - 5$
$-5 - 6 \overset{?}{=} -6 - 5$
$-11 = -11$

The solution is -1.

25. $8y + 2 - 6y = 3 + y - 10$
$2y + 2 = y - 7$
$2y + 2 - y = y - 7 - y$
$y + 2 = -7$
$y + 2 - 2 = -7 - 2$
$y = -9$

Check: $8y + 2 - 6y = 3 + y - 10$
$8(-9) + 2 - 6(-9) \overset{?}{=} 3 + (-9) - 10$
$-72 + 2 + 54 \overset{?}{=} 3 - 9 - 10$
$-16 = -16$

The solution is -9.

27. $13x - 9 + 2x - 5 = 12x - 1 + 2x$
$15x - 14 = 14x - 1$
$15x - 14 - 14x = 14x - 1 - 14x$
$x - 14 = -1$
$x - 14 + 14 = -1 + 14$
$x = 13$

Check: $13x - 9 + 2x - 5 = 12x - 1 + 2x$
$13(13) - 9 + 2(13) - 5 \overset{?}{=} 12(13) - 1 + 2(13)$
$169 - 9 + 26 - 5 \overset{?}{=} 156 - 1 + 26$
$181 = 181$

The solution is 13.

29. $-6.5 - 4x - 1.6 - 3x = -6x + 9.8$
$-7x - 8.1 = -6x + 9.8$
$-7x - 8.1 + 6x = -6x + 9.8 + 6x$
$-x - 8.1 = 9.8$
$-x - 8.1 + 8.1 = 9.8 + 8.1$
$-x = 17.9$
If $-x = 17.9$, then $x = -17.9$.

Check: $-6.5 - 4x - 1.6 - 3x = -6x + 9.8$
$-6.5 - 4(-17.9) - 1.6 - 3(-17.9) \overset{?}{=} -6(-17.9) + 9.8$
$-6.5 + 71.6 - 1.6 + 53.7 \overset{?}{=} 107.4 + 9.8$
$117.2 = 117.2$

The solution is -17.9.

31. $\dfrac{3}{8}x - \dfrac{1}{6} = -\dfrac{5}{8}x - \dfrac{2}{3}$

$\dfrac{3}{8}x - \dfrac{1}{6} - \dfrac{3}{8}x = -\dfrac{5}{8}x - \dfrac{2}{3} - \dfrac{3}{8}x$

$-\dfrac{1}{6} = -\dfrac{8}{8}x - \dfrac{2}{3}$

$-\dfrac{1}{6} + \dfrac{2}{3} = -x - \dfrac{2}{3} + \dfrac{2}{3}$

$-\dfrac{1}{6} + \dfrac{4}{6} = -x$

$\dfrac{3}{6} = -x$

$\dfrac{1}{2} = -x$

If $\dfrac{1}{2} = -x$, then $-\dfrac{1}{2} = x$.

   Check: $\dfrac{3}{8}x - \dfrac{1}{6} = -\dfrac{5}{8}x - \dfrac{2}{3}$

$\dfrac{3}{8}\left(-\dfrac{1}{2}\right) - \dfrac{1}{6} \overset{?}{=} -\dfrac{5}{8}\left(-\dfrac{1}{2}\right) - \dfrac{2}{3}$

$-\dfrac{3}{16} - \dfrac{1}{6} \overset{?}{=} \dfrac{5}{16} - \dfrac{2}{3}$

$-\dfrac{9}{48} - \dfrac{8}{48} \overset{?}{=} \dfrac{15}{48} - \dfrac{32}{48}$

$-\dfrac{17}{48} = -\dfrac{17}{48}$

   The solution is $-\dfrac{1}{2}$.

33. $2(x - 4) = x + 3$

$2(x) + 2(-4) = x + 3$

$2x - 8 = x + 3$

$2x - 8 - x = x + 3 - x$

$x - 8 = 3$

$x - 8 + 8 = 3 + 8$

$x = 11$

   Check: $2(x - 4) = x + 3$

$2(11 - 4) \overset{?}{=} 11 + 3$

$2(7) \overset{?}{=} 14$

$14 = 14$

The solution is 11.

35. $7(6 + w) = 6(2 + w)$

$7(6) + 7(w) = 6(2) + 6(w)$

$42 + 7w = 12 + 6w$

$42 + 7w - 6w = 12 + 6w - 6w$

$42 + w = 12$

$42 + w - 42 = 12 - 42$

$w = -30$

   Check: $7(6 + w) = 6(2 + w)$

$7(6 + (-30)) \overset{?}{=} 6(2 + (-30))$

$7(-24) \overset{?}{=} 6(-28)$

$-168 = -168$

The solution is –30.

37. $10 - (2x - 4) = 7 - 3x$
$10 - 1(2x - 4) = 7 - 3x$
$10 - 1(2x) - 1(-4) = 7 - 3x$
$10 - 2x + 4 = 7 - 3x$
$14 - 2x = 7 - 3x$
$14 - 2x + 3x = 7 - 3x + 3x$
$x + 14 = 7$
$x + 14 - 14 = 7 - 14$
$x = -7$

Check: $10 - (2x - 4) = 7 - 3x$
$10 - (2(-7) - 4) \stackrel{?}{=} 7 - 3(-7)$
$10 - (-14 - 4) \stackrel{?}{=} 7 + 21$
$10 - (-18) \stackrel{?}{=} 28$
$10 + 18 \stackrel{?}{=} 28$
$28 = 28$

The solution is -7.

39. $-5(n - 2) = 8 - 4n$
$-5(n) - 5(-2) = 8 - 4n$
$-5n + 10 = 8 - 4n$
$-5n + 10 + 5n = 8 - 4n + 5n$
$10 = 8 + n$
$10 - 8 = 8 + n - 8$
$2 = n$

Check: $-5(n - 2) = 8 - 4n$
$-5(2 - 2) \stackrel{?}{=} 8 - 4(2)$
$-5(0) \stackrel{?}{=} 8 - 8$
$0 = 0$

The solution is 2.

41. $-3(x - 4) = -4x$
$-3(x) - 3(-4) = -4x$
$-3x + 12 = -4x$
$-3x + 12 + 4x = -4x + 4x$
$x + 12 = 0$
$x + 12 - 12 = 0 - 12$
$x = -12$

Check: $-3(x - 4) = -4x$
$-3(-12 - 4) \stackrel{?}{=} -4(-12)$
$-3(-16) \stackrel{?}{=} 48$
$48 = 48$

The solution is -12.

43. $3(n - 5) - (6 - 2n) = 4n$
$3(n - 5) - 1(6 - 2n) = 4n$
$3(n) + 3(-5) - 1(6) - 1(-2n) = 4n$
$3n - 15 - 6 + 2n = 4n$
$5n - 21 = 4n$
$5n - 21 - 4n = 4n - 4n$
$n - 21 = 0$
$n - 21 + 21 = 0 + 21$
$n = 21$

Check: $3(n - 5) - (6 - 2n) = 4n$
$3(21 - 5) - (6 - 2(21)) \stackrel{?}{=} 4(21)$
$3(16) - (6 - 42) \stackrel{?}{=} 84$
$48 - (-36) \stackrel{?}{=} 84$
$48 + 36 \stackrel{?}{=} 84$
$84 = 84$

The solution is 21.

45. $-2(x + 6) + 3(2x - 5) = 3(x - 4) + 10$
$-2(x) - 2(6) + 3(2x) + 3(-5) = 3(x) + 3(-4) + 10$
$-2x - 12 + 6x - 15 = 3x - 12 + 10$
$4x - 27 = 3x - 2$
$4x - 27 - 3x = 3x - 2 - 3x$
$x - 27 = -2$
$x - 27 + 27 = -2 + 27$
$x = 25$

Check: $-2(x + 6) + 3(2x - 5) = 3(x - 4) + 10$
$-2(25 + 6) + 3(2(25) - 5) \stackrel{?}{=} 3(25 - 4) + 10$
$-2(31) + 3(50 - 5) \stackrel{?}{=} 3(21) + 10$
$-62 + 3(45) \stackrel{?}{=} 63 + 10$
$-62 + 135 \stackrel{?}{=} 73$
$73 = 73$

The solution is 25.

47. If the sum of two numbers is 20 and one number is p, we find the other number by subtracting p from 20. The other number is $20 - p$.

49. Since the sum of the lengths of the two pieces of board is 10 feet and one piece is x feet long, we can find the length of the other piece by subtracting x from 10. Therefore, the length of the other piece is $(10 - x)$ feet.

51. Since the sum of supplementary angles is 180° and one angle measures $x°$, we can find the measure of its supplement by subtracting x from 180. Therefore, the other angle is $(180 - x)°$.

53. If Joseph Brennan received n votes and Susan Collins received 30,898 more votes than Brennan, we find the number of votes that Collins received by adding 30,898 to n. Thus, Susan Collins received $(n + 30{,}898)$ votes.

55. The reciprocal or multiplicative inverse of $\frac{5}{8}$ is $\frac{8}{5}$ because $\frac{5}{8} \cdot \frac{8}{5} = 1$.

57. The reciprocal or multiplicative inverse of 2 is $\frac{1}{2}$ because $2 \cdot \frac{1}{2} = 1$.

59. The reciprocal or multiplicative inverse of $-\frac{1}{9}$ is -9 because $-\frac{1}{9} \cdot -9 = 1$.

61. $\frac{3x}{3} = \left(\frac{3}{3}\right)x = (1)x = x$

63. $-5\left(-\frac{1}{5}y\right) = \left(-5 \cdot -\frac{1}{5}\right)y = (1)y = y$

65. $\frac{3}{5}\left(\frac{5}{3}x\right) = \left(\frac{3}{5} \cdot \frac{5}{3}\right)x = (1)x = x$

67. $36.766 + x = -108.712$
$36.766 + x - 36.766 = -108.712 - 36.766$
$x = -108.712 - 36.766$
$x = -145.478$

69. Since the sum of the angles in a triangle is 180°, one angle measures $x°$, and a second angle measures $(2x + 7)°$, we find the measure of the third angle by subtracting x and $(2x + 7)$ from 180.
$180 - [x + (2x + 7)] = 180 - [x + 2x + 7]$
$= 180 - [3x + 7] = 180 - 1[3x + 7]$
$= 180 - 1(3x) - 1(7) = 180 - 3x - 7$
$= 173 - 3x$
The third angle measures $(173 - 3x)°$.

Section 2.3

Practice Problems

1. $\frac{3}{7}x = 9$

$\frac{7}{3}\left(\frac{3}{7}x\right) = \frac{7}{3} \cdot 9$

$1x = 21$
$x = 21$

2. $7x = 42$

$\frac{7x}{7} = \frac{42}{7}$

$1x = 6$
$x = 6$

3. $-4x = 52$

$\frac{-4x}{-4} = \frac{52}{-4}$

$1x = -13$
$x = -13$

4. $\frac{y}{5} = 13$

$5 \cdot \frac{1}{5}y = 5 \cdot 13$

$1y = 65$
$y = 65$

5. $2.6x = 13.52$

$\frac{2.6x}{2.6} = \frac{13.52}{2.6}$

$1x = 5.2$
$x = 5.2$

6. $-\dfrac{5}{6}y = -\dfrac{3}{5}$

$-\dfrac{6}{5}\left(-\dfrac{5}{6}y\right) = -\dfrac{6}{5}\left(-\dfrac{3}{5}\right)$

$1y = \dfrac{18}{25}$

$y = \dfrac{18}{25}$

7. $-x + 7 = -12$

$-x + 7 - 7 = -12 - 7$

$-x = -19$

$\dfrac{-x}{-1} = \dfrac{-19}{-1}$

$1x = 19$

$x = 19$

8. $-7x + 2x + 3 - 20 = -2$

$-5x - 17 = -2$

$-5x - 17 + 17 = -2 + 17$

$-5x = 15$

$\dfrac{-5x}{-5} = \dfrac{15}{-5}$

$1x = -3$

$x = -3$

9.   If x = the first integer, then $x + 1$ = the next integer. Their sum is

$x + (x + 1) = x + x + 1 = 2x + 1.$

Mental Math

1. $3a = 27$

$a = 9$

2. $9c = 54$

$c = 6$

3. $5b = 10$

$b = 2$

4. $7t = 14$

$t = 2$

5. $6x = -30$

$x = -5$

6. $8r = -64$

$r = -8$

Exercise Set 2.3

1. $-5x = 20$

$\dfrac{-5x}{-5} = \dfrac{20}{-5}$

$x = -4$

Check: $-5x = 20$

$-5(-4) \overset{?}{=} 20$

$20 = 20$

The solution is -4.

3. $3x = 0$

$\dfrac{3x}{3} = \dfrac{0}{3}$

$x = 0$

Check: $3x = 0$

$3(0) \overset{?}{=} 0$

$0 = 0$

The solution is 0.

5. $-x = -12$

$\dfrac{-x}{-1} = \dfrac{-12}{-1}$

$x = 12$

Check: $-x = -12$

$-(12) \overset{?}{=} -12$

$-12 = -12$

The solution is 12.

7. $\dfrac{2}{3}x = -8$

$\dfrac{3}{2}\left(\dfrac{2}{3}x\right) = \dfrac{3}{2}(-8)$

$x = -12$

Check: $\dfrac{2}{3}x = -8$

$\dfrac{2}{3}(-12) \overset{?}{=} -8$

$-8 = -8$

The solution is -12.

9. $\frac{1}{6}d = \frac{1}{2}$

$6\left(\frac{1}{6}d\right) = 6\left(\frac{1}{2}\right)$

$d = 3$

Check: $\frac{1}{6}d = \frac{1}{2}$

$\frac{1}{6}(3) \stackrel{?}{=} \frac{1}{2}$

$\frac{1}{2} = \frac{1}{2}$

The solution is 3.

11. $\frac{a}{2} = 1$

$2\left(\frac{a}{2}\right) = 2(1)$

$a = 2$

Check: $\frac{a}{2} = 1$

$\frac{2}{2} \stackrel{?}{=} 1$

$1 = 1$

The solution is 2.

13. $\frac{k}{-7} = 0$

$-7\left(\frac{k}{-7}\right) = -7(0)$

$k = 0$

Check: $\frac{k}{-7} = 0$

$\frac{0}{-7} \stackrel{?}{=} 0$

$0 = 0$

The solution is 0.

15. $1.7x = 10.71$

$\frac{1.7x}{1.7} = \frac{10.71}{1.7}$

$x = 6.3$

Check: $1.7x = 10.71$

$1.7(6.3) \stackrel{?}{=} 10.71$

$10.71 = 10.71$

The solution is 6.3.

17. $42 = 7x$

$\frac{42}{7} = \frac{7x}{7}$

$6 = x$

Check: $42 = 7x$

$42 \stackrel{?}{=} 7(6)$

$42 = 42$

The solution is 6.

19. $4.4 = -0.8x$

$\frac{4.4}{-0.8} = \frac{-0.8x}{-0.8}$

$-5.5 = x$

Check: $4.4 = -0.8x$

$4.4 \stackrel{?}{=} -0.8(-5.5)$

$4.4 = 4.4$

The solution is -5.5.

21. $-\frac{3}{7}p = -2$

$-\frac{7}{3}\left(-\frac{3}{7}p\right) = -\frac{7}{3}(-2)$

$p = \frac{14}{3}$

Check: $-\frac{3}{7}p = -2$

$-\frac{3}{7}\left(\frac{14}{3}\right) \stackrel{?}{=} -2$

$-2 = -2$

The solution is $\frac{14}{3}$.

23. $-\frac{4}{3}x = 12$

$-\frac{3}{4}\left(-\frac{4}{3}x\right) = -\frac{3}{4}(12)$

$x = -9$

Check: $-\frac{4}{3}x = 12$

$-\frac{4}{3}(-9) \stackrel{?}{=} 12$

$12 = 12$

The solution is -9.

25. $2x - 4 = 16$
$2x - 4 + 4 = 16 + 4$
$2x = 20$
$\dfrac{2x}{2} = \dfrac{20}{2}$
$x = 10$

    Check: $2x - 4 = 16$
             $2(10) - 4 \overset{?}{=} 16$
             $20 - 4 \overset{?}{=} 16$
             $16 = 16$
    The solution is 10.

27. $-x + 2 = 22$
$-x + 2 - 2 = 22 - 2$
$-x = 20$
$\dfrac{-x}{-1} = \dfrac{20}{-1}$
$x = -20$

    Check: $-x + 2 = 22$
             $-(-20) + 2 \overset{?}{=} 22$
             $20 + 2 \overset{?}{=} 22$
             $22 = 22$
    The solution is -20.

29. $6a + 3 = 3$
$6a + 3 - 3 = 3 - 3$
$6a = 0$
$\dfrac{6a}{6} = \dfrac{0}{6}$
$a = 0$

    Check: $6a + 3 = 3$
             $6(0) + 3 \overset{?}{=} 3$
             $0 + 3 \overset{?}{=} 3$
             $3 = 3$
    The solution is 0.

31. $6x + 10 = -20$
$6x + 10 - 10 = -20 - 10$
$6x = -30$
$\dfrac{6x}{6} = \dfrac{-30}{6}$
$x = -5$

    Check: $6x + 10 = -20$
             $6(-5) + 10 \overset{?}{=} -20$
             $-30 + 10 \overset{?}{=} -20$
             $-20 = -20$
    The solution is -5.

33. $5 - 0.3k = 5$
$5 - 0.3k - 5 = 5 - 5$
$-0.3k = 0$
$\dfrac{-0.3k}{-0.3} = \dfrac{0}{-0.3}$
$k = 0$

    Check: $5 - 0.3k = 5$
             $5 - 0.3(0) \overset{?}{=} 5$
             $5 - 0 \overset{?}{=} 5$
             $5 = 5$
    The solution is 0.

35. $-2x + \dfrac{1}{2} = \dfrac{7}{2}$
$-2x + \dfrac{1}{2} - \dfrac{1}{2} = \dfrac{7}{2} - \dfrac{1}{2}$
$-2x = \dfrac{6}{2}$
$-2x = 3$
$\dfrac{-2x}{-2} = \dfrac{3}{-2}$
$x = -\dfrac{3}{2}$

    Check:
$$-2x + \dfrac{1}{2} = \dfrac{7}{2}$$
$$-2\left(-\dfrac{3}{2}\right) + \dfrac{1}{2} \overset{?}{=} \dfrac{7}{2}$$
$$3 + \dfrac{1}{2} \overset{?}{=} \dfrac{7}{2}$$
$$\dfrac{6}{2} + \dfrac{1}{2} \overset{?}{=} \dfrac{7}{2}$$
$$\dfrac{7}{2} = \dfrac{7}{2}$$

    The solution is $-\dfrac{3}{2}$.

37. $\dfrac{x}{3} + 2 = -5$
$\dfrac{x}{3} + 2 - 2 = -5 - 2$
$\dfrac{x}{3} = -7$
$3\left(\dfrac{x}{3}\right) = 3(-7)$
$x = -21$

Check: $\dfrac{x}{3} + 2 = -5$

$$\dfrac{-21}{3} + 2 \stackrel{?}{=} -5$$

$$-7 + 2 \stackrel{?}{=} -5$$

$$-5 = -5$$

The solution is -21.

39. $10 = 2x - 1$

$10 + 1 = 2x - 1 + 1$

$11 = 2x$

$\dfrac{11}{2} = \dfrac{2x}{2}$

$\dfrac{11}{2} = x$

Check: $10 = 2x - 1$

$$10 \stackrel{?}{=} 2\left(\dfrac{11}{2}\right) - 1$$

$$10 \stackrel{?}{=} 11 - 1$$

$$10 = 10$$

The solution is $\dfrac{11}{2}$.

41. $6z - 8 - z + 3 = 0$

$5z - 5 = 0$

$5z - 5 + 5 = 0 + 5$

$5z = 5$

$\dfrac{5z}{5} = \dfrac{5}{5}$

$z = 1$

Check: $6z - 8 - z + 3 = 0$

$6(1) - 8 - 1 + 3 = 0$

$6 - 8 - 1 + 3 \stackrel{?}{=} 0$

$0 = 0$

The solution is 1.

43. $10 - 3x - 6 - 9x = 7$

$4 - 12x = 7$

$4 - 12x - 4 = 7 - 4$

$-12x = 3$

$\dfrac{-12x}{-12} = \dfrac{3}{-12}$

$x = -\dfrac{1}{4}$

Check: $10 - 3x - 6 - 9x = 7$

$$10 - 3\left(-\dfrac{1}{4}\right) - 6 - 9\left(-\dfrac{1}{4}\right) \stackrel{?}{=} 7$$

$$10 + \dfrac{3}{4} - 6 + \dfrac{9}{4} \stackrel{?}{=} 7$$

$$10 - 6 + \dfrac{3}{4} + \dfrac{9}{4} \stackrel{?}{=} 7$$

$$4 + \dfrac{12}{4} \stackrel{?}{=} 7$$

$$4 + 3 \stackrel{?}{=} 7$$

$$7 = 7$$

The solution is $-\dfrac{1}{4}$.

45. $1 = 0.4x - 0.6x - 5$

$1 = -0.2x - 5$

$1 + 5 = -0.2x - 5 + 5$

$6 = -0.2x$

$\dfrac{6}{-0.2} = \dfrac{-0.2x}{-0.2}$

$-30 = x$

Check: $1 = 0.4x - 0.6x - 5$

$1 \stackrel{?}{=} 0.4(-30) - 0.6(-30) - 5$

$1 \stackrel{?}{=} -12 + 18 - 5$

$1 = 1$

The solution is -30.

47. $z - 5z = 7z - 9 - z$

$-4z = 6z - 9$

$-4z - 6z = 6z - 9 - 6z$

$-10z = -9$

$\dfrac{-10z}{-10} = \dfrac{-9}{-10}$

$z = \dfrac{9}{10}$

Check: $z - 5z = 7z - 9 - z$

$$\dfrac{9}{10} - 5\left(\dfrac{9}{10}\right) \stackrel{?}{=} 7\left(\dfrac{9}{10}\right) - 9 - \dfrac{9}{10}$$

$$\dfrac{9}{10} - \dfrac{45}{10} \stackrel{?}{=} \dfrac{63}{10} - 9 - \dfrac{9}{10}$$

$$-\dfrac{36}{10} \stackrel{?}{=} \dfrac{54}{10} - 9$$

$$-\dfrac{36}{10} \stackrel{?}{=} \dfrac{54}{10} - \dfrac{90}{10}$$

$$-\dfrac{36}{10} = -\dfrac{36}{10}$$

The solution is $\dfrac{9}{10}$.

49. If x = the first odd integer, then $x + 2$ = the next odd integer. Their sum is
$x + (x + 2) = x + x + 2 = 2x + 2$

51. If x = the first integer, then $x + 1$ = the second consecutive integer, and $x + 2$ = the third consecutive integer. The sum of first and third integers is
$x + (x + 2) = x + x + 2 = 2x + 2$

53. $5x + 2(x - 6) = 5x + 2(x) + 2(-6)$
$= 5x + 2x - 12 = 7x - 12$

55. $6(2z + 4) + 20 = 6(2z) + 6(4) + 20$
$= 12z + 24 + 20 = 12z + 44$

57. $-(x - 1) + x = -1(x - 1) + x$
$= -(x) - 1(-1) + x = -x + 1 + x = 1$

59. $0.07x - 5.06 = -4.92$
$0.07x - 5.06 + 5.06 = -4.92 + 5.06$
$0.07x = 0.14$
$\dfrac{0.07x}{0.07} = \dfrac{0.14}{0.07}$
$x = 2$

61. Answers may vary. If we solve the equation for x, we obtain the following.
$3x + 6 = 2x + 10 + x - 4$
$3x + 6 = 3x + 6$
$3x + 6 - 6 = 3x + 6 - 6$
$3x = 3x$
$3x - 3x = 3x - 3x$
$0 = 0$

63. Answers may vary.

Section 2.4

Practice Problems

1. $5(3x - 1) + 2 = 12x + 6$
$15x - 5 + 2 = 12x + 6$
$15x - 3 = 12x + 6$
$15x - 3 - 12x = 12x + 6 - 12x$
$3x - 3 = 6$
$3x - 3 + 3 = 6 + 3$
$3x = 9$
$\dfrac{3x}{3} = \dfrac{9}{3}$
$x = 3$

2. $9(5 - x) = -3x$
$45 - 9x = -3x$
$45 - 9x + 9x = -3x + 9x$
$45 = 6x$
$\dfrac{45}{6} = \dfrac{6x}{6}$
$\dfrac{15}{2} = x$

3. $\dfrac{5}{2}x - 1 = \dfrac{3}{2}x - 4$
$2\left(\dfrac{5}{2}x - 1\right) = 2\left(\dfrac{3}{2}x - 4\right)$
$5x - 2 = 3x - 8$
$5x - 2 - 3x = 3x - 8 - 3x$
$2x - 2 = -8$
$2x - 2 + 2 = -8 + 2$
$2x = -6$
$\dfrac{2x}{2} = \dfrac{-6}{2}$
$x = -3$

4. $\dfrac{3(x - 2)}{5} = 3x + 6$
$5\left(\dfrac{3(x - 2)}{5}\right) = 5(3x + 6)$
$3(x - 2) = 5(3x + 6)$
$3x - 6 = 15x + 30$
$3x - 6 - 3x = 15x + 30 - 3x$
$-6 = 12x + 30$
$-6 - 30 = 12x + 30 - 30$
$-36 = 12x$
$\dfrac{-36}{12} = \dfrac{12x}{12}$
$-3 = x$

5. $0.06x - 0.10(x - 2) = -0.02(8)$
$100[0.06x - 0.10(x - 2)] = 100[-0.02(8)]$
$6x - 10(x - 2) = -2(8)$
$6x - 10x + 20 = -16$
$-4x + 20 = -16$
$-4x + 20 - 20 = -16 - 20$
$-4x = -36$
$\dfrac{-4x}{-4} = \dfrac{-36}{-4}$
$x = 9$

6. $5(2-x)+8x=3(x-6)$
$10-5x+8x=3x-18$
$10+3x=3x-18$
$10+3x-3x=3x-18-3x$
$10=-18$
There is no solution to this equation.

7. $-6(2x+1)-14=-10(x+2)-2x$
$-12x-6-14=-10x-20-2x$
$-12x-20=-12x-20$
$-12x-20+12x=-12x-20+12x$
$-20=-20$
Every real number is a solution.

Exercise Set 2.4

1. $-4y+10=-2(3y+1)$
$-4y+10=-6y-2$
$-4y+10+4y=-6y-2+4y$
$10=-2y-2$
$10+2=-2y-2+2$
$12=-2y$
$\dfrac{12}{-2}=\dfrac{-2y}{-2}$
$-6=y$

3. $9x-8=10+15x$
$9x-8-9x=10+15x-9x$
$-8=10+6x$
$-8-10=10+6x-10$
$-18=6x$
$\dfrac{-18}{6}=\dfrac{6x}{6}$
$-3=x$

5. $-2(3x-4)=2x$
$-6x+8=2x$
$-6x+8+6x=2x+6x$
$8=8x$
$\dfrac{8}{8}=\dfrac{8x}{8}$
$1=x$

7. $4(2n-1)=(6n+4)+1$
$8n-4=6n+4+1$
$8n-4=6n+5$
$8n-4-6n=6n+5-6n$
$2n-4=5$
$2n-4+4=5+4$
$2n=9$

$\dfrac{2n}{2}=\dfrac{9}{2}$
$n=\dfrac{9}{2}$

9. $5(2x-1)-2(3x)=1$
$10x-5-6x=1$
$4x-5=1$
$4x-5+5=1+5$
$4x=6$
$\dfrac{4x}{4}=\dfrac{6}{4}$
$x=\dfrac{3}{2}$

11. $6(x-3)+10=-8$
$6x-18+10=-8$
$6x-8=-8$
$6x-8+8=-8+8$
$6x=0$
$\dfrac{6x}{6}=\dfrac{0}{6}$
$x=0$

13. $8-2(a-1)=7+a$
$8-2a+2=7+a$
$10-2a=7+a$
$10-2a+2a=7+a+2a$
$10=7+3a$
$10-7=7+3a-7$
$3=3a$
$\dfrac{3}{3}=\dfrac{3a}{3}$
$1=a$

15. $4x+3=2x+11$
$4x+3-2x=2x+11-2x$
$2x+3=11$
$2x+3-3=11-3$
$2x=8$
$\dfrac{2x}{2}=\dfrac{8}{2}$
$x=4$

17. $-2y-10=5y+18$
$-2y-10+2y=5y+18+2y$
$-10=7y+18$
$-10-18=7y+18-18$
$-28=7y$
$\dfrac{-28}{7}=\dfrac{7y}{7}$
$-4=y$

19. $-3(t-5)+2t=5t-4$
$-3t+15+2t=5t-4$
$-t+15=5t-4$
$-t+15+t=5t-4+t$
$15=6t-4$
$15+4=6t-4+4$
$19=6t$
$\dfrac{19}{6}=\dfrac{6t}{6}$
$\dfrac{19}{6}=t$

21. $\dfrac{3}{4}x-\dfrac{1}{2}=1$
$4\left(\dfrac{3}{4}x-\dfrac{1}{2}\right)=4(1)$
$4\left(\dfrac{3}{4}x\right)+4\left(-\dfrac{1}{2}\right)=4$
$3x-2=4$
$3x-2+2=4+2$
$3x=6$
$\dfrac{3x}{3}=\dfrac{6}{3}$
$x=2$

23. $x+\dfrac{5}{4}=\dfrac{3}{4}x$
$4\left(x+\dfrac{5}{4}\right)=4\left(\dfrac{3}{4}x\right)$
$4(x)+4\left(\dfrac{5}{4}\right)=4\left(\dfrac{3}{4}x\right)$
$4x+5=3x$
$4x+5-4x=3x-4x$
$5=-x$
$\dfrac{5}{-1}=\dfrac{-x}{-1}$
$-5=x$

25. $\dfrac{x}{2}-1=\dfrac{x}{5}+2$
$10\left(\dfrac{x}{2}-1\right)=10\left(\dfrac{x}{5}+2\right)$
$10\left(\dfrac{x}{2}\right)+10(-1)=10\left(\dfrac{x}{5}\right)+10(2)$
$5x-10=2x+20$
$5x-10-2x=2x+20-2x$
$3x-10=20$
$3x-10+10=20+10$
$3x=30$
$\dfrac{3x}{3}=\dfrac{30}{3}$
$x=10$

27. $\dfrac{6(3-z)}{5}=-z$
$5\left(\dfrac{6(3-z)}{5}\right)=5(-z)$
$6(3-z)=-5z$
$6(3)+6(-z)=-5z$
$18-6z=-5z$
$18-6z+6z=-5z+6z$
$18=z$

29. $0.06-0.01(x+1)=-0.02(2-x)$
$100[0.06-0.01(x+1)]=100[-0.02(2-x)]$
$6-1(x+1)=-2(2-x)$
$6-x-1=-4+2x$
$5-x=-4+2x$
$5-x+x=-4+2x+x$
$5=-4+3x$
$5+4=-4+3x+4$
$9=3x$
$\dfrac{9}{3}=\dfrac{3x}{3}$
$3=x$

31. $\dfrac{3(x-5)}{2}=\dfrac{2(x+5)}{3}$
$6\left[\dfrac{3(x-5)}{2}\right]=6\left[\dfrac{2(x+5)}{3}\right]$
$9(x-5)=4(x+5)$
$9(x)+9(-5)=4(x)+4(5)$
$9x-45=4x+20$
$9x-45-4x=4x+20-4x$
$5x-45=20$
$5x-45+45=20+45$
$5x=65$
$\dfrac{5x}{5}=\dfrac{65}{5}$
$x=13$

33. $0.50x+0.15(70)=0.25(142)$
$100[0.50x+0.15(70)]=100[0.25(142)]$
$50x+15(70)=25(142)$
$50x+1050=3550$
$50x+1050-1050=3550-1050$
$50x=2500$
$\dfrac{50x}{50}=\dfrac{2500}{50}$
$x=50$

35. $0.12(y-6) + 0.06y = 0.08y - 0.07(10)$
$100[0.12(y-6) + 0.06y]$
$\qquad = 100[0.08y - 0.07(10)]$
$12(y-6) + 6y = 8y - 7(10)$
$12(y) + 12(-6) + 6y = 8y - 70$
$12y - 72 + 6y = 8y - 70$
$18y - 72 = 8y - 70$
$18y - 72 - 8y = 8y - 70 - 8y$
$10y - 72 = -70$
$10y - 72 + 72 = -70 + 72$
$10y = 2$
$\dfrac{10y}{10} = \dfrac{2}{10}$
$y = \dfrac{1}{5} = 0.2$

37. $\dfrac{2(x+1)}{4} = 3x - 2$
$4\left[\dfrac{2(x+1)}{4}\right] = 4(3x-2)$
$2(x+1) = 4(3x-2)$
$2(x) + 2(1) = 4(3x) + 4(-2)$
$2x + 2 = 12x - 8$
$2x + 2 - 2x = 12x - 8 - 2x$
$2 = 10x - 8$
$2 + 8 = 10x - 8 + 8$
$10 = 10x$
$\dfrac{10}{10} = \dfrac{10x}{10}$
$1 = x$

39. $x + \dfrac{7}{6} = 2x - \dfrac{7}{6}$
$6\left(x + \dfrac{7}{6}\right) = 6\left(2x - \dfrac{7}{6}\right)$
$6(x) + 6\left(\dfrac{7}{6}\right) = 6(2x) + 6\left(-\dfrac{7}{6}\right)$
$6x + 7 = 12x - 7$
$6x + 7 - 6x = 12x - 7 - 6x$
$7 = 6x - 7$
$7 + 7 = 6x - 7 + 7$
$14 = 6x$
$\dfrac{14}{6} = \dfrac{6x}{6}$
$\dfrac{14}{6} = x$
$\dfrac{7}{3} = x$

41. Answers may vary.

43. $5x - 5 = 2(x+1) + 3x - 7$
$5x - 5 = 2x + 2 + 3x - 7$
$5x - 5 = 5x - 5$
$5x - 5 + 5 = 5x - 5 + 5$
$5x = 5x$
$5x - 5x = 5x - 5x$
$0 = 0$
The equation $5x - 5 = 2(x+1) + 3x - 7$ is an identity and every real number is a solution.

45. $\dfrac{x}{4} + 1 = \dfrac{x}{4}$
$4\left(\dfrac{x}{4} + 1\right) = 4\left(\dfrac{x}{4}\right)$
$x + 4 = x$
$x + 4 - x = x - x$
$4 = 0$
There is no solution to the equation
$\dfrac{x}{4} + 1 = \dfrac{x}{4}$.

47. $3x - 7 = 3(x+1)$
$3x - 7 = 3x + 3$
$3x - 7 + 7 = 3x + 3 + 7$
$3x = 3x + 10$
$3x - 3x = 3x + 10 - 3x$
$0 = 10$
There is no solution to the equation
$3x - 7 = 3(x+1)$.

49. $2(x+3) - 5 = 5x - 3(1+x)$
$2x + 6 - 5 = 5x - 3 - 3x$
$2x + 1 = 2x - 3$
$2x + 1 - 1 = 2x - 3 - 1$
$2x = 2x - 4$
$2x - 2x = 2x - 4 - 2x$
$0 = -4$
There is no solution to the equation
$2(x+3) - 5 = 5x - 3(1+x)$.

51. Answers may vary.

53. The perimeter of the lot is the sum of the lengths of the sides.
$x + (2x-3) + (3x-5)$
$= x + 2x - 3 + 3x - 5$
$= 6x - 8$
The perimeter of the lot is $(6x - 8)$ meters.

55. $-8 - x$

57. $-3 + 2x$

59. $9(x + 20)$

61. $1000(7x - 10) = 50(412 + 100x)$
$7000x - 10,000 = 20,600 + 5000x$
$7000x - 10,000 + 10,000$
$\qquad = 20,600 + 5000x + 10,000$
$7000x = 30,600 + 5000x$
$7000x - 5000x$
$\qquad = 30,600 + 5000x - 5000x$
$2000x = 30,600$
$\dfrac{2000x}{2000} = \dfrac{30,600}{2000}$
$x = 15.3$

63. Clear the equation of decimals by multiplying both sides by 1000.
$0.035x + 5.112 = 0.010x + 5.107$
$1000(0.035x + 5.112)$
$\qquad = 1000(0.010x + 5.107)$
$35x + 5112 = 10x + 5107$
$35x + 5112 - 5112 = 10x + 5107 - 5112$
$35x = 10x - 5$
$35x - 10x = 10x - 5 - 10x$
$25x = -5$
$\dfrac{25x}{25} = \dfrac{-5}{25}$
$x = -0.2$

65. Since we know the perimeter of the pentagon is 28 cm,
$x + x + x + 2x + 2x = 28$
$(1 + 1 + 1 + 2 + 2)x = 28$
$7x = 28$
$\dfrac{7x}{7} = \dfrac{28}{7}$
$x = 4$
If $x = 4$ cm, then $2x = 2(4) = 8$ cm.

Integrated Review

1. $x - 10 = -4$
$x - 10 + 10 = -4 + 10$
$x = 6$

2. $y + 14 = -3$
$y + 14 - 14 = -3 - 14$
$y = -17$

3. $9y = 108$
$\dfrac{9y}{9} = \dfrac{108}{9}$
$y = 12$

4. $-3x = 78$
$\dfrac{-3x}{-3} = \dfrac{78}{-3}$
$x = -26$

5. $-6x + 7 = 25$
$-6x + 7 - 7 = 25 - 7$
$-6x = 18$
$\dfrac{-6x}{-6} = \dfrac{18}{-6}$
$x = -3$

6. $5y - 42 = -47$
$5y - 42 + 42 = -47 + 42$
$5y = -5$
$\dfrac{5y}{5} = \dfrac{-5}{5}$
$y = -1$

7. $\dfrac{2}{3}x = 9$
$\dfrac{3}{2}\left(\dfrac{2}{3}x\right) = \dfrac{3}{2}(9)$
$x = \dfrac{27}{2} = 13.5$

8. $\dfrac{4}{5}z = 10$
$\dfrac{5}{4}\left(\dfrac{4}{5}z\right) = \dfrac{5}{4}(10)$
$z = \dfrac{25}{2} = 12.5$

9. $\dfrac{r}{-4} = -2$
$-4\left(\dfrac{r}{-4}\right) = -4(-2)$
$r = 8$

10. $\dfrac{y}{-8} = 8$
$-8\left(\dfrac{y}{-8}\right) = -8(8)$
$y = -64$

11. $6 - 2x + 8 = 10$
$14 - 2x = 10$
$14 - 2x - 14 = 10 - 14$
$-2x = -4$
$\dfrac{-2x}{-2} = \dfrac{-4}{-2}$
$x = 2$

12. $-5 - 6y + 6 = 19$
$1 - 6y = 19$
$1 - 6y - 1 = 19 - 1$
$-6y = 18$
$\dfrac{-6y}{-6} = \dfrac{18}{-6}$
$y = -3$

13. $2x - 7 = 6x - 27$
$2x - 7 - 2x = 6x - 27 - 2x$
$-7 = 4x - 27$
$-7 + 27 = 4x - 27 + 27$
$20 = 4x$
$\dfrac{20}{4} = \dfrac{4x}{4}$
$5 = x$

14. $3 + 8y = 3y - 2$
$3 + 8y - 3 = 3y - 2 - 3$
$8y = 3y - 5$
$8y - 3y = 3y - 5 - 3y$
$5y = -5$
$\dfrac{5y}{5} = \dfrac{-5}{5}$
$y = -1$

15. $-3a + 6 + 5a = 7a - 8a$
$6 + 2a = -a$
$6 + 2a - 6 = -a - 6$
$2a = -a - 6$
$2a + a = -a - 6 + a$
$3a = -6$
$\dfrac{3a}{3} = \dfrac{-6}{3}$
$a = -2$

16. $4b - 8 - b = 10b - 3b$
$3b - 8 = 7b$
$3b - 8 + 8 = 7b + 8$
$3b = 7b + 8$
$3b - 7b = 7b + 8 - 7b$
$-4b = 8$
$\dfrac{-4b}{-4} = \dfrac{8}{-4}$
$b = -2$

17. $-\dfrac{2}{3}x = \dfrac{5}{9}$
$-\dfrac{3}{2}\left(-\dfrac{2}{3}x\right) = -\dfrac{3}{2}\left(\dfrac{5}{9}\right)$
$x = -\dfrac{5}{6}$

18. $-\dfrac{3}{8}y = -\dfrac{1}{16}$
$-\dfrac{8}{3}\left(-\dfrac{3}{8}y\right) = -\dfrac{8}{3}\left(-\dfrac{1}{16}\right)$
$y = \dfrac{1}{6}$

19. $10 = -6n + 16$
$10 - 16 = -6n + 16 - 16$
$-6 = -6n$
$\dfrac{-6}{-6} = \dfrac{-6n}{-6}$
$1 = n$

20. $-5 = -2m + 7$
$-5 - 7 = -2m + 7 - 7$
$-12 = -2m$
$\dfrac{-12}{-2} = \dfrac{-2m}{-2}$
$6 = m$

21. $3(5c - 1) - 2 = 13c + 3$
$15c - 3 - 2 = 13c + 3$
$15c - 5 = 13c + 3$
$15c - 5 + 5 = 13c + 3 + 5$
$15c = 13c + 8$
$15c - 13c = 13c + 8 - 13c$
$2c = 8$
$\dfrac{2c}{2} = \dfrac{8}{2}$
$c = 4$

22. $4(3t + 4) - 20 = 3 + 5t$
$12t + 16 - 20 = 3 + 5t$
$12t - 4 = 3 + 5t$
$12t - 4 + 4 = 3 + 5t + 4$
$12t = 5t + 7$
$12t - 5t = 5t + 7 - 5t$
$7t = 7$
$\dfrac{7t}{7} = \dfrac{7}{7}$
$t = 1$

23. $\dfrac{2(z + 3)}{3} = 5 - z$

$3\left[\dfrac{2(z + 3)}{3}\right] = 3(5 - z)$
$2(z + 3) = 3(5 - z)$
$2z + 6 = 15 - 3z$
$2z + 6 - 6 = 15 - 3z - 6$
$2z = 9 - 3z$
$2z + 3z = 9 - 3z + 3z$
$5z = 9$
$\dfrac{5z}{5} = \dfrac{9}{5}$
$z = \dfrac{9}{5}$

24. $\dfrac{3(w + 2)}{4} = 2w + 3$

$4\left[\dfrac{3(w + 2)}{4}\right] = 4(2w + 3)$
$3(w + 2) = 4(2w + 3)$
$3w + 6 = 8w + 12$
$3w + 6 - 6 = 8w + 12 - 6$
$3w = 8w + 6$
$3w - 8w = 8w + 6 - 8w$
$-5w = 6$
$\dfrac{-5w}{-5} = \dfrac{6}{-5}$
$w = -\dfrac{6}{5}$

25. $-2(2x - 5) = -3x + 7 - x + 3$
$-4x + 10 = -3x + 7 - x + 3$
$-4x + 10 = -4x + 10$
$-4x + 10 - 10 = -4x + 10 - 10$
$-4x = -4x$
$-4x + 4x = -4x + 4x$
$0 = 0$
$-2(2x - 5) = -3x + 7 - x + 3$ is an identity and every real number is a solution.

26. $-4(5x - 2) = -12x + 4 - 8x + 4$
$-20x + 8 = -12x + 4 - 8x + 4$
$-20x + 8 = -20x + 8$
$-20x + 8 - 8 = -20x + 8 - 8$
$-20x = -20x$
$-20x + 20x = -20x + 20x$
$0 = 0$
$-4(5x - 2) = -12x + 4 - 8x + 4$ is an identity and every real number is a solution.

27. $0.02(6t - 3) = 0.04(t - 2) + 0.02$
$100[0.02(6t - 3)] = 100[0.04(t - 2) + 0.02]$
$2(6t - 3) = 4(t - 2) + 2$
$12t - 6 = 4t - 8 + 2$
$12t - 6 = 4t - 6$
$12t - 6 + 6 = 4t - 6 + 6$
$12t = 4t$
$12t - 4t = 4t - 4t$
$8t = 0$
$\dfrac{8t}{8} = \dfrac{0}{8}$
$t = 0$

28. $0.03(m + 7) = 0.02(5 - m) + 0.03$
$100[0.03(m + 7)] = 100[0.02(5 - m) + 0.03]$
$3(m + 7) = 2(5 - m) + 3$
$3m + 21 = 10 - 2m + 3$
$3m + 21 = 13 - 2m$
$3m + 21 - 21 = 13 - 2m - 21$
$3m = -8 - 2m$
$3m + 2m = -8 - 2m + 2m$
$5m = -8$
$\dfrac{5m}{5} = \dfrac{-8}{5}$
$m = -\dfrac{8}{5} = -1.6$

Section 2.5

Practice Problems

1. Let x = the unknown number
$$3(x - 5) = 2x - 3$$
$$3x - 15 = 2x - 3$$
$$3x - 15 + 15 = 2x - 3 + 15$$
$$3x = 2x + 12$$
$$3x - 2x = 2x + 12 - 2x$$
$$x = 12$$
The number is 12.

2. Let x = length of the shorter piece. Then $5x$ = length of the longer piece.
$$x + 5x = 18$$
$$6x = 18$$
$$\frac{6x}{6} = \frac{18}{6}$$
$$x = 3$$
The shorter piece is 3 feet and the longer piece is 15 feet.

3. Let x = the number of Democrats. Then $x + 39$ = the number of Republicans.
$$x + (x + 39) = 433$$
$$x + x + 39 = 433$$
$$2x + 39 = 433$$
$$2x + 39 - 39 = 433 - 39$$
$$2x = 394$$
$$\frac{2x}{2} = \frac{394}{2}$$
$$x = 197$$
There were 197 Democrats and 236 Republicans.

4. Let x = the number of miles driven. Then $0.20x$ is the charge per mile driven.
$$34 + 0.20x = 104$$
$$34 + 0.20x - 34 = 104 - 34$$
$$0.20x = 70$$
$$\frac{0.20x}{0.20} = \frac{70}{0.20}$$
$$x = 350$$
350 miles were driven.

5. Let x = the measure of the smallest angle. Then $2x$ = the measure of the second angle and $3x$ = the measure of the third angle.
$$x + 2x + 3x = 180$$
$$6x = 180$$
$$\frac{6x}{6} = \frac{180}{6}$$
$$x = 30$$
The smallest angle is 30°; the second angle is 60°; and the third angle is 90°.

Exercise Set 2.5

1. Let x represent the number.
$$2x + \frac{1}{5} = 3x - \frac{4}{5}$$
$$5\left(2x + \frac{1}{5}\right) = 5\left(3x - \frac{4}{5}\right)$$
$$10x + 1 = 15x - 4$$
$$10x + 1 - 10x = 15x - 4 - 10x$$
$$1 = 5x - 4$$
$$1 + 4 = 5x - 4 + 4$$
$$5 = 5x$$
$$\frac{5}{5} = \frac{5x}{5}$$
$$1 = x$$
The number is 1.

3. Let x represent the number.
$$2(x - 8) = 3(x + 3)$$
$$2x - 16 = 3x + 9$$
$$2x - 16 - 2x = 3x + 9 - 2x$$
$$-16 = x + 9$$
$$-16 - 9 = x + 9 - 9$$
$$-25 = x$$
The number is −25.

5. Let x represent the number.
$$2x \cdot 3 = 5x - \frac{3}{4}$$
$$6x = 5x - \frac{3}{4}$$
$$6x - 5x = 5x - \frac{3}{4} - 5x$$
$$x = -\frac{3}{4}$$
The number is $-\frac{3}{4}$.

7. Let x represent the number.
    $3(x + 5) = 2x - 1$
    $3x + 15 = 2x - 1$
    $3x + 15 - 2x = 2x - 1 - 2x$
    $x + 15 = -1$
    $x + 15 - 15 = -1 - 15$
    $x = -16$
    The number is -16.

9. Let x = salary of the governor of Nebraska, then $2x$ = salary of the governor of New York.
    $x + 2x = 195,000$
    $3x = 195,000$
    $\dfrac{3x}{3} = \dfrac{195,000}{3}$
    $x = 65,000$
    The salary of the governor of Nebraska is $65,000. The salary of the governor of New York is $2 \cdot 65,000 = \$130,000$.

11. If x = length of the first piece, then $2x$ = length of the second piece, and $5x$ = length of the third piece.
    $x + 2x + 5x = 40$
    $8x = 40$
    $\dfrac{8x}{8} = \dfrac{40}{8}$
    $x = 5$
    The first piece is 5 inches long, the second piece is $2 \cdot 5 = 10$ inches long, and the third piece is $5 \cdot 5 = 25$ inches long.

13. The cost of renting the car is equal to the daily rental charge plus $0.29 per mile. Let x = number of miles.
    $2 \cdot 24.95 + 0.29x = 100$
    $49.90 + 0.29x = 100$
    $49.90 + 0.29x - 49.90 = 100 - 49.90$
    $0.29x = 50.10$
    $\dfrac{0.29x}{0.29} = \dfrac{50.10}{0.29}$
    $x = 172$
    You can drive 172 whole miles on a budget of $100.

15. Let x = measure of each of the two equal angles, then $2x + 30$ = measure of the third angle.
    $x + x + 2x + 30 = 180$
    $4x + 30 = 180$
    $4x + 30 - 30 = 180 - 30$
    $4x = 150$
    $\dfrac{4x}{4} = \dfrac{150}{4}$
    $x = 37.5$
    The angles measure $37.5°$, $37.5°$, and $105°$.

17. Let x = number of votes for Marc Little, then $x + 16,950$ = number of votes for Corrine Brown.
    $x + (x + 16,950) = 110,740$
    $x + x + 16,950 = 110,740$
    $2x + 16,950 = 110,740$
    $2x + 16,950 - 16,950 = 110,740 - 16,950$
    $2x = 93,790$
    $\dfrac{2x}{2} = \dfrac{93,790}{2}$
    $x = 46,895$
    Marc Little received 46,895 votes. Corrine Brown received $46,895 + 16,950 = 63,845$ votes.

19. If x = measure of the smaller angle, then $3x$ = measure of the supplementary angle.
    $x + 3x = 180$
    $4x = 180$
    $\dfrac{4x}{4} = \dfrac{180}{4}$
    $x = 45$
    The angles measure $45°$ and $135°$.

21. If x = length of the shorter piece, then $2x + 2$ = length of the longer piece.
    $x + (2x + 2) = 17$
    $x + 2x + 2 = 17$
    $3x + 2 = 17$
    $3x + 2 - 2 = 17 - 2$
    $3x = 15$
    $\dfrac{3x}{3} = \dfrac{15}{3}$
    $x = 5$
    The shorter piece is 5 feet long. The longer piece is 12 feet long.

23. Let x = height of the probe, then
$2x - 19$ = diameter of the probe.
$x + (2x - 19) = 83$
$x + 2x - 19 = 83$
$3x - 19 = 83$
$3x - 19 + 19 = 83 + 19$
$3x = 102$
$\dfrac{3x}{3} = \dfrac{102}{3}$
$x = 34$
The probe is 34 inches tall and has a
diameter of $2(34) - 19 = 49$ inches.

25. The tallest bar represents the most popular
name which is Midway.

27. If x = number of cities, towns, or villages
named Five Points, then
$x + 55 = 2x - 90$
$x + 55 - x = 2x - 90 - x$
$55 = x - 90$
$55 + 90 = x - 90 + 90$
$145 = x$
There are 145 cities, towns, or villages
named Five Points.
This is consistent with the graph.

29. $\dfrac{1}{2}(x - 1) = 37$

31. $\dfrac{3(x + 2)}{5} = 0$

33. $2W + 2L = 2(7) + 2(10) = 14 + 20 = 34$

35. $\pi r^2 = \pi\left(15^2\right) = \pi(225) = 225\pi$

37. Let x represent the first odd integer, then
$x + 2$ represents the next consecutive odd
integer.
$2(x + 2) = 3x + 15$
$2x + 4 = 3x + 15$
$2x + 4 - 2x = 3x + 15 - 2x$
$4 = x + 15$
$4 - 15 = x + 15 - 15$
$-11 = x$
The integers are -11 and -9.

39. Let x = the first even integer, then
$x + 2$ = the next consecutive even integer.

$x + (x + 2) = 178$
$x + x + 2 = 178$
$2x + 2 = 178$
$2x + 2 - 2 = 178 - 2$
$2x = 176$
$\dfrac{2x}{2} = \dfrac{176}{2}$
$x = 88$
The final scores were 88 for the Utah Jazz
and 90 for the Chicago Bulls.

41. Let x = the first integer, then
$x + 1$ = the next consecutive integer, and $x +$
2 = third consecutive integer.
$x + (x + 1) + (x + 2) = 2x + 13$
$x + x + 1 + x + 2 = 2x + 13$
$3x + 3 = 2x + 13$
$3x + 3 - 2x = 2x + 13 - 2x$
$x + 3 = 13$
$x + 3 - 3 = 13 - 3$
$x = 10$
The integers are 10, 11, and 12.

43. Let x = the first odd integer, then
$x + 2$ = the next consecutive odd integer, and
$x + 4$ = the third consecutive odd integer.
$x + (x + 2) + (x + 4) = 675$
$x + x + 2 + x + 4 = 675$
$3x + 6 = 675$
$3x + 6 - 6 = 675 - 6$
$3x = 669$
$\dfrac{3x}{3} = \dfrac{669}{3}$
$x = 223$
The codes are 223, 225, and 227 for Mali
Republic, Côte d'Ivoire, and Niger
respectively.

45. Let x = the first odd integer, then
$x + 2$ = the next consecutive odd integer.
$7x - 54 = 5(x + 2)$
$7x - 54 = 5x + 10$
$7x - 54 + 54 = 5x + 10 + 54$
$7x = 5x + 64$
$7x - 5x = 5x + 64 - 5x$
$2x = 64$
$\dfrac{2x}{2} = \dfrac{64}{2}$
$x = 32$
Since the solution is an even number, we
conclude that there are no odd integers
which satisfy these conditions.

47. Answers may vary.

Section 2.6

Practice Problems

1. Use the distance formula $d = rt$ with $d = 700$ and $r = 55$.

$d = rt$

$700 = 55t$

$\dfrac{700}{55} = \dfrac{55t}{55}$

$12.73 = t$

The trip will take approximately 12.73 hours.

2. The area of a rectangle is given by the formula $A = lw$. Let l = the length of the deck.

$450 = l \cdot 18$

$\dfrac{450}{18} = \dfrac{l \cdot 18}{18}$

$25 = l$

The deck will be 25 feet long.

3. $C = 2\pi r$

$\dfrac{C}{2\pi} = \dfrac{2\pi r}{2\pi}$

$\dfrac{C}{2\pi} = r$

4. $P = 2l + 2w$

$P - 2l = 2l + 2w - 2l$

$P - 2l = 2w$

$\dfrac{P - 2l}{2} = \dfrac{2w}{2}$

$\dfrac{P - 2l}{2} = w$

5. $A = \dfrac{a+b}{2}$

$2(A) = 2\left(\dfrac{a+b}{2}\right)$

$2A = a + b$

$2A - a = a + b - a$

$2A - a = b$

Exercise Set 2.6

1. $A = bh$

$45 = 15 \cdot h$

$\dfrac{45}{15} = \dfrac{15 \cdot h}{15}$

$3 = h$

3. $S = 4lw + 2wh$

$102 = 4(7)(3) + 2(3)h$

$102 = 84 + 6h$

$102 - 84 = 84 + 6h - 84$

$18 = 6h$

$\dfrac{18}{6} = \dfrac{6h}{6}$

$3 = h$

5. $A = \dfrac{1}{2}(B + b)h$

$180 = \dfrac{1}{2}(11 + 7)h$

$180 = \dfrac{1}{2}(18)h$

$180 = 9h$

$\dfrac{180}{9} = \dfrac{9h}{9}$

$20 = h$

7. $P = a + b + c$

$30 = 8 + 10 + c$

$30 = 18 + c$

$30 - 18 = 18 + c - 18$

$12 = c$

9. $C = 2\pi r$

$15.7 = 2\pi r$

$\dfrac{15.7}{2\pi} = \dfrac{2\pi r}{2\pi}$

$\dfrac{15.7}{6.28} = r$

$2.5 = r$

11. $I = PRT$

$3750 = (25,000)(0.05)T$

$3750 = 1250T$

$\dfrac{3750}{1250} = \dfrac{1250T}{1250}$

$3 = T$

13. $V = \frac{1}{3}\pi r^2 h$

$565.2 = \frac{1}{3}\pi\left(6^2\right)h$

$565.2 = \frac{1}{3}\pi(36)h$

$565.2 = 12\pi h$

$\frac{565.2}{12\pi} = \frac{12\pi h}{12\pi}$

$\frac{565.2}{37.68} = h$

$15 = h$

15. $A = bh$

$52{,}400 = 400 \cdot h$

$\frac{52{,}400}{400} = \frac{400 \cdot h}{400}$

$131 = h$

The width of the sign is 131 feet.

17. $d = rt$

$3000 = r \cdot 1.5$

$\frac{3000}{1.5} = \frac{r \cdot 1.5}{1.5}$

$2000 = r$

The plane travels at a rate of 2000 miles per hour.

19. Use the formula $C = \frac{5}{9}(F - 32)$ with $F = 14$.

$C = \left(\frac{5}{9}\right)(14 - 32) = \left(\frac{5}{9}\right)(-18) = -10$

14°F is the same as −10°C.

21. $V = lwh$

$V = (8)(3)(6)$

$V = 144$

Since the tank has a volume of 144 cubic feet, and each piranha requires 1.5 cubic feet, the tank can hold $\frac{144}{1.5} = 96$ piranhas.

23. $d = rt$

$135 = 60 \cdot t$

$\frac{135}{60} = \frac{60 \cdot t}{60}$

$2.25 = t$

It will take 2.25 hours.

25. $d = rt$

$25{,}000 = 4000 \cdot t$

$\frac{25{,}000}{4000} = \frac{4000 \cdot t}{4000}$

$6.25 = t$

It will take 6.25 hours.

27. $A = \frac{1}{2}(B + b)h$

$A = \frac{1}{2}(130 + 70)(60) = \frac{1}{2}(200)(60)$

$= 6000$

Since the area of the lawn is 6000 square feet and since each bag covers 4000 square feet, two bags must be purchased.

29. $V = lwh$

$V = (10)(8)(10)$

$V = 800$

The minimum volume of the box must be 800 cubic feet.

31. Use the formula for the area of a circle, $A = \pi r^2$, to solve for the number of square inches of pizza purchased in each case. For one 16-inch pizza use $r = 8$.

$A = \pi r^2 = \pi\left(8^2\right) = 64\pi$ square inches. For two 10-inch pizzas use $r = 5$ and multiply the result by 2.

$A = 2(\pi r^2) = 2\pi(5^2) = 2\pi(25) = 50\pi$ square inches

One 16-inch pizza gives more pizza for the price.

33. Use $F = \left(\frac{9}{5}\right)C + 32$ with $C = -78.5$.

$F = \left(\frac{9}{5}\right)C + 32$

$F = \left(\frac{9}{5}\right)(-78.5) + 32 = -141.3 + 32$

$= -109.3$

−78.5°C is the same as −109.3°F.

35. Use $d = rt$ with $d = 93{,}000{,}000$ and $r = 186{,}000$.

$d = rt$

$93{,}000{,}000 = 186{,}000 \cdot t$

$\dfrac{93{,}000{,}000}{186{,}000} = \dfrac{186{,}000 \cdot t}{186{,}000}$

$500 = t$

It takes 500 seconds or $8\frac{1}{3}$ minutes.

37. Use the formula for the volume of a sphere, $V = \frac{4}{3}\pi r^3$, with $r = 2000$.

$V = \dfrac{4}{3}\pi(2000)^3 = \dfrac{4}{3}\pi(8{,}000{,}000{,}000)$

$= 33{,}493{,}333{,}333$

The volume of the fireball was $33{,}493{,}333{,}333$ cubic miles.

39. Use the formula $d = rt$ with $d = 25{,}120$ and $r = 270{,}000$.

$d = rt$

$25{,}120 = 270{,}000 \cdot t$

$\dfrac{25{,}120}{270{,}000} = \dfrac{270{,}000 \cdot t}{270{,}000}$

$0.093 = t$

Thus, it takes 0.093 seconds for a bolt of lightning to travel around the world once. In one second it can travel around the world $\dfrac{1}{0.093} = 10.8$ times.

41. $\dfrac{20 \text{ miles}}{1 \text{ hour}} = \dfrac{20 \text{ miles}}{1 \text{ hour}} \cdot \dfrac{5280 \text{ feet}}{1 \text{ mile}} \cdot \dfrac{1 \text{ hour}}{3600 \text{ sec}}$

$= \dfrac{88}{3}$ feet per second

Use $d = rt$ with $d = 1300$ feet and $r = \dfrac{88}{3}$ feet per second.

$1300 = \dfrac{88}{3}t$

$\dfrac{3}{88}(1300) = \left(\dfrac{3}{88}\right)\left(\dfrac{88}{3}t\right)$

$44.3 = t$

It took about 44.3 seconds.

43. $f = 5gh$

$\dfrac{f}{5g} = \dfrac{5gh}{5g}$

$\dfrac{f}{5g} = h$

45. $V = LWH$

$\dfrac{V}{LH} = \dfrac{LWH}{LH}$

$\dfrac{V}{LH} = W$

47. $3x + y = 7$

$3x + y - 3x = 7 - 3x$

$y = 7 - 3x$

49. $A = p + PRT$

$A - p = p + PRT - p$

$A - p = PRT$

$\dfrac{A - p}{PT} = \dfrac{PRT}{PT}$

$\dfrac{A - p}{PT} = R$

51. $V = \dfrac{1}{3}Ah$

$3(V) = 3\left(\dfrac{1}{3}Ah\right)$

$3V = Ah$

$\dfrac{3V}{h} = \dfrac{Ah}{h}$

$\dfrac{3V}{h} = A$

53. $P = a + b + c$

$P - b = a + b + c - b$

$P - b = a + c$

$P - b - c = a + c - c$

$P - b - c = a$

55. $S = 2\pi rh + 2\pi r^2$

$S - 2\pi r^2 = 2\pi h + 2\pi r^2 - 2\pi r^2$

$S - 2\pi r^2 = 2\pi rh$

$\dfrac{S - 2\pi r^2}{2\pi r} = \dfrac{2\pi rh}{2\pi r}$

$\dfrac{S - 2\pi r^2}{2\pi r} = h$

57. $32\% = 0.32$

59. $200\% = 2.00$

61. $0.17 = 17\%$

63. $7.2 = 720\%$

65. $N = R + \dfrac{V}{G}$

$G(N) = G\left(R + \dfrac{V}{G}\right)$

$GN = GR + V$

$GN - GR = GR + V - GR$

$GN - GR = V$

$G(N - R) = V$

67. $V = (2 \cdot L)(2 \cdot W)(2 \cdot H) = 8LWH$
The volume is multiplied by 8.

69. Use the formula $C = \left(\dfrac{5}{9}\right)(F - 32)$ to find when $C = F$.

$C = \left(\dfrac{5}{9}\right)(F - 32)$

$F = \left(\dfrac{5}{9}\right)(F - 32)$

$9(F) = 9\left[\left(\dfrac{5}{9}\right)(F - 32)\right]$

$9F = 5(F - 32)$

$9F = 5F - 160$

$9F - 5F = 5F - 160 - 5F$

$4F = -160$

$\dfrac{4F}{4} = \dfrac{-160}{4}$

$F = -40$

$-40°F$ is the same as $-40°C$.

Section 2.7

Practice Problems

1. Let $x =$ the unknown percent.

$22 = x \cdot 40$

$\dfrac{22}{40} = \dfrac{x \cdot 40}{40}$

$0.55 = x$

$55\% = x$

The number 22 is 55% of 40.

2. Let $x =$ the unknown number.

$150 = 0.40x$

$\dfrac{150}{0.40} = \dfrac{0.40x}{0.40}$

$375 = x$

The number 150 is 40% of 375.

3. a. $38\% + 16\% = 54\%$
54% of homenowners spend
$250–$4999 per year on maintenance.

   b. $38\% + 16\% + 1\% = 55\%$
55% of homeowners spend $250 or
more per year on maintenance.

   c. 38% of $22,000 = 0.38(22,000) = 8360$
We might expect 8360 homeowners in
Fairview to spend $250–$999 on
maintenance.

4. a. $\dfrac{3}{7}$

   b. 3 hours $= 3 \cdot 60$ minutes $= 180$ minutes
The ratio of 40 minutes to 3 hours is
$\dfrac{40}{180} = \dfrac{2}{9}$.

5. $\dfrac{3}{8} = \dfrac{63}{x}$

$3x = 8 \cdot 63$

$\dfrac{3x}{3} = \dfrac{504}{3}$

$x = 168$

6. $\dfrac{2x + 1}{7} = \dfrac{x - 3}{5}$

$5(2x + 1) = 7(x - 3)$

$10x + 5 = 7x - 21$

$10x = 7x - 26$

$3x = -26$

$\dfrac{3x}{3} = \dfrac{-26}{3}$

$x = -\dfrac{26}{3}$

7. Let x = the number of people who are
uninsured.
$$\frac{39}{250} = \frac{x}{50,000}$$
$$39 \cdot 50,000 = 250x$$
$$\frac{1,950,000}{250} = \frac{250x}{250}$$
$$7800 = x$$
We would expect that 7800 people are
uninsured.

8. Compare unit prices.
8 ounces: $\frac{\$2.59}{8} = \$0.32375$

10 ounces: $\frac{\$3.11}{10} = \$0.311$
The 10-ounce size is the better buy.

Exercise Set 2.7

1. Let x = the unknown number.
$x = 0.16 \cdot 70$
$x = 11.2$
The number 11.2 is 16% of 70.

3. Let x = the unknown percent.
$28.6 = x \cdot 52$
$0.55 = x$
$55\% = x$
The number 28.6 is 55% of 52.

5. Let x = the unknown number.
$45 = 0.25 \cdot x$
$180 = x$
The number 45 is 25% of 180.

7. Let x = the unknown number.
$x = 0.23 \cdot 20$
$x = 4.6$
The number 4.6 is 23% of 20.

9. Let x = the unknown number.
$40 = 0.80 \cdot x$
$50 = x$
The number 40 is 80% of 50.

11. Let x = the unknown percent.
$144 = x \cdot 480$
$0.30 = x$
$30\% = x$
The number 144 is 30% of 480.

13. To find the decrease in price, we find 25%
of $156.
25% of $156 = 0.25(156) = 39
The coat is selling for $39 off the original
price. The sale price is $156 - 39 = \$117$.

15. To find how much further the men's record
throw is than the women's, we find 44.8% of
447.2 feet.
44.8% of $447.2 = 0.448(447.2)$
$= 200.3$
The men's record throw is 200.3 feet further
than the women's.
$447.2 + 200.3 = 647.5$.
The men's record is 647.5 feet.

17. 55.40% of those surveyed have used over-
the-counter drugs to combat the common
cold.

19. Since 23.70% of those surveyed have used
over-the-counter drugs for allergies, we find
23.70% of 230.
23.70% of $230 = 0.237(230) = 54.51$
54 people used over-the-counter drugs for
allergies.

21. No, because many people have used over-
the-counter drugs for more than one of the
categories listed.

23. Since 26% of men doze off, we find 26% of
121.
26% of $121 = 0.26(121) = 31.46$
We would expect 31 of the men to have
dozed off.

25. The percent of total operating income from
each division is the ratio of the operating
income of the division to the total.
Creative Content: $\frac{1596}{3333} = 0.48 = 48\%$

Theme Parks: $\frac{990}{3333} = 0.30 = 30\%$

Broadcasting: $\frac{747}{3333} = 22 = 22\%$
Total: $48\% + 30\% + 22\% = 100\%$

27. $\frac{2}{15}$

29. $\dfrac{10}{12} = \dfrac{5}{6}$

31. 3 gallons = 3 · 4 quarts = 12 quarts

The ratio of 5 quarts to 3 gallons is $\dfrac{5}{12}$.

33. 2 dollars = 2 · 20 nickels = 40 nickels

The ratio of 4 nickels to 2 dollars is

$\dfrac{4}{40} = \dfrac{1}{10}$.

35. 5 meters = 5 · 100 centimeters

= 500 centimeters

The ratio of 175 centimeters to 5 meters is

$\dfrac{175}{500} = \dfrac{7}{20}$.

37. 3 hours = 3 · 60 minutes = 180 minutes

The ratio of 190 minutes to 3 hours is

$\dfrac{190}{180} = \dfrac{19}{18}$.

39. Answers may vary.

41. $\dfrac{2}{3} = \dfrac{x}{6}$

$2 \cdot 6 = 3 \cdot x$

$\dfrac{12}{3} = \dfrac{3x}{3}$

$4 = x$

43. $\dfrac{x}{10} = \dfrac{5}{9}$

$x \cdot 9 = 10 \cdot 5$

$\dfrac{9x}{9} = \dfrac{50}{9}$

$x = \dfrac{50}{9}$

45. $\dfrac{4x}{6} = \dfrac{7}{2}$

$4x \cdot 2 = 6 \cdot 7$

$\dfrac{8x}{8} = \dfrac{42}{8}$

$x = \dfrac{21}{4}$

47. $\dfrac{x-3}{x} = \dfrac{4}{7}$

$7(x - 3) = x \cdot 4$

$7x - 21 = 4x$

$7x = 4x + 21$

$3x = 21$

$\dfrac{3x}{3} = \dfrac{21}{3}$

$x = 7$

49. $\dfrac{x+1}{2x+3} = \dfrac{2}{3}$

$3(x + 1) = 2(2x + 3)$

$3x + 3 = 4x + 6$

$3x = 4x + 3$

$-x = 3$

$\dfrac{-x}{-1} = \dfrac{3}{-1}$

$x = -3$

51. $\dfrac{9}{5} = \dfrac{12}{3x+2}$

$9(3x + 2) = 5 \cdot 12$

$27x + 18 = 60$

$27x = 42$

$\dfrac{27x}{27} = \dfrac{42}{27}$

$x = \dfrac{14}{9}$

53. $\dfrac{3}{x+1} = \dfrac{5}{2x}$

$3 \cdot 2x = 5(x + 1)$

$6x = 5x + 5$

$x = 5$

55. Let x = the elephant's weight on Pluto.

$\dfrac{100}{3} = \dfrac{4100}{x}$

$100x = 3(4100)$

$100x = 12{,}300$

$\dfrac{100x}{100} = \dfrac{12{,}300}{100}$

$x = 123$

The elephant's weight on Pluto is 123 lb.

57. Let x = number of calories in 42.6 grams.

$\dfrac{110}{28.4} = \dfrac{x}{42.6}$

$110(42.6) = 28.4x$

$4686 = 28.4x$

$\dfrac{4686}{28.4} = \dfrac{28.4x}{28.4}$

$165 = x$

There are 165 calories in 42.6 grams of the cereal.

59. Let x = number of women earning bigger paychecks.

$\dfrac{1}{6} = \dfrac{x}{23,000}$

$1(23,000) = 6x$

$23,000 = 6x$

$\dfrac{23,000}{6} = \dfrac{6x}{6}$

$3833 \approx x$

We would expect about 3833 women to earn bigger paychecks.

61. Let x = number of gallons of water needed for 36 teaspoons of weed killer.

$\dfrac{8}{2} = \dfrac{36}{x}$

$8x = 2(36)$

$8x = 72$

$\dfrac{8x}{8} = \dfrac{72}{8}$

$x = 9$

9 gallons of water are needed.

63. Compare unit prices.

110 ounces: $\dfrac{\$5.79}{110} \approx \$0.053$

240 ounces: $\dfrac{\$13.99}{240} \approx \$0.058$

The 110-ounce size is the better buy.

65. Compare unit prices.

6 ounces: $\dfrac{\$0.69}{6} = \$0.115$

8 ounces: $\dfrac{\$0.90}{8} \approx \$0.113$

16 ounces: $\dfrac{\$1.89}{16} \approx \$0.118$

The 8-ounce size is the best buy.

67. $-5 > -7$

69. Since $|-5| = 5$ and $-(-5) = 5$, $|-5| = -(-5)$.

71. Since $(-3)^2 = 9$ and $-3^2 = -9$,
$(-3)^2 > -3^2$.

73. $\dfrac{230}{2400} \approx 0.096 = 9.6\%$

About 9.6% of the daily value of sodium is contained in one serving.

75. Find the ratio of the calories from fat to the total calories.

$\dfrac{35}{130} \approx 0.269 = 26.9\%$

Since 26.9% < 30%, the food satisfies the recommendation.

77. Find the ratio of the calories from protein to the total calories. Each serving contains 12 g of protein.

$12g = 12 \cdot 4 = 48$ calories

$\dfrac{48}{280} \approx 0.171 = 17.1\%$

About 17.1% of the calories in each serving come from protein.

Section 2.8

Practice Problems

1. $x \geq -2$

2. $5 > x$

3. $x - 6 \geq -11$
$x - 6 + 6 \geq -11 + 6$
$x \geq -5$

4. $-3x \leq 12$
$\dfrac{-3x}{-3} \geq \dfrac{12}{-3}$
$x \geq -4$

5. $5x > -20$
$\dfrac{5x}{5} > \dfrac{-20}{5}$
$x > -4$

6. $-3x + 11 \leq -13$
$-3x + 11 - 11 \leq -13 - 11$
$-3x \leq -24$
$\dfrac{-3x}{-3} \geq \dfrac{-24}{-3}$
$x \geq 8$

7. $-6x - 3 > -4(x + 1)$
$-6x - 3 > -4x - 4$
$-6x - 3 + 3 > -4x - 4 + 3$
$-6x > -4x - 1$
$-6x + 4x > -4x - 1 + 4x$
$-2x > -1$
$\dfrac{-2x}{-2} < \dfrac{-1}{-2}$
$x < \dfrac{1}{2}$

8. $3(x + 5) - 1 \geq 5(x - 1) + 7$
$3x + 15 - 1 \geq 5x - 5 + 7$
$3x + 14 \geq 5x + 2$
$3x + 14 - 3x \geq 5x + 2 - 3x$
$14 \geq 2x + 2$
$14 - 2 \geq 2x + 2 - 2$
$12 \geq 2x$
$\dfrac{12}{2} \geq \dfrac{2x}{2}$
$6 \geq x$

9. Let x = the amount of sales.
Then $0.04x$ = the amount Alex earns from the sales.
$600 + 0.04x \geq 3000$
$600 + 0.04x - 600 \geq 3000 - 600$
$0.04x \geq 2400$
$\dfrac{0.04x}{0.04} \geq \dfrac{2400}{0.04}$
$x \geq 60{,}000$
At least \$60,000 in sales must be made.

Mental Math

1. $5x > 10$
$x > 2$

2. $4x < 20$
$x < 5$

3. $2x \geq 16$
$x \geq 8$

4. $9x \leq 63$
$x \leq 7$

5. -5 is not a solution to $x \geq -3$.

6. $|-6|$ is not a solution to $x < 6$.

7. 4.1 is not a solution to $x < 4.01$.

8. -4 is not a solution to $x \geq -3$.

Exercise Set 2.8

1. $x \leq -1$

3. $x > \dfrac{1}{2}$

5. $y < 4$

7. $-2 \leq m$

9. $x - 2 \geq -7$
$x - 2 + 2 \geq -7 + 2$
$x \geq -5$

11. $-9 + y < 0$
$-9 + y + 9 < 0 + 9$
$y < 9$

13. $3x - 5 > 2x - 8$
$3x - 5 + 5 > 2x - 8 + 5$
$3x > 2x - 3$
$3x - 2x > 2x - 3 - 2x$
$x > -3$

15. $4x - 1 \leq 5x - 2x$
$4x - 1 \leq 3x$
$4x - 1 - 3x \leq 3x - 3x$
$x - 1 \leq 0$
$x - 1 + 1 \leq 0 + 1$
$x \leq 1$

17. $2x < -6$

$$\frac{2x}{2} < \frac{-6}{2}$$

$$x < -3$$

19. $-8x \le 16$

$$\frac{-8x}{-8} \ge \frac{16}{-8}$$

$$x \ge -2$$

21. $-x > 0$

$$\frac{-x}{-1} < \frac{0}{-1}$$

$$x < 0$$

23. $\frac{3}{4} y \ge -2$

$$\frac{4}{3}\left(\frac{3}{4} y\right) \ge \frac{4}{3}(-2)$$

$$y \ge -\frac{8}{3}$$

25. $-0.6y < -1.8$

$$\frac{-0.6y}{-0.6} > \frac{-1.8}{-0.6}$$

$$y > 3$$

27. When we multiply or divide both sides of an inequality by a negative number, the direction of the inequality sign must be reversed.

29. $3x - 7 < 6x + 2$

$3x - 7 - 6x < 6x + 2 - 6x$

$-3x - 7 < 2$

$-3x - 7 + 7 < 2 + 7$

$-3x < 9$

$$\frac{-3x}{-3} > \frac{9}{-3}$$

$x > -3$

31. $5x - 7x \le x + 2$

$-2x \le x + 2$

$-2x - x \le 2$

$-3x \le 2$

$$\frac{-3x}{-3} \ge \frac{2}{-3}$$

$$x \ge -\frac{2}{3}$$

33. $-6x + 2 \ge 2(5 - x)$

$-6x + 2 \ge 10 - 2x$

$-6x + 2 + 2x \ge 10 - 2x + 2x$

$-4x + 2 \ge 10$

$-4x + 2 - 2 \ge 10 - 2$

$-4x \ge 8$

$$\frac{-4x}{-4} \le \frac{8}{-4}$$

$x \le -2$

35. $4(3x - 1) \le 5(2x - 4)$

$12x - 4 \le 10x - 20$

$12x - 4 - 10x \le 10x - 20 - 10x$

$2x - 4 \le -20$

$2x - 4 + 4 \le -20 + 4$

$2x \le -16$

$$\frac{2x}{2} \le \frac{-16}{2}$$

$x \le -8$

37. $3(x + 2) - 6 > -2(x - 3) + 14$

$3x + 6 - 6 > -2x + 6 + 14$

$3x > -2x + 20$

$3x + 2x > -2x + 20 + 2x$

$5x > 20$

$$\frac{5x}{5} > \frac{20}{5}$$

$x > 4$

39. $-2(x - 4) - 3x < -(4x + 1) + 2x$

$-2x + 8 - 3x < -4x - 1 + 2x$

$-5x + 8 < -2x - 1$

$-5x + 8 + 2x < -2x - 1 + 2x$

$-3x + 8 < -1$

$-3x + 8 - 8 < -1 - 8$

$-3x < -9$

$$\frac{-3x}{-3} > \frac{-9}{-3}$$

$x > 3$

41. $\frac{1}{2}(x-5) < \frac{1}{3}(2x-1)$

$6\left[\frac{1}{2}(x-5)\right] < 6\left[\frac{1}{3}(2x-1)\right]$

$3(x-5) < 2(2x-1)$

$3x - 15 < 4x - 2$

$3x - 15 - 4x < 4x - 2 - 4x$

$-x - 15 < -2$

$-x - 15 + 15 < -2 + 15$

$-x < 13$

$\frac{-x}{-1} > \frac{13}{-1}$

$x > -13$

43. $-5x + 4 \le -4(x-1)$

$-5x + 4 \le -4x + 4$

$-5x + 4 + 4x \le -4x + 4 + 4x$

$-x + 4 \le 4$

$-x + 4 - 4 \le 4 - 4$

$-x \le 0$

$\frac{-x}{-1} \ge \frac{0}{-1}$

$x \ge 0$

45. Let x = the unknown number.

$2x + 6 > -14$

$2x + 6 - 6 > -14 - 6$

$2x > -20$

$\frac{2x}{2} > \frac{-20}{2}$

$x > -10$

The statement is true for all numbers greater than -10.

47. If we let $l = x$ cm and $w = 15$ cm, the problem states

$2x + 2(15) \le 100$

$2x + 30 \le 100$

$2x - 30 \le 100 - 30$

$2x \le 70$

$\frac{2x}{2} \le \frac{70}{2}$

$x \le 35$

The maximum length is 35 cm.

49. Let x = score for the third game.

$\frac{146 + 201 + x}{3} \ge 180$

$\frac{347 + x}{3} \ge 180$

$3\left(\frac{347 + x}{3}\right) \ge 3(180)$

$347 + x \ge 540$

$347 + x - 347 \ge 540 - 347$

$x \ge 193$

He must score at least 193 in his third game.

51. Let x = his score on the final. Since the final counts as two tests, it is included twice in the average.

$\frac{75 + 83 + 85 + x + x}{5} \ge 80$

$\frac{243 + 2x}{5} \ge 80$

$5\left(\frac{243 + 2x}{5}\right) \ge 5(80)$

$243 + 2x \ge 400$

$243 + 2x - 243 \ge 400 - 243$

$2x \ge 157$

$\frac{2x}{2} \ge \frac{157}{2}$

$x \ge 78.5$

He must score 78.5 or higher on the final exam to receive a B in the course.

53. $(2)^3 = 2 \cdot 2 \cdot 2 = 8$

55. $(1)^{12} = 1$

57. $\left(\frac{4}{7}\right)^2 = \frac{4}{7} \cdot \frac{4}{7} = \frac{16}{49}$

59. About 37 million people were enrolled in Health Maintenance Organizations in 1992.

61. From the graph we can see that the greatest increase in the number of members occurred between the years 1992 and 1995. The membership increased by 3 million people each year.

63. Answers may vary.

Chapter 2 Review

1. $5x - x + 2x = (5 - 1 + 2)x = (4 + 2)x = 6x$

2. $0.2z - 4.6z - 7.4z = (0.2 - 4.6 - 7.4)z$
   $= (-4.4 - 7.4)z = -11.8z$

3. $\frac{1}{2}x + 3 + \frac{7}{2}x - 5 = \frac{1}{2}x + \frac{7}{2}x + (3 - 5)$
   $= \left(\frac{1}{2} + \frac{7}{2}\right)x + (3 - 5) = \frac{8}{2}x - 2 = 4x - 2$

4. $\frac{4}{5}y + 1 + \frac{6}{5}y + 2 = \frac{4}{5}y + \frac{6}{5}y + (1 + 2)$
   $= \left(\frac{4}{5} + \frac{6}{5}\right)y + (1 + 2) = \frac{10}{5}y + 3 = 2y + 3$

5. $2(n - 4) + n - 10 = 2(n) + 2(-4) + n - 10$
   $= 2n - 8 + n - 10 = 2n + n + (-8 - 10)$
   $= (2 + 1)n + (-8 - 10) = 3n - 18$

6. $3(w + 2) - (12 - w) = 3(w + 2) - 1(12 - w)$
   $= 3(w) + 3(2) - 1(12) - 1(-w)$
   $= 3w + 6 - 12 + w = 3w + w + (6 - 12)$
   $= (3 + 1)w + (6 - 12) = 4w - 6$

7. $(x + 5) - (7x - 2) = x + 5 - 7x + 2$
   $= x - 7x + 5 + 2 = -6x + 7$

8. $(y - 0.7) - (1.4y - 3) = y - 0.7 - 1.4y + 3$
   $= y - 1.4y - 0.7 + 3 = -0.4y + 2.3$

9. $3x - 7$

10. $2(x + 2.8) + 3x = 2x + 5.6 + 3x = 5x + 5.6$

11. $8x + 4 = 9x$
    $8x + 4 - 8x = 9x - 8x$
    $4 = x$

12. $5 - 3 = 6y$
    $5y - 3 - 5y = 6y - 5y$
    $-3 = y$

13. $\frac{2}{7}x + \frac{5}{7}x = 6$
    $\left(\frac{2}{7} + \frac{5}{7}\right)x = 6$
    $\frac{7}{7}x = 6$
    $x = 6$

14. $3x - 5 = 4x + 1$
    $3x - 5 - 3x = 4x + 1 - 3x$
    $-5 = x + 1$
    $-5 - 1 = x + 1 - 1$
    $-6 = x$

15. $2x - 6 = x - 6$
    $2x - 6 - x = x - 6 - x$
    $x - 6 = -6$
    $x - 6 + 6 = -6 + 6$
    $x = 0$

16. $4(x + 3) = 3(1 + x)$
    $4(x) + 4(3) = 3(1) + 3(x)$
    $4x + 12 = 3 + 3x$
    $4x + 12 - 3x = 3 + 3x - 3x$
    $x + 12 = 3$
    $x + 12 - 12 = 3 - 12$
    $x = -9$

17. $6(3 + n) = 5(n - 1)$
    $6(3) + 6(n) = 5(n) + 5(-1)$
    $18 + 6n = 5n - 5$
    $18 + 6n - 5n = 5n - 5 - 5n$
    $18 + n = -5$
    $18 + n - 18 = -5 - 18$
    $n = -23$

18. $5(2 + x) - 3(3x + 2) = -5(x - 6) + 2$
    $5(2) + 5(x) - 3(3x) - 3(2) = -5(x) - 5(-6) + 2$
    $10 + 5x - 9x - 6 = -5x + 30 + 2$
    $5x - 9x + 10 - 6 = -5x + 30 + 2$
    $-4x + 10 - 6 = -5x + 30 + 2$
    $-4x + 4 = -5x + 32$
    $-4x + 4 + 5x = -5x + 32 + 5x$
    $x + 4 = 32$
    $x + 4 - 4 = 32 - 4$
    $x = 28$

19. b; if the sum of two numbers is 10 and one number is x, we find the other number by subtracting x from 10. The other number is $10 - x$.

20. a; if Mandy is x inches tall and is 5 inches taller than Melissa, we find Melissa's height by subtracting 5 from x. Thus Melissa's height is $(x - 5)$ inches.

21. c; since the sum of supplementary angles is $180°$ and one angle measures $(x + 5)°$, we can find the measure of its supplement by subtracting $(x + 5)$ from 180. Thus, the measure of the supplemental angle is
$180 - (x + 5) = 180 - x - 5 = 180 - 5 - x$
$= 175 - x$
The supplement measures $(175 - x)°$.

22. $\dfrac{3}{4}x = -9$

$\dfrac{4}{3}\left(\dfrac{3}{4}x\right) = \dfrac{4}{3}(-9)$

$\dfrac{12}{12}x = -\dfrac{36}{3}$

$x = -12$

23. $\dfrac{x}{6} = \dfrac{2}{3}$

$6\left(\dfrac{x}{6}\right) = 6\left(\dfrac{2}{3}\right)$

$\dfrac{6x}{6} = \dfrac{12}{3}$

$x = 4$

24. $-5x = 0$

$\dfrac{-5x}{-5} = \dfrac{0}{-5}$

$x = 0$

25. $-y = 7$

$\dfrac{-y}{-1} = \dfrac{7}{-1}$

$y = -7$

26. $0.2x = 0.15$

$\dfrac{0.2x}{0.2} = \dfrac{0.15}{0.2}$

$x = 0.75$

27. $\dfrac{-x}{3} = 1$

$-3\left(\dfrac{-x}{3}\right) = -3(1)$

$\dfrac{3x}{3} = -3$

$x = -3$

28. $-3x + 1 = 19$
$-3x + 1 - 1 = 19 - 1$
$-3x = 18$
$\dfrac{-3x}{-3} = \dfrac{18}{-3}$
$x = -6$

29. $5x + 25 = 20$
$5x + 25 - 25 = 20 - 25$
$5x = -5$
$\dfrac{5x}{5} = \dfrac{-5}{5}$
$x = -1$

30. $5x - 6 + x = 4x$
$6x - 6 = 4x$
$6x - 6 - 6x = 4x - 6x$
$-6 = -2x$
$\dfrac{-6}{-2} = \dfrac{-2x}{-2}$
$3 = x$

31. $-y + 4y = -y$
$3y = -y$
$3y + y = -y + y$
$4y = 0$
$\dfrac{4y}{4} = \dfrac{0}{4}$
$y = 0$

32. $-5x + \dfrac{3}{7} = \dfrac{10}{7}$

$-5x + \dfrac{3}{7} - \dfrac{3}{7} = \dfrac{10}{7} - \dfrac{3}{7}$

$-5x = \dfrac{7}{7}$

$-5x = 1$

$\dfrac{-5x}{-5} = \dfrac{1}{-5}$

$x = -\dfrac{1}{5}$

33. If x = the first integer, then $x + 1$ = the next consecutive integer, and $x + 2$ = the third consecutive integer. Their sum is
$x + (x + 1) + (x + 2) = x + x + 1 + x + 2$
$= x + x + x + 1 + 2 = 3x + 3$

34. $\frac{5}{3}x + 4 = \frac{2}{3}x$

$3\left(\frac{5}{3}x + 4\right) = 3\left(\frac{2}{3}x\right)$

$5x + 12 = 2x$

$5x + 12 - 2x = 2x - 2x$

$3x + 12 = 0$

$3x + 12 - 12 = 0 - 12$

$3x = -12$

$\frac{3x}{3} = \frac{-12}{3}$

$x = -4$

35. $-(5x + 1) = -7x + 3$

$-5x - 1 = -7x + 3$

$-5x - 1 + 7x = -7x + 3 + 7x$

$2x - 1 = 3$

$2x - 1 + 1 = 3 + 1$

$2x = 4$

$\frac{2x}{2} = \frac{4}{2}$

$x = 2$

36. $-4(2x + 1) = -5x + 5$

$-8x - 4 = -5x + 5$

$-8x - 4 + 5x = -5x + 5 + 5x$

$-3x - 4 = 5$

$-3x - 4 + 4 = 5 + 4$

$-3x = 9$

$\frac{-3x}{-3} = \frac{9}{-3}$

$x = -3$

37. $-6(2x - 5) = -3(9 + 4x)$

$-12x + 30 = -27 - 12x$

$-12x + 30 + 12x = -27 - 12x + 12x$

$30 = -27$

There is no solution to the equation $-6(2x - 5) = -3(9 + 4x)$.

38. $3(8y - 1) = 6(5 + 4y)$

$24y - 3 = 30 + 24y$

$24y - 3 - 24y = 30 + 24y - 24y$

$-3 = 30$

There is no solution to the equation $3(8y - 1) = 6(5 + 4y)$.

39. $\frac{3(2 - z)}{5} = z$

$5\left[\frac{3(2 - z)}{5}\right] = 5(z)$

$3(2 - z) = 5z$

$6 - 3z = 5z$

$6 - 3z + 3z = 5z + 3z$

$6 = 8z$

$\frac{6}{8} = \frac{8z}{8}$

$\frac{3}{4} = z$

40. $\frac{4(n + 2)}{5} = -n$

$5\left[\frac{4(n + 2)}{5}\right] = 5(-n)$

$4(n + 2) = -5n$

$4n + 8 = -5n$

$4n + 8 - 4n = -5n - 4n$

$8 = -9n$

$\frac{8}{-9} = \frac{-9n}{-9}$

$-\frac{8}{9} = n$

41. $0.5(2n - 3) - 0.1 = 0.4(6 + 2n)$

$10[0.5(2n - 3) - 0.1] = 10[0.4(6 + 2n)]$

$5(2n - 3) - 1 = 4(6 + 2n)$

$10n - 15 - 1 = 24 + 8n$

$10n - 16 = 24 + 8n$

$10n - 16 - 8n = 24 + 8n - 8n$

$2n - 16 = 24$

$2n - 16 + 16 = 24 + 16$

$2n = 40$

$\frac{2n}{2} = \frac{40}{2}$

$n = 20$

42. $-9 - 5a = 3(6a - 1)$

$-9 - 5a = 18a - 3$

$-9 - 5a + 5a = 18a - 3 + 5a$

$-9 = 23a - 3$

$-9 + 3 = 23a - 3 + 3$

$-6 = 23a$

$\frac{-6}{23} = \frac{23a}{23a}$

$-\frac{6}{23} = a$

43. $\dfrac{5(c+1)}{6} = 2c - 3$

$6\left[\dfrac{5(c+1)}{6}\right] = 6(2c - 3)$

$5(c + 1) = 6(2c - 3)$

$5c + 5 = 12c - 18$

$5c + 5 - 5c = 12c - 18 - 5c$

$5 = 7c - 18$

$5 + 18 = 7c - 18 + 18$

$23 = 7c$

$\dfrac{23}{7} = \dfrac{7c}{7}$

$\dfrac{23}{7} = c$

44. $\dfrac{2(8-a)}{3} = 4 - 4a$

$3\left[\dfrac{2(8-a)}{3}\right] = 3(4 - 4a)$

$2(8 - a) = 3(4 - 4a)$

$16 - 2a = 12 - 12a$

$16 - 2a + 12a = 12 - 12a + 12a$

$16 + 10a = 12$

$16 + 10a - 16 = 12 - 16$

$10a = -4$

$\dfrac{10a}{10} = \dfrac{-4}{10}$

$a = -\dfrac{2}{5}$

45. $200(70x - 3560) = -179(150x - 19{,}300)$

$14{,}000x - 712{,}000 = -26{,}850x + 3{,}454{,}700$

$14{,}000x - 712{,}000 + 26{,}850x = -26{,}850x + 3{,}454{,}700 + 26{,}850x$

$40{,}850x - 712{,}000 = 3{,}454{,}700$

$40{,}850x - 712{,}000 + 712{,}000 = 3{,}454{,}700 + 712{,}000$

$40{,}850x = 4{,}166{,}700$

$\dfrac{40{,}850x}{40{,}850} = \dfrac{4{,}166{,}700}{40{,}850}$

$x = 102$

46. $1.72y - 0.04y = 0.42$

$1.68y = 0.42$

$\dfrac{1.68y}{1.68} = \dfrac{0.42}{1.68}$

$y = 0.25$

47. Let x = length of the base.
Then $3x + 68$ = height.

$x + (3x + 68) = 1380$

$x + 3x + 68 = 1380$

$4x + 68 = 1380$

$4x + 68 - 68 = 1380 - 68$

$4x = 1312$

$\dfrac{4x}{4} = \dfrac{1312}{4}$

$x = 328$

The base is 328 feet, the height is
$3(328) + 68 = 1052$ feet.

48. If x = length of the shorter piece, then
$2x$ = length of the longer piece.
$x + 2x = 12$
$3x = 12$
$\dfrac{3x}{3} = \dfrac{12}{3}$
$x = 4$
The shorter piece is 4 feet and the longer piece is $2(4) = 8$ feet.

49. Let x = one area code; then $3x + 34$ = the other area code.
$x + (3x + 34) = 1262$
$x + 3x + 34 = 1262$
$4x + 34 = 1262$
$4x + 34 - 34 = 1262 - 34$
$4x = 1228$
$\dfrac{4x}{4} = \dfrac{1228}{4}$
$x = 307$
The two area codes are 307 and $3(307) + 34 = 955$.

50. Let x represent the first integer, then $x + 1$ represents the next consecutive integer, and $x + 2$ represents the third consecutive integer.
$x + (x + 1) + (x + 2) = -114$
$x + x + 1 + x + 2 = -114$
$3x + 3 = -114$
$3x + 3 - 3 = -114 - 3$
$3x = -117$
$\dfrac{3x}{3} = \dfrac{-117}{3}$
$x = -39$
The numbers are -39, -38, and -37.

51. Let x = unknown number.
$\dfrac{x}{3} = x - 2$
$3\left(\dfrac{x}{3}\right) = 3(x - 2)$
$x = 3x - 6$
$x - 3x = 3x - 6 - 3x$
$-2x = -6$
$\dfrac{-2x}{-2} = \dfrac{-6}{-2}$
$x = 3$
The number is 3.

52. Let x = unknown number.
$2(x + 6) = -x$
$2x + 12 = -x$
$2x + 12 - 2x = -x - 2x$
$12 = -3x$
$\dfrac{12}{-3} = \dfrac{-3x}{-3}$
$-4 = x$
The number is -4.

53. $P = 2l + 2w$
$46 = 2(14) + 2w$
$46 = 28 + 2w$
$46 - 28 = 28 + 2w - 28$
$18 = 2w$
$\dfrac{18}{2} = \dfrac{2w}{2}$
$9 = w$

54. $V = lwh$
$192 = (8)(6)h$
$192 = 48h$
$\dfrac{192}{48} = \dfrac{48h}{48}$
$4 = h$

55. $y = mx + b$
$y - b = mx + b - b$
$y - b = mx$
$\dfrac{y - b}{x} = \dfrac{mx}{x}$
$\dfrac{y - b}{x} = m$

56. $r = vst - 5$
$r + 5 = vst - 5 + 5$
$r + 5 = vst$
$\dfrac{r + 5}{vt} = \dfrac{vst}{vt}$
$\dfrac{r + 5}{vt} = s$

57. $2y - 5x = 7$
$2y - 5x - 2y = 7 - 2y$
$-5x = 7 - 2y$
$\dfrac{-5x}{-5} = \dfrac{7 - 2y}{-5}$
$x = -\dfrac{7 - 2y}{5}$
$x = \dfrac{2y - 7}{5}$

58. $3x - 6y = -2$

$3x - 6y - 3x = -2 - 3x$

$-6y = -2 - 3x$

$\dfrac{-6y}{-6} = \dfrac{-2 - 3x}{-6}$

$y = -\dfrac{-2 - 3x}{6}$

$y = \dfrac{2 + 3x}{6}$

59. $C = \pi D$

$\dfrac{C}{D} = \dfrac{\pi D}{D}$

$\dfrac{C}{D} = \pi$

60. $C = 2\pi r$

$\dfrac{C}{2r} = \dfrac{2\pi r}{2r}$

$\dfrac{C}{2r} = \pi$

61. The volume of the rectangular swimming pool is given by the formula $V = lwh$.

$V = lwh$

$900 = 20 \cdot w \cdot 3$

$900 = 60w$

$\dfrac{900}{60} = \dfrac{60w}{60}$

$15 = w$

The width is 15 meters.

62. Use the formula $C = \dfrac{5}{9}(F - 32)$.

$C = \dfrac{5}{9}(F - 32)$

$C = \dfrac{5}{9}(104 - 32)$

$C = \dfrac{5}{9}(72)$

$C = 40$

104°F is the same as 40°C.

63. Use the formula $d = rt$.

$d = rt$

$10,000 = 125 \cdot t$

$\dfrac{10,000}{125} = \dfrac{125 \cdot t}{125}$

$80 = t$

It will take 80 minutes or 1 hour and 20 minutes.

64. Let x = unknown percent.

$9 = x \cdot 45$

$0.20 = x$

$20\% = x$

The number 9 is 20% of 45.

65. Let x = unknown percent.

$59.5 = x \cdot 85$

$0.70 = x$

$70\% = x$

The number 59.5 is 70% of 85.

66. Let x = unknown number.

$137.5 = 1.25 \cdot x$

$110 = x$

The number 137.5 is 125% of 110.

67. Let x = unknown number.

$768 = 0.60 \cdot x$

$1280 = x$

The number 768 is 60% of 1280.

68. Since 12.6% of the households do not have phones, find 12.6% of 50,000.

$12.6\% \text{ of } 50,000 = 0.126(50,000)$

$= 6300$

We would expect that 6300 of the households in the city are phoneless.

69. 6% of the business travelers surveyed relax by taking a nap.

70. The most popular way to relax according to the survey was to eat from the minibar.

71. Since 40% of those surveyed tend to relax by watching TV, find 40% of 300.

$40\% \text{ of } 300 = 0.40(300)$

$= 120$

We would expect that 120 of the travelers will relax by watching TV.

72. No; answers may vary.

73. 1 dollar = 100 cents

The ratio of 20 cents to 1 dollar is

$\dfrac{20}{100} = \dfrac{20}{5 \cdot 20} = \dfrac{1}{5}$.

74. $\dfrac{4}{6} = \dfrac{2 \cdot 2}{2 \cdot 3} = \dfrac{2}{3}$

75. $\dfrac{x}{2} = \dfrac{12}{4}$

$x \cdot 4 = 2 \cdot 12$

$\dfrac{4x}{4} = \dfrac{24}{4}$

$x = 6$

76. $\dfrac{20}{1} = \dfrac{x}{25}$

$20 \cdot 25 = 1 \cdot x$

$500 = x$

77. $\dfrac{32}{100} = \dfrac{100}{x}$

$32 \cdot x = 100 \cdot 100$

$\dfrac{32x}{32} = \dfrac{10,000}{32}$

$x = 312.5$

78. $\dfrac{20}{2} = \dfrac{c}{5}$

$20 \cdot 5 = 2 \cdot c$

$\dfrac{100}{2} = \dfrac{2c}{2}$

$50 = c$

79. $\dfrac{2}{x-1} = \dfrac{3}{x+3}$

$2(x+3) = 3(x-1)$

$2x + 6 = 3x - 3$

$6 = x - 3$

$9 = x$

80. $\dfrac{4}{y-3} = \dfrac{2}{y-3}$

$4(y-3) = 2(y-3)$

$4y - 12 = 2y - 6$

$2y - 12 = -6$

$2y = 6$

$\dfrac{2y}{2} = \dfrac{6}{2}$

$y = 3$

Since $y = 3$ makes the denominators in the original equation zero, there is no solution.

81. $\dfrac{y+2}{y} = \dfrac{5}{3}$

$3(y+2) = 5y$

$3y + 6 = 5y$

$6 = 2y$

$\dfrac{6}{2} = \dfrac{2y}{2}$

$3 = y$

82. $\dfrac{x-3}{3x+2} = \dfrac{2}{6}$

$6(x-3) = 2(3x+2)$

$6x - 18 = 6x + 4$

$-18 = 4$

There is no solution to the equation $\dfrac{x-3}{3x+2} = \dfrac{2}{6}$.

83. Compare unit prices.

10 ounces: $\dfrac{\$1.29}{10} = \$0.129$

16 ounces: $\dfrac{\$2.15}{16} \approx \$0.134$

The 10-ounce size is the better buy.

84. Compare unit prices.

8 ounces: $\dfrac{\$0.89}{8} \approx \$0.111$

15 ounces: $\dfrac{\$1.63}{15} \approx \$0.109$

20 ounces: $\dfrac{\$2.36}{20} = \$0.118$

The 15-ounce size is the best buy.

85. Let $x =$ number of parts that can be processed in 45 minutes.

$\dfrac{300}{20} = \dfrac{x}{45}$

$300 \cdot 45 = 20 \cdot x$

$13,500 = 20x$

$\dfrac{13,500}{20} = \dfrac{20x}{20}$

$675 = x$

Thus, 675 parts can be processed in 45 minutes.

86. Let $x =$ amount charged for 3 hours of consulting.

$\dfrac{90}{8} = \dfrac{x}{3}$

$90 \cdot 3 = 8 \cdot x$

$\dfrac{270}{8} = \dfrac{8x}{8}$

$33.75 = x$

He charges $33.75 for 3 hours.

87. Let x = number of letters he can address in 55 minutes.

$$\frac{100}{35} = \frac{x}{55}$$
$$100 \cdot 55 = 35 \cdot x$$
$$\frac{5500}{35} = \frac{35x}{35}$$
$$157.1 \approx x$$

Thus, he can address about 157 letters in 55 minutes.

88. $x \leq -2$

89. $x > 0$

90. $x - 5 \leq -4$
$x - 5 + 5 \leq -4 + 5$
$x \leq 1$

91. $x + 7 > 2$
$x + 7 - 7 > 2 - 7$
$x > -5$

92. $-2x \geq -20$
$\dfrac{-2x}{-2} \leq \dfrac{-20}{-2}$
$x \leq 10$

93. $-3x > 12$
$\dfrac{-3x}{-3} < \dfrac{12}{-3}$
$x < -4$

94. $5x - 7 > 8x + 5$
$5x - 7 - 5x > 8x + 5 - 5x$
$-7 > 3x + 5$
$-7 - 5 > 3x + 5 - 5$
$-12 > 3x$
$\dfrac{-12}{3} > \dfrac{3x}{3}$
$-4 > x$

95. $x + 4 \geq 6x - 16$
$x + 4 - x \geq 6x - 16 - x$
$4 \geq 5x - 16$
$4 + 16 \geq 5x - 16 + 16$
$20 \geq 5x$
$\dfrac{20}{5} \geq \dfrac{5x}{5}$
$4 \geq x$

96. $\dfrac{2}{3}y > 6$

$\dfrac{3}{2}\left(\dfrac{2}{3}y\right) > \dfrac{3}{2}(6)$

$y > 9$

97. $-0.5y \leq 7.5$
$\dfrac{-0.5y}{-0.5} \geq \dfrac{7.5}{-0.5}$
$y \geq -15$

98. $-2(x - 5) > 2(3x - 2)$
$-2x + 10 > 6x - 4$
$-2x + 10 + 2x > 6x - 4 + 2x$
$10 > 8x - 4$
$10 + 4 > 8x - 4 + 4$
$14 > 8x$
$\dfrac{14}{8} > \dfrac{8x}{8}$
$\dfrac{14}{8} > x$
$\dfrac{7}{4} > x$

99. $4(2x - 5) \leq 5x - 1$
$8x - 20 \leq 5x - 1$
$8x - 20 - 5x \leq 5x - 1 - 5x$
$3x - 20 \leq -1$
$3x - 20 + 20 \leq -1 + 20$
$3x \leq 19$
$\dfrac{3x}{3} \leq \dfrac{19}{3}$
$x \leq \dfrac{19}{3}$

100. Let x = amount of sales (in dollars).
$175 + 0.05x \geq 300$
$175 + 0.05x - 175 \geq 300 - 175$
$0.05x \geq 125$
$\dfrac{0.05x}{0.05} \geq \dfrac{125}{0.05}$
$x \geq 2500$
She must sell at least $2500 worth of goods.

101. Let $x =$ score on the next round.

$$\frac{76 + 82 + 79 + x}{4} < 80$$

$$\frac{237 + x}{4} < 80$$

$$4\left(\frac{237 + x}{4}\right) < 4(80)$$

$$237 + x < 320$$

$$237 + x - 237 < 320 - 237$$

$$x < 83$$

She must shoot less than 83 on her next round.

Chapter 2 Test

1. $2y - 6 - y - 4 = 2y - y + (-6 - 4)$
$\quad = (2 - 1)y - 10 = y - 10$

2. $2.7x + 6.1 + 3.2x - 4.9$
$\quad = 2.7x + 3.2x + (6.1 - 4.9)$
$\quad = (2.7 + 3.2)x + (6.1 - 4.9)$
$\quad = 5.9x + 1.2$

3. $4(x - 2) - 3(2x - 6)$
$\quad = 4(x) + 4(-2) - 3(2x) - 3(-6)$
$\quad = 4x - 8 - 6x + 18$
$\quad = 4x - 6x + (-8 + 18)$
$\quad = (4 - 6)x + (-8 + 18)$
$\quad = -2x + 10$

4. $-5(y + 1) + 2(3 - 5y)$
$\quad = -5(y) - 5(1) + 2(3) + 2(-5y)$
$\quad = -5y - 5 + 6 - 10y$
$\quad = -5y - 10y + (-5 + 6)$
$\quad = (-5 - 10)y + (-5 + 6)$
$\quad = -15y + 1$

5. $-\frac{4}{5}x = 4$
$\quad -\frac{5}{4}\left(-\frac{4}{5}x\right) = -\frac{5}{4}(4)$
$\quad x = -5$

6. $4(n - 5) = -(4 - 2n)$
$\quad 4(n - 5) = -1(4 - 2n)$
$\quad 4(n) + 4(-5) = -1(4) - 1(-2n)$
$\quad 4n - 20 = -4 + 2n$
$\quad 4n - 20 - 2n = -4 + 2n - 2n$
$\quad 2n - 20 = -4$
$\quad 2n - 20 + 20 = -4 + 20$
$\quad 2n = 16$

$$\frac{2n}{2} = \frac{16}{2}$$
$$n = 8$$

7. $5y - 7 + y = -(y + 3y)$
$\quad 5y + y - 7 = -4y$
$\quad 6y - 7 + 4y = -4y + 4y$
$\quad 6y + 4y - 7 + 7 = 0 + 7$
$\quad 10y = 7$
$\quad \frac{10}{10}y = \frac{7}{10}$
$\quad y = \frac{7}{10}$

8. $4z + 1 - z = 1 + z$
$\quad 3z + 1 = 1 + z$
$\quad 3z + 1 - z = 1 + z - z$
$\quad 2z + 1 = 1$
$\quad 2z + 1 - 1 = 1 - 1$
$\quad 2z = 0$
$\quad \frac{2z}{2} = \frac{0}{2}$
$\quad z = 0$

9. $\frac{2(x + 6)}{3} = x - 5$
$\quad 3\left(\frac{2(x + 6)}{3}\right) = 3(x - 5)$
$\quad 2(x + 6) = 3(x - 5)$
$\quad 2(x) + 2(6) = 3(x) + 3(-5)$
$\quad 2x + 12 = 3x - 15$
$\quad 2x + 12 - 2x = 3x - 15 - 2x$
$\quad 12 = x - 15$
$\quad 12 + 15 = x - 15 + 15$
$\quad 27 = x$

10. $\frac{4(y - 1)}{5} = 2y + 3$
$\quad 5\left(\frac{4(y - 1)}{5}\right) = 5(2y + 3)$
$\quad 4(y - 1) = 5(2y + 3)$
$\quad 4y - 4 = 10y + 15$
$\quad 4y - 4 - 15 = 10y + 15 - 15$
$\quad 4y - 19 = 10y$
$\quad 4y - 19 - 4y = 10y - 4y$
$\quad -19 = 6y$
$\quad \frac{-19}{6} = \frac{6y}{6}$
$\quad -\frac{19}{6} = y$

11. $\frac{1}{2} - x + \frac{3}{2} = x - 4$

$\frac{4}{2} - x = x - 4$

$2 - x = x - 4$

$2 - x + x = x - 4 + x$

$2 = 2x - 4$

$2 + 4 = 2x - 4 + 4$

$6 = 2x$

$\frac{6}{2} = \frac{2x}{2}$

$3 = 1x$

$3 = x$

12. $\frac{5}{y+1} = \frac{4}{y+2}$

$(y+2)(y+1)\frac{5}{y+1} = (y+2)(y+1)\frac{4}{y+2}$

$5(y+2) = 4(y+1)$

$5(y) + 5(2) = 4(y) + 4(1)$

$5y + 10 = 4y + 4$

$5y + 10 - 10 - 4y = 4y + 4 - 10 - 4y$

$y = 4 - 10$

$y = -6$

13. $\frac{1}{3}(y+3) = 4y$

$3\left[\frac{1}{3}(y+3)\right] = 3(4y)$

$y + 3 = 12y$

$y + 3 - y = 12y - y$

$3 = 11y$

$\frac{3}{11} = \frac{11y}{11}$

$\frac{3}{11} = 1y$

$\frac{3}{11} = y$

14. $-0.3(x-4) + x = 0.5(3-x)$

$10[-0.3(x-4)] + 10(x) = 10[0.5(3-x)]$

$-3(x-4) + 10x = 5(3-x)$

$-3x + 12 + 10x = 15 - 5x$

$7x + 12 = 15 - 5x$

$7x + 12 + 5x = 15 - 5x + 5x$

$12x + 12 = 15$

$12x + 12 - 12 = 15 - 12$

$12x = 3$

$\frac{12x}{12} = \frac{3}{12}$

$1x = \frac{1}{4}$

$x = \frac{1}{4}$ or $x = 0.25$

15. $-4(a+1) - 3a = -7(2a-3)$

$-4a - 4 - 3a = -14a + 21$

$-7a - 4 = -14a + 21$

$-7a - 4 + 7a = -14a + 21 + 7a$

$-4 = -7a + 21$

$-4 - 21 = -7a + 21 - 21$

$-25 = -7a$

$\frac{-25}{-7} = \frac{-7a}{-7}$

$\frac{25}{7} = 1a$

$\frac{25}{7} = a$

16. Let x = unknown number.

$x + \frac{2}{3}x = 35$

$3\left(x + \frac{2}{3}x\right) = 3(35)$

$3x + 2x = 105$

$5x = 105$

$\frac{5x}{5} = \frac{105}{5}$

$x = 21$

The number is 21.

17. The area of the rectangular deck is given by the formula $A = bh$. If $b = 20$ and $h = 35$, then

$A = bh$

$A = (20)(35)$

$A = 700$

Thus, the area of the deck is 700 square feet. Since 1 gallon covers 200 square feet, we can form the following proportion, where x = number of gallons needed to cover 700 square feet.

$\frac{1}{200} = \frac{x}{700}$

$1 \cdot 700 = 200 \cdot x$

$\frac{700}{200} = \frac{200x}{200}$

$3.5 = x$

Since we are painting two coats we will need twice as much or $2 \cdot 3.5 = 7$ gallons of water seal.

18. 6 ounces of $1.19 is $\dfrac{1.19}{6} \approx 20\cancel{c}$ per ounce

10 ounces for $2.15 is $\dfrac{2.15}{10} = 21.5\cancel{c}$ per ounce.

16 ounces for $3.25 is $\dfrac{3.25}{16} \approx 20.3\cancel{c}$ per ounce

6 ounces for $1.19 is the best buy.

19. We can set up the proportion
$$\dfrac{3 \text{ defective}}{85 \text{ bulbs}} = \dfrac{x \text{ defective}}{510 \text{ bulbs}}$$
$3(510) = 85x$
$1530 = 85x$
$\dfrac{1}{85}(1530) = \dfrac{1}{85}(85x)$
$x = 18$ bulbs
18 defective bulbs should be found among 510 bulbs.

20. $y = mx + b$
$-14 = -2x + (-2)$
$-14 = -2x - 2$
$-14 + 2 = -2x - 2 + 2$
$-12 = -2x$
$\dfrac{-12}{-2} = \dfrac{-2x}{-2}$
$6 = x$

21. $V = \pi r^2 h$
$\dfrac{1}{\pi r^2} \cdot V = \dfrac{1}{\pi r^2}(\pi r^2)h$
$\dfrac{V}{\pi r^2} = h$
$h = \dfrac{V}{\pi r^2}$

22. $3x - 4y = 10$
$3x - 4y - 3x = 10 - 3x$
$-4y = 10 - 3x$
$-\dfrac{4y}{4} = \dfrac{10 - 3x}{4}$
$y = -\dfrac{10 - 3x}{4}$
$y = \dfrac{3x - 10}{4}$

23. $3x - 5 > 7x + 3$
$3x - 5 - 3x > 7x + 3 - 3x$
$-5 > 4x + 3$
$-5 - 3 > 4x + 3 - 3$
$-8 > 4x$
$\dfrac{-8}{4} > \dfrac{4x}{4}$
$-2 > x$

24. $x + 6 > 4x - 6$
$x + 6 - x > 4x - 6 - x$
$6 > 3x - 6$
$6 + 6 > 3x - 6 + 6$
$12 > 3x$
$\dfrac{12}{3} > \dfrac{3x}{3}$
$4 > x$

25. $-0.3x \geq 2.4$
$-\dfrac{1}{0.3}(-0.3x) \leq -\dfrac{1}{0.3}(2.4)$
$x \leq -8$

26. $-5(x - 1) + 6 \leq -3(x + 4) + 1$
$-5x + 5 + 6 \leq -3x - 12 + 1$
$-5x + 11 \leq -3x - 11$
$-5x + 11 + 3x \leq -3x + 3x - 11$
$-2x + 11 - 11 \leq -11 - 11$
$-2x \leq -22$
$\dfrac{-2}{-2}x \geq \dfrac{-22}{-2}$
$x \geq 11$

27. $\dfrac{2(5x + 1)}{3} > 2$
$3\left[\dfrac{2(5x + 1)}{3}\right] > 3(2)$
$2(5x + 1) > 6$
$10x + 2 > 6$
$10x + 2 - 2 > 6 - 2$
$10x > 4$
$\dfrac{10x}{10} > \dfrac{4}{10}$
$x > \dfrac{2}{5}$

28. From the graph, we can see that 80.8% of charity income comes from individuals.

29. Since 5.1% of the income came from corporations, we find 5.1% of $143.84 billion.

5.1% of $143.84 billion
$= 0.051(143.84)$ billion
$= 7.33584$ billion

Thus, $7.33584 billion came from corporations.

30. $72 = \dfrac{x}{100} \cdot 180$

$72 = 1.8x$

$\dfrac{72}{1.8} = \dfrac{1.8}{1.8}x$

$x = 40$

40% of 180 is 72.

31. Let x = the number of public libraries in NY.
Let $x - 520$ = the number of public libraries in Indiana.
Then
$x + (x - 520) = 996$
$x + x - 520 = 996$
$2x - 520 + 520 = 996 + 520$
$2x = 1516$
$\dfrac{2}{2}x = \dfrac{1516}{2}$
$x = 758$
$x - 520 = 238$
There are 758 libraries in NY and 238 in Indiana.

Chapter 2 Cumulative Review

1. True, since 8 = 8 is true.

2. True, since 8 = 8 is true.

3. False, since 23 is larger than 0.

4. True

5. **a.** $|0| < 2$ since $|0| = 0$ and $0 < 2$.

    b. $|-5| = 5$

    c. $|-3| > |-2|$ since $|-3| = 3$ and $|-2| = 2$ and $3 > 2$.

    d. $|5| < |6|$ since $|5| = 5$ and $|6| = 6$ and $5 < 6$.

    e. $|-7| > |6|$ since $|-7| = 7$ and $|6| = 6$ and $7 > 6$.

6. $\dfrac{3 + |4 - 3| + 2^2}{6 - 3} = \dfrac{3 + |1| + 2^2}{6 - 3} = \dfrac{3 + 1 + 2^2}{3}$

$= \dfrac{3 + 1 + 4}{3} = \dfrac{8}{3}$

7. $(-8) + (-11) = -19$

8. $(-2) + 10 = 8$

9. $0.2 + (-0.5) = -0.3$

10. **a.** $-3 + [(-2 - 5) - 2]$
$= -3 + [(-2 + (-5)) - 2]$
$= -3 + [(-7) - 2]$
$= -3 + [-7 + (-2)]$
$= -3 + [-9] = -12$

    b. $2^3 - 10 + [-6 - (-5)]$
$= 2^3 - 10 + [-6 + 5]$
$= 2^3 - 10 + [-1]$
$= 8 - 10 + [-1]$
$= -2 + [-1]$
$= -3$

11. **a.** $7 \cdot 0(-6) = 7 \cdot 0 = 0$

    b. $(-2)(-3)(-4) = 6(-4) = -24$

    c. $(-1)(5)(-9) = (-5)(-9) = 45$

    d. $(-2)^3 = (-2)(-2)(-2) = 4(-2) = -8$

    e. $-4(-11) - 5(-2) = 44 + 10 = 54$

12. **a.** $-18 \div 3 = -6$

    b. $\dfrac{-14}{-2} = \dfrac{14}{2} = 7$

c. $\dfrac{20}{-4} = -5$

13. $-5(-3 + 2z) = -5(-3) + (-5)(2z) = 15 - 10z$

14. $4(3x + 7) + 10 = 4(3x) + 4(7) + 10$
$= 12x + 28 + 10 = 12x + 38$

15. **a.** The total cost of renting the truck if 100 miles are driven is $70.

b. 280 miles are driven if the total cost is $140.

16. **a.** Unlike

b. Like

c. Like, since $-3zy = -3yz$

d. Like

17. $(2x - 3) - (4x - 2) = 2x - 3 - 4x + 2$
$= -2x - 1$

18. $x - 7 = 10$
$x - 7 + 7 = 10 + 7$
$x = 17$

19. $-z - 4 = 6$
$-z - 4 + 4 = 6 + 4$
$-z = 10$
$\dfrac{-z}{-1} = \dfrac{10}{-1}$
$z = -10$

20. $\dfrac{2(a + 3)}{3} = 6a + 2$

$3\left(\dfrac{2(a + 3)}{3}\right) = 3(6a + 2)$

$2(a + 3) = 3(6a + 2)$
$2a + 6 = 18a + 6$
$2a + 6 - 6 = 18a + 6 - 6$
$2a = 18a$
$2a - 18a = 18a - 18a$
$-16a = 0$
$\dfrac{-16a}{-16} = \dfrac{0}{-16}$
$a = 0$

21. Let $x =$ the number of Democrats.
Then $x + 8 =$ the number of Republicans.
$x + (x + 8) = 100$
$x + x + 8 = 100$
$2x + 8 = 100$
$2x + 8 - 8 = 100 - 8$
$2x = 92$
$\dfrac{2x}{2} = \dfrac{92}{2}$
$x = 46$
There were 46 Democrats and 54 Republicans.

22. Use the distance formula $d = rt$. Let $d = 31{,}680$ and $r = 400$.
$d = rt$
$31{,}680 = 400t$
$\dfrac{31{,}680}{400} = \dfrac{400t}{400}$
$79.2 = t$
It takes 79.2 years.

23. Let $x =$ the unknown percent.
$63 = x \cdot 72$
$\dfrac{63}{72} = \dfrac{x \cdot 72}{72}$
$0.875 = x$
$87.5\% = x$
The number 63 is 87.5% of 72.

24. $\dfrac{45}{x} = \dfrac{5}{7}$
$45 \cdot 7 = x \cdot 5$
$\dfrac{315}{5} = \dfrac{5x}{5}$
$63 = x$

25. $-1 > x$

26. $2(x - 3) - 5 \le 3(x + 2) - 18$
$2x - 6 - 5 \le 3x + 6 - 18$
$2x - 11 \le 3x - 12$
$-x - 11 \le -12$
$-x \le -1$
$\dfrac{-x}{-1} \ge \dfrac{-1}{-1}$
$x \ge 1$

Chapter 3

1. $\left(-\dfrac{3}{4}\right)^2 = \left(-\dfrac{3}{4}\right)\left(-\dfrac{3}{4}\right) = \dfrac{9}{16}$

2. $(4y^6)(2y^7) = 4 \cdot 2 \cdot y^6 \cdot y^7 = 8y^{13}$

3. $\dfrac{a^9 b^{16}}{a^{12} b^5} = (a^{9-12})(b^{16-5}) = a^{-3} b^{11} = \dfrac{b^{11}}{a^3}$

4. $4^0 + 2x^0 = 1 + 2 \cdot 1 = 1 + 2 = 3$

5. $\left(-\dfrac{1}{6}\right)^{-3} = \left(\dfrac{-1}{6}\right)^{-3} = \dfrac{(-1)^{-3}}{6^{-3}} = \dfrac{6^3}{(-1)^3}$

$= \dfrac{216}{-1} = -216$

6. $\left(\dfrac{m^{-2} n}{m^6 n^{-8}}\right)^{-2} = \dfrac{m^4 n^{-2}}{m^{-12} n^{16}}$

$= (m^{4-(-12)})(n^{-2-16}) = m^{16} n^{-18} = \dfrac{m^{16}}{n^{18}}$

7. $12x^2 + 3x - 5 - 8x^2 + 7x$

$= 12x^2 - 8x^2 + 3x + 7x - 5$

$= 4x^2 + 10x - 5$

8. $0.000000814 = 8.14 \times 10^{-7}$

9. The degree of $8x - 4x^5 + 6x^3 + 10$ is 5.

10. $-3x^3 + 2x^2 - 4 = -3(-1)^3 + 2(-1)^2 - 4$

$= -3(-1) + 2(1) - 4 = 3 + 2 - 4 = 1$

11. $(4x^2 - 3x + 9) + (6x^2 + 3x - 8)$

$= 4x^2 - 3x + 9 + 6x^2 + 3x - 8$

$= 4x^2 + 6x^2 - 3x + 3x + 9 - 8$

$= 10x^2 + 1$

12. $(6y^2 - 4) - (-3y^2 + 5y - 1)$

$= 6y^2 - 4 + 3y^2 - 5y + 1$

$= 6y^2 + 3y^2 - 5y - 4 + 1$

$= 9y^2 - 5y - 4 + 1$

$= 9y^2 - 5y - 3$

13. $(2a^2 + 3ab - 7b^2) - (3a^2 + 3ab + 9b^2)$

$= 2a^2 + 3ab - 7b^2 - 3a^2 - 3ab - 9b^2$

$= 2a^2 - 3a^2 + 3ab - 3ab - 7b^2 - 9b^2$

$= -a^2 - 16b^2$

14. $\left(-\dfrac{2}{7} n^6\right)\left(\dfrac{21}{16} n^3\right) = -\dfrac{2}{7} \cdot \dfrac{21}{16} \cdot n^6 \cdot n^3$

$= -\dfrac{42}{112} n^9 = -\dfrac{3}{8} n^9$

15. $-2t^2(3t^5 + 4t^3 - 8)$

$= -2t^2(3t^5) + (-2t^2)(4t^3) - (-2t^2)(8)$

$= -6t^7 - 8t^5 + 16t^2$

16. $(2y - 1)(5y + 6)$

$= 2y(5y) + 5y(-1) + 2y(6) + (-1)(6)$

$= 10y^2 - 5y + 12y - 6$

$= 10y^2 + 7y - 6$

17. $(7a - 5)^2 = (7a - 5)(7a - 5)$

$= (7a)(7a) + (7a)(-5) + (-5)(7a) + (-5)(-5)$

$= 49a^2 - 35a - 35a + 25$

$= 49a^2 - 70a + 25$

18. $(4b + 9)(4b - 9)$

$= (4b)(4b) + (4b)(-9) + (9)(4b) + (9)(-9)$

$= 16b^2 - 36b + 36b - 81 = 16b^2 - 81$

19. $\dfrac{16p^4 - 8p^3 + 20p^2}{4p} = \dfrac{16p^4}{4p} - \dfrac{8p^3}{4p} + \dfrac{20p^2}{4p}$

$= 4p^3 - 2p^2 + 5p$

20.

$$\begin{array}{r} 5x+2 \\ x-6\overline{\smash{\big)}\,5x^2-28x-12} \\ \underline{5x^2-30x} \\ 2x-12 \\ \underline{2x-12} \\ 0 \end{array}$$

Thus, $\dfrac{5x^2-28x-12}{x-6}=5x+2$.

Section 3.1

Practice Problems

1. $3^4 = 3\cdot3\cdot3\cdot3 = 81$

2. $7^1 = 7$

3. $(-2)^3 = (-2)(-2)(-2) = -8$

4. $-2^3 = -(2\cdot2\cdot2) = -8$

5. $\left(\dfrac{2}{3}\right)^2 = \left(\dfrac{2}{3}\right)\left(\dfrac{2}{3}\right) = \dfrac{4}{9}$

6. $5\cdot6^2 = 5\cdot6\cdot6 = 180$

7. **a.** $3x^2 = 3(4)^2 = 3\cdot4\cdot4 = 48$

   b. $\dfrac{x^4}{-8} = \dfrac{(-2)^4}{-8} = \dfrac{(-2)(-2)(-2)(-2)}{-8}$
   $= \dfrac{16}{-8} = -2$

8. $7^3\cdot7^2 = 7^{3+2} = 7^5$

9. $x^4\cdot x^9 = x^{4+9} = x^{13}$

10. $r^5\cdot r = r^5\cdot r^1 = r^{5+1} = r^6$

11. $s^6\cdot s^2\cdot s^3 = s^{6+2+3} = s^{11}$

12. $(-3)^9(-3) = (-3)^9(-3)^1 = (-3)^{9+1} = (-3)^{10}$

13. $(6x^3)(-2x^9) = 6\cdot(-2)\cdot x^3\cdot x^9 = -12x^{12}$

14. $(9^4)^{10} = 9^{4\cdot10} = 9^{40}$

15. $(z^6)^3 = z^{6\cdot3} = z^{18}$

16. $(xy)^7 = x^7y^7$

17. $(3y)^4 = 3^4y^4 = 81y^4$

18.
$$(-2p^4q^2r)^3 = (-2)^3(p^4)^3(q^2)^3r^3$$
$$= -8p^{12}q^6r^3$$

19. $\left(\dfrac{r}{s}\right)^6 = \dfrac{r^6}{s^6},\ s\neq0$

20. $\left(\dfrac{5x^6}{9y^3}\right)^2 = \dfrac{5^2(x^6)^2}{9^2(y^3)^2} = \dfrac{25x^{12}}{81y^6},\ y\neq0$

21. $\dfrac{y^7}{y^3} = y^{7-3} = y^4$

22. $\dfrac{5^9}{5^6} = 5^{9-6} = 5^3 = 125$

23. $\dfrac{(-2)^{14}}{(-2)^{10}} = (-2)^{14-10} = (-2)^4 = 16$

24.
$$\dfrac{7a^4b^{11}}{ab} = 7\cdot\dfrac{a^4}{a^1}\cdot\dfrac{b^{11}}{b^1} = 7a^{4-1}b^{11-1}$$
$$= 7a^3b^{10}$$

25. $8^0 = 1$

26. $(2r^2s)^0 = 1$

27. $(-5)^0 = 1$

28. $-5^0 = -1\cdot5^0 = -1\cdot1 = -1$

29. **a.** $\dfrac{x^7}{x^4} = x^{7-4} = x^3$

b. $(3y^4)^4 = 3^4(y^4)^4 = 81y^{16}$

c. $\left(\dfrac{x}{4}\right)^3 = \dfrac{x^3}{4^3} = \dfrac{x^3}{64}$

Mental Math

1. 3^2
base: 3
exponent: 2

2. 5^4
base: 5
exponent: 4

3. $(-3)^6$
base: −3
exponent: 6

4. $(-3)^7$
base: 3
exponent: 7

5. -4^2
base: 4
exponent: 2

6. $(-4)^3$
base: −4
exponent: 3

7. $5 \cdot 3^4$
base: 5; exponent: 1
base: 3; exponent: 4

8. $9 \cdot 7^6$
base: 9; exponent: 1
base 7; exponent: 6

9. $5x^2$
base: 5; exponent: 1
base: x, exponent: 2

10. $(5x)^2$
base: $5x$
exponent: 2

Exercise Set 3.1

1. $7^2 = 7 \cdot 7 = 49$

3. $(-5)^1 = -5$

5. $-2^4 = -2 \cdot 2 \cdot 2 \cdot 2 = -16$

7. $(-2)^4 = (-2)(-2)(-2)(-2) = 16$

9. $\left(\dfrac{1}{3}\right)^3 = \left(\dfrac{1}{3}\right)\left(\dfrac{1}{3}\right)\left(\dfrac{1}{3}\right) = \dfrac{1}{27}$

11. $7 \cdot 2^4 = 7 \cdot 2 \cdot 2 \cdot 2 \cdot 2 = 112$

13. Answers may vary.

15. $x^2 = (-2)^2 = (-2)(-2) = 4$

17. $5x^3 = 5(3)^3 = 5 \cdot 3 \cdot 3 \cdot 3 = 135$

19. $2xy^2 = 2(3)(5)^2 = 2(3)(5)(5) = 150$

21. $\dfrac{2z^4}{5} = \dfrac{2(-2)^4}{5} = \dfrac{2(-2)(-2)(-2)(-2)}{5} = \dfrac{32}{5}$

23. $x^2 \cdot x^5 = x^{2+5} = x^7$

25. $(-3)^3 \cdot (-3)^9 = (-3)^{3+9} = (-3)^{12}$

27. $\left(5y^4\right)(3y) = 5(3)y^{4+1} = 15y^5$

29. $\left(4z^{10}\right)\left(-6z^7\right)\left(z^3\right) = 4(-6)z^{10+7+3} = -24z^{20}$

31. $\left(4x^2\right)\left(5x^3\right) = 4(5)x^{2+3} = 20x^5$ sq ft

33. $\left(x^9\right)^4 = x^{9 \cdot 4} = x^{36}$

35. $(pq)^7 = p^7q^7$

37. $\left(2a^5\right)^3 = 2^3 a^{5 \cdot 3} = 8a^{15}$

39. $\left(\dfrac{m}{n}\right)^9 = \dfrac{m^9}{n^9}$

41. $\left(x^2 y^3\right)^5 = x^{2 \cdot 5} y^{3 \cdot 5} = x^{10} y^{15}$

43. $\left(\dfrac{-2xz}{y^5}\right)^2 = \dfrac{(-2)^2 x^2 z^2}{y^{5 \cdot 2}} = \dfrac{4x^2 z^2}{y^{10}}$

45. $\left(8z^5\right)^2 = 8^2 z^{5 \cdot 2} = 64z^{10}$

The area is $64z^{10}$ sq. decimeters.

47. $\left(3y^4\right)^3 = 3^3 y^{4 \cdot 3} = 27y^{12}$

The volume is $27y^{12}$ cubic feet.

49. $\dfrac{x^3}{x} = \dfrac{x^3}{x^1} = x^{3-1} = x^2$

51. $\dfrac{(-2)^5}{(-2)^3} = (-2)^{5-3} = (-2)^2 = 4$

53. $\dfrac{p^7 q^{20}}{pq^{15}} = p^{7-1} q^{20-15} = p^6 q^5$

55. $\dfrac{7x^2 y^6}{14x^2 y^3} = \dfrac{7}{14} x^{2-2} y^{6-3} = \dfrac{1}{2} x^0 y^3 = \dfrac{y^3}{2}$

57. $(2x)^0 = 1$

59. $-2x^0 = -2(1) = -2$

61. $5^0 + y^0 = 1 + 1 = 2$

63. Answers may vary.

65. $-5^2 = -5 \cdot 5 = -25$

67. $\left(\dfrac{1}{4}\right)^3 = \dfrac{1^3}{4^3} = \dfrac{1}{64}$

69. $\dfrac{z^{12}}{z^4} = z^{12-4} = z^8$

71. $(9xy)^2 = 9^2 x^2 y^2 = 81x^2 y^2$

73. $(6b)^0 = 1$

75. $2^3 + 2^5 = 8 + 32 = 40$

77. $b^4 b^2 = b^{4+2} = b^6$

79. $a^2 a^3 a^4 = a^{2+3+4} = a^9$

81. $\left(2x^3\right)\left(-8x^4\right) = 2(-8)x^{3+4} = -16x^7$

83. $(4a)^3 = 4^3 a^3 = 64a^3$

85. $\left(-6xyz^3\right)^2 = (-6)^2 x^2 y^2 z^{3 \cdot 2} = 36x^2 y^2 z^6$

87. $\left(\dfrac{3y^5}{6x^4}\right)^3 = \dfrac{3^3 y^{5 \cdot 3}}{6^3 x^{4 \cdot 3}} = \dfrac{27y^{15}}{216x^{12}} = \dfrac{y^{15}}{8x^{12}}$

89. $\dfrac{3x^5}{x^4} = 3x^{5-4} = 3x$

91. $\dfrac{2x^3 y^2 z}{xyz} = 2x^{3-1} y^{2-1} z^{1-1} = 2x^2 y$

93. $5 - 7 = 5 + (-7) = -2$

95. $3 - (-2) = 3 + 2 = 5$

97. $-11 - (-4) = -11 + 4 = -7$

99. $V = x^3 = 7^3 = 7 \cdot 7 \cdot 7 = 343$
The volume is 343 cubic meters.

101. We use the volume formula.

103. $x^{5a} x^{4a} = x^{5a+4a} = x^{9a}$

105. $\left(a^b\right)^5 = a^{b \cdot 5} = a^{5b}$

107. $\dfrac{x^{9a}}{x^{4a}} = x^{9a-4a} = x^{5a}$

109. $A = P\left(1 + \dfrac{r}{12}\right)^6$

$A = 1000\left(1 + \dfrac{0.09}{12}\right)^6$

$A = 1000(1.0075)^6$

$A = 1045.85$

You need \$1045.85 to pay off the loan.

Section 3.2

Practice Problems

1. $5^{-3} = \dfrac{1}{5^3} = \dfrac{1}{125}$

2. $7x^{-4} = 7 \cdot \dfrac{1}{x^4} = \dfrac{7}{x^4}$

3. $5^{-1} + 3^{-1} = \dfrac{1}{5^1} + \dfrac{1}{3^1} = \dfrac{3}{15} + \dfrac{5}{15} = \dfrac{8}{15}$

4. $(-3)^{-4} = \dfrac{1}{(-3)^4} = \dfrac{1}{81}$

5. $\left(\dfrac{6}{7}\right)^{-2} = \dfrac{6^{-2}}{7^{-2}} = \dfrac{7^2}{6^2} = \dfrac{49}{36}$

6. $\dfrac{x}{x^{-4}} = \dfrac{x^1}{x^{-4}} = x^{1-(-4)} = x^5$

7. $\dfrac{y^{-9}}{z^{-5}} = \dfrac{z^5}{y^9}$

8. $\dfrac{y^{-4}}{y^6} = y^{-4-6} = y^{-10} = \dfrac{1}{y^{10}}$

9. $\dfrac{(x^5)^3 x}{x^4} = \dfrac{x^{15} x^1}{x^4} = x^{15+1-4} = x^{12}$

10. $\left(\dfrac{9x^3}{y}\right)^{-2} = \dfrac{9^{-2} x^{-6}}{y^{-2}} = \dfrac{y^2}{9^2 x^6} = \dfrac{y^2}{81x^6}$

11. $(a^{-4} b^7)^{-5} = a^{20} b^{-35} = \dfrac{a^{20}}{b^{35}}$

12. $\dfrac{(2x)^4}{x^8} = \dfrac{2^4 x^4}{x^8} = 2^4 x^{4-8} = 16x^{-4} = \dfrac{16}{x^4}$

13. $\dfrac{y^{-10}}{(y^5)^4} = \dfrac{y^{-10}}{y^{20}} = y^{-10-20} = y^{-30} = \dfrac{1}{y^{30}}$

14. $(4a^2)^{-3} = 4^{-3} a^{-6} = \dfrac{1}{4^3 a^6} = \dfrac{1}{64a^6}$

15. a. $420,000 = 4.2 \times 10^5$

    b. $0.00017 = 1.7 \times 10^{-4}$

    c. $9,060,000,000 = 9.06 \times 10^9$

    d. $0.000007 = 7 \times 10^{-6}$

16. a. $3.062 \times 10^{-4} = 0.0003062$

    b. $5.21 \times 10^4 = 52,100$

    c. $9.6 \times 10^{-5} = 0.000096$

    d. $6.002 \times 10^6 = 6,002,000$

17. a. $(9 \times 10^7)(4 \times 10^{-9}) = 9 \cdot 4 \cdot 10^7 \cdot 10^{-9}$
$= 36 \times 10^{-2} = 0.36$

    b. $\dfrac{8 \times 10^4}{2 \times 10^{-3}} = \dfrac{8}{2} \times 10^{4-(-3)} = 4 \times 10^7$

        $= 40,000,000$

Mental Math

1. $5x^{-2} = \dfrac{5}{x^2}$

2. $3x^{-3} = \dfrac{3}{x^3}$

3. $\dfrac{1}{y^{-6}} = y^6$

4. $\dfrac{1}{x^{-3}} = x^3$

5. $\dfrac{4}{y^{-3}} = 4y^3$

6. $\dfrac{16}{y^{-7}} = 16y^7$

Exercise Set 3.2

1. $4^{-3} = \dfrac{1}{4^3} = \dfrac{1}{64}$

3. $7x^{-3} = 7 \cdot \dfrac{1}{x^3} = \dfrac{7}{x^3}$

5. $\left(-\dfrac{1}{4}\right)^{-3} = \dfrac{(-1)^{-3}}{(4)^{-3}} = \dfrac{4^3}{(-1)^3} = \dfrac{64}{-1} = -64$

7. $3^{-1} + 2^{-1} = \dfrac{1}{3} + \dfrac{1}{2} = \dfrac{2}{6} + \dfrac{3}{6} = \dfrac{5}{6}$

9. $\dfrac{1}{p^{-3}} = p^3$

11. $\dfrac{p^{-5}}{q^{-4}} = \dfrac{q^4}{p^5}$

13. $\dfrac{x^{-2}}{x} = x^{-2-1} = x^{-3} = \dfrac{1}{x^3}$

15. $\dfrac{z^{-4}}{z^{-7}} = z^{-4-(-7)} = z^3$

17. $2^0 + 3^{-1} = 1 + \dfrac{1}{3} = \dfrac{3}{3} + \dfrac{1}{3} = \dfrac{4}{3}$

19. $(-3)^{-2} = \dfrac{1}{(-3)^2} = \dfrac{1}{9}$

21. $\dfrac{-1}{p^{-4}} = -1\left(p^4\right) = -p^4$

23. $-2^0 - 3^0 = -1(1) - 1 = -2$

25. $\dfrac{x^2 x^5}{x^3} = x^{2+5-3} = x^4$

27. $\dfrac{p^2 p}{p^{-1}} = p^{2+1-(-1)} = p^{2+1+1} = p^4$

29. $\dfrac{(m^5)^4 m}{m^{10}} = m^{5(4)+1-10} = m^{20+1-10} = m^{11}$

31. $\dfrac{r}{r^{-3} r^{-2}} = r^{1-(-3)-(-2)} = r^{1+3+2} = r^6$

33. $\left(x^5 y^3\right)^{-3} = x^{5(-3)} y^{3(-3)} = x^{-15} y^{-9} = \dfrac{1}{x^{15} y^9}$

35. $\dfrac{\left(x^2\right)^3}{x^{10}} = \dfrac{x^6}{x^{10}} = x^{6-10} = x^{-4} = \dfrac{1}{x^4}$

37. $\dfrac{\left(a^5\right)^2}{\left(a^3\right)^4} = \dfrac{a^{10}}{a^{12}} = a^{10-12} = a^{-2} = \dfrac{1}{a^2}$

39. $\dfrac{8k^4}{2k} = \dfrac{8}{2} \cdot k^{4-1} = 4k^3$

41. $\dfrac{-6m^4}{-2m^3} = \dfrac{-6}{-2} \cdot m^{4-3} = 3m$

43. $\dfrac{-24a^6 b}{6ab^2} = \dfrac{-24}{6} \cdot a^{6-1} b^{1-2} = -4a^5 b^{-1}$

$= -\dfrac{4a^5}{b}$

45 $\dfrac{6x^2 y^3}{-7xy^5} = -\dfrac{6}{7} x^{2-1} y^{3-5} = -\dfrac{6}{7} x^1 y^{-2}$

$= -\dfrac{6x}{7y^2}$

47. $\left(a^{-5} b^2\right)^{-6} = a^{-5(-6)} b^{2(-6)}$

$= a^{30} b^{-12} = \dfrac{a^{30}}{b^{12}}$

49. $\left(\dfrac{x^{-2}y^4}{x^3y^7}\right)^2 = \dfrac{x^{-2(2)}y^{4(2)}}{x^{3(2)}y^{7(2)}} = \dfrac{x^{-4}y^8}{x^6y^{14}}$

$= x^{-4-6}y^{8-14} = x^{-10}y^{-6} = \dfrac{1}{x^{10}y^6}$

51. $\dfrac{4^2z^{-3}}{4^3z^{-5}} = 4^{2-3}z^{-3-(-5)} = 4^{-1}z^2 = \dfrac{z^2}{4}$

53. $\dfrac{2^{-3}x^{-4}}{2^2x} = 2^{-3-2}x^{-4-1} = 2^{-5}x^{-5}$

$= \dfrac{1}{2^5x^5} = \dfrac{1}{32x^5}$

55. $\dfrac{7ab^{-4}}{7^{-1}a^{-3}b^2} = 7^{1-(-1)}a^{1-(-3)}b^{-4-2}$

$= 7^2a^4b^{-6} = \dfrac{49a^4}{b^6}$

57. $\left(\dfrac{a^{-5}b}{ab^3}\right)^{-4} = \dfrac{a^{-5(-4)}b^{-4}}{a^{-4}b^{3(-4)}} = \dfrac{a^{20}b^{-4}}{a^{-4}b^{-12}}$

$= a^{20-(-4)}b^{-4-(-12)} = a^{24}b^8$

59. $\dfrac{\left(xy^3\right)^5}{(xy)^{-4}} = \dfrac{x^5y^{3(5)}}{x^{-4}y^{-4}} = \dfrac{x^5y^{15}}{x^{-4}y^{-4}}$

$= x^{5-(-4)}y^{15-(-4)} = x^9y^{19}$

61. $\dfrac{\left(-2xy^{-3}\right)^{-3}}{\left(xy^{-1}\right)^{-1}} = \dfrac{(-2)^{-3}x^{-3}y^9}{x^{-1}y^1}$

$= (-2)^{-3}x^{-3-(-1)}y^{9-1} = -\dfrac{y^8}{8x^2}$

63. $\left(\dfrac{3x^{-2}}{z}\right)^3 = \dfrac{3^3x^{-6}}{z^3} = \dfrac{27}{x^6z^3}$

The volume is $\dfrac{27}{x^6z^3}$ cubic inches.

65. $78,000 = 7.8 \times 10^4$

67. $0.00000167 = 1.67 \times 10^{-6}$

69. $0.00635 = 6.35 \times 10^{-3}$

71. $1,160,000 = 1.16 \times 10^6$

73. $20,000,000 = 2.0 \times 10^7$

75. $93,000,000 = 9.3 \times 10^7$

77. $120,000,000 = 1.2 \times 10^8$

79. $8.673 \times 10^{-10} = 0.0000000008673$

81. $3.3 \times 10^{-2} = 0.033$

83. $2.032 \times 10^4 = 20,320$

85. $6.25 \times 10^{18} = 6,250,000,000,000,000,000$

87. $9.460 \times 10^{12} = 9,460,000,000,000$

89. $\left(1.2 \times 10^{-3}\right)\left(3 \times 10^{-2}\right) = 1.2 \cdot 3 \cdot 10^{-3} \cdot 10^{-2}$

$= 3.6 \times 10^{-5}$

$= 0.000036$

91. $\left(4 \times 10^{-10}\right)\left(7 \times 10^{-9}\right) = 4 \cdot 7 \cdot 10^{-10} \cdot 10^{-9}$

$= 28 \times 10^{-19}$

$= 0.0000000000000000028$

93. $\dfrac{8 \times 10^{-1}}{16 \times 10^5} = \dfrac{8}{16} \times 10^{-1-5} = 0.5 \times 10^{-6}$

$= 0.0000005$

95. $\dfrac{1.4 \times 10^{-2}}{7 \times 10^{-8}} = \dfrac{1.4}{7} \times 10^{-2-(-8)} = 0.2 \times 10^6$

$= 200,000$

97. $3600 \times 4.2 \times 10^6 = 3.6 \times 10^3 \times 4.2 \times 10^6$

$= 3.6 \times 4.2 \times 10^3 \times 10^6 = 15.12 \times 10^9$

$= 1.512 \times 10^{10}$ cubic feet

99. $3x - 5x + 7 = (3-5)x + 7 = -2x + 7$

101. $y - 10 + y = (1 + 1)y - 10 = 2y - 10$

103. $7x + 2 - 8x - 6 = (7 - 8)x + (2 - 6)$
$= -x - 4$

105. $\left(2.63 \times 10^{12}\right)\left(-1.5 \times 10^{-10}\right)$
$= 2.63 \cdot (-1.5) \cdot 10^{12} \cdot 10^{-10}$
$= -3.945 \times 10^{2} = -394.5$

107. $d = r \cdot t$
$238,857 = \left(1.86 \times 10^{5}\right)t$
$t = \dfrac{238,857}{1.86 \times 10^{5}}$
$t = \dfrac{2.38857}{1.86} \times 10^{5-5}$
$t = 1.3$ seconds

109. $a^{-4m} \cdot a^{5m} = a^{-4m+5m} = a^{m}$

111. $\left(3y^{2z}\right)^{3} = 3^{3}y^{2z \cdot 3} = 27y^{6z}$

113. Answers may vary.

115. Answers may vary.

Section 3.3

Practice Problems

1. $-6x^{6} + 4x^{5} + 7x^{3} - 9x^{2} - 1$

Term	Coefficient
$7x^{3}$	7
$-9x^{2}$	-9
$-6x^{6}$	-6
$4x^{5}$	4
-1	-1

2. $-15x^{3} + 2x^{2} - 5$
The term $-15x^{3}$ has degree 3.
The term $2x^{2}$ has degree 2.
The term -5 has degree 0 because -5 is $-5x^{0}$.

3. a. The degree of the binomial $-6x + 14$ is 1 because $-6x$ is $-6x^{1}$.

   b. The degree of the polynomial $9x - 3x^{6} + 5x^{4} + 2$ is 6, the greatest degree of any of its terms. It is not a monomial, binomial, or trinomial.

   c. The degree of the trinomial $10x^{2} - 6x - 6$ is 2, the greatest degree of any of its terms.

4. a. $-2x + 10 = -2(-1) + 10 = 2 + 10 = 12$

   b. $6x^{2} + 11x - 20 = 6(-1)^{2} + 11(-1) - 20$
   $= 6 - 11 - 20 = -25$

5. $-16t^{2} + 1821 = -16(3)^{2} + 1821$
$= -16(9) + 1821 = -144 + 1821 = 1677$
The height of the object at 3 seconds is 16777 feet.
$-16t^{2} + 1821 = -16(7)^{2} + 1821$
$= -16(49) + 1821$
$= 1037$
The height of the object at 7 seconds is 1037 feet.

6. $-6y + 8y = (-6 + 8)y = 2y$

7. $14y^{2} + 3 - 10y^{2} - 9 = 14y^{2} - 10y^{2} + 3 - 9$
$= 4y^{2} - 6$

8. $7x^{3} + x^{3} = 7x^{3} + 1x^{3} = 8x^{3}$

9. $23x^{2} - 6x - x - 15 = 23x^{2} - 7x - 15$

10. $\dfrac{2}{7}x^3 - \dfrac{1}{4}x + 2 - \dfrac{1}{2}x^3 + \dfrac{3}{8}x$

$= \left(\dfrac{2}{7} - \dfrac{1}{2}\right)x^3 + \left(-\dfrac{1}{4} + \dfrac{3}{8}\right)x + 2$

$= \left(\dfrac{4}{14} - \dfrac{7}{14}\right)x^3 + \left(-\dfrac{2}{8} + \dfrac{3}{8}\right)x + 2$

$= -\dfrac{3}{14}x^3 + \dfrac{1}{8}x + 2$

11. Area $= 5 \cdot x + x \cdot x + 4 \cdot 5 + x \cdot x + 8 \cdot x$
$= 5x + x^2 + 20 + x^2 + 8x$
$= 2x^2 + 13x + 20$

12. $-2x^3y^2 + 4 - 8xy + 3x^3y + 5xy^2$

Terms	Degree	Degree of Polynomial
$-2x^3y^2$	3 + 2 or 5	5 (highest degree)
4	0	
$-8xy$	1 + 1 = 2	
$3x^3y$	3 + 1 or 4	
$5xy^2$	1 + 2 or 3	

13. $11ab - 6a^2 - ba + 8b^2$
$= (11 - 1)ab - 6a^2 + 8b^2$
$= 10ab - 6a^2 + 8b^2$

14. $7x^2y^2 + 2y^2 - 4y^2x^2 + x^2 - y^2 + 5x^2$
$= (7 - 4)x^2y^2 + (2 - 1)y^2 + (1 + 5)x^2$
$= 3x^2y^2 + y^2 + 6x^2$

Exercise Set 3.3

1. $x^2 - 3x + 5$

Term	Coefficient
x^2	1
$-3x$	-3
5	5

3. $-5x^4 + 3.2x^2 + x - 5$

Term	Coefficient
$-5x^4$	-5
$3.2x^2$	3.2
x	1
-5	-5

5. $x + 2$
The degree is 1 since x is x^1. It is a binomial because it has two terms.

7. $9m^3 - 5m^2 + 4m - 8$
The degree is 3, the greatest degree of any of its terms. It is none of these because it has more than three terms.

9. $12x^4 - x^2 - 12x^2 = 12x^4 - 13x^2$
The degree is 4, the greatest degree of any of its terms. It is a binomial because the simplified form has two terms.

11. $3z - 5$
The degree is 1 because $3z$ is $3z^1$. It is a binomial because it has two terms.

13. Answers may vary.

15. Answers may vary.

17. a. $x + 6 = 0 + 6 = 6$

  b. $x + 6 = -1 + 6 = 5$

19. a. $x^2 - 5x - 2 = 0^2 - 5(0) - 2 = -2$

b. $x^2 - 5x - 2 = (-1)^2 - 5(-1) - 2$
     $= 1 + 5 - 2 = 4$

21. a. $x^3 - 15 = 0^3 - 15 = -15$

  b. $x^3 - 15 = (-1)^3 - 15 = -1 - 15 = -16$

23. $-16t^2 + 200t = -16(1)^2 + 200(1)$
    $= -16 + 200 = 184$ feet

25. $-16t^2 + 200t = -16(7.6)^2 + 200(7.6)$
    $= -924.16 + 1520 = 595.84$ feet

27. $0.15x^2 + 0.15x + 7.2$
    $x = 1998 - 1993 = 5$
    $0.15(5)^2 + 0.15(5) + 7.2$
    $= 3.75 + 0.75 + 7.2 = 11.7$
    There were 11.7 million departures.

29. $14x^2 + 9x^2 = (14 + 9)x^2 = 23x^2$

31. $15x^2 - 3x^2 - y$
    $(15 - 3)x^2 - y = 12x^2 - y$

33. $8s - 5s + 4s = (8 - 5 + 4)s = 7s$

35. $0.1y^2 - 1.2y^2 + 6.7 - 1.9$
    $= (0.1 - 1.2)y^2 + (6.7 - 1.9)$
    $= -1.1y^2 + 4.8$

37. $5x + 3 + 4x + 3 + 2x + 6 + 3x + 7x$
    $= (5x + 4x + 2x + 3x + 7x) + (3 + 3 + 6)$
    $= 21x + 12$

39. $(2x)^2 + 7x + x^2 + 5x = 4x^2 + x^2 + 7x + 5x$
    $= 5x^2 + 12x$

41. $9ab - 6a + 5b - 3$

Terms	Degree	Degree of Polynomial
$9ab$	1 + 1 or 2	2 (highest degree)
$-6a$	1	
$5b$	1	
-3	0	

43. $x^3y - 6 + 2x^2y^2 + 5y^3$

Terms	Degree	Degree of Polynomial
x^3y	3 + 1 or 4	4
-6	0	
$2x^2y^2$	2+2 or 4	
$5y^3$	3	

45. $3ab - 4a + 6ab - 7a = (3 + 6)ab + (-4 - 7)a$
    $= 9ab - 11a$

47. $4x^2 - 6xy + 3y^2 - xy$
    $= 4x^2 + (-6 - 1)xy + 3y^2$
    $= 4x^2 - 7xy + 3y^2$

49. $5x^2y + 6xy^2 - 5yx^2 + 4 - 9y^2x$
    $= (5 - 5)x^2y + (6 - 9)xy^2 + 4$
    $= -3xy^2 + 4$

51. $14y^3 - 9 + 3a^2b^2 - 10 - 19b^2a^2$
    $= 14y^3 + (-9 - 10) + (3 - 19)a^2b^2$
    $= 14y^3 - 19 - 16a^2b^2$

53. $4 + 5(2x + 3) = 4 + 10x + 15 = 10x + 19$

55. $2(x - 5) + 3(5 - x) = 2x - 10 + 15 - 3x$
    $= -x + 5$

57. Answers may vary.

59. $1.85x^2 - 3.76x + 9.25x^2 + 10.76 - 4.21x$
$= (1.85 + 9.25)x^2 + (-3.76 - 4.21)x + 10.76$
$= 11.1x^2 - 7.97x + 10.76$

Section 3.4

Practice Problems

1. $(3x^5 - 7x^3 + 2x - 1) + (3x^3 - 2x)$
$= 3x^5 - 7x^3 + 2x - 1 + 3x^3 - 2x$
$= 3x^5 + (-7x^3 + 3x^3) + (2x - 2x) - 1$
$= 3x^5 - 4x^3 - 1$

2. $(5x^2 - 2x + 1) + (-6x^2 + x - 1)$
$= 5x^2 - 2x + 1 - 6x^2 + x - 1$
$= (5x^2 - 6x^2) + (-2x + x) + (1 - 1)$
$= -x^2 - x$

3. $\quad 9y^2 - 6y + 55$
$\quad\quad\quad\quad \underline{4y + 3}$
$\quad 9y^2 - 2y + 8$

4. $(9x + 5) - (4x - 3) = (9x + 5) + [-(4x - 3)]$
$= (9x + 5) + (-4x + 3) = 5x + 8$

5. $(4x^3 - 10x^2 + 1) - (-4x^3 + x^2 - 11)$
$= (4x^3 - 10x^2 + 1) + (4x^3 - x^2 + 11)$
$= 4x^3 + 4x^3 - 10x^2 - x^2 + 1 + 11$
$= 8x^3 - 11x^2 + 12$

6. $\quad\quad 2y^2 - 2y + 7$
$\quad\quad \underline{-(6y^2 - 3y + 2)}$

$\quad\quad 2y^2 - 2y + 7$
$\quad\quad \underline{-6y^2 + 3y - 2}$
$\quad\quad -4y^2 + y + 5$

7. $[(4x - 3) + (12x - 5)] - (3x + 1)$
$= 4x - 3 + 12x - 5 - 3x - 1$
$= 4x + 12x - 3x - 3 - 5 - 1$
$= 13x - 9$

8. $(2a^2 - ab + 6b^2) - (-3a^2 + ab - 7b^2)$
$= 2a^2 - ab + 6b^2 + 3a^2 - ab + 7b^2$
$= 5a^2 - 2ab + 13b^2$

9. $(5x^2y^2 + 3 - 9x^2y + y^2) - (-x^2y^2 + 7 - 8xy^2 + 2y^2)$
$= 5x^2y^2 + 3 - 9x^2y + y^2 + x^2y^2 - 7 + 8xy^2 - 2y^2$
$= 6x^2y^2 - 4 - 9x^2y - y^2 + 8xy^2$

Exercise Set 3.4

1. $(3x + 7) + (9x + 5) = 3x + 7 + 9x + 5$
$= (3x + 9x) + (7 + 5) = 12x + 12$

3. $(-7x + 5) + \left(-3x^2 + 7x + 5\right)$
$= -7x + 5 + \left(-3x^2\right) + 7x + 5$
$= -3x^2 + (-7x + 7x) + (5 + 5)$
$= -3x^2 + 10$

5. $\left(-5x^2+3\right)+\left(2x^2+1\right)$

$= -5x^2+3+2x^2+1$

$= \left(-5x^2+2x^2\right)+(3+1)$

$= -3x^2+4$

7. $\left(-3y^2-4y\right)+\left(2y^2+y-1\right)$

$= -3y^2-4y+2y^2+y-1$

$= \left(-3y^2+2y^2\right)+(-4y+y)-1$

$= -y^2-3y-1$

9.
$$3t^2+4$$
$$\underline{+\ 5t^2-8}$$
$$8t^2-4$$

11.
$$10a^3-8a^2+9$$
$$\underline{+\ 5a^3+9a^2+7}$$
$$15a^3+a^2+16$$

13. $(2x+5)-(3x-9) = (2x+5)+(-3x+9)$
$= 2x+5+(-3x)+9 = (2x-3x)+(5+9)$
$= -x+14$

15. $3x-(5x-9) = 3x+(-5x+9)$
$= 3x+(-5x)+9 = -2x+9$

17. $\left(2x^2+3x-9\right)-(-4x+7)$

$= \left(2x^2+3x-9\right)+(4x-7)$

$= 2x^2+3x-9+4x-7$

$= 2x^2+(3x+4x)+(-9-7)$

$= 2x^2+7x-16$

19. $\left(-7y^2+5\right)-\left(-8y^2+12\right)$

$= \left(-7y^2+5\right)+\left(8y^2-12\right)$

$= -7y^2+5+8y^2-12$

$= \left(-7y^2+8y^2\right)+(5-12)$

$= y^2-7$

21. $(5x+8)-\left(-2x^2-6x+8\right)$

$= (5x+8)+\left(2x^2+6x-8\right)$

$= 5x+8+2x^2+6x-8$

$= 2x^2+(5x+6x)+(8-8)$

$= 2x^2+11x$

23.
$$4z^2-8z+3$$
$$\underline{-\left(6z^2+8z-3\right)}$$

$$4z^2-8z+3$$
$$\underline{+\left(-6z^2-8z+3\right)}$$
$$-2z^2-16z+6$$

25.
$$5u^5-4u^2+3u-7$$
$$\underline{-(3u^5+6u^2-8u+2)}$$
$$5u^5-4u^2+3u-7$$
$$\underline{+(-3u^5-6u^2+8u-2)}$$
$$2u^5-10u^2+11u-9$$

27. $(3x+5)+(2x-14) = 3x+5+2x-14$
$= 5x-9$

29. $(7y+7)-(y-6) = 7y+7-y+6$
$= 6y+13$

31. $\left(x^2+2x+1\right)-\left(3x^2-6x+2\right)$

$= x^2+2x+1-3x^2+6x-2$

$= -2x^2+8x-1$

33. $\left(3x^2+5x-8\right)+\left(5x^2+9x+12\right)-\left(x^2-14\right)$

$= 3x^2+5x-8+5x^2+9x+12-x^2+14$

$= 7x^2+14x+18$

35. $(7x-3)-4x = 7x-3-4x = 3x-3$

37. $\left(4x^2-6x+1\right)+\left(3x^2+2x+1\right)$

$= 4x^2-6x+1+3x^2+2x+1$

$= 7x^2-4x+2$

39. $\left(7x^2 + 3x + 9\right) - (5x + 7)$

$= 7x^2 + 3x + 9 - 5x - 7$

$= 7x^2 - 2x + 2$

41. $\left[\left(8y^2 + 7\right) + (6y + 9)\right] - \left(4y^2 - 6y - 3\right)$

$= 8y^2 + 7 + 6y + 9 - 4y^2 + 6y + 3$

$= 4y^2 + 12y + 19$

43. $(9a + 6b - 5) + (-11a - 7b + 6)$

$= 9a + 6b - 5 - 11a - 7b + 6$

$= -2a - b + 1$

45. $\left(4x^2 + y^2 + 3\right) - \left(x^2 + y^2 - 2\right)$

$= 4x^2 + y^2 + 3 - x^2 - y^2 + 2$

$= 3x^2 + 5$

47. $\left(x^2 + 2xy - y^2\right) + \left(5x^2 - 4xy + 20y^2\right)$

$= x^2 + 2xy - y^2 + 5x^2 - 4xy + 20y^2$

$= 6x^2 - 2xy + 19y^2$

49. $\left(11r^2s + 16rs - 3 - 2r^2s^2\right) - \left(3sr^2 + 5 - 9r^2s^2\right)$

$= 11r^2s + 16rs - 3 - 2r^2s^2 - 3sr^2 - 5 + 9r^2s^2$

$= 8r^2s + 16rs + 7r^2s^2 - 8$

51. $3x(2x) = 3 \cdot 2 \cdot x \cdot x = 6x^2$

53. $\left(12x^3\right)\left(-x^5\right) = \left(12x^3\right)\left(-1x^5\right)$

$= (12)(-1)\left(x^3\right)\left(x^5\right) = -12x^8$

55. $10x^2\left(20xy^2\right) = 10 \cdot 20x^2 \cdot x \cdot y^2 = 200x^3y^2$

57. $\left(-x^2 + 3x\right) + \left(2x^2 + 5\right) + (4x - 1)$

$= -x^2 + 3x + 2x^2 + 5 + 4x - 1$

$= \left(x^2 + 7x + 4\right)$ feet

59. $(-x + 4) + 5x + \left(x^2 - 6x - 2\right) + x^2$

$= -x + 4 + 5x + x^2 - 6x - 2 + x^2$

$= \left(2x^2 - 2x + 2\right)$ cm

61. $\left[\left(1.2x^2 - 3x + 9.1\right) - \left(7.8x^2 - 3.1 + 8\right)\right] + (1.2x - 6) = 1.2x^2 - 3x + 9.1 - 7.8x^2 + 3.1 - 8 + 1.2x - 6$

$= -6.6x^2 - 1.8x - 1.8$

63. $\left(-250.5x^2 + 1587.5x + 23,049\right) + \left(-214x^2 + 801x + 17,710\right)$

$= -250.5x^2 + 1587.5x + 23,049 - 214x^2 + 801x + 17,710$

$= \left(-464.5x^2 + 2388.5x + 40,759\right)$ millions of pounds

Section 3.5

Practice Problems

1. $10x \cdot 9x = (10 \cdot 9)(x \cdot x) = 90x^2$

2. $8x^3(-11x^7) = 8 \cdot (-11)(x^3 \cdot x^7) = -88x^{10}$

3. $(-5x^4)(-x) = (-5x^4)(-1x)$
$= (-5)(-1)(x^4 \cdot x) = 5x^5$

4. $4x(x^2 + 4x + 3) = 4x(x^2) + 4x(4x) + 4x(3)$
$= 4x^3 + 16x^2 + 12x$

5. $8x(7x^4 + 1) = 8x(7x^4) + 8x(1)$
$= 56x^5 + 8x$

6. $-2x^3(3x^2 - x + 2)$
$= -2x^3(3x^2) - 2x^3(-x) - 2x^3(2)$
$= -6x^5 + 2x^4 - 4x^3$

7. $(4x + 5)(3x - 4)$
$= 4x(3x) + 4x(-4) + 5(3x) + 5(-4)$
$= 12x^2 - 16x + 15x - 20 = 12x^2 - x - 20$

8. $(3x - 2y)^2 = (3x - 2y)(3x - 2y)$
$= 3x(3x) + 3x(-2y) - 2y(3x) - 2y(-2y)$
$= 9x^2 - 6xy - 6xy + 4y^2$
$= 9x^2 - 12xy + 4y^2$

9. $(x + 3)(2x^2 - 5x + 4) = x(2x^2) + x(-5x) + x(4) + 3(2x^2) + 3(-5x) + 3(4)$
$= 2x^3 - 5x^2 + 4x + 6x^2 - 15x + 12 = 2x^3 + x^2 - 11x + 12$

10.
$$\begin{array}{r} y^2 - 4y + 5 \\ 3y^2 \qquad\quad + 1 \\ \hline y^2 - 4y + 5 \\ 3y^4 - 12y^3 + 15y^2 \\ \hline 3y^4 - 12y^3 + 16y^2 - 4y + 5 \end{array}$$

11.
$$\begin{array}{r} 4x^2 \quad - x - 1 \\ 3x^2 + 6x - 2 \\ \hline -8x^2 + 2x + 2 \\ 24x^3 - 6x^2 - 6x \\ 12x^4 - 3x^3 - 3x^2 \\ \hline 12x^4 + 21x^3 - 17x^2 - 4x + 2 \end{array}$$

Mental Math

1. $x^3 \cdot x^5 = x^8$

2. $x^2 \cdot x^6 = x^8$

3. $y^4 y = y^5$

4. $y^9 \cdot y = y^{10}$

5. $x^7 \cdot x^7 = x^{14}$

6. $x^{11} \cdot x^{11} = x^{22}$

Exercise Set 3.5

1. $8x^2 \cdot 3x = (8 \cdot 3)\left(x^2 \cdot x\right) = 24x^3$

3. $\left(-3.1x^3\right)\left(4x^9\right) = (-3.1 \cdot 4)\left(x^3 \cdot x^9\right)$
$= -12.4x^{12}$

5. $\left(-x^3\right)(-x) = (-1)(-1)\left(x^3 \cdot x\right) = x^4$

7. $\left(-\frac{1}{3}y^2\right)\left(\frac{2}{5}y\right) = \left(-\frac{1}{3} \cdot \frac{2}{5}\right)\left(y^2 \cdot y\right) = -\frac{2}{15}y^3$

9. $(2x)\left(-3x^2\right)\left(4x^5\right) = (2)(-3)(4)\left(x \cdot x^2 \cdot x^5\right)$
$\quad = -24x^8$

11. $3x(2x + 5) = 3x(2x) + 3x(5) = 6x^2 + 15x$

13. $7x\left(x^2 + 2x - 1\right)$
$\quad = 7x\left(x^2\right) + 7x(2x) + 7x(-1)$
$\quad = 7x^3 + 14x^2 - 7x$

15. $-2a(a + 4) = -2a(a) + (-2a)(4)$
$\quad = -2a^2 - 8a$

17. $3x\left(2x^2 - 3x + 4\right)$
$\quad = 3x\left(2x^2\right) + 3x(-3x) + 3x(4)$
$\quad = 6x^3 - 9x^2 + 12x$

19. $3a\left(a^2 + 2\right) = 3a\left(a^2\right) + 3a(2) = 3a^3 + 6a$

21. $-2a^2\left(3a^2 - 2a + 3\right)$
$\quad = -2a^2\left(3a^2\right) - 2a^2(-2a) - 2a^2(3)$
$\quad = -6a^4 + 4a^3 - 6a^2$

23. $3x^2y\left(2x^3 - x^2y^2 + 8y^3\right)$
$\quad = 3x^2y\left(2x^3\right) + 3x^2y\left(-x^2y^2\right) + 3x^2y\left(8y^3\right)$
$\quad = 6x^5y - 3x^4y^3 + 24x^2y^4$

25. $x^2 + 3x = x(x + 3)$

27. $(x + 4)(x + 3) = x(x) + x(3) + 4(x) + 4(3)$
$\quad = x^2 + 3x + 4x + 12 = x^2 + 7x + 12$

29. $(a + 7)(a - 2) = a(a) + a(-2) + 7(a) + 7(-2)$
$\quad = a^2 - 2a + 7a - 14 = a^2 + 5a - 14$

31. $\left(x + \frac{2}{3}\right)\left(x - \frac{1}{3}\right)$
$\quad = x(x) + x\left(-\frac{1}{3}\right) + \frac{2}{3}(x) + \frac{2}{3}\left(-\frac{1}{3}\right)$
$\quad = x^2 - \frac{1}{3}x + \frac{2}{3}x - \frac{2}{9} = x^2 + \frac{1}{3}x - \frac{2}{9}$

33. $\left(3x^2 + 1\right)\left(4x^2 + 7\right)$
$\quad = 3x^2\left(4x^2\right) + 3x^2(7) + 1\left(4x^2\right) + 1(7)$
$\quad = 12x^4 + 21x^2 + 4x^2 + 7$
$\quad = 12x^4 + 25x^2 + 7$

35. $(4x - 3)(3x - 5)$
$\quad = 4x(3x) + 4x(-5) - 3(3x) - 3(-5)$
$\quad = 12x^2 - 20x - 9x + 15$
$\quad = 12x^2 - 29x + 15$

37. $(1 - 3a)(1 - 4a)$
$\quad = 1(1) + 1(-4a) - 3a(1) - 3a(-4a)$
$\quad = 1 - 4a - 3a + 12a^2$
$\quad = 1 - 7a + 12a^2$

39. $(2y - 4)^2 = (2y - 4)(2y - 4)$
$\quad = 2y(2y) + 2y(-4) - 4(2y) - 4(-4)$
$\quad = 4y^2 - 8y - 8y + 16$
$\quad = 4y^2 - 16y + 16$

41. $(x - 2)\left(x^2 - 3x + 7\right)$
$\quad = x\left(x^2\right) + x(-3x) + x(7) - 2\left(x^2\right) - 2(-3x) - 2(7)$
$\quad = x^3 - 3x^2 + 7x - 2x^2 + 6x - 14$
$\quad = x^3 - 5x^2 + 13x - 14$

43. $(x+5)\left(x^3-3x+4\right)=x\left(x^3\right)+x(-3x)+x(4)+5\left(x^3\right)+5(-3x)+5(4)$

$\quad=x^4-3x^2+4x+5x^3-15x+20$

$\quad=x^4+5x^3-3x^2-11x+20$

45. $(2a-3)\left(5a^2-6a+4\right)=2a\left(5a^2\right)+2a(-6a)+2a(4)-3\left(5a^2\right)-3(-6a)-3(4)$

$\quad=10a^3-12a^2+8a-15a^2+18a-12$

$\quad=10a^3-27a^2+26a-12$

47. $(7xy-y)^2=(7xy-y)(7xy-y)$

$\quad=7xy(7xy)+7xy(-y)-y(7xy)-y(-y)$

$\quad=49x^2y^2-7xy^2-7xy^2+y^2$

$\quad=49x^2y^2-14xy^2+y^2$

49. $x^2+2x+3x+2(3)=x^2+5x+6$

51.

$$\begin{array}{r} 2x-11 \\ 6x+1 \\ \hline 2x-11 \\ 12x^2-66x \\ \hline 12x^2-64x-11 \end{array}$$

53.

$$\begin{array}{r} 2x^2+4x-1 \\ x+3 \\ \hline 6x^2+12x-3 \\ 2x^3+\ 4x^2-\ \ x \\ \hline 2x^3+10x^2+11x-3 \end{array}$$

55.

$$\begin{array}{r} x^2\ \ +5x\ -7 \\ x^2\ \ -7x\ -9 \\ \hline -9x^2-45x+63 \\ -7x^3-35x^2+49x \\ x^4+5x^3-\ 7x^2 \\ \hline x^4-2x^3-51x^2\ +4x\ +63 \end{array}$$

57. $(5x)^2=5^2\,x^2=25x^2$

59. $\left(-3y^3\right)^2=(-3)^2\,y^{3\cdot2}=9y^6$

61. At $t=0$, value $=\$7000$

63. At $t=0$, value $=\$7000$
At $t=1$, value $=\$6500$
$\$7000-\$6500=\$500$

65. Answers may vary.

67. $(2x-5)(2x+5)$

$\quad=2x(2x)+2x(5)-5(2x)-5(5)$

$\quad=4x^2+10x-10x-25=4x^2-25$

$\quad\left(4x^2-25\right)$ square yards

69. $\frac{1}{2}(3x-2)(4x)=2x(3x-2)$

$\quad=2x(3x)+2x(-2)=6x^2-4x$

$\quad\left(6x^2-4x\right)$ square inches

71. a. $(3x+5)+(3x+7)=3x+5+3x+7$
$\quad=6x+12$
Answers may vary.

b. $(3x+5)(3x+7)$
$\quad=3x(3x)+3x(7)+5(3x)+5(7)$
$\quad=9x^2+21x+15x+35$
$\quad=9x^2+36x+35$
Answers may vary.

73. a. $(a+b)(a-b)=a^2-ab+ab-b^2$
$\quad=a^2-b^2$

b. $(2x+3y)(2x-3y)$
$\quad=(2x)^2-6xy+6xy-(3y)^2$
$\quad=4x^2-9y^2$

c. $(4x+7)(4x-7)$
$\quad=(4x)^2-28x+28x-7^2$
$\quad=16x^2-49$
$\quad(x+y)(x-y)=x^2-y^2$

Section 3.6

Practice Problems

1. $(x + 7)(x - 5)$
$= (x)(x) + (x)(-5) + (7)(x) + (7)(-5)$
$= x^2 - 5x + 7x - 35 = x^2 + 2x - 35$

2. $(6x - 1)(x - 4)$
$= 6x(x) + 6x(-4) + (-1)(x) + (-1)(-4)$
$= 6x^2 - 24x - x + 4$
$= 6x^2 - 25x + 4$

3. $(2y^2 + 3)(y - 4)$
$= (2y^2)(y) + (2y^2)(-4) + (3)(y) + (3)(-4)$
$= 2y^3 - 8y^2 + 3y - 12$

4. $(2x + 9)^2 = (2x + 9)(2x + 9)$
$= 2x(2x) + 2x(9) + 9(2x) + 9(9)$
$= 4x^2 + 18x + 18x + 81$
$= 4x^2 + 36x + 81$

5. $(y + 3)^2 = y^2 + 2(y)(3) + 3^2$
$= y^2 + 6y + 9$

6. $(r - s)^2 = r^2 - 2rs + s^2$

7. $(6x + 5)^2 = (6x)^2 + 2(6x)(5) + 5^2$
$= 36x^2 + 60x + 25$

8. $(x^2 - 3y)^2 = (x^2)^2 - 2(x^2)(3y) + (3y)^2$
$= x^4 - 6x^2 y + 9y^2$

9. $(x + 7)(x - 7) = x^2 - 7^2 = x^2 - 49$

10. $(4y + 5)(4y - 5) = (4y)^2 - 5^2$
$= 16y^2 - 25$

11. $\left(x - \dfrac{1}{3}\right)\left(x + \dfrac{1}{3}\right) = x^2 - \left(\dfrac{1}{3}\right)^2 = x^2 - \dfrac{1}{9}$

12. $(3a - b)(3a + b) = (3a)^2 - b^2 = 9a^2 - b^2$

13. $(2x^2 - 6y)(2x^2 + 6y) = (2x^2)^2 - (6y)^2$
$= 4x^4 - 36y^2$

14. $(7x - 1)^2 = (7x)^2 - 2(7x)(1) + 1^2$
$= 49x^2 - 14x + 1$

15. $(5y + 3)(2y - 5)$
$= 5y(2y) + 5y(-5) + 3(2y) + 3(-5)$
$= 10y^2 - 25y + 6y - 15$
$= 10y^2 - 19y - 15$

16. $(2a - 1)(2a + 1) = (2a)^2 - 1^2 = 4a^2 - 1$

Exercise Set 3.6

1. $(x + 3)(x + 4) = x^2 + 4x + 3x + 12$
$= x^2 + 7x + 12$

3. $(x - 5)(x + 10) = x^2 + 10x - 5x - 50$
$= x^2 + 5x - 50$

5. $(5x - 6)(x + 2) = 5x^2 + 10x - 6x - 12$
$= 5x^2 + 4x - 12$

7. $(y - 6)(4y - 1) = 4y^2 - 1y - 24y + 6$
$= 4y^2 - 25y + 6$

9. $(2x + 5)(3x - 1) = 6x^2 - 2x + 15x - 5$
$= 6x^2 + 13x - 5$

11. $\left(y^2 + 7\right)(6y + 4) = 6y^3 + 4y^2 + 42y + 28$

13. $\left(x - \dfrac{1}{3}\right)\left(x + \dfrac{2}{3}\right) = x^2 + \dfrac{2}{3}x - \dfrac{1}{3}x - \dfrac{2}{9}$
$= x^2 + \dfrac{1}{3}x - \dfrac{2}{9}$

15. $(4 - 3a)(2 - 5a) = 8 - 20a - 6a + 15a^2$
$= 8 - 26a + 15a^2$

17. $(x + 5y)(2x - y) = 2x^2 - xy + 10xy - 5y^2$
$= 2x^2 + 9xy - 5y^2$

19. $(x+2)^2 = x^2 + 2(x)(2) + 2^2$
$\qquad = x^2 + 4x + 4$

21. $(2x-1)^2 = (2x)^2 - 2(2x)(1) + (1)^2$
$\qquad = 4x^2 - 4x + 1$

23. $(3a-5)^2 = (3a)^2 - 2(3a)(5) + 5^2$
$\qquad = 9a^2 - 30a + 25$

25. $\left(x^2 + 5\right)^2 = \left(x^2\right)^2 + 2\left(x^2\right)(5) + 5^2$
$\qquad = x^4 + 10x^2 + 25$

27. $\left(y - \frac{2}{7}\right)^2 = y^2 - 2(y)\left(\frac{2}{7}\right) + \left(\frac{2}{7}\right)^2$
$\qquad = y^2 - \frac{4}{7}y + \frac{4}{49}$

29. $(2a-3)^2 = (2a)^2 - 2(2a)(3) + 3^2$
$\qquad = 4a^2 - 12a + 9$

31. $(5x+9)^2 = (5x)^2 + 2(5x)(9) + 9^2$
$\qquad = 25x^2 + 90x + 81$

33. $(3x-7y)^2 = (3x)^2 - 2(3x)(7y) + (7y)^2$
$\qquad = 9x^2 - 42xy + 49y^2$

35. $(4m+5n)^2 = (4m)^2 + 2(4m)(5n) + (5n)^2$
$\qquad = 16m^2 + 40mn + 25n^2$

37. Answers may vary.

39. $(a-7)(a+7) = a^2 - 7^2 = a^2 - 49$

41. $(x+6)(x-6) = x^2 - 6^2 = x^2 - 36$

43. $(3x-1)(3x-1) = (3x)^2 - 1^2 = 9x^2 - 1$

45. $\left(x^2 + 5\right)\left(x^2 - 5\right) = \left(x^2\right)^2 - 5^2 = x^4 - 25$

47. $\left(2y^2 - 1\right)\left(2y^2 + 1\right) = \left(2y^2\right)^2 - 1^2 = 4y^4 - 1$

49. $(4-7x)(4+7x) = 4^2 - (7x)^2 = 16 - 49x^2$

51. $\left(3x - \frac{1}{2}\right)\left(3x + \frac{1}{2}\right) = (3x)^2 - \left(\frac{1}{2}\right)^2 = 9x^2 - \frac{1}{4}$

53. $(9x+y)(9x-y) = (9x)^2 - y^2 = 81x^2 - y^2$

55. $(2m+5n)(2m-5n) = (2m)^2 - (5n)^2$
$\qquad = 4m^2 - 25n^2$

57. $(a+5)(a+4) = a^2 + 4a + 5a + 20$
$\qquad = a^2 + 9a + 20$

59. $(a-7)^2 = a^2 - 2(a)(7) + 7^2$
$\qquad = a^2 - 14a + 49$

61. $(4a+1)(3a-1) = 12a^2 - 4a + 3a - 1$
$\qquad = 12a^2 - a - 1$

63. $(x+2)(x-2) = x^2 - 2^2 = x^2 - 4$

65. $(3a+1)^2 = (3a)^2 + 2(3a)(1) + 1^2$
$\qquad = 9a^2 + 6a + 1$

67. $(x+y)(4x-y) = 4x^2 - xy + 4xy - y^2$
$\qquad = 4x^2 + 3xy - y^2$

69. $\left(a - \frac{1}{2}y\right)\left(a + \frac{1}{2}y\right) = a^2 - \left(\frac{1}{2}y\right)^2$
$\qquad = a^2 - \frac{1}{4}y^2$

71. $(3b+7)(2b-5) = 6b^2 - 15b + 14b - 35$
$\qquad = 6b^2 - b - 35$

73. $\left(x^2 + 10\right)\left(x^2 - 10\right) = \left(x^2\right)^2 - (10)^2$
$\qquad = x^4 - 100$

75. $(4x+5)(4x-5) = (4x)^2 - 5^2$
$\qquad = 16x^2 - 25$

77. $(5x-6y)^2 = (5x)^2 - 2(5x)(6y) + (6y)^2$
$\qquad = 25x^2 - 60xy + 36y^2$

79. $(2r - 3s)(2r + 3s) = (2r)^2 - (3s)^2$
$$= 4r^2 - 9s^2$$

81. $\dfrac{50b^{10}}{70b^5} = \dfrac{10 \cdot 5 \cdot b^5 \cdot b^5}{10 \cdot 7 \cdot b^5} = \dfrac{5b^5}{7}$

83. $\dfrac{8a^{17}b^5}{-4a^7b^{10}} = \dfrac{4 \cdot 2 \cdot a^7 \cdot a^{10} \cdot b^5}{4 \cdot (-1) \cdot a^7 \cdot b^5 \cdot b^5} = -\dfrac{2a^{10}}{b^5}$

85. $\dfrac{2x^4 y^{12}}{3x^4 y^4} = \dfrac{2 \cdot x^4 \cdot y^4 \cdot y^8}{3 \cdot x^4 \cdot y^4} = \dfrac{2y^8}{3}$

87. $(2x + 1)^2 = (2x)^2 + 2(2x)(1) + 1^2$
$$= 4x^2 + 4x + 1$$
$\left(4x^2 + 4x + 1\right)$ square feet

89. $(5x - 3)^2 - (x + 1)^2$
$$= \left[(5x)^2 - 2(5x)(3) + 3^2\right] - \left[x^2 + 2(x)(1) + 1^2\right]$$

$$= \left(25x^2 - 30x + 9\right) - \left(x^2 + 2x + 1\right)$$
$$= 25x^2 - 30x + 9 - x^2 - 2x - 1$$
$$= \left(24x^2 - 32x + 8\right) \text{ square meters}$$

Chapter 3 Integrated Review

1. $\left(5x^2\right)\left(7x^3\right)$
$$= (5 \cdot 7)\left(x^2 \cdot x^3\right)$$
$$= 35x^5$$

2. $\left(4y^2\right)\left(8y^7\right)$
$$= (4 \cdot 8)\left(y^2 \cdot y^7\right)$$
$$= 32y^9$$

3. $(x - 5)(2x + 1)$
$$= x(2x) + x(1) - 5(2x) - 5(1)$$
$$= 2x^2 + x - 10x - 5$$
$$= 2x^2 - 9x - 5$$

4. $(3x - 2)(x + 5)$
$$= 3x(x) + 3x(5) - 2(x) - 2(5)$$
$$= 3x^2 + 15x - 2x - 10$$
$$= 3x^2 + 13x - 10$$

5. $(x - 5) + (2x + 1)$
$$= x - 5 + 2x + 1$$
$$= 3x - 4$$

6. $(3x - 2) + (x + 5)$
$$= 3x - 2 + x + 5$$
$$= 4x + 3$$

7. $(4y - 3)(4y + 3)$
$$= (4y)^2 - 3^2$$
$$= 16y^2 - 9$$

8. $(7x - 1)(7x + 1)$
$$= (7x)^2 - 1^2$$
$$= 49x^2 - 1$$

9. $\left(7x^2 - 2x + 3\right) - \left(5x^2 + 9\right)$
$$= 7x^2 - 2x + 3 - 5x^2 - 9$$
$$= 2x^2 - 2x - 6$$

10. $\left(10x^2 + 7x - 9\right) - \left(4x^2 - 6x + 2\right)$
$$= 10x^2 + 7x - 9 - 4x^2 + 6x - 2$$
$$= 6x^2 + 13x - 11$$

11. $(x + 4)^2$
$$= (x + 4)(x + 4)$$
$$= x^2 + 2(x)(4) + 4^2$$
$$= x^2 + 8x + 16$$

12. $(y - 9)^2$
$$= (y - 9)(y - 9)$$
$$= y^2 - 2(y)(9) + 9^2$$
$$= y^2 - 18y + 81$$

13. $(x-3)\left(x^2+5x-1\right)$

$= x\left(x^2\right)+x(5x)+x(-1)-3\left(x^2\right)-3(5x)-3(-1)$

$= x^3+5x^2-x-3x^2-15x+3$

$= x^3+2x^2-16x+3$

14. $(x+1)\left(x^2-3x-2\right)$

$= x\left(x^2\right)+x(-3x)+x(-2)+1\left(x^2\right)+1(-3x)+1(-2)$

$= x^3-3x^2-2x+x^2-3x-2$

$= x^3-2x^2-5x-2$

Section 3.7

Practice Problems

1. $\dfrac{25x^3+5x^2}{5x^2}=\dfrac{25x^3}{5x^2}+\dfrac{5x^2}{5x^2}=5x+1$

2. $\dfrac{30x^7+10x^2-5x}{5x^2}=\dfrac{30x^7}{5x^2}+\dfrac{10x^2}{5x^2}-\dfrac{5x}{5x^2}$

$= 6x^5+2-\dfrac{1}{x}$

3. $\dfrac{12x^3y^3-18xy+6y}{3xy}$

$= \dfrac{12x^3y^3}{3xy}-\dfrac{18xy}{3xy}+\dfrac{6y}{3xy}$

$= 4x^2y^2-6+\dfrac{2}{x}$

4.
$$
\begin{array}{r}
x+7 \\
x+5 \overline{\smash{\big)}\ x^2+12x+35} \\
\underline{x^2+\ 5x} \\
7x+35 \\
\underline{7x+35} \\
0
\end{array}
$$

The quotient is $x+7$.

5.
$$
\begin{array}{r}
3x+5 \\
2x-1 \overline{\smash{\big)}\ 6x^2+7x-5} \\
\underline{6x^2-3x} \\
10x-5 \\
\underline{10x-5} \\
0
\end{array}
$$

The quotient is $3x+5$.

6.
$$
\begin{array}{r}
3x^2-2x+1 \\
3x+2 \overline{\smash{\big)}\ 9x^3+0x^2-\ x+5} \\
\underline{9x^3+6x^2} \\
-6x^2-\ x \\
\underline{-6x^2-4x} \\
3x+5 \\
\underline{3x+2} \\
3
\end{array}
$$

$\dfrac{5-x+9x^3}{3x+2}=3x^2-2x+1+\dfrac{3}{3x+2}$

Mental Math

1. $\dfrac{a^6}{a^4}=a^2$

2. $\dfrac{y^2}{y}=y$

3. $\dfrac{a^3}{a}=a^2$

4. $\dfrac{p^8}{p^3} = p^5$

5. $\dfrac{k^5}{k^2} = k^3$

6. $\dfrac{k^7}{k^5} = k^2$

Exercise Set 3.7

1. $\dfrac{20x^2 + 5x + 9}{5} = \dfrac{20x^2}{5} + \dfrac{5x}{5} + \dfrac{9}{5}$

$\qquad = 4x^2 + x + \dfrac{9}{5}$

3. $\dfrac{12x^4 + 3x^2}{x} = \dfrac{12x^4}{x} + \dfrac{3x^2}{x} = 12x^3 + 3x$

5. $\dfrac{15p^3 + 18p^2}{3p} = \dfrac{15p^3}{3p} + \dfrac{18p^2}{3p} = 5p^2 + 6p$

7. $\dfrac{-9x^4 + 18x^5}{6x^5} = \dfrac{-9x^4}{6x^5} + \dfrac{18x^5}{6x^5} = -\dfrac{3}{2x} + 3$

9. $\dfrac{-9x^5 + 3x^4 - 12}{3x^3} = \dfrac{-9x^5}{3x^3} + \dfrac{3x^4}{3x^3} - \dfrac{12}{3x^3}$

$\qquad = -3x^2 + x - \dfrac{4}{x^3}$

11. $\dfrac{4x^4 - 6x^3 + 7}{-4x^4} = \dfrac{4x^4}{-4x^4} - \dfrac{6x^3}{-4x^4} + \dfrac{7}{-4x^4}$

$\qquad = -1 + \dfrac{3}{2x} - \dfrac{7}{4x^4}$

13. $\dfrac{a^2b^2 - ab^3}{ab} = \dfrac{a^2b^2}{ab} - \dfrac{ab^3}{ab} = ab - b^2$

15. $\dfrac{2x^2y + 8x^2y^2 - xy^2}{2xy}$

$\qquad = \dfrac{2x^2y}{2xy} + \dfrac{8x^2y^2}{2xy} - \dfrac{xy^2}{2xy}$

$\qquad = x + 4xy - \dfrac{y}{2}$

17.
$$\begin{array}{r} x+1 \\ x+3\,\overline{)\,x^2 + 4x + 3} \\ \underline{x^2 + 3x} \\ x+3 \\ \underline{x+3} \\ 0 \end{array}$$

$\dfrac{x^2 + 4x + 3}{x+3} = x + 1$

19.
$$\begin{array}{r} 2x+3 \\ x+5\,\overline{)\,2x^2 + 13x + 15} \\ \underline{2x^2 + 10x} \\ 3x+15 \\ \underline{3x+15} \\ 0 \end{array}$$

$\dfrac{2x^2 + 13x + 15}{x+5} = 2x + 3$

21.
$$\begin{array}{r} 2x+1 \\ x-4\,\overline{)\,2x^2 - 7x + 3} \\ \underline{2x^2 - 8x} \\ x+3 \\ \underline{x-4} \\ 7 \end{array}$$

$\dfrac{2x^2 - 7x + 3}{x-4} = 2x + 1 + \dfrac{7}{x-4}$

23.
$$\begin{array}{r} 4x+9 \\ 2x-3\,\overline{)\,8x^2 + 6x - 27} \\ \underline{8x^2 - 12x} \\ 18x-27 \\ \underline{18x-27} \\ 0 \end{array}$$

$\dfrac{8x^2 + 6x - 27}{2x-3} = 4x + 9$

25.
$$\begin{array}{r} 3a^2 - 3a + 1 \\ 3a+2\,\overline{)\,9a^3 - 3a^2 - 3a + 4} \\ \underline{9a^3 + 6a^2} \\ -9a^2 - 3a \\ \underline{-9a^2 - 6a} \\ 3a+4 \\ \underline{3a+2} \\ 2 \end{array}$$

$\dfrac{9a^3 - 3a^2 - 3a + 4}{3a+2} = 3a^2 - 3a + 1 + \dfrac{2}{3a+2}$

27.

$$
\begin{array}{r}
2b^2 + b + 2 \\
b+4\overline{\smash{\big)}\,2b^3 + 9b^2 + 6b - 4} \\
\underline{2b^3 + 8b^2} \\
b^2 + 6b \\
\underline{b^2 + 4b} \\
2b - 4 \\
\underline{2b + 8} \\
-12
\end{array}
$$

$$\frac{2b^3 + 9b^2 + 6b - 4}{b+4} = 2b^2 + b + 2 - \frac{12}{b+4}$$

29.

$$
\begin{array}{r}
4x + 3 \\
2x+1\overline{\smash{\big)}\,8x^2 + 10x + 1} \\
\underline{8x^2 + \ 4x} \\
6x + 1 \\
\underline{6x + 3} \\
-2
\end{array}
$$

$$\frac{8x^2 + 10x + 1}{2x+1} = 4x + 3 - \frac{2}{2x+1}$$

31.

$$
\begin{array}{r}
2x^2 + 6x - 5 \\
x-2\overline{\smash{\big)}\,2x^3 + 2x^2 - 17x + 8} \\
\underline{2x^3 - 4x^2} \\
6x^2 - 17x \\
\underline{6x^2 - 12x} \\
-5x + \ 8 \\
\underline{-5x + 10} \\
-2
\end{array}
$$

$$\frac{2x^3 + 2x^2 - 17x + 8}{x-2} = 2x^2 + 6x - 5 - \frac{2}{x-2}$$

33.

$$
\begin{array}{r}
x^2 + 3x + 9 \\
x-3\overline{\smash{\big)}\,x^3 + 0x^2 + 0x - 27} \\
\underline{x^3 - 3x^2} \\
3x^2 + 0x \\
\underline{3x^2 - 9x} \\
9x - 27 \\
\underline{9x - 27} \\
0
\end{array}
$$

$$\frac{x^3 - 27}{x-3} = x^2 + 3x + 9$$

35.

$$
\begin{array}{r}
-3x + 6 \\
x+2\overline{\smash{\big)}\,-3x^2 + 0x + 1} \\
\underline{-3x^2 - 6x} \\
6x + \ 1 \\
\underline{6x + 12} \\
-11
\end{array}
$$

$$\frac{1 - 3x^2}{x+2} = -3x + 6 - \frac{11}{x+2}$$

37.

$$
\begin{array}{r}
2b - 1 \\
2b-1\overline{\smash{\big)}\,4b^2 - 4b - 5} \\
\underline{4b^2 - 2b} \\
-2b - 5 \\
\underline{-2b + 1} \\
-6
\end{array}
$$

$$\frac{-4b + 4b^2 - 5}{2b-1} = 2b - 1 - \frac{6}{2b-1}$$

39. $12 = 4 \cdot 3$

41. $20 = -5 \cdot (-4)$

43. $9x^2 = 3x \cdot 3x$

45. $36x^2 = 4x \cdot 9x$

47.

$$
\begin{array}{r}
x^3 - x^2 + x \\
x^2+x\overline{\smash{\big)}\,x^5 + 0x^4 + 0x^3 + x^2} \\
\underline{x^5 + \ x^4} \\
-x^4 + 0x^3 \\
\underline{-x^4 - \ x^3} \\
x^3 + x^2 \\
\underline{x^3 + x^2} \\
0
\end{array}
$$

$$\frac{x^5 + x^2}{x^2 + x} = x^3 - x^2 + x$$

49. $\dfrac{12x^3 + 4x - 16}{4} = \dfrac{12x^3}{4} + \dfrac{4x}{4} - \dfrac{16}{4}$

$= 3x^3 + x - 4$

Each side is $\left(3x^3 + x - 4\right)$ feet

51.

$$5x+3 \overline{)\begin{array}{r} 2x+5 \\ 10x^2+31x+15 \\ \underline{10x^2+\ 6x} \\ 25\ +15 \\ \underline{25\ +15} \\ 0 \end{array}}$$

The height is $(2x + 5)$ meters.

53. Answers may vary.

Chapter 3 Review

1. 3^2
base: 3
exponent: 2

2. $(-5)^4$
base: -5
exponent: 4

3. -5^4
base: 5
exponent: 4

4. x^6
base: x
exponent: 6

5. $8^3 = 8 \cdot 8 \cdot 8 = 512$

6. $(-6)^2 = (-6)(-6) = 36$

7. $-6^2 = -6 \cdot 6 = -36$

8. $-4^3 - 4^0 = -4 \cdot 4 \cdot 4 - 1 = -65$

9. $(3b)^0 = 1$

10. $\dfrac{8b}{8b} = 1$

11. $y^2 \cdot y^7 = y^{2+7} = y^9$

12. $x^9 \cdot x^5 = x^{9+5} = x^{14}$

13. $\left(2x^5\right)\left(-3x^6\right) = 2(-3) \cdot \left(x^5 \cdot x^6\right) = -6x^{11}$

14. $\left(-5y^3\right)\left(4y^4\right) = (-5 \cdot 4)\left(y^3 \cdot y^4\right) = -20y^7$

15. $\left(x^4\right)^2 = x^{4 \cdot 2} = x^8$

16. $\left(y^3\right)^5 = y^{3 \cdot 5} = y^{15}$

17. $\left(3y^6\right)^4 = 3^4 y^{6 \cdot 4} = 81y^{24}$

18. $\left(2x^3\right)^3 = 2^3 x^{3 \cdot 3} = 8x^9$

19. $\dfrac{x^9}{x^4} = x^{9-4} = x^5$

20. $\dfrac{z^{12}}{z^5} = z^{12-5} = z^7$

21. $\dfrac{a^5 b^4}{ab} = a^{5-1} b^{4-1} = a^4 b^3$

22. $\dfrac{x^4 y^6}{xy} = x^{4-1} y^{6-1} = x^3 y^5$

23. $\dfrac{12xy^6}{3x^4 y^{10}}$
$= \dfrac{12}{3} \cdot x^{1-4} \cdot y^{6-10}$
$= 4x^{-3} y^{-4}$
$= \dfrac{4}{x^3 y^4}$

24. $\dfrac{2x^7 y^8}{8xy^2}$
$= \dfrac{2}{8} \cdot x^{7-1} y^{8-2}$
$= \dfrac{x^6 y^6}{4}$

25. $5a^7\left(2a^4\right)^3$

$= 5a^7\left(2^3 a^{4\cdot3}\right)$

$= 5a^7\left(8a^{12}\right)$

$= 5\cdot 8a^{7+12}$

$= 40a^{19}$

26. $(2x)^2(9x)$

$= \left(2^2\cdot x^2\right)(9x)$

$= 4x^2\cdot 9x$

$= 4\cdot 9\cdot x^{2+1}$

$= 36x^3$

27. $(-5a)^0 + 7^0 + 8^0$

$= 1 + 1 + 1$

$= 3$

28. $8x^0 + 9x^0$

$= 8\cdot 1 + 1$

$= 9$

29. $\left(\dfrac{3x^4}{4y}\right)^3$

$= \dfrac{3^3 x^{4\cdot3}}{4^3 y^3}$

$= \dfrac{27x^{12}}{64y^3}$

Answer: b

30. $\left(\dfrac{5a^6}{b^3}\right)^2$

$= \dfrac{5^2 a^{6\cdot2}}{b^{3\cdot2}}$

$= \dfrac{25a^{12}}{b^6}$

Answer: c

31. $7^{-2} = \dfrac{1}{7^2} = \dfrac{1}{49}$

32. $-7^{-2} = -\dfrac{1}{7^2} = -\dfrac{1}{49}$

33. $2x^{-4} = \dfrac{2}{x^4}$

34. $(2x)^{-4} = \dfrac{1}{(2x)^4} = \dfrac{1}{16x^4}$

35. $\left(\dfrac{1}{5}\right)^{-3} = \dfrac{1^{-3}}{5^{-3}} = \dfrac{5^3}{1^3} = 125$

36. $\left(\dfrac{-2}{3}\right)^{-2} = \dfrac{(-2)^{-2}}{3^{-2}} = \dfrac{3^2}{(-2)^2} = \dfrac{9}{4}$

37. $2^0 + 2^{-4} = 1 + \dfrac{1}{2^4} = \dfrac{16}{16} + \dfrac{1}{16} = \dfrac{17}{16}$

38. $6^{-1} - 7^{-1} = \dfrac{1}{6} - \dfrac{1}{7} = \dfrac{7}{42} - \dfrac{6}{42} = \dfrac{1}{42}$

39. $\dfrac{x^5}{x^{-3}} = x^{5-(-3)} = x^8$

40. $\dfrac{z^4}{z^{-4}} = z^{4-(-4)} = z^8$

41. $\dfrac{r^{-3}}{r^{-4}} = r^{-3-(-4)} = r^1 = r$

42. $\dfrac{y^{-2}}{y^{-5}} = y^{-2-(-5)} = y^3$

43. $\left(\dfrac{bc^{-2}}{bc^{-3}}\right)^4 = \dfrac{b^4 c^{-8}}{b^4 c^{-12}} = b^{4-4} c^{-8-(-12)} = c^4$

44. $\left(\dfrac{x^{-3}y^{-4}}{x^{-2}y^{-5}}\right)^{-3} = \dfrac{x^9 y^{12}}{x^6 y^{15}} = x^{9-6} y^{12-15} = \dfrac{x^3}{y^3}$

45. $\dfrac{x^{-4}y^{-6}}{x^2 y^7} = x^{-4-2} y^{-6-7} = x^{-6} y^{-13} = \dfrac{1}{x^6 y^{13}}$

46. $\dfrac{a^5 b^{-5}}{a^{-5} b^5}$

$= a^{5-(-5)} b^{-5-5}$

$= a^{10} b^{-10}$

$= \dfrac{a^{10}}{b^{10}}$

47. $0.00027 = 2.7 \times 10^{-4}$

48. $0.8868 = 8.868 \times 10^{-1}$

49. $80,800,000 = 8.08 \times 10^7$

50. $-868,000 = -8.68 \times 10^5$

51. $31,880,000 = 3.188 \times 10^7$

52. $4000 = 4.0 \times 10^3$

53. $8.67 \times 10^5 = 867,000$

54. $3.86 \times 10^{-3} = 0.00386$

55. $8.6 \times 10^{-4} = 0.00086$

56. $8.936 \times 10^5 = 893,600$

57. $1 \times 10^{20} = 100,000,000,000,000,000,000$

58. 3×10^{-25}
$= 0.0000000000000000000000003$

59. $\left(8 \times 10^4\right)\left(2 \times 10^{-7}\right)$

$= (8 \times 2) \times \left(10^4 \times 10^{-7}\right)$

$= 16 \times 10^{-3}$

$= 0.016$

60. $\dfrac{8 \times 10^4}{2 \times 10^{-7}}$

$= \dfrac{8}{2} \times \left(10^{4-(-7)}\right)$

$= 4 \times 10^{11}$

$= 400,000,000,000$

61. The degree is 5 because y^5 is the term with highest degree.

62. The degree is 2 because $9y^2$ is the term with the highest degree.

63. The degree is 5 because $-28x^2 y^3$ is the term with the highest degree.

64. The degree is 6 because $6x^2 y^2 z^2$ is the term with the highest degree.

65. $2(1)^2 + 20(1) = 22$
$2(3)^2 + 20(3) = 78$
$2(5.1)^2 + 20(5.1) = 154.02$
$2(10)^2 + 20(10) = 400$

x	1	3	5.1	10
$2x^2 + 20x$	22	78	154.02	400

66. $7a^2 - 4a^2 - a^2$
$= (7 - 4 - 1)a^2$
$= 2a^2$

67. $9y + y - 14y$
$= (9 + 1 - 14)y$
$= -4y$

68. $6a^2 - 4a + 9a^2$
$= (6 + 9)a^2 + 4a$
$= 15a^2 + 4a$

69. $21x^2 + 3x + x^2 + 6$
$= (21 + 1)x^2 + 3x + 6$
$= 22x^2 + 3x + 6$

70. $4a^2 b - 3b^2 - 8q^2 - 10a^2 b + 7q^2$
$= \left(4a^2 b - 10a^2 b\right) - 3b^2 + \left(-8q^2 + 7q^2\right)$
$= -6a^2 b - 3b^2 - q^2$

71. $2s^{14} + 3s^{13} + 12s^{12} - s^{10}$
Cannot be combined.

72. $\left(3x^2 + 2x + 6\right) + \left(5x^2 + x\right)$
$= 3x^2 + 2x + 6 + 5x^2 + x$
$= 8x^2 + 3x + 6$

73. $\left(2x^5 + 3x^4 + 4x^3 + 5x^2\right) + \left(4x^2 + 7x + 6\right)$
$= 2x^5 + 3x^4 + 4x^3 + 5x^2 + 4x^2 + 7x + 6$
$= 2x^5 + 3x^4 + 4x^3 + 9x^2 + 7x + 6$

74. $\left(-5y^2 + 3\right) - \left(2y^2 + 4\right)$
$= -5y^2 + 3 - 2y^2 - 4$
$= -7y^2 - 1$

75. $\left(2m^7 + 3x^4 + 7m^6\right) - \left(8m^7 + 4m^2 + 6x^4\right)$
$= 2m^7 + 3x^4 + 7m^6 - 8m^7 - 4m^2 - 6x^4$
$= -6m^7 - 3x^4 + 7m^6 - 4m^2$

76. $\left(3x^2 - 7xy + 7y^2\right) - \left(4x^2 - xy + 9y^2\right)$
$= 3x^2 - 7xy + 7y^2 - 4x^2 + xy - 9y^2$
$= -x^2 - 6xy - 2y^2$

77. $\left(-9x^2 + 6x + 2\right) + \left(4x^2 - x - 1\right)$
$= -9x^2 + 6x + 2 + 4x^2 - x - 1$
$= -5x^2 + 5x + 1$

78. $\left[\left(x^2 + 7x + 9\right) + \left(x^2 + 4\right)\right] - \left(4x^2 + 8x - 7\right)$
$= x^2 + 7x + 9 + x^2 + 4 - 4x^2 - 8x + 7$
$= -2x^2 - x + 20$

79. $6(x + 5)$
$= 6x + 6(5)$
$= 6x + 30$

80. $9(x - 7)$
$= 9x - 9(7)$
$= 9x - 63$

81. $4(2a + 7)$
$= 4(2a) + 4(7)$
$= 8a + 28$

82. $9(6a - 3)$
$= 9(6a) - 9(3)$
$= 54a - 27$

83. $-7x\left(x^2 + 5\right)$
$= -7\left(x^2\right) - 7x(5)$
$= -7x^3 - 35x$

84. $-8y\left(4y^2 - 6\right)$
$= -8y\left(4y^2\right) - 8y(-6)$
$= -32y^3 + 48y$

85. $-2\left(x^3 - 9x^2 + x\right)$
$= -2\left(x^3\right) - 2\left(-9x^2\right) - 2(x)$
$= -2x^3 + 18x^2 - 2x$

86. $-3a\left(a^2b + ab + b^2\right)$
$= -3a\left(a^2b\right) - 3a(ab) - 3a\left(b^2\right)$
$= -3a^3b - 3a^2b - 3ab^2$

87. $\left(3a^3 - 4a + 1\right)(-2a)$
$= 3a^3(-2a) - 4a(-2a) + 1(-2a)$
$= -6a^4 + 8a^2 - 2a$

88. $\left(6b^3 - 4b + 2\right)(7b)$
$= 6b^3(7b) - 4b(7b) + 2(7b)$
$= 42b^4 - 28b^2 + 14b$

89. $(2x + 2)(x - 7)$
$= 2x(x) + 2x(-7) + 2(x) + 2(-7)$
$= 2x^2 - 14x + 2x - 14$
$= 2x^2 - 12x - 14$

90. $(2x-5)(3x+2) = 2x(3x) + 2x(2) - 5(3x) - 5(2)$
$= 6x^2 + 4x - 15x - 10$
$= 6x^2 - 11x - 10$

91. $(4a-1)(a+7) = 4a^2 + 28a - a - 7$
$= 4a^2 + 27a - 7$

92. $(6a-1)(7a+3) = 42a^2 + 18a - 7a - 3$
$= 42a^2 + 11a - 3$

93. $(x+7)\left(x^3 + 4x - 5\right) = x^4 + 4x^2 - 5x + 7x^3 + 28x - 35$
$= x^4 + 7x^3 + 4x^2 + 23x - 35$

94. $(x+2)\left(x^5 + x + 1\right) = x^6 + x^2 + x + 2x^5 + 2x + 2$
$= x^6 + 2x^5 + x^2 + 3x + 2$

95. $\left(x^2 + 2x + 4\right)\left(x^2 + 2x - 4\right) = x^4 + 2x^3 - 4x^2 + 2x^3 + 4x^2 - 8x + 4x^2 + 8x - 16$
$= x^4 + 4x^3 + 4x^2 - 16$

96. $\left(x^3 + 4x + 4\right)\left(x^3 + 4x - 4\right) = x^6 + 4x^4 - 4x^3 + 4x^4 + 16x^2 - 16x + 4x^3 + 16x - 16$
$= x^6 + 8x^4 + 16x^2 - 16$

97. $(x+7)^3$
$= (x+7)(x+7)(x+7)$
$= \left(x^2 + 7x + 7x + 49\right)(x+7)$
$= \left(x^2 + 14x + 49\right)(x+7)$
$= x^3 + 7x^2 + 14x^2 + 98x + 49x + 343$
$= x^3 + 21x^2 + 147x + 343$

98. $(2x-5)^3$
$= (2x-5)(2x-5)(2x-5)$
$= \left(4x^2 - 10x - 10x + 25\right)(2x-5)$
$= \left(4x^2 - 20x + 25\right)(2x-5)$
$= 8x^3 - 20x^2 - 40x^2 + 100x + 50x - 125$
$= 8x^3 - 60x^2 + 150x - 125$

99. $(x+7)^2$
$= x^2 + 2(x)(7) + 7^2$
$= x^2 + 14x + 49$

100. $(x-5)^2$
$= x^2 - 2(x)(5) + 5^2$
$= x^2 - 10x + 25$

101. $(3x-7)^2$
$= (3x)^2 - 2(3x)(7) + 7^2$
$= 9x^2 - 42x + 49$

102. $(4x+2)^2$
$= (4x)^2 + 2(4x)(2) + 2^2$
$= 16x^2 + 16x + 4$

103. $(5x-9)^2$
$$= (5x)^2 - 2(5x)(9) + 9^2$$
$$= 25x^2 - 90x + 81$$

104. $(5x+1)(5x-1)$
$$= (5x)^2 - 1^2$$
$$= 25x^2 - 1$$

105. $(7x+4)(7x-4)$
$$= (7x)^2 - 4^2$$
$$= 49x^2 - 16$$

106. $(a+2b)(a-2b)$
$$= a^2 - (2b)^2$$
$$= a^2 - 4b^2$$

107. $(2x-6)(2x+6)$
$$= (2x)^2 - 6^2$$
$$= 4x^2 - 36$$

108. $\left(4a^2 - 2b\right)\left(4a^2 + 2b\right)$
$$= \left(4a^2\right)^2 - (2b)^2$$
$$= 16a^4 - 4b^2$$

109. $\dfrac{x^2 + 21x + 49}{7x^2}$
$$= \frac{x^2}{7x^2} + \frac{21x}{7x^2} + \frac{49}{7x^2}$$
$$= \frac{1}{7} + \frac{3}{x} + \frac{7}{x^2}$$

110. $\dfrac{5a^3b - 15ab^2 + 20ab}{-5ab}$
$$= \frac{5a^3b}{-5ab} - \frac{15ab^2}{-5ab} + \frac{20ab}{-5ab}$$
$$= -a^2 + 3b - 4$$

111.
$$\begin{array}{r}
a+1 \\
a-2\overline{\smash{\big)}\,a^2 -\ a+4} \\
\underline{a^2 - 2a} \\
a+4 \\
\underline{a-2} \\
6
\end{array}$$
$$\left(a^2 - a + 4\right) \div (a-2) = a + 1 + \frac{6}{a-2}$$

112.
$$\begin{array}{r}
4x \\
x+5\overline{\smash{\big)}\,4x^2 + 20x + 7} \\
\underline{4x^2 + 20x} \\
7
\end{array}$$
$$(4x^2 + 20x + 7) \div (x+5) = 4x + \frac{7}{x+5}$$

113.
$$\begin{array}{r}
a^2 + 3a + 8 \\
a-2\overline{\smash{\big)}\,a^3 +\ a^2 + 2a + 6} \\
\underline{a^3 - 2a^2} \\
3a^2 + 2a \\
\underline{3a^2 - 6a} \\
8a + 6 \\
\underline{8a - 16} \\
22
\end{array}$$
$$\frac{a^3 + a^2 + 2a + 6}{a-2} = a^2 + 3a + 8 + \frac{22}{a-2}$$

114.
$$\begin{array}{r}
3b^2 - 4b \\
3b-2\overline{\smash{\big)}\,9b^3 - 18b^2 + 8b - 1} \\
\underline{9b^3 -\ 6b^2} \\
-12b^2 + 8b \\
\underline{-12b^2 + 8b} \\
-1
\end{array}$$
$$\frac{9b^3 - 18b^2 + 8b - 1}{3b-2} = 3b^2 - 4b - \frac{1}{3b-2}$$

115.

$$2x-1 \overline{)\,4x^4 - 4x^3 + x^2 + 4x - 3\,}$$

quotient: $2x^3 - x^2 + 2$

$$\underline{4x^4 - 2x^3}$$
$$-2x^3 + x^2$$
$$\underline{-2x^2 + x^2}$$
$$4x - 3$$
$$\underline{4x - 2}$$
$$-1$$

$$\frac{4x^4 - 4x^3 + x^2 + 4x - 3}{2x - 1}$$
$$= 2x^3 - x^2 + 2 - \frac{1}{2x - 1}$$

116.

$$x-6 \overline{)\,-x^3 - 10x^2 - 21x + 18\,}$$

quotient: $-x^2 - 16x - 117$

$$\underline{-x^3 + 6x^2}$$
$$-16x^2 - 21x$$
$$\underline{-16x^2 + 96x}$$
$$-117x + 18$$
$$\underline{-117x + 702}$$
$$-684$$

$$\frac{-10x^2 - x^3 - 21x + 18}{x - 6}$$
$$= -x^2 - 16x - 117 - \frac{684}{x - 6}$$

Chapter 3 Test

1. $2^5 = 2 \cdot 2 \cdot 2 \cdot 2 \cdot 2 = 32$

2. $(-3)^4 = (-3)(-3)(-3)(-3) = 81$

3. $-3^4 = -3 \cdot 3 \cdot 3 \cdot 3 = -81$

4. $4^{-3} = \frac{1}{4^3} = \frac{1}{64}$

5. $\left(3x^2\right)\left(-5x^9\right)$
$$= (3)(-5)\left(x^2 \cdot x^9\right)$$
$$= -15x^{11}$$

6. $\frac{y^7}{y^2} = y^{7-2} = y^5$

7. $\frac{r^{-8}}{r^{-3}} = r^{-8-(-3)} = r^{-5} = \frac{1}{r^5}$

8. $\left(\frac{x^2 y^3}{x^3 y^{-4}}\right)^2$
$$= \frac{x^4 y^6}{x^6 y^{-8}}$$
$$= x^{4-6} y^{6-(-8)}$$
$$= x^{-2} y^{14}$$
$$= \frac{y^{14}}{x^2}$$

9. $\frac{6^2 x^{-4} y^{-1}}{6^3 x^{-3} y^7}$
$$= 6^{2-3} x^{-4-(-3)} y^{-1-7}$$
$$= 6^{-1} x^{-1} y^{-8}$$
$$= \frac{1}{6xy^8}$$

10. $563,000 = 5.63 \times 10^5$

11. $0.0000863 = 8.63 \times 10^{-5}$

12. $1.5 \times 10^{-3} = 0.0015$

13. $6.23 \times 10^4 = 62,300$

14. $\left(1.2 \times 10^5\right)\left(3 \times 10^{-7}\right)$
$$= (1.2)(3) \times 10^{5-7}$$
$$= 3.6 \times 10^{-2}$$
$$= 0.036$$

15. The degree is 5 because $9x^3 yz$ or $9x^3 y^1 z^1$ is the term with the highest degree.

16. $5x^2 + 4x - 7x^2 + 11 + 8x$
$$= \left(5x^2 - 7x^2\right) + (4x + 8x) + 11$$
$$= -2x^2 + 12x + 11$$

17. $\left(8x^3 + 7x^2 + 4x - 7\right) + \left(8x^3 - 7x - 6\right)$

$= 8x^3 + 7x^2 + 4x - 7 + 8x^3 - 7x - 6$

$= 16x^3 + 7x^2 - 3x - 13$

18.
$$5x^3 + x^2 + 5x - 2$$
$$\underline{-\left(8x^3 - 4x^2 + x - 7\right)}$$

$$5x^3 + x^2 + 5x - 2$$
$$\underline{-8x^3 + 4x^2 - x + 7}$$
$$-3x^3 + 5x^2 + 4x + 5$$

19. $\left[\left(8x^2 + 7x + 5\right) + \left(x^3 - 8\right)\right] - (4x + 2)$

$= 8x^2 + 7x + 5 + x^3 - 8 - 4x - 2$

$= x^3 + 8x^2 + 3x - 5$

20. $(3x + 7)\left(x^2 + 5x + 2\right)$

$= 3x^3 + 15x^2 + 6x + 7x^2 + 35x + 14$

$= 3x^3 + 22x^2 + 41x + 14$

21.
$$x^3 \quad - x^2 \quad + x + 1$$
$$\underline{2x^2 - 3x + 7}$$
$$7x^3 - 7x^2 + 7x + 7$$
$$-3x^4 + \ 3x^3 - 3x^2 \ - 3x$$
$$\underline{2x^5 - 2x^4 + \ 2x^3 + 2x^2}$$
$$2x^5 - 5x^4 + 12x^3 - 8x^2 + 4x + 7$$

22. $(x + 7)(3x - 5)$

$= 3x^2 - 5x + 21x - 35$

$= 3x^2 + 16x - 35$

23. $(3x - 7)(3x + 7)$

$= (3\hat{x})^2 - 7^2$

$= 9x^2 - 49$

24. $(4x - 2)^2$

$= (4x)^2 - 2(4x)(2) + 2^2$

$= 16x^2 - 16x + 4$

25. $(8x + 3)^2$

$= (8x)^2 + 2(8x)(3) + 3^2$

$= 64x^2 + 48x + 9$

26. $\left(x^2 - 9b\right)\left(x^2 + 9b\right)$

$= \left(x^2\right)^2 - (9b)^2$

$= x^4 - 81b^2$

27. $-16(0)^2 + 1001 = 1001$

$-16(1)^2 + 1001 = 985$

$-16(3)^2 + 1001 = 857$

$-16(5)^2 + 1001 = 601$

t	0 seconds	1 second	3 seconds	5 seconds
$-16t^2 + 1001$	1001 ft	985 ft	857 ft	601 ft

Answer only: 1001, 985, 857, 601

28. $\dfrac{4x^2 + 2xy - 7x}{8xy}$

$= \dfrac{4x^2}{8xy} + \dfrac{2xy}{8xy} - \dfrac{7x}{8xy}$

$= \dfrac{x}{2y} + \dfrac{1}{4} - \dfrac{7}{8y}$

29.
$$
\begin{array}{r}
x+2 \\
x+5{\overline{\smash{\big)}\,x^2 + 7x + 10 }} \\
\underline{x^2 + 5x } \\
2x + 10 \\
\underline{2x + 10} \\
0
\end{array}
$$

$\left(x^2 + 7x + 10\right) \div (x+5) = x+2$

30.
$$
\begin{array}{r}
9x^2 - 6x + 4 \\
3x+2{\overline{\smash{\big)}\,27x^3 + 0x^2 + 0x - 8}} \\
\underline{27x^3 + 18x^2 } \\
-18x^2 + 0x \\
\underline{-18x^2 - 12x } \\
12x - 8 \\
\underline{12x + 8} \\
-16
\end{array}
$$

$\dfrac{27x^3 - 8}{3x + 2} = 9x^2 - 6x + 4 - \dfrac{16}{3x + 2}$

Chapter 3 Cumulative Review

1. a. The natural numbers are 11 and 112.

b. The whole numbers are 0, 11, and 112.

c. The integers are −3, −2, 0, 11, and 112.

d. The rational numbers are −3, −2, 0, $\dfrac{1}{4}$, 11, and 112.

e. The irrational number is $\sqrt{2}$.

f. The real numbers are all the numbers in the given set.

2. a. $3^2 = 3 \cdot 3 = 9$

b. $5^3 = 5 \cdot 5 \cdot 5 = 125$

c. $2^4 = 2 \cdot 2 \cdot 2 \cdot 2 = 16$

d. $7^1 = 7$

e. $\left(\dfrac{3}{7}\right)^2 = \left(\dfrac{3}{7}\right)\left(\dfrac{3}{7}\right) = \dfrac{9}{49}$

3. $\dfrac{3}{2} \cdot \dfrac{1}{2} - \dfrac{1}{2} = \dfrac{3 \cdot 1}{2 \cdot 2} - \dfrac{1}{2} = \dfrac{3}{4} - \dfrac{1}{2}$

$= \dfrac{3}{4} - \dfrac{1 \cdot 2}{2 \cdot 2} = \dfrac{3}{4} - \dfrac{2}{4} = \dfrac{3-2}{4} = \dfrac{1}{4}$

4. a. $x + 3$

b. $3x$

c. $2x$

d. $10 - x$

e. $5x + 7$

5. $11.4 + (-4.7) = 6.7$

6. a. $\dfrac{x - y}{12 + x} = \dfrac{2 - (-5)}{12 + 2} = \dfrac{2 + 5}{12 + 2}$

$= \dfrac{7}{14} = \dfrac{7}{2 \cdot 7} = \dfrac{1}{2}$

b. $x^2 - y = 2^2 - (-5) = 4 - (-5)$
$= 4 + 5 = 9$

7. $\dfrac{-30}{-10} = \dfrac{-3 \cdot 2 \cdot 5}{-2 \cdot 5} = 3$

8. $\dfrac{42}{-0.6} = -70$

9. $5(x + 2) = 5(x) + 5(2) = 5x + 10$

10. $-2(y + 0.3z - 1)$
$= -2(y) + (-2)(0.3z) - (-2)(1)$
$= -2y - 0.6z + 2$

11. $-(x + y - 2z + 6) = -1(x + y - 2z + 6)$
$= -1(x) + (-1)(y) - (-1)(2z) + (-1)(6)$
$= -x - y + 2z - 6$

12. $6(2a-1)-(11a+6)=7$
$12a-6-11a-6=7$
$12a-11a-6-6=7$
$a-12=7$
$a-12+12=7+12$
$a=19$

13. $\dfrac{y}{7}=20$

$7\left(\dfrac{y}{7}\right)=7(20)$

$y=140$

14. $0.25x+0.10(x-3)=0.05(22)$
$100[0.25x+0.10(x-3)]=100[0.05(22)]$
$25x+10(x-3)=5(22)$
$25x+10x-30=110$
$35x-30=110$
$35x-30+30=110+30$
$35x=140$
$\dfrac{35x}{35}=\dfrac{140}{35}$
$x=4$

15. Let $x=$ the unknown number.
$2(x+4)=4x-12$
$2x+8=4x-12$
$2x+8-2x=4x-12-2x$
$8=2x-12$
$8+12=2x-12+12$
$20=2x$
$\dfrac{20}{2}=\dfrac{2x}{2}$
$10=x$
The number is 10.

16. The perimeter of a rectangle is given by the formula $P=2l+2w$. Let
$l=$ the length of the garden.
$P=2l+2w$
$140=2l+2(30)$
$140=2l+60$
$140-60=2l+60-60$
$80=2l$
$\dfrac{80}{2}=\dfrac{2l}{2}$
$40=l$
The length of the garden is 40 feet.

17. Let $x=$ the unknown number.
$120=(0.15)(x)$
$\dfrac{120}{0.15}=\dfrac{0.15x}{0.15}$
$800=x$

18. $-4x+7\geq-9$
$-4x+7-7\geq-9-7$
$-4x\geq-16$
$\dfrac{-4x}{-4}\leq\dfrac{-16}{-4}$
$x\leq4$

19. a. $x^7\cdot x^4=x^{7+4}=x^{11}$

 b. $\left(\dfrac{1}{2}\right)^4=\dfrac{1^4}{2^4}=\dfrac{1}{16}$

 c. $(9y^5)^2=9^2(y^5)^2=81y^{10}$

20. $\left(\dfrac{3a^2}{b}\right)^{-3}=\dfrac{3^{-3}a^{-6}}{b^{-3}}=\dfrac{b^3}{3^3a^6}=\dfrac{b^3}{27a^6}$

21. $(5y^3)^{-2}=5^{-2}(y^3)^{-2}=5^{-2}y^{-6}$
$=\dfrac{1}{5^2y^6}=\dfrac{1}{25y^6}$

22. $9x^3+x^3=9x^3+1x^3=10x^3$

23. $5x^2+6x-9x-3=5x^2-3x-3$

24. $7x(x^2+2x+5)=7x(x^2)+7x(2x)+7x(5)$
$=7x^3+14x^2+35x$

25. $\dfrac{9x^5-12x^2+3x}{3x^2}=\dfrac{9x^5}{3x^2}-\dfrac{12x^2}{3x^2}+\dfrac{3x}{3x^2}$
$=3x^3-4+\dfrac{1}{x}$

Chapter 4

Pretest

1. $2x^3y - 6x^2y^2 = 2x^2y(x - 3y)$

2. $xy + 6x - 4y - 24 = (xy + 6x) + (-4y - 24)$
$= x(y + 6) - 4(y + 6) = (y + 6)(x - 4)$

3. $a^2 + 8a + 12 = (a + 6)(a + 2)$

4. $m^2 + 4m - 3$ is a prime polynomial.

5. $3x^3 - 18x^2 + 15x = 3x(x^2 - 6x + 5)$
$= 3x(x - 5)(x - 1)$

6. $2x^2 + 5x - 12 = (2x - 3)(x + 4)$

7. $14x^2 + 63x + 70 = 7(2x^2 + 9x + 10)$
$= 7(2x + 5)(x + 2)$

8. $24b^2 - 25b + 6 = 24b^2 - 16b - 9b + 6$
$= 8b(3b - 2) - 3(3b - 2) = (3b - 2)(8b - 3)$

9. $15y^2 + 38y + 7 = 15y^2 + 35y + 3y + 7$
$= 5y(3y + 7) + 1(3y + 7) = (3y + 7)(5y + 1)$

10. $x^2 + 24x + 144 = (x)^2 + 2(x)(12) + (12)^2$
$= (x + 12)(x + 12) = (x + 12)^2$

11. $4x^2 - 12xy + 9y^2$
$= (2x)^2 - 2(2x)(3y) + (3y)^2$
$= (2x - 3y)(2x - 3y) = (2x - 3y)^2$

12. $a^2 - 49b^2 = a^2 - (7b)^2 = (a + 7b)(a - 7b)$

13. $1 - 64t^2 = 1^2 - (8t)^2 = (1 + 8t)(1 - 8t)$

14. $25b^2 + 4$ is a prime polynomial.

15. $x^2 + 18x + 81$ is a perfect square trinomial.

16. $(x - 12)(x + 5) = 0$
$x - 12 = 0$ or $x + 5 = 0$
$x = 12$ $x = -5$
The solutions are 12 and –5.

17. $y^2 - 13y = 0$
$y(y - 13) = 0$
$y = 0$ or $y - 13 = 0$
$y = 13$
The solutions are 0 and 13.

18. $2m^3 - 2m^2 - 24m = 0$
$2m(m^2 - m - 12) = 0$
$2m(m - 4)(m + 3) = 0$
$2m = 0$ or $m - 4 = 0$ or $m + 3 = 0$
$m = 0$ $m = 4$ $m = -3$
The solutions are 0, 4, and –3.

19. Let $x =$ the width. Then $x + 7 =$ the length.
$A = lw$
$120 = (x + 7)(x)$
$120 = x^2 + 7x$
$0 = x^2 + 7x - 120$
$0 = (x + 15)(x - 8)$
$x + 15 = 0$ or $x - 8 = 0$
$x = -15$ $x = 8$
Since the width cannot be negative, we discard the result –15. The width is 8 inches and the length is $8 + 7 = 15$ inches.

20. Let $x =$ the number.
$x + x^2 = 240$
$x^2 + x - 240 = 0$
$(x + 16)(x - 15) = 0$
$x + 16 = 0$ or $x - 15 = 0$
$x = -16$ $x = 15$
The number is –16 or 15.

Section 4.1

Practice Problems

1. a. $6x^2 = 2 \cdot 3 \cdot x^2$
$9x^4 = 3 \cdot 3 \cdot x^4$
$-12x^5 = -2 \cdot 2 \cdot 3 \cdot x^5$
$\text{GCF} = 3 \cdot x^2 = 3x^2$

b. $-16y = -2 \cdot 2 \cdot 2 \cdot 2 \cdot y$
$-20y^6 = -2 \cdot 2 \cdot 5 \cdot y^6$
$40y^4 = 2 \cdot 2 \cdot 2 \cdot 5 \cdot y^4$
GCF: $= 2 \cdot 2 \cdot y = 4y$

c. The GCF of a^5, a, and a^3 is a.
The GCF of b^4, b^3 and b^2 is b^2.
The GCF of a^5b^4, ab^3, and a^3b^2 is ab^2

2. a. $10y + 25 = 5(2y + 5)$

b. $x^4 - x^9 = x^4\left(1 - x^5\right)$

3. $-10x^3 + 8x^2 - 2x = -2x\left(5x^2 - 4x + 1\right)$

4. $4x^3 + 12x = 4x\left(x^2 + 3\right)$

5. $\dfrac{2}{5}a^5 - \dfrac{4}{5}a^3 + \dfrac{1}{5}a^2 = \dfrac{1}{5}a^2\left(2a^3 - 4a + 1\right)$

6.
$6a^3b + 3a^3b^2 + 9a^2b^4$
$= 3a^2b\left(2a + ab + 3b^3\right)$

7. $7(p + 2) + q(p + 2) = (p + 2)(7 + q)$

8. $ab + 7a + 2b + 14 = a(b + 7) + 2(b + 7)$
$= (b + 7)(a + 2)$

9. $28x^3 - 7x^2 + 12x - 3$
$= 7x^2(4x - 1) + 3(4x - 1)$
$= (4x - 1)\left(7x^2 + 3\right)$

10. $2xy + 5y^2 - 4x - 10y$
$= y(2x + 5y) - 2(2x + 5y)$
$= (2x + 5y)(y - 2)$

11. $4x^3 + x - 20x^2 - 5$
$= x\left(4x^2 + 1\right) - 5\left(4x^2 + 1\right)$
$= \left(4x^2 + 1\right)(x - 5)$

12. $2x - 2 + x^3 - 3x^2 = 2(x - 1) + x^2(x - 3)$
The polynomial is not factorable by grouping.

13. $3xy - 4 + x - 12y = 3xy + x - 4 - 12y$
$= x(3y + 1) - 4(1 + 3y)$
$= (3y + 1)(x - 4)$

Mental Math

1. $2 = 2$
$16 = 2 \cdot 2 \cdot 2 \cdot 2$
GCF $= 2$

2. $3 = 3$
$18 = 2 \cdot 3 \cdot 3$
GCF $= 3$

3. $6 = 2 \cdot 3$
$15 = 3 \cdot 5$
GCF $= 3$

4. $20 = 2 \cdot 2 \cdot 5$
$15 = 3 \cdot 5$
GCF $= 5$

5. $14 = 2 \cdot 7$
$35 = 5 \cdot 7$
GCF $= 7$

6. $27 = 3 \cdot 3 \cdot 3$
$36 = 2 \cdot 2 \cdot 3 \cdot 3$
GCF $= 3 \cdot 3 = 9$

Exercise Set 4.1

1. y^2

3. xy^2

5. $8x = 2 \cdot 2 \cdot 2 \cdot x$
$4 = 2 \cdot 2$
GCF $= 2 \cdot 2 = 4$

7. $12y^4 = 2 \cdot 2 \cdot 3 \cdot y^4$
$20y^3 = 2 \cdot 2 \cdot 5 \cdot y^3$
GCF $= 2 \cdot 2 \cdot y^3 = 4y^3$

9. $-10x^2 = -2 \cdot 5 \cdot x^2$
$15x^3 = 3 \cdot 5 \cdot x^3$
$\text{GCF} = 5 \cdot x^2 = 5x^2$

11. $12x^3 = 2 \cdot 2 \cdot 3 \cdot x^3$
$-6x^4 = -2 \cdot 3 \cdot x^4$
$3x^5 = 3 \cdot x^5$
$\text{GCF} = 3 \cdot x^3 = 3x^3$

13. $-18x^2 y = -2 \cdot 3 \cdot 3 \cdot x^2 \cdot y$
$9x^3 y^3 = 3 \cdot 3 \cdot x^3 \cdot y^3$
$36x^3 y = 2 \cdot 2 \cdot 3 \cdot 3 \cdot x^3 \cdot y$
$\text{GCF} = 3 \cdot 3 \cdot x^2 \cdot y = 9x^2 y$

15. $3a + 6 = 3(a + 2)$

17. $30x - 15 = 15(2x - 1)$

19. $x^3 + 5x^2 = x^2(x + 5)$

21. $6y^4 - 2y = 2y\left(3y^3 - 1\right)$

23. $32xy - 18x^2 = 2x(16y - 9x)$

25. $4x - 8y + 4 = 4(x - 2y + 1)$

27. $6x^3 - 9x^2 + 12x = 3x\left(2x^2 - 3x + 4\right)$

29. $a^7 b^6 - a^3 b^2 + a^2 b^5 - a^2 b^2$
$= a^2 b^2\left(a^5 b^4 - a + b^3 - 1\right)$

31. $5x^3 y - 15x^2 y + 10xy = 5xy\left(x^2 - 3x + 2\right)$

33. $8x^5 + 16x^4 - 20x^3 + 12$
$= 4\left(2x^5 + 4x^4 - 5x^3 + 3\right)$

35. $\dfrac{1}{3}x^4 + \dfrac{2}{3}x^3 - \dfrac{4}{3}x^5 + \dfrac{1}{3}x$
$= \dfrac{1}{3}x\left(x^3 + 2x^2 - 4x^4 + 1\right)$

37. $y(x + 2) + 3(x + 2) = (x + 2)(y + 3)$

39. $8(x + 2) - y(x + 2) = (x + 2)(8 - y)$

41. Answers may vary.

43. $x^3 + 2x^2 + 5x + 10 = x^2(x + 2) + 5(x + 2)$
$= \left(x^2 + 5\right)(x + 2)$

45. $5x + 15 + xy + 3y = 5(x + 3) + y(x + 3)$
$= (x + 3)(5 + y)$

47. $6x^3 - 4x^2 + 15x - 10$
$= 2x^2(3x - 2) + 5(3x - 2)$
$= \left(2x^2 + 5\right)(3x - 2)$

49. $2y - 8 + xy - 4x = 2(y - 4) + x(y - 4)$
$= (y - 4)(2 + x)$

51. $2x^3 + x^2 + 8x + 4 = x^2(2x + 1) + 4(2x + 1)$
$= (2x + 1)\left(x^2 + 4\right)$

53. $4x^2 - 8xy - 3x + 6y$
$= 4x(x - 2y) - 3(x - 2y)$
$= (x - 2y)(4x - 3)$

55. Answers may vary.

57. $(x + 2)(x + 5) = x^2 + 2x + 5x + 10$
$= x^2 + 7x + 10$

59. $(b + 1)(b - 4) = b^2 + b - 4b - 4$
$= b^2 - 3b - 4$

61. The two numbers are 2 and 6.
$2 \cdot 6 = 12; \ 2 + 6 = 8$

63. The two numbers are -1 and -8.
$-1 \cdot (-8) = 8; \ -1 + (-8) = -9$

65. The two numbers are -2 and 5.
$-2 \cdot 5 = -10; \ -2 + 5 = 3$

67. The two numbers are -8 and 3.
$-8 \cdot 3 = -24; \ -8 + 3 = -5$

69. $12x^2y - 42x^2 - 4y + 14$

$= 2\left(6x^2y - 21x^2 - 2y + 7\right)$

$= 2\left(3x^2(2y - 7) - 1(2y - 7)\right)$

$= 2\left(3x^2 - 1\right)(2y - 7)$

71. Subtract the area of the inner rectangle from the area of the outer rectangle.
Outer rectangle: $A = l \cdot w$
$$A = 12x \cdot x^2 = 12x^3$$
Inner rectangle: $A = l \cdot w$
$$A = 2 \cdot x = 2x$$
The area of the shaded region is given by the expression $12x^3 - 2x = 2x\left(6x^2 - 1\right)$.

73. Let l = length of the rectangle.
$A = l \cdot w$

$4n^4 - 24n = 4n \cdot l$

$4n\left(n^3 - 6\right) = 4n \cdot l$

$$\dfrac{4n\left(n^3 - 6\right)}{4n} = \dfrac{4n \cdot l}{4n}$$

$n^3 - 6 = l$

The length is $\left(n^3 - 6\right)$ units.

75. a. $3x^2 - 21x + 42 = 3\left(4^2\right) - 21(4) + 42$
$= 3(16) - 21(4) + 42 = 48 - 84 + 42$
$= 6$
6 million CD singles were sold in 1994.

  b. $3x^2 - 21x + 42 = 3\left(5^2\right) - 21(5) + 42$
$= 3(25) - 21(5) + 42 = 75 - 105 + 42$
$= 12$
12 million CD singles were sold in 1995.

  c. $3x^2 - 21x + 42 = 3\left(x^2 - 7x + 14\right)$

Section 4.2

Practice Problems

1. $x^2 + 9x + 20 = (x + 4)(x + 5)$

2. a. $x^2 - 13x + 22 = (x - 11)(x - 2)$

  b. $x^2 - 27x + 50 = (x - 25)(x - 2)$

3. $x^2 + 5x - 36 = (x + 9)(x - 4)$

4. a. $q^2 - 3q - 40 = (q - 8)(q + 5)$

  b. $y^2 + 2y - 48 = (y + 8)(y - 6)$

5. $x^2 + 6x + 15$ is a prime polynomial.

6. a. $x^2 + 6xy + 8y^2 = (x + 4y)(x + 2y)$

  b. $a^2 - 13ab + 30b^2 = (a - 10b)(a - 3b)$

7. $x^4 + 8x^2 + 12 = \left(x^2 + 6\right)\left(x^2 + 2\right)$

8. a. $x^3 + 3x^2 - 4x = x\left(x^2 + 3x - 4\right)$
$= x(x + 4)(x - 1)$

  b. $4x^2 - 24x + 36 = 4\left(x^2 - 6x + 9\right)$
$= 4(x - 3)(x - 3)$

Mental Math

1. $x^2 + 9x + 20 = (x + 4)(x + 5)$

2. $x^2 + 12x + 35 = (x + 5)(x + 7)$

3. $x^2 - 7x + 12 = (x - 4)(x - 3)$

4. $x^2 - 13x + 22 = (x - 2)(x - 11)$

5. $x^2 + 4x + 4 = (x + 2)(x + 2)$

6. $x^2 + 10x + 24 = (x + 6)(x + 4)$

Exercise Set 4.2

1. $x^2 + 7x + 6 = (x + 6)(x + 1)$

3. $x^2 - 10x + 9 = (x - 9)(x - 1)$

5. $x^2 - 3x - 18 = (x - 6)(x + 3)$

7. $x^2 + 3x - 70 = (x + 10)(x - 7)$

9. $x^2 + 5x + 2$ is a prime polynomial.

11. $x^2 + 8xy + 15y^2 = (x + 5y)(x + 3y)$

13. $a^4 - 2a^2 - 15 = \left(a^2 - 5\right)\left(a^2 + 3\right)$

15. $(x - 3)(x + 8) = x^2 - 3x + 8x - 24$
$\qquad = x^2 + 5x - 24$

17. Answers may vary.

19. $2z^2 + 20z + 32 = 2\left(z^2 + 10z + 16\right)$
$\qquad = 2(z + 8)(z + 2)$

21. $2x^3 - 18x^2 + 40x = 2x\left(x^2 - 9x + 20\right)$
$\qquad = 2x(x - 5)(x - 4)$

23. $x^2 - 3xy - 4y^2 = (x - 4y)(x + y)$

25. $x^2 + 15x + 36 = (x + 12)(x + 3)$

27. $x^2 - x - 2 = (x - 2)(x + 1)$

29. $r^2 - 16r + 48 = (r - 12)(r - 4)$

31. $x^2 + xy - 2y^2 = (x + 2y)(x - y)$

33. $3x^2 + 9x - 30 = 3\left(x^2 + 3x - 10\right)$
$\qquad = 3(x + 5)(x - 2)$

35. $3x^2 - 60x + 108 = 3\left(x^2 - 20x + 36\right)$
$\qquad = 3(x - 18)(x - 2)$

37. $x^2 - 18x - 144 = (x - 24)(x + 6)$

39. $r^2 - 3r + 6$ is a prime polynomial.

41. $x^2 - 8x + 15 = (x - 5)(x - 3)$

43. $6x^3 + 54x^2 + 120x = 6x\left(x^2 + 9x + 20\right)$
$\qquad = 6x(x + 4)(x + 5)$

45. $4x^2y + 4xy - 12y = 4y\left(x^2 + x - 3\right)$

47. $x^2 - 4x - 21 = (x - 7)(x + 3)$

49. $x^2 + 7xy + 10y^2 = (x + 5y)(x + 2y)$

51. $64 + 24t + 2t^2 = 2t^2 + 24t + 64$
$\qquad = 2\left(t^2 + 12t + 32\right) = 2(t + 8)(t + 4)$

53. $x^3 - 2x^2 - 24x = x\left(x^2 - 2x - 24\right)$
$\qquad = x(x - 6)(x + 4)$

55. $2t^5 - 14t^4 + 24t^3 = 2t^3\left(t^2 - 7t + 12\right)$
$\qquad = 2t^3(t - 4)(t - 3)$

57. $5x^3y - 25x^2y^2 - 120xy^3$
$\qquad = 5xy\left(x^2 - 5xy - 24y^2\right)$
$\qquad = 5xy(x - 8y)(x + 3y)$

59. $(2x + 1)(x + 5) = 2x^2 + x + 10x + 5$
$\qquad = 2x^2 + 11x + 5$

61. $(5y - 4)(3y - 1) = 15y^2 - 12y - 5y + 4$
$\qquad = 15y^2 - 17y + 4$

63. $(a + 3)(9a - 4) = 9a^2 + 27a - 4a - 12$
$\qquad = 9a^2 + 23a - 12$

65. $P = 2l + 2w$
$l = x^2 + 10x$ and $w = 4x + 33$, so
$P = 2\left(x^2 + 10x\right) + 2(4x + 33)$

$\qquad = 2x^2 + 20x + 8x + 66$
$\qquad = 2x^2 + 28x + 66 = 2\left(x^2 + 14x + 33\right)$

$\qquad = 2(x + 11)(x + 3)$
The perimeter of the rectangle is given by
the polynomial $2x^2 + 28x + 66$ which
factors as $2(x + 11)(x + 3)$.

67. $y^2(x+1) - 2y(x+1) - 15(x+1)$

$= (x+1)\left(y^2 - 2y - 15\right)$

$= (x+1)(y-5)(y+3)$

69. $y^2 - 4y + c$ if factorable when c is 3 or 4.

71. $x^2 + bx + 15$ is factorable when b is 8 or 16.

73. $x^{2n} + 5x^n + 6 = \left(x^n + 2\right)\left(x^n + 3\right)$

Section 4.3

Practice Problems

1. a. $4x^2 + 12x + 5 = (2x+5)(2x+1)$

    b. $5x^2 + 27x + 10 = (5x+2)(x+5)$

2. a. $6x^2 - 5x + 1 = (3x-1)(2x-1)$

    b. $2x^2 - 11x + 12 = (2x-3)(x-4)$

3. a. $35x^2 + 4x - 4 = (5x+2)(7x-2)$

    b. $4x^2 + 3x - 7 = (4x+7)(x-1)$

4. a. $14x^2 - 3xy - 2y^2 = (7x+2y)(2x-y)$

    b. $12a^2 - 16ab - 3b^2 = (6a+b)(2a-3b)$

5. a. $3x^3 + 17x^2 + 10x = x\left(3x^2 + 17x + 10\right)$

        $= x(3x+2)(x+5)$

    b. $6xy^2 + 33xy - 18x = 3x\left(2y^2 + 11y - 6\right)$

        $= 3x(2y-1)(y+6)$

Exercise Set 4.3

1. $5x^2 + 22x + 8 = (5x+2)(x+4)$

3. $50x^2 + 15x - 2 = (5x+2)(10x-1)$

5. $20x^2 - 7x - 6 = (5x+2)(4x-3)$

7. $2x^2 + 13x + 15 = (2x+3)(x+5)$

9. $8y^2 - 17y + 9 = (y-1)(8y-9)$

11. $2x^2 - 9x - 5 = (2x+1)(x-5)$

13. $20r^2 + 27r - 8 = (4r-1)(5r+8)$

15. $10x^2 + 17x + 3 = (5x+1)(2x+3)$

17. $3x^2 + x - 2 = (3x-2)(x+1)$

19. $6x^2 - 13xy + 5y^2 = (3x-5y)(2x-y)$

21. $15x^2 - 16x - 15 = (3x-5)(5x+3)$

23. $x^2 - 9x + 20 = (x-4)(x-5)$

25. $2x^2 - 7x - 99 = (2x+11)(x-9)$

27. $-27t + 7t^2 - 4 = 7t^2 - 27t - 4$

    $= (7t+1)(t-4)$

29. $3a^2 + 10ab + 3b^2 = (3a+b)(a+3b)$

31. $49x^2 - 7x - 2 = (7x+1)(7x-2)$

33. $18x^2 - 9x - 14 = (6x-7)(3x+2)$

35. $12x^3 + 11x^2 + 2x = x\left(12x^2 + 11x + 2\right)$

    $= x(3x+2)(4x+1)$

37. $21x^2 - 48x - 45 = 3\left(7x^2 - 16x - 15\right)$

    $= 3(7x+5)(x-3)$

39. $12x^2 + 7x - 12 = (3x+4)(4x-3)$

41. $6x^2y^2 - 2xy^2 - 60y^2 = 2y^2\left(3x^2 - x - 30\right)$

    $= 2y^2(3x-10)(x+3)$

43. $4x^2 - 8x - 21 = (2x-7)(2x+3)$

45. $3x^2 - 42x + 63 = 3\left(x^2 - 14x + 21\right)$

47. $8x^2 + 6x - 27 = (4x + 9)(2x - 3)$

49. $4x^3 - 9x^2 - 9x = x\left(4x^2 - 9x - 9\right)$
$= x(4x + 3)(x - 3)$

51. $24x^2 - 58x + 9 = (4x - 9)(6x - 1)$

53. $40a^2b + 9ab - 9b = b\left(40a^2 + 9a - 9\right)$
$= b(8a - 3)(5a + 3)$

55. $15x^4 + 19x^2 + 6 = \left(3x^2 + 2\right)\left(5x^2 + 3\right)$

57. $6y^3 - 8y^2 - 30y = 2y\left(3y^2 - 4y - 15\right)$
$= 2y(3y + 5)(y - 3)$

59. $10x^3 + 25x^2y - 15xy^2$
$= 5x\left(2x^2 + 5xy - 3y^2\right)$
$= 5x(2x - y)(x + 3y)$

61. The greatest percentage of households having a personal computer corresponds to households having an income of $60,000 and above.

63. Answers may vary.

65. $4x^2(y - 1)^2 + 10x(y - 1)^2 + 25(y - 1)^2$
$= (y - 1)^2\left(4x^2 + 10x + 25\right)$

67. $-12x^3y^2 + 3x^2y^2 + 15xy^2$
$= -3xy^2\left(4x^2 - x - 5\right)$
$= -3xy^2(4x - 5)(x + 1)$

69. $2z^2 + bz - 7$ is factorable when b is 5 or 13.

71. $3x^2 - 8x + c$ is factorable when c is 4 or 5.

Section 4.4

Practice Problems

1. a. $3x^2 + 14x + 8 = 3x^2 + 12x + 2x + 8$
$= 3x(x + 4) + 2(x + 4)$
$= (x + 4)(3x + 2)$

b. $12x^2 + 19x + 5 = 12x^2 + 15x + 4x + 5$
$= 3x(4x + 5) + 1(4x + 5)$
$= (4x + 5)(3x + 1)$

2. a. $6x^2y - 7xy - 5y = y\left(6x^2 - 7x - 5\right)$
$= y\left(6x^2 - 10x + 3x - 5\right)$
$= y[2x(3x - 5) + 1(3x - 5)]$
$= y(3x - 5)(2x + 1)$

b. $30x^2 - 26x + 4 = 2\left(15x^2 - 13x + 2\right)$
$= 2\left(15x^2 - 10x - 3x + 2\right)$
$= 2[5x(3x - 2) - 1(3x - 2)]$
$= 2(3x - 2)(5x - 1)$

Exercise Set 4.4

1. $x^2 + 3x + 2x + 6 = x(x + 3) + 2(x + 3)$
$= (x + 3)(x + 2)$

3. $x^2 - 4x + 7x - 28 = x(x - 4) + 7(x - 4)$
$= (x - 4)(x + 7)$

5. $y^2 + 8y - 2y - 16 = y(y + 8) - 2(y + 8)$
$= (y + 8)(y - 2)$

7. $3x^2 + 4x + 12x + 16 = x(3x + 4) + 4(3x + 4)$
$= (3x + 4)(x + 4)$

9. $8x^2 - 5x - 24x + 15 = x(8x - 5) - 3(8x - 5)$
$= (8x - 5)(x - 3)$

11. $5x^4 - 3x^2 + 25x^2 - 15$
$= x^2\left(5x^2 - 3\right) + 5\left(5x^2 - 3\right)$
$= \left(5x^2 - 3\right)\left(x^2 + 5\right)$

13. a. The numbers are 9 and 2.
$$9 \cdot 2 = 18$$
$$9 + 2 = 11$$

   b. $9x + 2x = 11x$

   c. $6x^2 + 11x + 3 = 6x^2 + 9x + 2x + 3$
$$= 3x(2x + 3) + 1(2x + 3)$$
$$= (2x + 3)(3x + 1)$$

15. a. The numbers are -20 and -3.
$$-20 \cdot (-3) = 60$$
$$-20 + (-3) = -23$$

   b. $-20x - 3x = -23x$

   c. $15x^2 - 23x + 4 = 15x^2 - 20x - 3x + 4$
$$= 5x(3x - 4) - 1(3x - 4)$$
$$= (3x - 4)(5x - 1)$$

17. $21y^2 + 17y + 2 = 21y^2 + 3y + 14y + 2$
$$= 3y(7y + 1) + 2(7y + 1)$$
$$= (3y + 2)(7y + 1)$$

19. $7x^2 - 4x - 11 = 7x^2 - 11x + 7x - 11$
$$= x(7x - 11) + 1(7x - 11)$$
$$= (7x - 11)(x + 1)$$

21. $10x^2 - 9x + 2 = 10x^2 - 5x - 4x + 2$
$$= 5x(2x - 1) - 2(2x - 1)$$
$$= (2x - 1)(5x - 2)$$

23. $2x^2 - 7x + 5 = 2x^2 - 5x - 2x + 5$
$$= x(2x - 5) - 1(2x - 5) = (2x - 5)(x - 1)$$

25. $4x^2 + 12x + 9 = 4x^2 + 6x + 6x + 9$
$$= 2x(2x + 3) + 3(2x + 3)$$
$$= (2x + 3)(2x + 3)$$
$$= (2x + 3)^2$$

27. $4x^2 - 8x - 21 = 4x^2 - 14x + 6x - 21$
$$= 2x(2x - 7) + 3(2x - 7)$$
$$= (2x - 7)(2x + 3)$$

29. $10x^2 - 23x + 12 = 10x^2 - 15x - 8x + 12$
$$= 5x(2x - 3) - 4(2x - 3)$$
$$= (2x - 3)(5x - 4)$$

31. $2x^3 + 13x^2 + 15x = x\left(2x^2 + 13x + 15\right)$
$$= x\left(2x^2 + 10x + 3x + 15\right)$$
$$= x[2x(x + 5) + 3(x + 5)] = x(x + 5)(2x + 3)$$

33. $16y^2 - 34y + 18 = 2\left(8y^2 - 17y + 9\right)$
$$= 2\left(8y^2 - 8y - 9y + 9\right)$$
$$= 2[(8y(y - 1) - 9(y - 1)] = 2(y - 1)(8y - 9)$$

35. $6x^2 - 13x + 6 = 6x^2 - 9x - 4x + 6$
$$= 3x(2x - 3) - 2(2x - 3)$$
$$= (2x - 3)(3x - 2)$$

37. $54a^2 - 9a - 30 = 3\left(18a^2 - 3a - 10\right)$
$$= 3\left(18a^2 - 15a + 12a - 10\right)$$
$$= 3[3a(6a - 5) + 2(6a - 5)]$$
$$= 3(6a - 5)(3a + 2)$$

39. $20a^3 + 37a^2 + 8a = a\left(20a^2 + 37a + 8\right)$
$$= a\left(20a^2 + 5a + 32a + 8\right)$$
$$= a[5a(4a + 1) + 8(4a + 1)]$$
$$= a(4a + 1)(5a + 8)$$

41. $12x^3 - 27x^2 - 27x = 3x\left(4x^2 - 9x - 9\right)$
$$= 3x\left(4x^2 - 12x + 3x - 9\right)$$
$$= 3x[4x(x - 3) + 3(x - 3)] = 3x(x - 3)(4x + 3)$$

43. $(x - 2)(x + 2) = x^2 - 2x + 2x - 4 = x^2 - 4$

45. $(y + 4)(y + 4) = y^2 + 4y + 4y + 16$
$$= y^2 + 8y + 16$$

47. $(9z + 5)(9z - 5) = 81z^2 + 45z - 45z - 25$
$$= 81z^2 - 25$$

49. $(4x - 3)^2 = 16x^2 + 2(4x)(-3) + 9$
$$= 16x^2 - 24x + 9$$

51. $x^{2n} + 2x^n + 3x^n + 6$

$\quad = x^n\left(x^n + 2\right) + 3\left(x^n + 2\right)$

$\quad = \left(x^n + 2\right)\left(x^n + 3\right)$

53. $3x^{2n} + 16x^n - 35 = 3x^{2n} - 5x^n + 21x^n - 35$

$\quad = x^n\left(3x^n - 5\right) + 7\left(3x^n - 5\right)$

$\quad = \left(3x^n - 5\right)\left(x^n + 7\right)$

Section 4.5

Practice Problems

1. a. Yes; two terms, x^2 and 36, are squares $\left(36 = 6^2\right)$ and the third term of the trinomial, $12x$, is twice the product of x and 6 $(2 \cdot x \cdot 6 = 12x)$.

    b. Yes; two terms, x^2 and 100, are squares $\left(100 = 10^2\right)$, and the third term of the trinomial, $20x$, is twice the product of x and 10 $(2 \cdot x \cdot 10 = 20x)$.

2. a. No; the two terms $9x^2$ and 25 $\left(9x^2 = (3x)^2 \text{ and } 25 = 5^2\right)$ are squares, but the third term, $20x$, is not twice the product of $3x$ and 5, or its opposite.

    b. No; only one of the terms, $4x^2$, is a square.

3. a. Yes; two terms, $25x^2$ and 1, are squares ($25x^2 = (5x)^2$ and $1 = 1^2$), and the third term of the trinomial, $-10x$, is the opposite of twice the product of $5x$ and 1 $(-2 \cdot 5x \cdot 1 = -10x)$.

    b. Yes; two terms, $9x^2$ and 49 are squares $(9x^2 = (3x)^2$, and $49 = 7^2)$, and the third term, $-42x$, is the opposite of twice the product of $3x$ and 7 $(-2 \cdot 3x \cdot 7 = -42x)$.

4. $x^2 + 16x + 64 = x^2 + 2 \cdot x \cdot 8 + 8^2 = (x+8)^2$

5. $9r^2 + 24rs + 16s^2 = (3r)^2 + 2 \cdot 3r \cdot 4s + (4s)^2$

$\quad\quad = (3r + 4s)^2$

6. $9n^2 - 6n + 1 = (3n)^2 - 2 \cdot 3n \cdot 1 + 1^2$

$\quad\quad = (3n - 1)^2$

7. $9x^2 + 15x + 4 = 9x^2 + 12x + 3x + 4$

$\quad\quad = 3x(3x + 4) + 1(3x + 4)$

$\quad\quad = (3x + 4)(3x + 1)$

8. $12x^3 - 84x^2 + 147x = 3x(4x^2 - 28x + 49)$

$\quad\quad = 3x[(2x)^2 - 2 \cdot 2x \cdot 7 + 7^2] = 3x(2x - 7)^2$

9. $x^2 - 9 = x^2 - 3^2 = (x+3)(x-3)$

10. $a^2 - 16 = a^2 - 4^2 = (a+4)(a-4)$

11. $c^2 - \dfrac{9}{25} = c^2 - \left(\dfrac{3}{5}\right)^2 = \left(c + \dfrac{3}{5}\right)\left(c - \dfrac{3}{5}\right)$

12. $s^2 + 9 = s^2 + 3^2$

This is not a difference of squares, it is a prime polynomial.

13. $9s^2 - 1 = (3s)^2 - 1^2 = (3s+1)(3s-1)$

14. $16x^2 - 49y^2 = (4x)^2 - (7y)^2$

$\quad\quad = (4x + 7y)(4x - 7y)$

15. $p^4 - 81 = \left(p^2\right)^2 - 9^2 = \left(p^2 + 9\right)\left(p^2 - 9\right)$

$\quad\quad = \left(p^2 + 9\right)\left(p^2 - 3^2\right)$

$\quad\quad = \left(p^2 + 9\right)(p+3)(p-3)$

16. $9x^3 - 25x = x\left(9x^2 - 25\right)$

$\quad\quad = x[(3x)^2 - 5^2] = x(3x + 5)(3x - 5)$

17. $48x^4 - 3 = 3(16x^4 - 1) = 3\left[(4x^2)^2 - 1^2\right]$

$\qquad = 3(4x^2 + 1)(4x^2 - 1)$

$\qquad = 3(4x^2 + 1)[(2x)^2 - 1^2]$

$\qquad = 3(4x^2 + 1)(2x + 1)(2x - 1)$

Mental Math

1. $1 = 1^2$

2. $25 = 5^2$

3. $81 = 9^2$

4. $64 = 8^2$

5. $9 = 3^2$

6. $100 = 10^2$

7. $9x^2 = (3x)^2$

8. $16y^2 = (4y)^2$

9. $25a^2 = (5a)^2$

10. $81b^2 = (9b)^2$

11. $36p^4 = \left(6p^2\right)^2$

12. $4q^4 = \left(2q^2\right)^2$

Exercise Set 4.5

1. Yes; two terms, x^2 and 64, are squares $\left(64 = 8^2\right)$, and the third term of the trinomial, $16x$, is twice the product of x and 8 $(2 \cdot x \cdot 8 = 16x)$.

3. No; the two terms, y^2 and 25, are squares $\left(25 = 5^2\right)$, but the third term of the trinomial, $5y$, is not twice the product of y and 5, or its opposite.

5. Yes; two terms, m^2 and 1, are squares $\left(1 = 1^2\right)$, and the third term of the trinomial, $-2m$, is the opposite of twice the product of m and $1(-(2 \cdot m \cdot 1) = -2m)$.

7. No; the two terms, a^2 and 49, are squares $\left(49 = 7^2\right)$, but the third term of the trinomial, $-16a$, is not twice the product of a and 7, or its opposite.

9. No; if we first factor out the GCF, 4, we find that only one of the terms, x^2, is a square.

11. Yes, two of the terms, $25a^2$ and $16b^2$, are squares ($25a^2 = (5a)^2$ and $16b^2 = (4b)^2$), and the third term of the trinomial, $-40ab$, is the opposite of twice the product of $5a$ and $4b$ $(-(2 \cdot 5a \cdot 4b) = -40ab)$.

13. $x^2 + 8x + 16$ is a perfect square trinomial because, x^2 and 16 are squares $\left(16 = 4^2\right)$, and $8x$ is twice the product of x and 4 $(2 \cdot x \cdot 4 = 8x)$.

15. $x^2 + 22x + 121 = x^2 + 2 \cdot x \cdot 11 + 11^2$

$\qquad = (x + 11)^2$

17. $x^2 - 16x + 64 = x^2 - 2 \cdot x \cdot 8 + 8^2$

$\qquad = (x - 8)^2$

19. $16a^2 - 24a + 9 = (4a)^2 - 2 \cdot 4a \cdot 3 + 3^2$

$\qquad = (4a - 3)^2$

21. $x^4 + 4x^2 + 4 = \left(x^2\right)^2 + 2 \cdot x^2 \cdot 2 + 2^2$

$\qquad = \left(x^2 + 2\right)^2$

23. $2n^2 - 28n + 98 = 2\left(n^2 - 14n + 49\right)$

$= 2\left(n^2 - 2 \cdot n \cdot 7 + 7^2\right) = 2(n-7)^2$

25. $16y^2 + 40y + 25 = (4y)^2 + 2 \cdot 4y \cdot 5 + 5^2$

$= (4y+5)^2$

27. $x^2 y^2 - 10xy + 25 = (xy)^2 - 2 \cdot xy \cdot 5 + 5^2$

$= (xy-5)^2$

29. $m^3 + 18m^2 + 81m = m\left(m^2 + 18m + 81\right)$

$= m\left(m^2 + 2 \cdot m \cdot 9 + 9^2\right)$

$= m(m+9)^2$

31. The trinomial $1 + 6x^2 + x^4 = x^4 + 6x^2 + 1$ is not factorable with integers, and is, therefore, a prime polynomial.

33. $9x^2 - 24xy + 16y^2$

$= (3x)^2 - 2 \cdot 3x \cdot 4y + (4y)^2$

$= (3x-4y)^2$

35. $x^2 + 14xy + 49y^2 = x^2 + 2 \cdot x \cdot 7y + (7y)^2$

$= (x+7y)^2$

37. Answers may vary.

39. $x^2 - 4 = x^2 - 2^2 = (x+2)(x-2)$

41. $81 - p^2 = 9^2 - p^2 = (9+p)(9-p)$

43. $4r^2 - 1 = (2r)^2 - 1^2 = (2r+1)(2r-1)$

45. $9x^2 - 16 = (3x)^2 - 4^2 = (3x+4)(3x-4)$

47. $16r^2 + 1$ is the sum of two squares, $(4r)^2 + 1^2$, not the difference of two squares. $16r^2 + 1$ is a prime polynomial.

49. $-36 + x^2 = -\left(6^2\right) + x^2 = (-6+x)(6+x)$

51. $m^4 - 1 = \left(m^2\right)^2 - 1^2 = \left(m^2 + 1\right)\left(m^2 - 1\right)$

$= \left(m^2 + 1\right)\left(m^2 - 1^2\right)$

$= \left(m^2 + 1\right)(m+1)(m-1)$

53. $x^2 - 169y^2 = x^2 - (13y)^2$

$= (x+13y)(x-13y)$

55. $18r^2 - 8 = 2\left(9r^2 - 4\right) = 2\left((3r)^2 - 2^2\right)$

$= 2(3r+2)(3r-2)$

57. $9xy^2 - 4x = x\left(9y^2 - 4\right) = x\left((3y)^2 - 2^2\right)$

$= x(3y+2)(3y-2)$

59. $25y^4 - 100y^2 = 25y^2\left(y^2 - 4\right)$

$= 25y^2\left(y^2 - 2^2\right) = 25y^2(y+2)(y-2)$

61. $x^3 y - 4xy^3 = xy\left(x^2 - 4y^2\right)$

$= xy\left(x^2 - (2y)^2\right)$

$= xy(x+2y)(x-2y)$

63. $225a^2 - 81b^2 = 9\left(25a^2 - 9b^2\right)$

$= 9\left((5a)^2 - (3b)^2\right)$

$= 9(5a+3b)(5a-3b)$

65. $12x^2 - 27 = 3\left(4x^2 - 9\right) = 3\left((2x)^2 - 3^2\right)$

$= 3(2x+3)(2x-3)$

67. $49a^2 - 16 = (7a)^2 - 4^2 = (7a+4)(7a-4)$

69. $169a^2 - 49b^2 = (13a)^2 - (7b)^2$

$= (13a+7b)(13a-7b)$

71. $16 - a^2 b^2 = 4^2 - (ab)^2 = (4+ab)(4-ab)$

73. $y^2 - \dfrac{1}{16} = y^2 - \left(\dfrac{1}{4}\right)^2 = \left(y + \dfrac{1}{4}\right)\left(y - \dfrac{1}{4}\right)$

75. $100 - \frac{4}{81} n^2 = 10^2 - \left(\frac{2}{9} n\right)^2$

$= \left(10 + \frac{2}{9} n\right)\left(10 - \frac{2}{9} n\right)$

77. $5 - y$, since

$(5 - y)(5 + y) = 25 - 5y + 5y - y^2$

$= 25 - y^2 = 5^2 - y^2$

79. $y + 5 = 0$

$y + 5 - 5 = 0 - 5$

$y = -5$

81. $3x - 9 = 0$

$3x - 9 + 9 = 0 + 9$

$3x = 9$

$\frac{3x}{3} = \frac{9}{3}$

$x = 3$

83. $4a + 2 = 0$

$4a + 2 - 2 = 0 - 2$

$4a = -2$

$\frac{4a}{4} = \frac{-2}{4}$

$a = -\frac{1}{2}$

85. The sail is shaped like a triangle. The area of a triangle is given by $A = \frac{1}{2} bh$. Use $b = 10$ feet and $h = x$ feet. Then,

$A = \frac{1}{2} bh$

$25 = \frac{1}{2} \cdot 10 \cdot x$

$25 = 5x$

$\frac{25}{5} = \frac{5x}{5}$

$5 = x$

The height, x, is 5 feet.

87. $(y - 6)^2 - z^2 = (y - 6 + z)(y - 6 - z)$

89. $m^2(n + 8) - 9(n + 8) = (n + 8)\left(m^2 - 9\right)$

$= (n + 8)\left(m^2 - 3^2\right) = (n + 8)(m + 3)(m - 3)$

91. $\left(x^2 + 2x + 1\right) - 36y^2$

$= [(x + 1)(x + 1)] - 36y^2$

$= (x + 1)^2 - (6y)^2 = (x + 1 + 6y)(x + 1 - 6y)$

93. $x^{2n} - 81 = \left(x^n\right)^2 - 9^2 = \left(x^n + 9\right)\left(x^n - 9\right)$

95. The formula for factoring a perfect square trinomial

97. a. Let $t = 1$.

$529 - 16t^2 = 529 - 16\left(1^2\right) = 529 -$

$16(1)$

$= 529 - 16 = 513$

After 1 second the height of the bolt is 513 feet.

b. Let $t = 4$.

$529 - 16t^2 = 529 - 16\left(4^2\right)$

$= 529 - 16(16)$

$= 529 - 256 = 273$

After 4 seconds the height of the bolt is 273 feet.

c. When the object hits the ground, its height is zero feet. Thus, to find the time, t, when the object's height is zero feet above the ground, we set the expression $529 - 16t^2$ equal to 0 and solve for t.

$529 - 16t^2 = 0$

$529 - 16t^2 + 16t^2 = 0 + 16t^2$

$529 = 16t^2$

$\frac{529}{16} = \frac{16t^2}{16}$

$33.0625 = t^2$

$\sqrt{33.0625} = \sqrt{t^2}$

$5.75 = t$

Thus, the object will hit the ground after approximately 6 seconds.

d. $529 - 16t^2 = 23^2 - (4t)^2$

$= (23 + 4t)(23 - 4t)$

Integrated Review

1. $x^2 + x - 12 = (x+4)(x-3)$

2. $x^2 - 10x + 16 = (x-2)(x-8)$

3. $x^2 - x - 6 = (x+2)(x-3)$

4. $x^2 + 2x + 1 = (x+1)(x+1) = (x+1)^2$

5. $x^2 - 6x + 9 = (x-3)(x-3) = (x-3)^2$

6. $x^2 + x - 2 = (x+2)(x-1)$

7. $x^2 + x - 6 = (x+3)(x-2)$

8. $x^2 + 7x + 12 = (x+4)(x+3)$

9. $x^2 - 7x + 10 = (x-5)(x-2)$

10. $x^2 - x - 30 = (x-6)(x+5)$

11. $2x^2 - 98 = 2(x^2 - 49) = 2(x^2 - 7^2)$
    $= 2(x+7)(x-7)$

12. $3x^2 - 75 = 3(x^2 - 25) = 3(x^2 - 5^2)$
    $= 3(x+5)(x-5)$

13. $x^2 + 3x + 5x + 15 = x(x+3) + 5(x+3)$
    $= (x+3)(x+5)$

14. $3y - 21 + xy - 7x = 3(y-7) + x(y-7)$
    $= (y-7)(3+x)$

15. $x^2 + 6x - 16 = (x+8)(x-2)$

16. $x^2 - 3x - 28 = (x-7)(x+4)$

17. $4x^3 + 20x^2 - 56x = 4x(x^2 + 5x - 14)$
    $= 4x(x+7)(x-2)$

18. $6x^3 - 6x^2 - 120x = 6x(x^2 - x - 20)$
    $= 6x(x-5)(x+4)$

19. $12x^2 + 34x + 24 = 2(6x^2 + 17x + 12)$
    $= 2(6x^2 + 9x + 8x + 12)$
    $= 2[3x(2x+3) + 4(2x+3)]$
    $= 2(2x+3)(3x+4)$

20. $8a^2 + 6ab - 5b^2 = 8a^2 + 10ab - 4ab - 5b^2$
    $= 2a(4a+5b) - b(4a+5b)$
    $= (4a+5b)(2a-b)$

21. $4a^2 - b^2 = (2a)^2 - b^2 = (2a+b)(2a-b)$

22. $x^2 - 25y^2 = x^2 - (5y)^2 = (x+5y)(x-5y)$

23. $28 - 13x - 6x^2 = 28 - 21x + 8x - 6x^2$
    $= 7(4-3x) + 2x(4-3x) = (4-3x)(7+2x)$

24. $20 - 3x - 2x^2 = 20 - 8x + 5x - 2x^2$
    $= 4(5-2x) + x(5-2x)$
    $= (5-2x)(4+x)$

25. $x^2 - 2x + 4$ is a prime polynomial.

26. $a^2 + a - 3$ is a prime polynomial.

27. $6y^2 + y - 15 = 6y^2 + 10y - 9y - 15$
    $= 2y(3y+5) - 3(3y+5) = (3y+5)(2y-3)$

28. $4x^2 - x - 5 = 4x^2 - 5x + 4x - 5$
    $= x(4x-5) + 1(4x-5) = (4x-5)(x+1)$

29. $18x^3 - 63x^2 + 9x = 9x(2x^2 - 7x + 1)$

30. $12a^3 - 24a^2 + 4a = 4a(3a^2 - 6a + 1)$

31. $16a^2 - 56a + 49 = (4a)^2 - 2 \cdot 4a \cdot 7 + 7^2$
    $= (4a-7)^2$

32. $25p^2 - 70p + 49 = (5p)^2 - 2 \cdot 5p \cdot 7 + 7^2$
    $= (5p-7)^2$

33. $14 + 5x - x^2 = (7-x)(2+x)$

34. $3 - 2x - x^2 = (3+x)(1-x)$

35. $3x^4y + 6x^3y - 72x^2y = 3x^2y\left(x^2 + 2x - 24\right)$

$= 3x^2y(x+6)(x-4)$

36.

$2x^3y + 8x^2y^2 - 10xy^3$

$= 2xy\left(x^2 + 4xy - 5y^2\right)$

$= 2xy(x+5y)(x-y)$

37. $12x^3y + 243xy = 3xy\left(4x^2 + 81\right)$

38. $6x^3y^2 + 8xy^2 = 2xy^2\left(3x^2 + 4\right)$

39. $2xy - 72x^3y = 2xy\left(1 - 36x^2\right)$

$= 2xy\left(1^2 - (6x)^2\right) = 2xy(1+6x)(1-6x)$

40. $2x^3 - 18x = 2x\left(x^2 - 9\right) = 2x\left(x^2 - 3^2\right)$

$= 2x(x+3)(x-3)$

41. $x^3 + 6x^2 - 4x - 24 = x^2(x+6) - 4(x+6)$

$= (x+6)\left(x^2 - 4\right) = (x+6)\left(x^2 - 2^2\right)$

$= (x+6)(x+2)(x-2)$

42.

$x^3 - 2x^2 - 36x + 72$

$= x^2(x-2) - 36(x-2)$

$= (x-2)\left(x^2 - 36\right) = (x-2)\left(x^2 - 6^2\right)$

$= (x-2)(x+6)(x-6)$

43. $6a^3 + 10a^2 = 2a^2(3a+5)$

44. $4n^2 - 6n = 2n(2n-3)$

45. $3x^3 - x^2 + 12x - 4 = x^2(3x-1) + 4(3x-1)$

$= (3x-1)\left(x^2 + 4\right)$

46. $x^3 - 2x^2 + 3x - 6 = x^2(x-2) + 3(x-2)$

$= (x-2)\left(x^2 + 3\right)$

47. $6x^2 + 18xy + 12y^2 = 6\left(x^2 + 3xy + 2y^2\right)$

$= 6(x+2y)(x+y)$

48. $12x^2 + 46xy - 8y^2 = 2\left(6x^2 + 23xy - 4y^2\right)$

$= 2\left(6x^2 + 24xy - xy - 4y^2\right)$

$= 2[6x(x+4y) - y(x+4y)]$

$= 2(x+4y)(6x-y)$

49. $5(x+y) + x(x+y) = (x+y)(5+x)$

50. $7(x-y) + y(x-y) = (x-y)(7+y)$

51. $14t^2 - 9t + 1 = 14t^2 - 7t - 2t + 1$

$= 7t(2t-1) - 1(2t-1)$

$= (2t-1)(7t-1)$

52. $3t^2 - 5t + 1$ is a prime polynomial.

53. $3x^2 + 2x - 5 = 3x^2 + 5x - 3x - 5$

$= x(3x+5) - 1(3x+5) = (3x+5)(x-1)$

54. $7x^2 + 19x - 6 = 7x^2 + 21x - 2x - 6$

$= 7x(x+3) - 2(x+3) = (x+3)(7x-2)$

55. $1 - 8a - 20a^2 = 1 - 10a + 2a - 20a^2$

$= 1(1-10a) + 2a(1-10a) = (1-10a)(1+2a)$

56. $1 - 7a - 60a^2 = 1 - 12a + 5a - 60a^2$

$= 1(1-12a) + 5a(1-12a)$

$= (1-12a)(1+5a)$

57. $x^4 - 10x^2 + 9 = \left(x^2 - 9\right)\left(x^2 - 1\right)$

$= \left(x^2 - 3^2\right)\left(x^2 - 1^2\right)$

$= (x+3)(x-3)(x+1)(x-1)$

58. $x^4 - 13x^2 + 36 = \left(x^2 - 9\right)\left(x^2 - 4\right)$

$= \left(x^2 - 3^2\right)\left(x^2 - 2^2\right)$

$= (x+3)(x-3)(x+2)(x-2)$

59. $x^2 - 23x + 120 = (x-15)(x-8)$

60. $y^2 + 22y + 96 = (y+16)(y+6)$

61. Answers may vary.

62. Yes; $9x^2 + 81y^2 = 9\left(x^2 + 9y^2\right)$

Section 4.6

Practice Problems

1. $(x - 7)(x + 2) = 0$
$x - 7 = 0$ or $x + 2 = 0$
     $x = 7$ $x = -2$
The solutions are 7 and –2.

2. $(x - 10)(3x + 1) = 0$
$x - 10 = 0$ or $3x + 1 = 0$
    $x = 10$ $3x = -1$
                       $x = -\dfrac{1}{3}$
The solutions are 10 and $-\dfrac{1}{3}$.

3. a. $y(y + 3) = 0$
    $y = 0$ or $y + 3 = 0$
                   $y = -3$
    The solutions are 0 and –3.

   b. $x(4x - 3) = 0$
      $x = 0$ or $4x - 3 = 0$
                     $4x = 3$
                      $x = \dfrac{3}{4}$
    The solutions are 0 and $\dfrac{3}{4}$.

4. $x^2 - 3x - 18 = 0$
$(x - 6)(x + 3) = 0$
$x - 6 = 0$ or $x + 3 = 0$
    $x = 6$ $x = -3$
The solutions are 6 and –3.

5. $x^2 - 14x = -24$
$x^2 - 14x + 24 = 0$
$(x - 12)(x - 2) = 0$
$x - 12 = 0$ or $x - 2 = 0$
     $x = 12$ $x = 2$
The solutions are 12 and 2.

6. a. $x(x - 4) = 5$
     $x^2 - 4x = 5$
     $x^2 - 4x - 5 = 0$
     $(x - 5)(x + 1) = 0$
     $x - 5 = 0$ or $x + 1 = 0$
        $x = 5$ $x = -1$
     The solutions are 5 and –1.

   b. $x(3x + 7) = 6$
     $3x^2 + 7x = 6$
     $3x^2 + 7x - 6 = 0$
     $(3x - 2)(x + 3) = 0$
     $3x - 2 = 0$ or $x + 3 = 0$
        $3x = 2$ $x = -3$
         $x = \dfrac{2}{3}$
     The solutions are $\dfrac{2}{3}$ and –3.

7. $2x^3 - 18x = 0$
$2x\left(x^2 - 9\right) = 0$
$2x(x + 3)(x - 3) = 0$
$2x = 0$ or $x + 3 = 0$ or $x - 3 = 0$
  $x = 0$ $x = -3$ $x = 3$
The solutions are 0, –3, and 3.

8. $(x + 3)\left(3x^2 - 20x - 7\right) = 0$
$(x + 3)(3x + 1)(x - 7) = 0$
$x + 3 = 0$ or $3x + 1 = 0$ or $x - 7 = 0$
   $x = -3$ $3x = -1$ $x = 7$
                  $x = -\dfrac{1}{3}$
The solutions are –3, $-\dfrac{1}{3}$, and 7.

Mental Math

1. $(a - 3)(a - 7) = 0$
$a - 3 = 0$ or $a - 7 = 0$
   $a = 3$ $a = 7$
The solutions are 3 and 7.

2. $(a - 5)(a - 2) = 0$
$a - 5 = 0$ or $a - 2 = 0$
   $a = 5$ $a = 2$
The solutions are 5 and 2.

3. $(x + 8)(x + 6) = 0$
$x + 8 = 0$ or $x + 6 = 0$
$x = -8$ $x = -6$
The solutions are -8 and -6.

4. $(x + 2)(x + 3) = 0$
$x + 2 = 0$ or $x + 3 = 0$
$x = -2$ $x = -3$
The solutions are -2 and -3.

5. $(x + 1)(x - 3) = 0$
$x + 1 = 0$ or $x - 3 = 0$
$x = -1$ $x = 3$
The solutions are -1 and 3.

6. $(x - 1)(x + 2) = 0$
$x - 1 = 0$ or $x + 2 = 0$
$x = 1$ $x = -2$
The solutions are 1 and -2.

Exercise Set 4.6

1. $(x - 2)(x + 1) = 0$
$x - 2 = 0$ or $x + 1 = 0$
$x = 2$ $x = -1$
The solutions are 2 and -1.

3. $(x - 6)(x - 7) = 0$
$x - 6 = 0$ or $x - 7 = 0$
$x = 6$ $x = 7$
The solutions are 6 and 7.

5. $(x + 9)(x + 17) = 0$
$x + 9 = 0$ or $x + 17 = 0$
$x = -9$ $x = -17$
The solutions are -9 and -17.

7. $x(x + 6) = 0$
$x = 0$ or $x + 6 = 0$
                    $x = -6$
The solutions are 0 and -6.

9. $3x(x - 8) = 0$
$3x = 0$ or $x - 8 = 0$
$x = 0$ $x = 8$
The solutions are 0 and 8.

11. $(2x + 3)(4x - 5) = 0$
$2x + 3 = 0$ or $4x - 5 = 0$
$2x = -3$ $4x = 5$
$x = -\dfrac{3}{2}$ $x = \dfrac{5}{4}$
The solutions are $-\dfrac{3}{2}$ and $\dfrac{5}{4}$.

13. $(2x - 7)(7x + 2) = 0$
$2x - 7 = 0$ or $7x + 2 = 0$
$2x = 7$ $7x = -2$
$x = \dfrac{7}{2}$ $x = -\dfrac{2}{7}$
The solutions are $\dfrac{7}{2}$ and $-\dfrac{2}{7}$.

15. $\left(x - \dfrac{1}{2}\right)\left(x + \dfrac{1}{3}\right) = 0$
$x - \dfrac{1}{2} = 0$ or $x + \dfrac{1}{3} = 0$
$x = \dfrac{1}{2}$ $x = -\dfrac{1}{3}$
The solutions are $\dfrac{1}{2}$ and $-\dfrac{1}{3}$.

17. $(x + 0.2)(x + 1.5) = 0$
$x + 0.2 = 0$ or $x + 1.5 = 0$
$x = -0.2$ $x = -1.5$
The solutions are -0.2 and -1.5.

19. If $x = 6$ and $x = -1$ are the solutions, then
$x = 6$ or $x = -1$
$x - 6 = 0$ $x + 1 = 0$
$(x - 6)(x + 1) = 0$

21. $x^2 - 13x + 36 = 0$
$(x - 9)(x - 4) = 0$
$x - 9 = 0$ or $x - 4 = 0$
$x = 9$ $x = 4$
The solutions are 9 and 4.

23. $x^2 + 2x - 8 = 0$
$(x + 4)(x - 2) = 0$
$x + 4 = 0$ or $x - 2 = 0$
$x = -4$ $x = 2$
The solutions are -4 and 2.

25. $x^2 - 7x = 0$
$x(x - 7) = 0$
$x = 0$ or $x - 7 = 0$
                $x = 7$
The solutions are 0 and 7.

27. $x^2 + 20x = 0$
$x(x + 20) = 0$
$x = 0$ or $x + 20 = 0$
                $x = -20$
The solutions are 0 and -20.

29. $x^2 = 16$
$x^2 - 16 = 0$
$x^2 - 4^2 = 0$
$(x + 4)(x - 4) = 0$
$x + 4 = 0$ or $x - 4 = 0$
   $x = -4$ $x = 4$
The solutions are -4 and 4.

31. $x^2 - 4x = 32$
$x^2 - 4x - 32 = 0$
$(x - 8)(x + 4) = 0$
$x - 8 = 0$ or $x + 4 = 0$
   $x = 8$ $x = -4$
The solutions are 8 and -4.

33. $x(3x - 1) = 14$
$3x^2 - x = 14$
$3x^2 - x - 14 = 0$
$(3x - 7)(x + 2) = 0$
$3x - 7 = 0$ or $x + 2 = 0$
   $3x = 7$ $x = -2$
   $x = \dfrac{7}{3}$
The solutions are $\dfrac{7}{3}$ and -2.

35. $3x^2 + 19x - 72 = 0$
$(3x - 8)(x + 9) = 0$
$3x - 8 = 0$ or $x + 9 = 0$
$3x = 8$ $x = -9$
$x = \dfrac{8}{3}$
The solutions are $\dfrac{8}{3}$ and -9.

37. $4x^3 - x = 0$
$x(4x^2 - 1) = 0$
$x(2x + 1)(2x - 1) = 0$
$x = 0$ or $2x + 1 = 0$ or $2x - 1 = 0$
            $2x = -1$ $2x = 1$
            $x = -\dfrac{1}{2}$ $x = \dfrac{1}{2}$
The solutions are 0, $-\dfrac{1}{2}$, and $\dfrac{1}{2}$.

39. $4(x - 7) = 6$
$4x - 28 = 6$
$4x = 34$
$x = \dfrac{34}{4}$
$x = \dfrac{17}{2}$
The solution is $\dfrac{17}{2}$.

41. $(4x - 3)(16x^2 - 24x + 9) = 0$
$(4x - 3)(4x - 3)^2 = 0$
$(4x - 3)^3 = 0$
$4x - 3 = 0$
$4x = 3$
$x = \dfrac{3}{4}$
The solution is $\dfrac{3}{4}$.

43. $4y^2 - 1 = 0$
$(2y + 1)(2y - 1) = 0$
$2y + 1 = 0$ or $2y - 1 = 0$
$2y = -1$ $2y = 1$
$y = -\dfrac{1}{2}$ $y = \dfrac{1}{2}$
The solutions are $-\dfrac{1}{2}$ and $\dfrac{1}{2}$.

45. $(2x + 3)(2x^2 - 5x - 3) = 0$
$(2x + 3)(2x + 1)(x - 3) = 0$
$2x + 3 = 0$ or $2x + 1 = 0$ or $x - 3 = 0$
$2x = -3$ $2x = -1$ $x = 3$
$x = -\dfrac{3}{2}$ $x = -\dfrac{1}{2}$
The solutions are $-\dfrac{3}{2}$, $-\dfrac{1}{2}$, and 3.

47. $x^2 - 15 = -2x$
$x^2 + 2x - 15 = 0$
$(x + 5)(x - 3) = 0$
$x + 5 = 0$ or $x - 3 = 0$
$x = -5$ $x = 3$
The solutions are -5 and 3.

49. $x^2 - 16x = 0$
$x(x - 16) = 0$
$x = 0$ or $x - 16 = 0$
                       $x = 16$
The solutions are 0 and 16.

51. $x^2 - x = 30$
$x^2 - x - 30 = 0$
$(x - 6)(x + 5) = 0$
$x - 6 = 0$ or $x + 5 = 0$
$x = 6$ $x = -5$
The solutions are 6 and -5.

53. $6y^2 - 22y - 40 = 0$
$2(3y^2 - 11y - 20) = 0$
$2(3y + 4)(y - 5) = 0$
$3y + 4 = 0$ or $y - 5 = 0$
$3y = -4$ $y = 5$
$y = -\dfrac{4}{3}$

The solutions are $-\dfrac{4}{3}$ and 5.

55. $(y - 2)(y + 3) = 6$
$y^2 - 2y + 3y - 6 = 6$
$y^2 + y - 12 = 0$
$(y + 4)(y - 3) = 0$
$y + 4 = 0$ or $y - 3 = 0$
$y = -4$ $y = 3$
The solutions are -4 and 3.

57. $x^3 - 12x^2 + 32x = 0$
$x(x^2 - 12x + 32) = 0$
$x(x - 8)(x - 4) = 0$
$x = 0$ or $x - 8 = 0$ or $x - 4 = 0$
             $x = 8$ $x = 4$
The solutions are 0, 8, and 4.

59. If the solutions are $x = 5$ and $x = 7$, then, by the zero factor property,
$x = 5$ or $x = 7$
$x - 5 = 0$ $x - 7 = 0$
$(x - 5)(x - 7) = 0$
$x^2 - 5x - 7x + 35 = 0$
$x^2 - 12x + 35 = 0$

61. $\dfrac{3}{5} + \dfrac{4}{9} = \dfrac{3 \cdot 9}{5 \cdot 9} + \dfrac{4 \cdot 5}{9 \cdot 5} = \dfrac{27}{45} + \dfrac{20}{45}$
$= \dfrac{27 + 20}{45} = \dfrac{47}{45}$

63. $\dfrac{7}{10} - \dfrac{5}{12} = \dfrac{7 \cdot 6}{10 \cdot 6} - \dfrac{5 \cdot 5}{12 \cdot 5} = \dfrac{42}{60} - \dfrac{25}{60}$
$= \dfrac{42 - 25}{60} = \dfrac{17}{60}$

65. $\dfrac{4}{5} \cdot \dfrac{7}{8} = \dfrac{4 \cdot 7}{5 \cdot 8} = \dfrac{4 \cdot 7}{5 \cdot 2 \cdot 4} = \dfrac{7}{10}$

67. The equation is not written in standard form.

69. a. When $x = 0$:
$y = -16x^2 + 20x + 300$
$y = -16(0^2) + 20(0) + 300$
$= -16(0) + 20(0) + 300$
$= 0 + 0 + 300 = 300$
When $x = 1$:
$y = -16x^2 + 20x + 300$
$y = -16(1)^2 + 20(1) + 300$
$= -16(1) + 20(1) + 300$
$= -16 + 20 + 300 = 304$
When $x = 2$:
$y = -16x^2 + 20x + 300$
$y = -16(2^2) + 20(2) + 300$
$= -16(4) + 20(2) + 300$
$= -64 + 40 + 300$
$= 276$
When $x = 3$:
$y = -16x^2 + 20x + 300$
$y = -16(3^2) + 20(3) + 300$
$= -16(9) + 20(3) + 300$
$= -144 + 60 + 300$
$= 216$

When $x = 4$:

$$y = -16x^2 + 20x + 300$$
$$y = -16(4^2) + 20(4) + 300$$
$$= -16(16) + 20(4) + 300$$
$$= -256 + 80 + 300$$
$$= 124$$

When $x = 5$:

$$y = -16x^2 + 20x + 300$$
$$y = -16(5^2) + 20(5) + 300$$
$$= -16(25) + 20(5) + 300$$
$$= -400 + 100 + 300$$
$$= 0$$

When $x = 6$:

$$y = -16x^2 + 20x + 300$$
$$y = -16(6^2) + 20(6) + 300$$
$$= -16(36) + 20(6) + 300$$
$$= -576 + 120 + 300$$
$$= -156$$

b. The compass strikes the ground after 5 seconds, when the height, y, is zero feet.

c. The maximum height was approximately 304 feet.

71. $(x - 3)(3x + 4) = (x + 2)(x - 6)$

$$3x^2 - 9x + 4x - 12 = x^2 + 2x - 6x - 12$$
$$3x^2 - 5x - 12 = x^2 - 4x - 12$$
$$2x^2 - x = 0$$
$$x(2x - 1) = 0$$

$x = 0$ or $\quad 2x - 1 = 0$
$\qquad\qquad\qquad\qquad 2x = 1$
$\qquad\qquad\qquad\qquad x = \dfrac{1}{2}$

The solutions are 0 and $\dfrac{1}{2}$.

73. $(2x - 3)(x + 8) = (x - 6)(x + 4)$

$$2x^2 - 3x + 16x - 24 = x^2 - 6x + 4x - 24$$
$$2x^2 + 13x - 24 = x^2 - 2x - 24$$
$$x^2 + 15x = 0$$
$$x(x + 15) = 0$$

$x = 0$ or $\quad x + 15 = 0$
$\qquad\qquad\qquad\qquad x = -15$

The solutions are 0 and -15.

Section 4.7

Practice Problems

1. Find t when $h = 0$.

$$h = -16t^2 + 144$$
$$0 = -16t^2 + 144$$
$$0 = -16\left(t^2 - 9\right)$$
$$0 = -16(t + 3)(t - 3)$$

$t + 3 = 0$ $\quad$ or $\quad$ $t - 3 = 0$
$\quad t = -3$ $\qquad\qquad\quad t = 3$

Since the time cannot be negative, the solution is 3 seconds.

2. Let $x =$ the unknown number.

$$x^2 - 2x = 63$$
$$x^2 - 2x - 63 = 0$$
$$(x - 9)(x + 7) = 0$$

$x - 9 = 0$ $\quad$ or $\quad$ $x + 7 = 0$
$\quad x = 9$ $\qquad\qquad\quad x = -7$

There are two numbers. They are 9 and -7.

3. Let $x =$ the width of the rectangle. Then $x + 5 =$ the length of the rectangle.

$$A = lw$$
$$176 = (x + 5)(x)$$
$$176 = x^2 + 5x$$
$$0 = x^2 + 5x - 176$$
$$0 = (x + 16)(x - 11)$$

$x + 16 = 0$ $\quad$ or $\quad$ $x - 11 = 0$
$\quad x = -16$ $\qquad\qquad x = 11$

Since the dimensions cannot be negative, we discard $x = -16$. The width is 11 feet and the length is $11 + 5 = 16$ feet.

4. a. Let $x =$ the first integer. Then $x + 2 =$ the next consecutive odd integer.

$$x(x + 2) - 23 = x + (x + 2)$$
$$x^2 + 2x - 23 = 2x + 2$$
$$x^2 - 25 = 0$$
$$(x + 5)(x - 5) = 0$$

$x + 5 = 0$ $\quad$ or $\quad$ $x - 5 = 0$
$\quad x = -5$ $\qquad\qquad x = 5$

The integers are -5 and -3 or 5 and 7.

b. Let x = the length of the shorter leg. Then,
$x + 7$ = the length of the longer leg. By the Pythagorean theorem

$$x^2 + (x+7)^2 = 13^2$$
$$x^2 + x^2 + 14x + 49 = 169$$
$$2x^2 + 14x + 49 = 169$$
$$2x^2 + 14x - 120 = 0$$
$$2\left(x^2 + 7x - 60\right) = 0$$
$$2(x + 12)(x - 5) = 0$$
$$x + 12 = 0 \quad \text{or} \quad x - 5 = 0$$
$$x = -12 \qquad\qquad x = 5$$

Since the length cannot be negative, we discard $x = -12$. The legs are 5 meters and
$5 + 7 = 12$ meters.

Exercise Set 4.7

1. Let x = the width, then $x + 4$ = the length.

3. Let x = the first odd integer, then
$x + 2$ = the next consecutive odd integer.

5. Let x = the base, then $4x + 1$ = the height.

7. Let x = the length of one side.
$$A = x^2$$
$$121 = x^2$$
$$0 = x^2 - 121$$
$$0 = x^2 - 11^2$$
$$0 = (x + 11)(x - 11)$$
$$x + 11 = 0 \quad \text{or} \quad x - 11 = 0$$
$$x = -11 \qquad\qquad x = 11$$

Since the length cannot be negative, the sides are 11 units long.

9. The perimeter is the sum of the lengths of the sides.

$$120 = (x+5) + (x^2 - 3x) + (3x - 8) + (x+3)$$
$$120 = x + 5 + x^2 - 3x + 3x - 8 + x + 3$$
$$120 = x^2 + 2x$$
$$0 = x^2 + 2x - 120$$
$$x^2 + 2x - 120 = 0$$
$$(x + 12)(x - 10) = 0$$
$$x + 12 = 0 \quad \text{or} \quad x - 10 = 0$$
$$x = -12 \qquad\qquad x = 10$$

Since the dimensions cannot be negative, the lengths of the sides are: $10 + 5 = 15$ cm,
$10^2 - 3(10) = 70$ cm, $3(10) - 8 = 22$ cm, and
$10 + 3 = 13$ cm.

11. $x + 5$ = the base and $x - 5$ = the height.
$$A = bh$$
$$96 = (x + 5)(x - 5)$$
$$96 = x^2 + 5x - 5x - 25$$
$$96 = x^2 - 25$$
$$0 = x^2 - 121$$
$$x^2 - 121 = 0$$
$$(x + 11)(x - 11) = 0$$
$$x + 11 = 0 \quad \text{or} \quad x - 11 = 0$$
$$x = -11 \qquad\qquad x = 11$$

Since the dimensions cannot be negative,
$x = 11$. The base is $11 + 5 = 16$ miles, and the height is $11 - 5 = 6$ miles.

13. Find t when $h = 0$.
$$h = -16t^2 + 64t + 80$$
$$0 = -16t^2 + 64t + 80$$
$$0 = -16(t^2 - 4t - 5)$$
$$0 = -16(t - 5)(t + 1)$$
$$t - 5 = 0 \quad \text{or} \quad t + 1 = 0$$
$$t = 5 \qquad\qquad t = -1$$

Since the time t cannot be negative, the object hits the ground after 5 seconds.

15. Let x = the width, then $2x - 7$ = the length.
$$A = lw$$
$$30 = (2x - 7)(x)$$
$$30 = 2x^2 - 7x$$
$$0 = 2x^2 - 7x - 30$$
$$0 = (2x + 5)(x - 6)$$
$$2x + 5 = 0 \quad \text{or} \quad x - 6 = 0$$
$$2x = -5 \qquad\qquad x = 6$$
$$x = -\frac{5}{2}$$

Since the dimensions cannot be negative, the width is 6 cm and the length is
$2(6) - 7 = 5$ cm.

17. Let $n = 12$.

$$D = \frac{1}{2}n(n-3)$$

$$D = \frac{1}{2} \cdot 12(12-3) = 6(9) = 54$$

A polygon with 12 sides has 54 diagonals.

19. Let $D = 35$ and solve for n.

$$D = \frac{1}{2}n(n-3)$$

$$35 = \frac{1}{2}n(n-3)$$

$$35 = \frac{1}{2}n^2 - \frac{3}{2}n$$

$$0 = \frac{1}{2}n^2 - \frac{3}{2}n - 35$$

$$0 = \frac{1}{2}(n^2 - 3n - 70)$$

$$0 = \frac{1}{2}(n-10)(n+7)$$

$n - 10 = 0$ or $n + 7 = 0$
$n = 10$ $n = -7$

The polygon has 10 sides.

21. Let $x =$ the unknown number.

$$x + x^2 = 132$$
$$x^2 + x - 132 = 0$$
$$(x + 12)(x - 11) = 0$$

$x + 12 = 0$ or $x - 11 = 0$
$x = -12$ $x = 11$

There are two numbers. They are -12 and 11.

23. Let $x =$ the rate (in mph) of the slower boat, then $x + 7 =$ the rate (in mph) of the faster boat. After one hour, the slower boat has traveled x miles and the faster boat has traveled $x + 7$ miles. By the Pythagorean theorem,

$$x^2 + (x+7)^2 = 17^2$$
$$x^2 + x^2 + 14x + 49 = 289$$
$$2x^2 + 14x + 49 = 289$$
$$2x^2 + 14x - 240 = 0$$
$$2(x^2 + 7x - 120) = 0$$
$$2(x + 15)(x - 8) = 0$$

$x + 15 = 0$ or $x - 8 = 0$
$x = -15$ $x = 8$

Since the rate cannot be negative, the slower boat travels at 8 mph. The faster boat travels at $8 + 7 = 15$ mph.

25. Let $x =$ the first number, then $20 - x =$ the other number.

$$x^2 + (20 - x)^2 = 218$$
$$x^2 + 400 - 40x + x^2 = 218$$
$$2x^2 - 40x + 400 = 218$$
$$2x^2 - 40x + 182 = 0$$
$$2(x^2 - 20x + 91) = 0$$
$$2(x - 13)(x - 7) = 0$$

$x - 13 = 0$ or $x - 7 = 0$
$x = 13$ $x = 7$

The numbers are 13 and 7.

27. Let $x =$ the length of a side of the original square. Then $x + 3 =$ the length of a side of the larger square.

$$64 = (x + 3)^2$$
$$64 = x^2 + 6x + 9$$
$$0 = x^2 + 6x - 55$$
$$0 = (x + 11)(x - 5)$$

$x + 11 = 0$ or $x - 5 = 0$
$x = -11$ $x = 5$

Since the length cannot be negative, the sides of the original square are 5 inches long.

29. Let $x =$ the length of the shorter leg. Then $x + 4 =$ the length of the longer leg and $x + 8 =$ the length of the hypotenuse. By the Pythagorean theorem

$$x^2 + (x+4)^2 = (x+8)^2$$
$$x^2 + x^2 + 8x + 16 = x^2 + 16x + 64$$
$$2x^2 + 8x + 16 = x^2 + 16x + 64$$
$$x^2 - 8x - 48 = 0$$
$$(x - 12)(x + 4) = 0$$

$x - 12 = 0$ or $x + 4 = 0$
$x = 12$ $x = -4$

Since the length cannot be negative, the sides of the triangle are 12 mm, $12 + 4 = 16$ mm, and $12 + 8 = 20$ mm.

31. Let x = the height of the triangle, then $2x$ = the base.

$$A = \frac{1}{2}bh$$

$$100 = \frac{1}{2}(2x)(x)$$

$$100 = x^2$$

$$0 = x^2 - 100$$

$$0 = (x + 10)(x - 10)$$

$x + 10 = 0$ or $x - 10 = 0$

$x = -10$ $x = 10$

Since the altitude cannot be negative, the height of the triangle is 10 km.

33. Let x = the length of the shorter leg, then $x + 12$ = the length of the longer leg and $2x - 12$ = the length of the hypotenuse. By the Pythagorean theorem

$$x^2 + (x + 12)^2 = (2x - 12)^2$$

$$x^2 + x^2 + 24x + 144 = 4x^2 - 48x + 144$$

$$2x^2 + 24x + 144 = 4x^2 - 48x + 144$$

$$0 = 2x^2 - 72x$$

$$0 = 2x(x - 36)$$

$2x = 0$ or $x - 36 = 0$

$x = 0$ $x = 36$

Since the length cannot be zero feet, the shorter leg is 36 feet long.

35. Find t when $h = 0$.

$$h = -16t^2 + 625$$

$$0 = -16t^2 + 625$$

$$0 = -(16t^2 - 625)$$

$$0 = -(4t + 25)(4t - 25)$$

$4t + 25 = 0$ or $4t - 25 = 0$

$4t = -25$ $4t = 25$

$t = -\dfrac{25}{4}$ $t = \dfrac{25}{4}$

$t = -6.25$ $t = 6.25$

Since the time cannot be negative, the object will reach the ground after 6.25 seconds.

37. The size of the average farm in 1940 was approximately 175 acres.

39. The number of farms in 1940 was approximately 6.25 million.

41. The lines intersect at approximately 1966.

43. Answers may vary.

45. Let x = the width of the pool, then $x + 6$ = the length of the pool, $x + 8$ = the combined width of the pool and the walk, and $x + 14$ = the combined length of the pool and the walk.

The area of the pool is $(x + 6)(x) = x^2 + 6x$. The total area is

$$(x + 14)(x + 8) = x^2 + 22x + 112.$$

$$x^2 + 22x + 112 = \left(x^2 + 6x\right) + 576$$

$$16x = 464$$

$$x = 29$$

The pool is 29 meters wide and $29 + 6 = 35$ meters long.

47. Let x = the first integer, then $x + 1$ = the next consecutive integer.

$$x(x + 1) = 870$$

$$x^2 + x = 870$$

$$x^2 + x - 870 = 0$$

$$(x + 30)(x - 29) = 0$$

$x + 30 = 0$ or $x - 29 = 0$

     $x = -30$ $x = 29$

Since the digits are positive integers, the ZIP codes are 60129 and 60130.

60129: Esmond, IL

60130: Forest Park, IL

Chapter 4 Review

1. $6x^2 - 15x = 3x(2x - 5)$

2. $4x^5 + 2x - 10x^4 = 2x\left(2x^4 + 1 - 5x^3\right)$

3. $5m + 30 = 5(m + 6)$

4. $20x^3 + 12x^2 + 24x = 4x\left(5x^2 + 3x + 6\right)$

5. $3x(2x + 3) - 5(2x + 3) = (2x + 3)(3x - 5)$

6. $5x(x + 1) - (x + 1) = (x + 1)(5x - 1)$

7. $3x^2 - 3x + 2x - 2 = 3x(x - 1) + 2(x - 1)$
$ = (x - 1)(3x + 2)$

8. $6x^2 + 10x - 3x - 5 = 2x(3x+5) - 1(3x+5)$
$= (3x+5)(2x-1)$

9. $3a^2 + 9ab + 3b^2 + ab$
$= 3a(a+3b) + b(3b+a)$
$= 3a(a+3b) + b(a+3b) = (a+3b)(3a+b)$

10. $x^2 + 6x + 8 = (x+4)(x+2)$

11. $x^2 - 11x + 24 = (x-8)(x-3)$

12. $x^2 + x + 2$ is a prime polynomial.

13. $x^2 - 5x - 6 = (x-6)(x+1)$

14. $x^2 + 2x - 8 = (x+4)(x-2)$

15. $x^2 + 4xy - 12y^2 = (x+6y)(x-2y)$

16. $x^2 + 8xy + 15y^2 = (x+5y)(x+3y)$

17. $72 - 18x - 2x^2 = 2\left(36 - 9x - x^2\right)$
$= 2(3-x)(12+x)$

18. $32 + 12x - 4x^2 = 4\left(8 + 3x - x^2\right)$

19. $5y^3 - 50y^2 + 120y = 5y\left(y^2 - 10y + 24\right)$
$= 5y(y-6)(y-4)$

20. To factor $x^2 + 2x - 48$, think of two numbers whose product is −48 and whose sum is 2.

21. Factor out the GCF, which is 3.

22. $2x^2 + 13x + 6 = 2x^2 + 12x + x + 6$
$= 2x(x+6) + 1(x+6)$
$= (x+6)(2x+1)$

23. $4x^2 + 4x - 3 = 4x^2 + 6x - 2x - 3$
$= 2x(2x+3) - 1(2x+3)$
$= (2x+3)(2x-1)$

24. $6x^2 + 5xy - 4y^2 = 6x^2 + 8xy - 3xy - 4y^2$
$= 2x(3x+4y) - y(3x+4y) = (3x+4y)(2x - y)$

25. $x^2 - x + 2$ is a prime polynomial.

26. $2x^2 - 23x - 39 = 2x^2 - 26x + 3x - 39$
$= 2x(x-13) + 3(x-13) = (x-13)(2x+3)$

27. $18x^2 - 9xy - 20y^2$
$= 18x^2 - 24xy + 15xy - 20y^2$
$= 6x(3x-4y) + 5y(3x-4y)$
$= (3x-4y)(6x+5y)$

28. $10y^3 + 25y^2 - 60y = 5y\left(2y^2 + 5y - 12\right)$
$= 5y\left(2y^2 + 8y - 3y - 12\right)$
$= 5y[2y(y+4) - 3(y+4)]$
$= 5y(y+4)(2y-3)$

29. The perimeter is the sum of the lengths of the sides.
$P = \left(x^2 - 2\right) + \left(x^2 - 4x\right) + \left(3x^2 - 5x\right)$
$= x^2 - 2 + x^2 - 4x + 3x^2 - 5x$
$= 5x^2 - 9x - 2$
$= (5x+1)(x-2)$

30. $l = 6x^2 - 14x$ and $w = 2x^2 + 3$.
$P = 2l + 2w$
$P = 2\left(6x^2 - 14x\right) + 2\left(2x^2 + 3\right)$
$= 12x^2 - 28x + 4x^2 + 6$
$= 16x^2 - 28x + 6$
$= 2\left(8x^2 - 14x + 3\right)$
$= 2(4x-1)(2x-3)$

31. Yes; two terms, x^2 and 9, are squares $\left(9 = 3^2\right)$ and the third term of the trinomial, $6x$, is twice the product of x and 3 $(2 \cdot x \cdot 3 = 6x)$.

32. No; the two terms x^2 and 64 are squares $\left(64 = 8^2\right)$, but the third term of the trinomial, $8x$, is not twice the product of x and 8, or its opposite.

33. No; the two terms $9m^2$ and 16 are squares $(9m^2 = (3m)^2$ and $16 = 4^2)$, but the third term of the trinomial, $-12m$, is not twice the product of $3m$ and 4, or its opposite.

34. Yes; two terms, $4y^2$ and 49, are squares $(4y^2 = (2y)^2$ and $49 = 7^2)$ and the third term of the trinomial, $-28y$, is the opposite of twice the product of $2y$ and 7.

35. Yes, $x^2 - 9 = x^2 - 3^2$ is the difference of squares.

36. No; $x^2 + 16$ is the sum of two squares, $x^2 + 16 = x^2 + 4^2$.

37. Yes; $4x^2 - 25y^2 = (2x)^2 - (5y)^2$ is the difference of two squares.

38. No; only one of the terms, 1, is a square.

39. $x^2 - 81 = x^2 - 9^2 = (x + 9)(x - 9)$

40. $x^2 + 12x + 36 = (x + 6)(x + 6) = (x + 6)^2$

41. $4x^2 - 9 = (2x)^2 - 3^2 = (2x + 3)(2x - 3)$

42. $9t^2 - 25s^2 = (3t)^2 - (5s)^2$
$= (3t + 5s)(3t - 5s)$

43. $16x^2 + y^2$ is a prime polynomial.

44. $n^2 - 18n + 81 = (n - 9)(n - 9) = (n - 9)^2$

45. $3r^2 + 36r + 108 = 3\left(r^2 + 12r + 36\right)$
$= 3(r + 6)(r + 6) = 3(r + 6)^2$

46. $9y^2 - 42y + 49 = (3y - 7)(3y - 7)$
$= (3y - 7)^2$

47. $5m^8 - 5m^6 = 5m^6\left(m^2 - 1\right) = 5m^6\left(m^2 - 1^2\right)$
$= 5m^6(m + 1)(m - 1)$

48. $4x^2 - 28xy + 49y^2 = (2x - 7y)(2x - 7y)$
$= (2x - 7y)^2$

49. $3x^2y + 6xy^2 + 3y^3 = 3y\left(x^2 + 2xy + y^2\right)$
$= 3y(x + y)(x + y) = 3y(x + y)^2$

50. $16x^4 - 1 = \left(4x^2\right)^2 - 1^2 = \left(4x^2 + 1\right)\left(4x^2 - 1\right)$
$= \left(4x^2 + 1\right)\left((2x)^2 - 1^2\right)$
$= \left(4x^2 + 1\right)(2x + 1)(2x - 1)$

51. $(x + 6)(x - 2) = 0$
$x + 6 = 0 \quad$ or $\quad x - 2 = 0$
$x = -6 \qquad\qquad x = 2$
The solutions are –6 and 2.

52. $3x(x + 1)(7x - 2) = 0$
$3x = 0 \quad$ or $\quad x + 1 = 0 \quad$ or $\quad 7x - 2 = 0$
$x = 0 \qquad\qquad x = -1 \qquad\qquad 7x = 2$
$\qquad\qquad\qquad\qquad\qquad\qquad\qquad x = \frac{2}{7}$

The solutions are 0, –1, and $\frac{2}{7}$.

53. $4(5x + 1)(x + 3) = 0$
$5x + 1 = 0 \quad$ or $\quad x + 3 = 0$
$5x = -1 \qquad\qquad x = -3$
$x = -\frac{1}{5}$

The solutions are $-\frac{1}{5}$ and –3.

54. $x^2 + 8x + 7 = 0$
$(x + 7)(x + 1) = 0$
$x + 7 = 0 \quad$ or $\quad x + 1 = 0$
$x = -7 \qquad\qquad x = -1$
The solutions are –7 and –1.

55. $x^2 - 2x - 24 = 0$
$(x - 6)(x + 4) = 0$
$x - 6 = 0 \quad$ or $\quad x + 4 = 0$
$x = 6 \qquad\qquad x = -4$
The solutions are 6 and –4.

56. $x^2 + 10x = -25$
$x^2 + 10x + 25 = 0$
$(x + 5)(x + 5) = 0$
$x + 5 = 0 \quad$ or $\quad x + 5 = 0$
$\quad\quad x = -5 \quad\quad\quad\quad x = -5$
The solution is -5.

57. $x(x - 10) = -16$
$x^2 - 10x = -16$
$x^2 - 10x + 16 = 0$
$(x - 8)(x - 2) = 0$
$x - 8 = 0 \quad$ or $\quad x - 2 = 0$
$\quad x = 8 \quad\quad\quad\quad\quad x = 2$
The solutions are 8 and 2.

58. $(3x - 1)\left(9x^2 + 3x + 1\right) = 0$

$3x - 1 = 0 \quad\quad$ or $\quad\quad 9x^2 + 3x + 1 = 0$
$9x^2 + 3x + 1$ is a prime polynomial.
$3x - 1 = 0$
$3x = 1$
$x = \dfrac{1}{3}$

59. $56x^2 - 5x - 6 = 0$
$56x^2 + 16x - 21x - 6 = 0$
$8x(7x + 2) - 3(7x + 2) = 0$
$(7x + 2)(8x - 3) = 0$
$7x + 2 = 0 \quad$ or $\quad 8x - 3 = 0$
$7x = -2 \quad\quad\quad\quad 8x = 3$
$x = -\dfrac{2}{7} \quad\quad\quad x = \dfrac{3}{8}$
The solutions are $-\dfrac{2}{7}$ and $\dfrac{3}{8}$.

60. $m^2 = 6m$
$m^2 - 6m = 0$
$m(m - 6) = 0$
$m = 0 \quad$ or $\quad m - 6 = 0$
$\quad\quad\quad\quad\quad\quad m = 6$
The solutions are 0 and 6.

61. $r^2 = 25$
$r^2 - 25 = 0$
$r^2 - 5^2 = 0$
$(r + 5)(r - 5) = 0$
$r + 5 = 0 \quad$ or $\quad r - 5 = 0$
$\quad r = -5 \quad\quad\quad\quad r = 5$
The solutions are -5 and 5.

62. If $x = 4$ and $x = 5$ are the solutions, then, by the zero factor property
$x = 4 \quad$ or $\quad\quad x = 5$
$x - 4 = 0 \quad\quad\quad x - 5 = 0$
$(x - 4)(x - 5) = 0$
$x^2 - 4x - 5x + 20 = 0$
$x^2 - 9x + 20 = 0$

63. Let $x =$ the width, then $2x =$ the length.
$P = 2l + 2w$
$24 = 2(2x) + 2x$
$24 = 4x + 2x$
$24 = 6x$
$4 = x$
The width is 4 inches and the length is $2 \cdot 4 = 8$ inches. Thus, c is the correct answer.

64. Let $x =$ the width, then $3x + 1 =$ the length.
$A = lw$
$80 = (3x + 1)(x)$
$80 = 3x^2 + x$
$0 = 3x^2 + x - 80$
$0 = (3x + 16)(x - 5)$
$3x + 16 = 0 \quad$ or $\quad x - 5 = 0$
$3x = -16 \quad\quad\quad\quad x = 5$
$x = -\dfrac{16}{3}$
Since the width cannot be negative, the width is 5 meters and the length is $3(5) + 1 = 16$ meters. Thus d is the correct answer.

65. $x^2 = 81$
$x^2 - 81 = 0$
$(x + 9)(x - 9) = 0$
$x + 9 = 0 \quad\quad$ or $\quad\quad x - 9 = 0$
$x = -9 \quad\quad\quad\quad\quad\quad x = 9$
Since the length cannot be negative, the sides are 9 units long.

66. The perimeter is the sum of the lengths of the sides.
$47 = (2x + 3) + (3x + 1) + \left(x^2 - 3x\right) + (x + 3)$
$47 = 2x + 3 + 3x + 1 + x^2 - 3x + x + 3$
$47 = x^2 + 3x + 7$
$0 = x^2 + 3x - 40$
$0 = (x + 8)(x - 5)$

$$x + 8 = 0 \quad \text{or} \quad x - 5 = 0$$
$$x = -8 \qquad\qquad x = 5$$

Since the lengths cannot be negative, $x = 5$.
The lengths of the sides are $2(5) + 3 = 13$ units,

$3(5) + 1 = 16$ units, $5^2 - 3(5) = 10$ units, and
$5 + 3 = 8$ units.

67. Let $x =$ the width of the flag. Then
$2x - 15 =$ the length of the flag.
$A = lw$
$$500 = (2x - 15)(x)$$
$$500 = 2x^2 - 15x$$
$$0 = 2x^2 - 15x - 500$$
$$0 = (2x + 25)(x - 20)$$
$$2x + 25 = 0 \quad \text{or} \quad x - 20 = 0$$
$$2x = -25 \qquad\qquad x = 20$$
$$x = -\frac{25}{2}$$

Since the dimensions cannot be negative, the
width is 20 inches and the length is
$2(20) - 15 = 25$ inches.

68. Let $x =$ the height of the sail. Then
$4x =$ the base of the sail.
$A = \frac{1}{2}bh$
$$162 = \frac{1}{2} \cdot 4x \cdot x$$
$$162 = 2x^2$$
$$0 = 2x^2 - 162$$
$$0 = 2\left(x^2 - 81\right)$$
$$0 = 2(x + 9)(x - 9)$$
$$x + 9 = 0 \quad \text{or} \quad x - 9 = 0$$
$$x = -9 \qquad\qquad x = 9$$

Since the dimensions cannot be negative, the
height is 9 yards and the base is $4 \cdot 9 = 36$
yards.

69. Let $x =$ the first integer. Then
$x + 1 =$ the next consecutive integer.
$$x(x + 1) = 380$$
$$x^2 + x = 380$$
$$x^2 + x - 380 = 0$$
$$(x + 20)(x - 19) = 0$$
$$x + 20 = 0 \quad \text{or} \quad x - 19 = 0$$
$$x = -21 \qquad\qquad x = 19$$

The integers are 19 and 20.

70. a. Let $h = 2800$ and solve for t.
$$h = -16t^2 + 440t$$
$$2800 = -16t^2 + 440t$$
$$0 = -16t^2 + 440t - 2800$$
$$0 = -8\left(2t^2 - 55t + 350\right)$$
$$0 = -8(2t - 35)(t - 10)$$
$$2t - 35 = 0 \quad \text{or} \quad t - 10 = 0$$
$$2t = 35 \qquad\qquad t = 10$$
$$t = \frac{35}{2}$$
$$t = 17.5$$

The solutions are 17.5 and 10. Answers
may vary.

b. Find t when $h = 0$.
$$h = -16t^2 + 440t$$
$$0 = 16t^2 + 440t$$
$$0 = -8t(2t - 55)$$
$$-8t = 0 \quad \text{or} \quad 2t - 55 = 0$$
$$t = 0 \qquad\qquad 2t = 55$$
$$t = \frac{55}{2}$$
$$t = 27.5$$

27.5 seconds after being fired, the
rocket will reach the ground again.

71. Let $x =$ the length of the longer leg. Then,
$x - 8 =$ the length of the shorter leg and
$x + 8 =$ the length of the hypotenuse. By the
Pythagorean theorem,
$$x^2 + (x - 8)^2 = (x + 8)^2$$
$$x^2 + x^2 - 16x + 64 = x^2 + 16x + 64$$
$$x^2 - 32x = 0$$
$$x(x - 32) = 0$$
$$x = 0 \quad \text{or} \quad x = 32$$

Since the length cannot be zero cm, the
length of the longer leg is 32 cm.

Chapter 4 Test

1. $9x^2 - 3x = 3x(3x - 1)$

2. $x^2 + 11x + 28 = (x + 7)(x + 4)$

3. $49 - m^2 = 7^2 - m^2 = (7 + m)(7 - m)$

4. $y^2 + 22y + 121 = (y + 11)(y + 11) = (y + 11)^2$

5. $x^4 - 16 = (x^2)^2 - 4^2 = \left(x^2 + 4\right)\left(x^2 - 4\right)$

       $= (x^2 + 4)(x^2 - 2^2) = \left(x^2 + 4\right)(x + 2)(x - 2)$

6. $4(a + 3) - y(a + 3) = (a + 3)(4 - y)$

7. $x^2 + 4$ is a prime polynomial.

8. $y^2 - 8y - 48 = (y - 12)(y + 4)$

9. $3a^2 + 3ab - 7a - 7b = 3a(a + b) - 7(a + b)$

       $= (a + b)(3a - 7)$

10. $3x^2 - 5x + 2 = (3x - 2)(x - 1)$

11. $180 - 5x^2 = 5\left(36 - x^2\right) = 5(6^2 - x^2)$

       $= 5(6 + x)(6 - x)$

12. $3x^3 - 21x^2 + 30x = 3x\left(x^2 - 7x + 10\right)$

       $= 3x(x - 5)(x - 2)$

13. $6t^2 - t - 5 = (6t + 5)(t - 1)$

14. $xy^2 - 7y^2 - 4x + 28 = y^2(x - 7) - 4(x - 7)$

       $= (x - 7)\left(y^2 - 4\right) = (x - 7)(y^2 - 2^2)$

       $= (x - 7)(y + 2)(y - 2)$

15. $x - x^5 = x\left(1 - x^4\right) = x(1 - (x^2)^2)$

       $= x\left(1 + x^2\right)\left(1 - x^2\right)$

       $= x\left(1 + x^2\right)(1 + x)(1 - x)$

16. $x^2 + 14xy + 24y^2 = (x + 12y)(x + 2y)$

17. $(x - 3)(x + 9) = 0$

      $x - 3 = 0$ or $x + 9 = 0$

         $x = 3$ $x = -9$

      The solutions are 3 and –9.

18. $x^2 + 10x + 24 = 0$

      $(x + 6)(x + 4) = 0$

      $x + 6 = 0$ or $x + 4 = 0$

        $x = -6$ $x = -4$

      The solutions are –6 and –4.

19. $x^2 + 5x = 14$

      $x^2 + 5x - 14 = 0$

      $(x + 7)(x - 2) = 0$

      $x + 7 = 0$ or $x - 2 = 0$

        $x = -7$ $x = 2$

      The solutions are –7 and 2.

20. $3x(2x - 3)(3x + 4) = 0$

      $3x = 0$ or $2x - 3 = 0$ or $3x + 4 = 0$

       $x = 0$ $2x = 3$ $3x = -4$

                      $x = \dfrac{3}{2}$ $x = -\dfrac{4}{3}$

      The solutions are 0, $\dfrac{3}{2}$, and $-\dfrac{4}{3}$.

21. $5t^3 - 45t = 0$

      $5t\left(t^2 - 9\right) = 0$

      $5t(t + 3)(t - 3) = 0$

      $5t = 0$ or $t + 3 = 0$ or $t - 3 = 0$

       $t = 0$ $t = -3$ $t = 3$

      The solutions are 0, –3, and 3.

22. $3x^2 = -12x$

      $3x^2 + 12x = 0$

      $3x(x + 4) = 0$

      $3x = 0$ or $x + 4 = 0$

       $x = 0$ $x = -4$

      The solutions are 0 and –4.

23. $t^2 - 2t - 15 = 0$

      $(t - 5)(t + 3) = 0$

      $t - 5 = 0$ or $t + 3 = 0$

        $t = 5$ $t = -3$

      The solutions are 5 and –3.

24. $(x - 1)\left(3x^2 - x - 2\right) = 0$

      $(x - 1)(3x + 2)(x - 1) = 0$

      $(x - 1)^2(3x + 2) = 0$

      $x - 1 = 0$ or $3x + 2 = 0$

        $x = 1$ $3x = -2$

                     $x = -\dfrac{2}{3}$

      The solutions are 1 and $-\dfrac{2}{3}$.

25. $x + 2 =$ the length of the rectangle and
$x - 1 =$ the width of the rectangle.
$A = lw$
$54 = (x + 2)(x - 1)$
$54 = x^2 + x - 2$
$0 = x^2 + x - 56$
$0 = (x + 8)(x - 7)$
$x + 8 = 0$ or $x - 7 = 0$
     $x = -8$ $x = 7$
Since the dimensions cannot be negative,
length of the rectangle is $7 + 2 = 9$ units, and
the width is $7 - 1 = 6$ units.

26. Let $x =$ the height of the triangle. Then
$x + 9 =$ the base of the triangle.
$A = \frac{1}{2} bh$
$68 = \frac{1}{2}(x + 9)(x)$
$68 = \frac{1}{2} x^2 + \frac{9}{2} x$
$0 = \frac{1}{2} x^2 + \frac{9}{2} x - 68$
$0 = \frac{1}{2}\left(x^2 + 9x - 136\right)$
$0 = \frac{1}{2}(x + 17)(x - 8)$
$x + 17 = 0$ or $x - 8 = 0$
     $x = -17$ $x = 8$
Since the length of the base cannot be
negative, the base is $8 + 9 = 17$ feet.

27. Let $x =$ the first number, then
$17 - x =$ the other number.
$x^2 + (17 - x)^2 = 145$
$x^2 + 289 - 34x + x^2 = 145$
$2x^2 - 34x + 144 = 0$
$2\left(x^2 - 17x + 72\right) = 0$
$2(x - 9)(x - 8) = 0$
$x - 9 = 0$ or $x - 8 = 0$
    $x = 9$ $x = 8$
The numbers are 8 and 9.

28. Find t when $h = 0$.
$h = -16t^2 + 784$
$0 = -16t^2 + 784$
$0 = -16\left(t^2 - 49\right)$
$0 = -16(t + 7)(t - 7)$

$t + 7 = 0$ or $t - 7 = 0$
   $t = -7$ $t = 7$
Since the time cannot be negative, the object
reaches the ground after 7 seconds.

29. Let $x =$ the length of the shorter leg. Then
$x + 10 =$ the length of the hypotenuse, and
$x + 5 =$ the length of the longer leg. By the
Pythagorean theorem
$x^2 + (x + 5)^2 = (x + 10)^2$
$x^2 + x^2 + 10x + 25 = x^2 + 20x + 100$
$x^2 - 10x - 75 = 0$
$(x - 15)(x + 5) = 0$
$x - 15 = 0$ or $x + 5 = 0$
    $x = 15$ $x = -5$
Since the lengths cannot be negative, the
length of the shorter leg is 15 cm, the longer
leg is
$15 + 5 = 20$ cm, and the hypotenuse is
$5 + 10 = 25$ cm.

Chapter 4 Cumulative Review

1. a. $9 \le 11$

    b. $8 > 1$

    c. $3 \ne 4$

2. Replace x with 2 and see if a true statement
results.
$3x + 10 = 8x$
$3(2) + 10 \overset{?}{=} 8(2)$
$6 + 10 \overset{?}{=} 16$
$16 = 16$
Since $16 = 16$ is a true statement, 2 is a
solution of the equation $3x + 10 = 8x$.

3. $-4 - 8 = -4 + (-8) = -12$

4. a. $5x - y = 5(-2) - (-4) = -10 + 4 = -6$

    b. $x^3 - y^2 = (-2)^3 - (-4)^2 = -8 - 16$
            $= -24$

5. $2x + 3x + 5 + 2 = (2 + 3)x + (5 + 2) = 5x + 7$

6. $-5a - 3 + a + 2 = -5a + a + (-3 + 2)$
     $= -4a - 1$

7. $2.3x + 5x - 6 = (2.3 + 5)x - 6 = 7.3x - 6$

8. $-3x = 33$

$$\frac{-3x}{-3} = \frac{33}{-3}$$

$$x = -11$$

9. $3(x - 4) = 3x - 12$

$3x - 12 = 3x - 12$

$3x - 12 + 12 = 3x - 12 + 12$

$3x = 3x$

$3x - 3x = 3x - 3x$

$0 = 0$

Since $0 = 0$ is true for every value of x, every real number is a solution.

10. $V = lwh$

$$\frac{V}{wh} = \frac{lwh}{wh}$$

$$\frac{V}{wh} = l$$

11. $\dfrac{x - 5}{3} = \dfrac{x + 2}{5}$

$5(x - 5) = 3(x + 2)$

$5x - 25 = 3x + 6$

$5x = 3x + 31$

$2x = 31$

$$\frac{2x}{2} = \frac{31}{2}$$

$$x = \frac{31}{2}$$

12. $(5^3)^6 = 5^{3 \cdot 6} = 5^{18}$

13. $(y^8)^2 = y^{8 \cdot 2} = y^{16}$

14. $\dfrac{(x^3)^4 x}{x^7} = \dfrac{x^{12} \cdot x}{x^7} = \dfrac{x^{12+1}}{x^7} = \dfrac{x^{13}}{x^7}$

$= x^{13-7} = x^6$

15. $(y^{-3}z^6)^{-6} = (y^{-3})^{-6}(z^6)^{-6}$

$= y^{18}z^{-36} = \dfrac{y^{18}}{z^{36}}$

16. $\dfrac{x^{-7}}{(x^4)^3} = \dfrac{x^{-7}}{x^{12}} = x^{-7-12} = x^{-19} = \dfrac{1}{x^{19}}$

17. $-3x + 7x = (-3 + 7)x = 4x$

18. $11x^2 + 5 + 2x^2 - 7 = 11x^2 + 2x^2 + 5 - 7$

$= 13x^2 - 2$

19. $(2x - y)^2 = (2x - y)(2x - y)$

$= 2x(2x) + 2x(-y) + (-y)(2x) + (-y)(-y)$

$= 4x^2 - 2xy - 2xy + y^2 = 4x^2 - 4xy + y^2$

20. $(t + 2)^2 = t^2 + 2(t)(2) + 2^2 = t^2 + 4t + 4$

21. $(x^2 - 7y)^2 = (x^2)^2 - 2(x^2)(7y) + (7y)^2$

$= x^4 - 14x^2y + 49y^2$

22. $\dfrac{8x^2y^2 - 16xy + 2x}{4xy} = \dfrac{8x^2y^2}{4xy} - \dfrac{16xy}{4xy} + \dfrac{2x}{4xy}$

$= 2xy - 4 + \dfrac{1}{2y}$

23. $5(x + 3) + y(x + 3) = (x + 3)(5 + y)$

24. $x^4 + 5x^2 + 6 = (x^2 + 2)(x^2 + 3)$

25. $6x^2 - 2x - 20 = 2(3x^2 - x - 10)$

$= 2(3x^2 - 6x + 5x - 10)$

$= 2[3x(x - 2) + 5(x - 2)]$

$= 2(x - 2)(3x + 5)$

Chapter 5

1. Find values for x that make the denominator 0.
$$x^2 - 9x - 10 = 0$$
$$(x - 10)(x + 1) = 0$$
$$x - 10 = 0 \quad \text{or} \quad x + 1 = 0$$
$$x = 10 \qquad\qquad x = -1$$

   The rational expression $\dfrac{x+2}{x^2 - 9x - 10}$ is undefined when $x = 10$ or when $x = -1$.

2. $\dfrac{4x + 32}{x^2 + 10x + 16} = \dfrac{4(x+8)}{(x+8)(x+2)} = \dfrac{4}{x+2}$

3. Factor each denominator.
$$5x + 10 = 5(x + 2)$$
$$2x^2 + 10x + 12 = 2(x^2 + 5x + 6)$$
$$= 2(x + 3)(x + 2)$$
$$\text{LCD} = 2 \cdot 5(x+2)(x+3) = 10(x+2)(x+3)$$

4. $\dfrac{y^2 - 8y + 7}{2y - 14} \cdot \dfrac{6y + 18}{y^2 + 2y - 3}$
$$= \dfrac{(y-7)(y-1)}{2(y-7)} \cdot \dfrac{2 \cdot 3(y+3)}{(y+3)(y-1)}$$
$$= \dfrac{(y-7)(y-1) \cdot 2 \cdot 3(y+3)}{2(y-7) \cdot (y+3)(y-1)} = 3$$

5. $\dfrac{5x^3}{x^2 - 25} \div \dfrac{x^6}{(x+5)^2} = \dfrac{5x^3}{x^2 - 25} \cdot \dfrac{(x+5)^2}{x^6}$
$$= \dfrac{5x^3}{(x+5)(x-5)} \cdot \dfrac{(x+5)^2}{x^3 \cdot x^3}$$
$$= \dfrac{5x^3 \cdot (x+5)^2}{(x+5)(x-5) \cdot x^3 \cdot x^3} = \dfrac{5(x+5)}{x^3(x-5)}$$

6. $\dfrac{b}{b^2 - 9b - 22} + \dfrac{2}{b^2 - 9b - 22}$
$$= \dfrac{b+2}{b^2 - 9b - 22} = \dfrac{b+2}{(b-11)(b+2)} = \dfrac{1}{b-11}$$

7. $\dfrac{3}{x-1} - 4 = \dfrac{3}{x-1} - \dfrac{4(x-1)}{1(x-1)} = \dfrac{3 - 4(x-1)}{x-1}$
$$= \dfrac{3 - 4x + 4}{x-1} = \dfrac{7 - 4x}{x-1}$$

8. $\dfrac{2}{x-5} - \dfrac{7}{5-x} = \dfrac{2}{x-5} - \dfrac{7}{-(x-5)}$
$$= \dfrac{2}{x-5} - \dfrac{-7}{x-5} = \dfrac{2 - (-7)}{x-5} = \dfrac{9}{x-5}$$

9. $\dfrac{x}{x^2 - 16} + \dfrac{3}{x^2 - 7x + 12}$
$$= \dfrac{x}{(x+4)(x-4)} + \dfrac{3}{(x-4)(x-3)}$$
$$= \dfrac{x(x-3)}{(x+4)(x-4)(x-3)} + \dfrac{3(x+4)}{(x+4)(x-4)(x-3)}$$
$$= \dfrac{x(x-3) + 3(x+4)}{(x+4)(x-4)(x-3)}$$
$$= \dfrac{x^2 - 3x + 3x + 12}{(x+4)(x-4)(x-3)}$$
$$= \dfrac{x^2 + 12}{(x+4)(x-4)(x-3)}$$

10. $\dfrac{5}{b} + \dfrac{3}{5} = \dfrac{4}{5b}$
$$5b\left(\dfrac{5}{b} + \dfrac{3}{5}\right) = 5b\left(\dfrac{4}{5b}\right)$$
$$5b\left(\dfrac{5}{b}\right) + 5b\left(\dfrac{3}{5}\right) = 4$$
$$25 + 3b = 4$$
$$3b = -21$$
$$b = -7$$

11. $9 + \dfrac{7}{d-7} = \dfrac{d}{d-7}$

$(d-7)\left(9 + \dfrac{7}{d-7}\right) = (d-7)\left(\dfrac{d}{d-7}\right)$

$(d-7)(9) + (d-7)\left(\dfrac{7}{d-7}\right) = d$

$9d - 63 + 7 = d$

$-56 = -8d$

$d = 7$

However, $d = 7$ makes the denominators 0 in the original equation. The equation has no solution.

12. $\dfrac{4y+5}{y^2+5y+6} + \dfrac{3}{y+3} = \dfrac{2}{y+2}$

$y^2 + 5y + 6 = (y+3)(y+2)$

$(y+3)(y+2)\left(\dfrac{4y+5}{(y+3)(y+2)} + \dfrac{3}{y+3}\right) = (y+3)(y+2)\left(\dfrac{2}{y+2}\right)$

$(y+3)(y+2)\left(\dfrac{4y+5}{(y+3)(y+2)}\right) + (y+3)(y+2)\left(\dfrac{3}{y+3}\right) = 2(y+3)$

$4y + 5 + 3(y+2) = 2(y+3)$

$4y + 5 + 3y + 6 = 2y + 6$

$5y = -5$

$y = -1$

13. $\dfrac{2A}{b} = h$

$b\left(\dfrac{2A}{b}\right) = b(h)$

$2A = bh$

$\dfrac{2A}{h} = \dfrac{bh}{h}$

$\dfrac{2A}{h} = b$

14. $\dfrac{\frac{12m^3}{5n^2}}{\frac{4m^6}{25n^8}} = \dfrac{12m^3}{5n^2} \cdot \dfrac{25n^8}{4m^6} = \dfrac{12m^3 \cdot 25n^8}{5n^2 \cdot 4m^6}$

$= \dfrac{3 \cdot 4 \cdot 5 \cdot 5 m^3 \cdot n^6 \cdot n^2}{5 \cdot 4 m^3 \cdot m^3 \cdot n^2} = \dfrac{3 \cdot 5 n^6}{m^3}$

$= \dfrac{15n^6}{m^3}$

15. $\dfrac{16 - \frac{1}{a^2}}{\frac{4}{a} + \frac{1}{a^2}} = \dfrac{a^2\left(16 - \frac{1}{a^2}\right)}{a^2\left(\frac{4}{a} + \frac{1}{a^2}\right)} = \dfrac{16a^2 - 1}{4a + 1}$

$= \dfrac{(4a+1)(4a-1)}{(4a+1)} = 4a - 1$

16. $\dfrac{3}{x} = \dfrac{9}{15}$

$45 = 9x$

$5 = x$

17. Let $x =$ the unknown number.

$\left(10 \cdot \dfrac{1}{x}\right) + x = 7$

$\dfrac{10}{x} + x = 7$

$x\left(\dfrac{10}{x} + x\right) = x(7)$

$10 + x^2 = 7x$

$x^2 - 7x + 10 = 0$

$(x - 5)(x - 2) = 0$

$x - 5 = 0$ or $x - 2 = 0$

$x = 5$ $x = 2$

The number is 2 or 5.

18. Let $x =$ the time in hours it takes if they work together. Then $\dfrac{1}{x} =$ the part of the job they complete in 1 hour.

Since Sonya completes $\dfrac{1}{5}$ of the job in

1 hour, and her daughter completes $\dfrac{1}{8}$ of the job in one hour,

$\dfrac{1}{5} + \dfrac{1}{8} = \dfrac{1}{x}$

$40x\left(\dfrac{1}{5}\right) + 40x\left(\dfrac{1}{8}\right) = 40x\left(\dfrac{1}{x}\right)$

$8x + 5x = 40$

$13x = 40$

$x = \dfrac{40}{13}$

$x = 3\dfrac{1}{13}$

It will take $3\dfrac{1}{13}$ hr if they work together.

19. Let $r =$ the rate of the plane in still air. Then $r + 25 =$ the rate with a tail wind and $r - 25 =$ the rate against the wind. Since $d = rt$, $t = \dfrac{d}{r}$.

With a tail wind:

$\dfrac{495}{r + 25} = t$

Against the wind:

$\dfrac{405}{r - 25} = t$

$\dfrac{495}{r + 25} = \dfrac{405}{r - 25}$

$(r + 25)(r - 25)\left(\dfrac{495}{r + 25}\right)$

$= (r + 25)(r - 25)\left(\dfrac{405}{r - 25}\right)$

$495(r - 25) = 405(r + 25)$

$495r - 12{,}375 = 405r + 10{,}125$

$90r = 22{,}500$

$r = 250$

The rate of the plane in still air is 250 mph.

Section 5.1

Practice Problems

1. **a.** $\dfrac{x - 3}{5x + 1} = \dfrac{4 - 3}{5(4) + 1} = \dfrac{1}{20 + 1} = \dfrac{1}{21}$

    b. $\dfrac{x - 3}{5x + 1} = \dfrac{-3 - 3}{5(-3) + 1} = \dfrac{-6}{-15 + 1} = \dfrac{-6}{-14}$

        $= \dfrac{-2 \cdot 3}{-2 \cdot 7} = \dfrac{3}{7}$

2. **a.** $x + 2 = 0$

       $x = -2$

    The expression is undefined when $x = -2$.

    b. $x^2 + 5x + 4 = 0$

       $(x + 4)(x + 1) = 0$

       $x + 4 = 0$ or $x + 1 = 0$

       $x = -4$ $x = -1$

    The expression is undefined when $x = -4$ or $x = -1$.

    c. The denominator of $\dfrac{x^2 - 3x + 2}{5}$ is never zero, so there are no values of x for which the expression is undefined.

3. $\dfrac{x^4 + x^3}{5x + 5} = \dfrac{x^3(x + 1)}{5(x + 1)} = \dfrac{x^3}{5}$

4. $\dfrac{x^2 + 11x + 18}{x^2 + x - 2} = \dfrac{(x + 2)(x + 9)}{(x + 2)(x - 1)} = \dfrac{x + 9}{x - 1}$

5. $\dfrac{x^2 + 10x + 25}{x^2 + 5x} = \dfrac{(x + 5)(x + 5)}{x(x + 5)} = \dfrac{x + 5}{x}$

6. $\dfrac{x + 5}{x^2 - 25} = \dfrac{x + 5}{(x - 5)(x + 5)} = \dfrac{1}{x - 5}$

7. a. $\dfrac{x+4}{4+x} = \dfrac{x+4}{x+4} = 1$

b. $\dfrac{x-4}{4-x} = \dfrac{x-4}{(-1)(x-4)} = \dfrac{1}{-1} = -1$

Mental Math

1. $x = 0$

2. $x = 3$

3. $x = 0,\ x = 1$

4. $x = 5,\ x = 6$

Exercise Set 5.1

1. $\dfrac{x+5}{x+2} = \dfrac{2+5}{2+2} = \dfrac{7}{4}$

3. $\dfrac{y^3}{y^2-1} = \dfrac{(-2)^3}{(-2)^2-1} = \dfrac{-8}{4-1} = \dfrac{-8}{3} = -\dfrac{8}{3}$

5. $\dfrac{x^2+8x+2}{x^2-x-6} = \dfrac{2^2+8(2)+2}{2^2-2-6}$

$= \dfrac{4+16+2}{4-8}$

$= \dfrac{22}{-4}$

$= \dfrac{11\cdot 2}{-2\cdot 2}$

$= -\dfrac{11}{2}$

7. a. $\dfrac{150x^2}{x^2+3} = \dfrac{150(1)^2}{1^2+3} = \dfrac{150}{4} = 37.5$

The revenue is approximately \$37.5 million at the end of the first year.

b. $\dfrac{150x^2}{x^2+3} = \dfrac{150(2)^2}{2^2+3} = \dfrac{150(4)}{4+3}$

$= \dfrac{600}{7} \approx 85.7$

The revenue is approximately \$85.7 million at the end of the second year.

c. $\$85.7 - \$37.5 = \$48.2$ million

9. $2x = 0$
$x = 0$
The expressioin is undefined when $x = 0$.

11. $x + 2 = 0$
$x = -2$
The expression is undefined when $x = -2$.

13. $2x - 8 = 0$
$2x = 8$
$x = 4$
The expression is undefined when $x = 4$.

15. $15x + 30 = 0$
$15x = -30$
$x = -2$
The expression is undefined when $x = -2$.

17. The denominator is never zero so there are no values for which $\dfrac{x^2-5x-2}{4}$ is undefined.

19. Answers may vary.

21. $\dfrac{2}{8x+16} = \dfrac{2}{8(x+2)} = \dfrac{2}{2\cdot 4(x+2)} = \dfrac{1}{4(x+2)}$

23. $\dfrac{x-2}{x^2-4} = \dfrac{x-2}{(x+2)(x-2)} = \dfrac{1}{x+2}$

25. $\dfrac{2x-10}{3x-30} = \dfrac{2(x-5)}{3(x-10)}$; does not simplify

27. $\dfrac{x+7}{7+x} = \dfrac{x+7}{x+7} = 1$

29. $\dfrac{x-7}{7-x} = \dfrac{x-7}{-1(x-7)} = \dfrac{1}{-1} = -1$

31. $\dfrac{-5a-5b}{a+b} = \dfrac{-5(a+b)}{a+b} = -5$

33. $\dfrac{x+5}{x^2-4x-45} = \dfrac{x+5}{(x-9)(x+5)} = \dfrac{1}{x-9}$

35. $\dfrac{5x^2+11x+2}{x+2} = \dfrac{(5x+1)(x+2)}{x+2}$

$= 5x+1$

37. $\dfrac{x+7}{x^2+5x-14} = \dfrac{x+7}{(x-2)(x+7)}$

$= \dfrac{1}{x-2}$

39. $\dfrac{2x^2+3x-2}{2x-1} = \dfrac{(2x-1)(x+2)}{2x-1}$

$= x+2$

41. $\dfrac{x^2+7x+10}{x^2-3x-10} = \dfrac{(x+5)(x+2)}{(x-5)(x+2)}$

$= \dfrac{x+5}{x-5}$

43. $\dfrac{3x^2+7x+2}{3x^2+13x+4} = \dfrac{(x+2)(3x+1)}{(x+4)(3x+1)}$

$= \dfrac{x+2}{x+4}$

45. $\dfrac{2x^2-8}{4x-8} = \dfrac{2(x^2-4)}{4(x-2)}$

$= \dfrac{2(x+2)(x-2)}{2\cdot 2(x-2)}$

$= \dfrac{x+2}{2}$

47. $\dfrac{11x^2-22x^3}{6x-12x^2} = \dfrac{11x^2(1-2x)}{6x(1-2x)}$

$= \dfrac{11x\cdot x(1-2x)}{6\cdot x(1-2x)}$

$= \dfrac{11x}{6}$

49. $\dfrac{2-x}{x-2} = \dfrac{-1(x-2)}{x-2} = -1$

51. $\dfrac{x^2-1}{x^2-2x+1} = \dfrac{(x-1)(x+1)}{(x-1)(x-1)}$

$= \dfrac{x+1}{x-1}$

53. $\dfrac{m^2-6m+9}{m^2-9} = \dfrac{(m-3)(m-3)}{(m+3)(m-3)}$

$= \dfrac{m-3}{m+3}$

55. $\dfrac{1}{3}\cdot\dfrac{9}{11} = \dfrac{1\cdot 9}{3\cdot 11} = \dfrac{3\cdot 3}{3\cdot 11} = \dfrac{3}{11}$

57. $\dfrac{5}{6}\cdot\dfrac{10}{11}\cdot\dfrac{2}{3} = \dfrac{5\cdot 10\cdot 2}{6\cdot 11\cdot 3} = \dfrac{5\cdot 2\cdot 5\cdot 2}{3\cdot 2\cdot 11\cdot 3}$

$= \dfrac{5\cdot 5\cdot 2}{3\cdot 11\cdot 3} = \dfrac{50}{99}$

59. $\dfrac{1}{3}\div\dfrac{1}{4} = \dfrac{1}{3}\cdot\dfrac{4}{1} = \dfrac{4}{3}$

61. $\dfrac{13}{20}\div\dfrac{2}{9} = \dfrac{13}{20}\cdot\dfrac{9}{2} = \dfrac{13\cdot 9}{20\cdot 2} = \dfrac{117}{40}$

63. $\dfrac{x^2+xy+2x+2y}{x+2} = \dfrac{x(x+y)+2(x+y)}{x+2}$

$= \dfrac{(x+y)(x+2)}{x+2} = x+y$

65. $\dfrac{5x+15-xy-3y}{2x+6} = \dfrac{5(x+3)-y(x+3)}{2(x+3)}$

$= \dfrac{(x+3)(5-y)}{2(x+3)}$

$= \dfrac{5-y}{2}$

67. Answers may vary.

69. $C = \dfrac{DA}{A+12} = \dfrac{(1000)(8)}{8+12} = \dfrac{8000}{20} = \dfrac{400\cdot 20}{20}$

$= 400$

The child should receive 400 mg.

71. 5 feet 6 inches = 66 inches

$B = \dfrac{705w}{h^2}$

$= \dfrac{705(148)}{66^2}$

$= \dfrac{104,340}{4356}$

≈ 23.95

No, the person does not need to lose weight.

Section 5.2

Practice Problems

1. a. $\dfrac{16y}{3} \cdot \dfrac{1}{x^2} = \dfrac{16y \cdot 1}{3 \cdot x^2} = \dfrac{16y}{3x^2}$

b. $\dfrac{-5a^3}{3b^3} \cdot \dfrac{2b^2}{15a} = \dfrac{-5a^3 \cdot 2b^2}{3b^3 \cdot 15a}$

$\quad = \dfrac{-1 \cdot 5 \cdot 2 \cdot a \cdot a^2 \cdot b^2}{3 \cdot 3 \cdot 5 \cdot a \cdot b \cdot b^2} = -\dfrac{2a^2}{9b}$

2. $\dfrac{6x+6}{7} \cdot \dfrac{14}{x^2-1} = \dfrac{6(x+1)}{7} \cdot \dfrac{2 \cdot 7}{(x-1)(x+1)}$

$\quad = \dfrac{6(x+1) \cdot 2 \cdot 7}{7 \cdot (x-1)(x+1)} = \dfrac{12}{x-1}$

3. $\dfrac{4x+8}{7x^2-14x} \cdot \dfrac{3x^2-5x-2}{9x^2-1}$

$\quad = \dfrac{4(x+2)}{7x(x-2)} \cdot \dfrac{(3x+1)(x-2)}{(3x+1)(3x-1)}$

$\quad = \dfrac{4(x+2)(3x+1)(x-2)}{7x(x-2)(3x+1)(3x-1)}$

$\quad = \dfrac{4(x+2)}{7x(3x-1)}$

4. $\dfrac{7x^2}{6} \div \dfrac{x}{2y} = \dfrac{7x^2}{6} \cdot \dfrac{2y}{x} = \dfrac{7 \cdot x \cdot 2y}{2 \cdot 3 \cdot x} = \dfrac{7xy}{3}$

5. $\dfrac{(2x+3)(x-4)}{6} \div \dfrac{3x-12}{2}$

$\quad = \dfrac{(2x+3)(x-4)}{6} \cdot \dfrac{2}{3x-12}$

$\quad = \dfrac{(2x+3)(x-4) \cdot 2}{2 \cdot 3 \cdot 3(x-4)} = \dfrac{2x+3}{9}$

6. $\dfrac{10x+4}{x^2-4} \div \dfrac{5x^3+2x^2}{x+2} = \dfrac{10x+4}{x^2-4} \cdot \dfrac{x+2}{5x^3+2x^2}$

$\quad = \dfrac{2(5x+2) \cdot (x+2)}{(x-2)(x+2) \cdot x^2(5x+2)} = \dfrac{2}{x^2(x-2)}$

7. $\dfrac{3x^2-10x+8}{7x-14} \div \dfrac{9x-12}{21}$

$\quad = \dfrac{3x^2-10x+8}{7x-14} \cdot \dfrac{21}{9x-12}$

$\quad = \dfrac{(3x-4)(x-2) \cdot 3 \cdot 7}{7(x-2) \cdot 3(3x-4)} = \dfrac{1}{1} = 1$

8. a. $\dfrac{x+3}{x} \cdot \dfrac{7}{x+3} = \dfrac{(x+3) \cdot 7}{x \cdot (x+3)} = \dfrac{7}{x}$

b. $\dfrac{x+3}{x} \div \dfrac{7}{x+3} = \dfrac{x+3}{x} \cdot \dfrac{x+3}{7}$

$\quad = \dfrac{(x+3) \cdot (x+3)}{x \cdot 7} = \dfrac{(x+3)^2}{7x}$

9. $40,000 \text{ sq ft} = 40,000 \text{ sq ft} \cdot \dfrac{1 \text{ sq yd}}{9 \text{ sq ft}}$

$\quad = \dfrac{40,000}{9} \text{ sq yd} \approx 4444.44 \text{ sq yd}$

10. 39.5 ft/sec

$\quad = \dfrac{39.5 \text{ feet}}{1 \text{ second}} \cdot \dfrac{3600 \text{ seconds}}{1 \text{ hour}} \cdot \dfrac{1 \text{ mile}}{5280 \text{ feet}}$

$\quad = \dfrac{39.5 \cdot 3600}{5280} \text{ miles/hour}$

$\quad \approx 26.9 \text{ miles/hour (rounded to the nearest tenth)}$

Mental Math

1. $\dfrac{2}{y} \cdot \dfrac{x}{3} = \dfrac{2x}{3y}$

2. $\dfrac{3x}{4} \cdot \dfrac{1}{y} = \dfrac{3x}{4y}$

3. $\dfrac{5}{7} \cdot \dfrac{y^2}{x^2} = \dfrac{5y^2}{7x^2}$

4. $\dfrac{x^5}{11} \cdot \dfrac{4}{z^3} = \dfrac{4x^5}{11z^3}$

5. $\dfrac{9}{x} \cdot \dfrac{x}{5} = \dfrac{9x}{5x} = \dfrac{9}{5}$

6. $\dfrac{y}{7} \cdot \dfrac{3}{y} = \dfrac{3y}{7y} = \dfrac{3}{7}$

Exercise Set 5.2

1. $\dfrac{3x}{y^2} \cdot \dfrac{7y}{4x} = \dfrac{3x \cdot 7y}{y^2 \cdot 4x} = \dfrac{3 \cdot 7 \cdot x \cdot y}{4 \cdot x \cdot y \cdot y} = \dfrac{3 \cdot 7}{4 \cdot y} = \dfrac{21}{4y}$

3. $\dfrac{8x}{2} \cdot \dfrac{x^5}{4x^2} = \dfrac{8x \cdot x^5}{2 \cdot 4x^2} = \dfrac{2 \cdot 4 \cdot x \cdot x \cdot x^4}{2 \cdot 4 \cdot x \cdot x} = x^4$

5. $-\dfrac{5a^2b}{30a^2b^2} \cdot b^3 = -\dfrac{5a^2b \cdot b^3}{30a^2b^2}$

$= -\dfrac{5 \cdot a^2 \cdot b \cdot b \cdot b^2}{5 \cdot 6 \cdot a^2 \cdot b^2} = -\dfrac{b \cdot b}{6} = -\dfrac{b^2}{6}$

7. $\dfrac{x}{2x-14} \cdot \dfrac{x^2 - 7x}{5} = \dfrac{x \cdot (x^2 - 7x)}{(2x-14) \cdot 5}$

$= \dfrac{x \cdot x(x-7)}{2(x-7) \cdot 5} = \dfrac{x \cdot x}{2 \cdot 5} = \dfrac{x^2}{10}$

9. $\dfrac{6x+6}{5} \cdot \dfrac{10}{36x+36} = \dfrac{(6x+6) \cdot 10}{5 \cdot (36x+36)}$

$= \dfrac{6(x+1) \cdot 2 \cdot 5}{5 \cdot 36(x+1)} = \dfrac{6 \cdot 5 \cdot 2 \cdot (x+1)}{6 \cdot 5 \cdot 2 \cdot 3 \cdot (x+1)} = \dfrac{1}{3}$

11. $\dfrac{m^2 - n^2}{m+n} \cdot \dfrac{m}{m^2 - mn} = \dfrac{(m^2 - n^2) \cdot m}{(m+n) \cdot (m^2 - mn)}$

$= \dfrac{(m-n)(m+n) \cdot m}{(m+n) \cdot m \cdot (m-n)} = 1$

13. $\dfrac{x^2 - 25}{x^2 - 3x - 10} \cdot \dfrac{x+2}{x} = \dfrac{(x^2 - 25) \cdot (x+2)}{(x^2 - 3x - 10) \cdot x}$

$= \dfrac{(x-5)(x+5) \cdot (x+2)}{(x-5)(x+2) \cdot x} = \dfrac{x+5}{x}$

15. $A = \dfrac{2x}{x^2 - 25} \cdot \dfrac{x+5}{9x} = \dfrac{2x \cdot (x+5)}{(x^2 - 25) \cdot 9x}$

$= \dfrac{2 \cdot x \cdot (x+5)}{9 \cdot x \cdot (x+5)(x-5)} = \dfrac{2}{9(x-5)}$

The area is $\dfrac{2}{9(x-5)}$ square feet.

17. $\dfrac{5x^7}{2x^5} \div \dfrac{10x}{4x^3} = \dfrac{5x^7}{2x^5} \cdot \dfrac{4x^3}{10x}$

$= \dfrac{5 \cdot x^2 \cdot x^5 \cdot 2 \cdot 2x \cdot x^2}{2x^5 \cdot 2 \cdot 5 \cdot x}$

$= x^4$

19. $\dfrac{8x^2}{y^3} \div \dfrac{4x^2y^3}{6} = \dfrac{8x^2}{y^3} \cdot \dfrac{6}{4x^2y^3} = \dfrac{2 \cdot 4 \cdot x^2 \cdot 6}{y^3 \cdot 4x^2y^3}$

$= \dfrac{12}{y^6}$

21. $\dfrac{(x-6)(x+4)}{4x} \div \dfrac{2x-12}{8x^2}$

$= \dfrac{(x-6)(x+4)}{4x} \cdot \dfrac{8x^2}{2x-12}$

$= \dfrac{(x-6)(x+4) \cdot 2 \cdot 4 \cdot x \cdot x}{4x \cdot 2(x-6)}$

$= x(x+4)$

23. $\dfrac{3x^2}{x^2 - 1} \div \dfrac{x^5}{(x+1)^2} = \dfrac{3x^2}{x^2 - 1} \cdot \dfrac{(x+1)^2}{x^5}$

$= \dfrac{3x^2 \cdot (x+1)(x+1)}{(x-1)(x+1) \cdot x^2 \cdot x^3}$

$= \dfrac{3(x+1)}{x^3(x-1)}$

25. $\dfrac{m^2 - n^2}{m+n} \div \dfrac{m}{m^2 + nm} = \dfrac{m^2 - n^2}{m+n} \cdot \dfrac{m^2 + nm}{m}$

$= \dfrac{(m-n)(m+n) \cdot m(m+n)}{(m+n) \cdot m}$

$= (m-n)(m+n) = m^2 - n^2$

27. $\dfrac{x+2}{7-x} \div \dfrac{x^2 - 5x + 6}{x^2 - 9x + 14} = \dfrac{x+2}{7-x} \cdot \dfrac{x^2 - 9x + 14}{x^2 - 5x + 6}$

$= \dfrac{(x+2) \cdot (x-7)(x-2)}{-1(x-7) \cdot (x-3)(x-2)} = -\dfrac{x+2}{x-3}$

29. $\dfrac{x^2+7x+10}{x-1} \div \dfrac{x^2+2x-15}{x-1}$

$= \dfrac{x^2+7x+10}{x-1} \cdot \dfrac{x-1}{x^2+2x-15}$

$= \dfrac{(x+5)(x+2)\cdot(x-1)}{(x-1)\cdot(x+5)(x-3)} = \dfrac{x+2}{x-3}$

31. $\dfrac{5x-10}{12} \div \dfrac{4x-8}{8} = \dfrac{5x-10}{12} \cdot \dfrac{8}{4x-8}$

$= \dfrac{5(x-2)\cdot 2\cdot 4}{6\cdot 2\cdot 4(x-2)} = \dfrac{5}{6}$

33. $\dfrac{x^2+5x}{8} \cdot \dfrac{9}{3x+15} = \dfrac{x(x+5)\cdot 3\cdot 3}{8\cdot 3(x+5)} = \dfrac{3x}{8}$

35. $\dfrac{7}{6p^2+q} \div \dfrac{14}{18p^2+3q} = \dfrac{7}{6p^2+q} \cdot \dfrac{18p^2+3q}{14}$

$= \dfrac{7\cdot 3(6p^2+q)}{(6p^2+q)\cdot 7\cdot 2} = \dfrac{3}{2}$

37. $\dfrac{3x+4y}{x^2+4xy+4y^2} \cdot \dfrac{x+2y}{2}$

$= \dfrac{(3x+4y)\cdot(x+2y)}{(x+2y)(x+2y)\cdot 2}$

$= \dfrac{3x+4y}{2(x+2y)}$

39. $\dfrac{(x+2)^2}{x-2} \div \dfrac{x^2-4}{2x-4} = \dfrac{(x+2)^2}{x-2} \cdot \dfrac{2x-4}{x^2-4}$

$= \dfrac{(x+2)(x+2)\cdot 2(x-2)}{(x-2)\cdot(x+2)(x-2)} = \dfrac{2(x+2)}{x-2}$

41. $\dfrac{a^2+7a+12}{a^2+5a+6} \cdot \dfrac{a^2+8a+15}{a^2+5a+4}$

$= \dfrac{(a+3)(a+4)\cdot(a+5)(a+3)}{(a+3)(a+2)\cdot(a+4)(a+1)}$

$= \dfrac{(a+5)(a+3)}{(a+2)(a+1)}$

43. 1 square foot is 12 inches by 12 inches or 144 square inches.

10 sq ft $\cdot \dfrac{144 \text{ sq in.}}{1 \text{ sq ft}} = 1440$ sq in.

45. $\dfrac{50 \text{ miles}}{1 \text{ hour}} \cdot \dfrac{1 \text{ hour}}{3600 \text{ seconds}} \cdot \dfrac{5280 \text{ feet}}{1 \text{ mile}}$

$= \dfrac{50\cdot 5280}{3600}$ feet/sec ≈ 73 feet/sec

47. $\dfrac{5023 \text{ feet}}{1 \text{ second}} \cdot \dfrac{3600 \text{ seconds}}{1 \text{ hour}} \cdot \dfrac{1 \text{ mile}}{5280 \text{ feet}}$

$= \dfrac{5023\cdot 3600}{5280}$ miles/hour

≈ 3424.8 miles/hour

49. $\dfrac{1000 \text{ yen}}{1} \cdot \dfrac{1 \text{ US dollar}}{94.06 \text{ yen}}$

$= \dfrac{1000}{94.06}$ U.S. dollars

$= \$10.63$

51. $\dfrac{1}{5} + \dfrac{4}{5} = \dfrac{5}{5} = 1$

53. $\dfrac{9}{9} - \dfrac{19}{9} = -\dfrac{10}{9}$

55. $\dfrac{6}{5} + \left(\dfrac{1}{5} - \dfrac{8}{5}\right) = \dfrac{6}{5} + \left(-\dfrac{7}{5}\right) = -\dfrac{1}{5}$

57. $\left(\dfrac{x^2-y^2}{x^2+y^2} \div \dfrac{x^2-y^2}{3x}\right) \cdot \dfrac{x^2+y^2}{6}$

$= \dfrac{x^2-y^2}{x^2+y^2} \cdot \dfrac{3x}{x^2-y^2} \cdot \dfrac{x^2+y^2}{6}$

$= \dfrac{(x^2-y^2)\cdot 3x\cdot(x^2+y^2)}{(x^2+y^2)\cdot(x^2-y^2)\cdot 2\cdot 3} = \dfrac{x}{2}$

59. $\left(\dfrac{2a+b}{b^2} \cdot \dfrac{3a^2-2ab}{ab+2b^2}\right) \div \dfrac{a^2-3ab+2b^2}{5ab-10b^2}$

$= \dfrac{2a+b}{b^2} \cdot \dfrac{3a^2-2ab}{ab+2b^2} \cdot \dfrac{5ab-10b^2}{a^2-3ab+2b^2}$

$= \dfrac{(2a+b)\cdot(3a^2-2ab)\cdot(5ab-10b^2)}{b^2\cdot(ab+2b^2)\cdot(a^2-3ab+2b^2)}$

$= \dfrac{(2a+b)\cdot a(3a-2b)\cdot 5b(a-2b)}{b^2\cdot b(a+2b)\cdot(a-2b)(a-b)}$

$= \dfrac{5a(2a+b)(3a-2b)}{b^2(a+2b)(a-b)}$

61. Answers may vary.

63. $\dfrac{3.125 \text{ gallons}}{1 \text{ urn}} \cdot \dfrac{64 \text{ fl oz}}{0.5 \text{ gallons}} \cdot \dfrac{1 \text{ cup}}{8 \text{ fl oz}}$

$= \dfrac{3.125 \cdot 64}{0.5 \cdot 8}$ cups/urn

$= 50$ cups/urn

Section 5.3

Practice Problems

1. $\dfrac{8x}{3y} + \dfrac{x}{3y} = \dfrac{8x+x}{3y} = \dfrac{9x}{3y} = \dfrac{3x}{y}$

2. $\dfrac{3x}{3x-7} - \dfrac{7}{3x-7} = \dfrac{3x-7}{3x-7} = \dfrac{1}{1} = 1$

3. $\dfrac{2x^2+5x}{x+2} - \dfrac{4x+6}{x+2} = \dfrac{2x^2+5x-(4x+6)}{x+2}$

$= \dfrac{2x^2+5x-4x-6}{x+2} = \dfrac{2x^2+x-6}{x+2}$

$= \dfrac{(2x-3)(x+2)}{x+2} = 2x-3$

4. a. $9 = 3\cdot3 = 3^2$ and $15 = 3\cdot5$

$\text{LCD} = 3^2\cdot5 = 9\cdot5 = 45$

    b. $6x^3 = 2\cdot3\cdot x^3$ and $8x^5 = 2^3\cdot x^5$

$\text{LCD} = 2^3\cdot3\cdot x^5 = 8\cdot3\cdot x^5 = 24x^5$

5. Since $a+5$ and $a-5$ are completely factored and each factor appears once, the LCD $= (a+5)(a-5)$.

6. $(x-4)^2 = (x-4)^2$

$3x-12 = 3(x-4)$

$\text{LCD} = 3(x-4)^2$

7. $y^2+2y-3 = (y+3)(y-1)$

$y^2-3y+2 = (y-2)(y-1)$

$\text{LCD} = (y+3)(y-1)(y-2)$

8. Since $x-4$ and $4-x$ are opposites, LCD $= x-4$ or LCD $= 4-x$.

9. $\dfrac{2x}{5y} = \dfrac{2x(4x^2y)}{5y(4x^2y)} = \dfrac{8x^3y}{20x^2y^2}$

10. $\dfrac{3}{x^2-25} = \dfrac{3}{(x+5)(x-5)}$

$= \dfrac{3(x-3)}{(x+5)(x-5)(x-3)}$

$= \dfrac{3x-9}{(x+5)(x-5)(x-3)}$

Mental Math

1. $\dfrac{2}{3} + \dfrac{1}{3} = \dfrac{3}{3} = 1$

2. $\dfrac{5}{11} + \dfrac{1}{11} = \dfrac{6}{11}$

3. $\dfrac{3x}{9} + \dfrac{4x}{9} = \dfrac{7x}{9}$

4. $\dfrac{3y}{8} + \dfrac{2y}{8} = \dfrac{5y}{8}$

5. $\dfrac{8}{9} - \dfrac{7}{9} = \dfrac{1}{9}$

6. $\dfrac{14}{12} - \dfrac{3}{12} = \dfrac{11}{12}$

7. $\dfrac{7y}{5} + \dfrac{10y}{5} = \dfrac{17y}{5}$

8. $\dfrac{12x}{7} - \dfrac{4x}{7} = \dfrac{8x}{7}$

Exercise Set 5.3

1. $\dfrac{a}{13} + \dfrac{9}{13} = \dfrac{a+9}{13}$

3. $\dfrac{4m}{3n} + \dfrac{5m}{3n} = \dfrac{4m+5m}{3n} = \dfrac{9m}{3n} = \dfrac{3m}{n}$

5. $\dfrac{4m}{m-6} - \dfrac{24}{m-6} = \dfrac{4m-24}{m-6} = \dfrac{4(m-6)}{m-6} = 4$

7. $\dfrac{9}{3+y} + \dfrac{y+1}{3+y} = \dfrac{9+y+1}{3+y} = \dfrac{y+10}{3+y}$

9. $\dfrac{5x+4}{x-1} - \dfrac{2x+7}{x-1} = \dfrac{5x+4-(2x+7)}{x-1}$

$= \dfrac{5x+4-2x-7}{x-1} = \dfrac{3x-3}{x-1} = \dfrac{3(x-1)}{x-1} = 3$

11. $\dfrac{a}{a^2+2a-15} - \dfrac{3}{a^2+2a-15} = \dfrac{a-3}{a^2+2a-15}$

$= \dfrac{a-3}{(a+5)(a-3)} = \dfrac{1}{a+5}$

13. $\dfrac{2x+3}{x^2-x-30} - \dfrac{x-2}{x^2-x-30}$

$= \dfrac{2x+3-(x-2)}{x^2-x-30}$

$= \dfrac{2x+3-x+2}{x^2-x-30} = \dfrac{x+5}{x^2-x-30}$

$= \dfrac{x+5}{(x-6)(x+5)} = \dfrac{1}{x-6}$

15. $P = \dfrac{5}{x-2} + \dfrac{5}{x-2} + \dfrac{5}{x-2} + \dfrac{5}{x-2}$

$= \dfrac{5+5+5+5}{x-2} = \dfrac{20}{x-2}$

The perimeter is $\dfrac{20}{x-2}$ meters.

17. Answers may vary.

19. $2x = 2 \cdot x$

$4x^3 = 2^2 \cdot x^3$

$\text{LCD} = 2^2 \cdot x^3 = 4x^3$

21. $8x = 2^3 \cdot x$

$2x + 4 = 2(x+2)$

$\text{LCD} = 2^3 \cdot x \cdot (x+2) = 8x(x+2)$

23. $x + 3 = x + 3$

$x - 2 = x - 2$

$\text{LCD} = (x+3)(x-2)$

25. $x + 6 = x + 6$

$3x + 18 = 3 \cdot (x+6)$

$\text{LCD} = 3(x+6)$

27. $3x + 3 = 3 \cdot (x+1)$

$2x^2 + 4x + 2 = 2(x^2+2x+1) = 2 \cdot (x+1)^2$

$\text{LCD} = 2 \cdot 3(x+1)^2$

$= 6(x+1)^2$

29. $x - 8 = x - 8$

$8 - x = -(x-8)$

$\text{LCD} = x - 8 \text{ or } 8 - x$

31. $x^2 + 3x - 4 = (x+4)(x-1)$

$x^2 + 2x - 3 = (x+3)(x-1)$

$\text{LCD} = (x+4)(x+3)(x-1)$

33. Answers may vary.

35. $\dfrac{3}{2x} = \dfrac{3(2x)}{2x(2x)} = \dfrac{6x}{4x^2}$

37. $\dfrac{6}{3a} = \dfrac{6(4b^2)}{3a(4b^2)} = \dfrac{24b^2}{12ab^2}$

39. $\dfrac{9}{x+3} = \dfrac{9(2)}{(x+3)(2)} = \dfrac{18}{2(x+3)}$

41. $\dfrac{9a+2}{5a+10} = \dfrac{9a+2}{5(a+2)} = \dfrac{(9a+2)(b)}{5(a+2)(b)}$

$= \dfrac{9ab+2b}{5b(a+2)}$

43. $\dfrac{x}{x^3 + 6x^2 + 8x} = \dfrac{x}{x(x^2 + 6x + 8)}$

$= \dfrac{x}{x(x+4)(x+2)} = \dfrac{x(x+1)}{x(x+4)(x+2)(x+1)}$

$= \dfrac{x^2 + x}{x(x+4)(x+2)(x+1)}$

45. $\dfrac{9y-1}{15x^2 - 30} = \dfrac{(9y-1)(2)}{(15x^2 - 30)2} = \dfrac{18y-2}{30x^2 - 60}$

47. LCD $= 21$

$\dfrac{2}{3} + \dfrac{5}{7} = \dfrac{2(7)}{3(7)} + \dfrac{5(3)}{7(3)} = \dfrac{14}{21} + \dfrac{15}{21} = \dfrac{29}{21}$

49. Since $6 = 2 \cdot 3$ and $4 = 2^2$,

LCD $= 2^2 \cdot 3 = 12$.

$\dfrac{2}{6} - \dfrac{3}{4} = \dfrac{2(2)}{6(2)} - \dfrac{3(3)}{4(3)} = \dfrac{4}{12} - \dfrac{9}{12} = \dfrac{4-9}{12}$

$= -\dfrac{5}{12}$

51. Since $12 = 2^2 \cdot 3$ and $20 = 2^2 \cdot 5$,

LCD $= 2^2 \cdot 3 \cdot 5 = 60$.

$\dfrac{1}{12} + \dfrac{3}{20} = \dfrac{1(5)}{12(5)} + \dfrac{3(3)}{20(3)} = \dfrac{5}{60} + \dfrac{9}{60} = \dfrac{14}{60}$

$= \dfrac{7(2)}{30(2)} = \dfrac{7}{30}$

53. Since $8 = 2^3$ and $12 = 2^2 \cdot 3$, the least common multiple of 8 and 12 is $2^3 \cdot 3 = 24$. Since $8 \cdot 3 = 24$ and $12 \cdot 2 = 24$, buy three packages of hot dogs and two packages of buns.

55. Answers may vary.

Section 5.4

Practice Problems

1. a. LCD $= 5 \cdot 3 = 15$

$\dfrac{y}{5} - \dfrac{3y}{15} = \dfrac{y(3)}{5(3)} - \dfrac{3y}{15} = \dfrac{3y}{15} - \dfrac{3y}{15}$

$= \dfrac{3y - 3y}{15} = \dfrac{0}{15} = 0$

b. $8x = 2^3 \cdot x$

$10x^2 = 2 \cdot 5 \cdot x^2$

LCD $= 2^3 \cdot 5 \cdot x^2 = 8 \cdot 5 \cdot x^2 = 40x^2$

$\dfrac{5}{8x} + \dfrac{11}{10x^2} = \dfrac{5(5x)}{8x(5x)} + \dfrac{11(4)}{10x^2(4)}$

$= \dfrac{25x}{40x^2} + \dfrac{44}{40x^2} = \dfrac{25x + 44}{40x^2}$

2. Since $x^2 - 9 = (x+3)(x-3)$, the LCD $= (x+3)(x-3)$.

$\dfrac{10x}{x^2 - 9} - \dfrac{5}{x+3}$

$= \dfrac{10x}{(x+3)(x-3)} - \dfrac{5(x-3)}{(x+3)(x-3)}$

$= \dfrac{10x - 5(x-3)}{(x+3)(x-3)} = \dfrac{10x - 5x + 15}{(x+3)(x-3)}$

$= \dfrac{5x + 15}{(x+3)(x-3)} = \dfrac{5(x+3)}{(x+3)(x-3)} = \dfrac{5}{x-3}$

3. $\dfrac{5}{7x} + \dfrac{2}{x+1} = \dfrac{5(x+1)}{7x(x+1)} + \dfrac{2(7x)}{7x(x+1)}$

$= \dfrac{5(x+1) + 2(7x)}{7x(x+1)} = \dfrac{5x + 5 + 14x}{7x(x+1)}$

$= \dfrac{19x + 5}{7x(x+1)}$

4. $\dfrac{10}{x-6} - \dfrac{15}{6-x} = \dfrac{10}{x-6} - \dfrac{15}{-(x-6)}$

$= \dfrac{10}{x-6} - \dfrac{-15}{x-6} = \dfrac{10 - (-15)}{x-6} = \dfrac{25}{x-6}$

5. $2 + \dfrac{x}{x+5} = \dfrac{2}{1} + \dfrac{x}{x+5} = \dfrac{2(x+5)}{1(x+5)} + \dfrac{x}{x+5}$

$= \dfrac{2x + 10 + x}{x+5} = \dfrac{3x + 10}{x+5}$

6. $\dfrac{4}{3x^2 + 2x} - \dfrac{3x}{12x + 8} = \dfrac{4}{x(3x+2)} - \dfrac{3x}{4(3x+2)}$

$= \dfrac{4(4)}{x(3x+2)(4)} - \dfrac{3x(x)}{4(3x+2)(x)} = \dfrac{16 - 3x^2}{4x(3x+2)}$

7. $\dfrac{6x}{x^2+4x+4}+\dfrac{x}{x^2-4}$

$=\dfrac{6x}{(x+2)^2}+\dfrac{x}{(x+2)(x-2)}$

$=\dfrac{6x(x-2)}{(x+2)^2(x-2)}+\dfrac{x(x+2)}{(x+2)(x-2)(x+2)}$

$=\dfrac{6x^2-12x+x^2+2x}{(x+2)^2(x-2)}$

$=\dfrac{7x^2-10x}{(x+2)^2(x-2)}=\dfrac{x(7x-10)}{(x+2)^2(x-2)}$

Exercise Set 5.4

1. LCD $=2\cdot3\cdot x=6x$

$\dfrac{4}{2x}+\dfrac{9}{3x}=\dfrac{4(3)}{2x(3)}+\dfrac{9(2)}{3x(2)}=\dfrac{12}{6x}+\dfrac{18}{6x}$

$=\dfrac{30}{6x}=\dfrac{5(6)}{6x}=\dfrac{5}{x}$

3. LCD $=5b$

$\dfrac{15a}{b}+\dfrac{6b}{5}=\dfrac{15a(5)}{b(5)}+\dfrac{6b(b)}{5(b)}=\dfrac{75a}{5b}+\dfrac{6b^2}{5b}$

$=\dfrac{75a+6b^2}{5b}$

5. LCD $=2x^2$

$\dfrac{3}{x}+\dfrac{5}{2x^2}=\dfrac{3(2x)}{x(2x)}+\dfrac{5}{2x^2}=\dfrac{6x}{2x^2}+\dfrac{5}{2x^2}$

$=\dfrac{6x+5}{2x^2}$

7. $2x+2=2(x+1)$

LCD $=2(x+1)$

$\dfrac{6}{x+1}+\dfrac{10}{2x+2}=\dfrac{6}{x+1}+\dfrac{10}{2(x+1)}$

$=\dfrac{6(2)}{(x+1)(2)}+\dfrac{10}{2(x+1)}=\dfrac{12}{2(x+1)}+\dfrac{10}{2(x+1)}$

$=\dfrac{12+10}{2(x+1)}=\dfrac{22}{2(x+1)}=\dfrac{2\cdot11}{2(x+1)}=\dfrac{11}{x+1}$

9. $2x-4=2(x-2)$

$x^2-4=(x-2)(x+2)$

LCD $=2(x-2)(x+2)$

$\dfrac{15}{2x-4}+\dfrac{x}{x^2-4}=\dfrac{15}{2(x-2)}+\dfrac{x}{(x-2)(x+2)}$

$=\dfrac{15(x+2)}{2(x-2)(x+2)}+\dfrac{x(2)}{(x-2)(x+2)(2)}$

$=\dfrac{15x+30}{2(x-2)(x+2)}+\dfrac{2x}{2(x-2)(x+2)}$

$=\dfrac{15x+30+2x}{2(x-2)(x+2)}=\dfrac{17x+30}{2(x-2)(x+2)}$

11. LCD $=4x(x-2)$

$\dfrac{3}{4x}+\dfrac{8}{x-2}=\dfrac{3(x-2)}{4x(x-2)}+\dfrac{8(4x)}{(x-2)(4x)}$

$=\dfrac{3x-6}{4x(x-2)}+\dfrac{32x}{4x(x-2)}=\dfrac{3x-6+32x}{4x(x-2)}$

$=\dfrac{35x-6}{4x(x-2)}$

13. $\dfrac{6}{x-3}+\dfrac{8}{3-x}=\dfrac{6}{x-3}+\dfrac{8}{-(x-3)}$

$=\dfrac{6}{x-3}+\dfrac{-8}{x-3}=\dfrac{6+(-8)}{x-3}=-\dfrac{2}{x-3}$

15. $\dfrac{-8}{x^2-1}-\dfrac{7}{1-x^2}=\dfrac{8}{-(x^2-1)}-\dfrac{7}{1-x^2}$

$=\dfrac{8}{1-x^2}-\dfrac{7}{1-x^2}=\dfrac{8-7}{1-x^2}$

$=\dfrac{1}{1-x^2}$ or $-\dfrac{1}{x^2-1}$

17. $\dfrac{5}{x}+2=\dfrac{5}{x}+\dfrac{2}{1}=\dfrac{5}{x}+\dfrac{2(x)}{1(x)}=\dfrac{5+2x}{x}$

19. $\dfrac{5}{x-2}+6=\dfrac{5}{x-2}+\dfrac{6}{1}=\dfrac{5}{x-2}+\dfrac{6(x-2)}{1(x-2)}$

$=\dfrac{5}{x-2}+\dfrac{6x-12}{x-2}=\dfrac{5+6x-12}{x-2}=\dfrac{6x-7}{x-2}$

21. $\dfrac{y+2}{y+3}-2=\dfrac{y+2}{y+3}-\dfrac{2}{1}=\dfrac{y+2}{y+3}-\dfrac{2(y+3)}{y+3}$

$=\dfrac{y+2}{y+3}-\dfrac{2y+6}{y+3}=\dfrac{y+2-(2y+6)}{y+3}$

$=\dfrac{y+2-2y-6}{y+3}=\dfrac{-y-4}{y+3}=\dfrac{-(y+4)}{y+3}$

$=-\dfrac{y+4}{y+3}$

23. Answers may vary.

25. $\dfrac{5x}{x+2} - \dfrac{3x-4}{x+2} = \dfrac{5x-(3x-4)}{x+2}$

$= \dfrac{5x-3x+4}{x+2} = \dfrac{2x+4}{x+2} = \dfrac{2(x+2)}{x+2} = 2$

27. $\dfrac{3x^4}{x} - \dfrac{4x^2}{x^2} = \dfrac{3x^4(x)}{x(x)} - \dfrac{4x^2}{x^2} = \dfrac{3x^5}{x^2} - \dfrac{4x^2}{x^2}$

$= \dfrac{3x^5-4x^2}{x^2} = \dfrac{x^2(3x^3-4)}{x^2} = 3x^3 - 4$

29. $\dfrac{1}{x+3} - \dfrac{1}{(x+3)^2} = \dfrac{1(x+3)}{(x+3)(x+3)} - \dfrac{1}{(x+3)^2}$

$= \dfrac{x+3}{(x+3)^2} - \dfrac{1}{(x+3)^2} = \dfrac{x+3-1}{(x+3)^2} = \dfrac{x+2}{(x+3)^2}$

31. $\dfrac{4}{5b} + \dfrac{1}{b-1} = \dfrac{4(b-1)}{5b(b-1)} + \dfrac{1(5b)}{(b-1)(5b)}$

$= \dfrac{4b-4}{5b(b-1)} + \dfrac{5b}{5b(b-1)} = \dfrac{4b-4+5b}{5b(b-1)}$

$= \dfrac{9b-4}{5b(b-1)}$

33. $\dfrac{2}{m} + 1 = \dfrac{2}{m} + \dfrac{1}{1} = \dfrac{2}{m} + \dfrac{1(m)}{1(m)} = \dfrac{2+m}{m}$

35. $\dfrac{6}{1-2x} - \dfrac{4}{2x-1} = \dfrac{6}{1-2x} - \dfrac{4}{-(1-2x)}$

$= \dfrac{6}{1-2x} - \dfrac{-4}{1-2x} = \dfrac{6-(-4)}{1-2x} = \dfrac{10}{1-2x}$

37. $\dfrac{7}{(x+1)(x-1)} + \dfrac{8}{(x+1)^2}$

$= \dfrac{7(x+1)}{(x+1)(x-1)(x+1)} + \dfrac{8(x-1)}{(x+1)^2(x-1)}$

$= \dfrac{7x+7}{(x+1)^2(x-1)} + \dfrac{8x-8}{(x+1)^2(x-1)}$

$= \dfrac{7x+7+8x-8}{(x+1)^2(x-1)} = \dfrac{15x-1}{(x+1)^2(x-1)}$

39. $\dfrac{x}{x^2-1} - \dfrac{2}{x^2-2x+1}$

$= \dfrac{x}{(x-1)(x+1)} - \dfrac{2}{(x-1)^2}$

$= \dfrac{x(x-1)}{(x-1)(x+1)(x-1)} - \dfrac{2(x+1)}{(x-1)^2(x+1)}$

$= \dfrac{x^2-x}{(x-1)^2(x+1)} - \dfrac{2x+2}{(x-1)^2(x+1)}$

$= \dfrac{x^2-x-(2x+2)}{(x-1)^2(x+1)} = \dfrac{x^2-x-2x-2}{(x-1)^2(x+1)}$

$= \dfrac{x^2-3x-2}{(x-1)^2(x+1)}$

41. $\dfrac{3a}{2a+6} - \dfrac{a-1}{a+3} = \dfrac{3a}{2(a+3)} - \dfrac{a-1}{a+3}$

$= \dfrac{3a}{2(a+3)} - \dfrac{(a-1)(2)}{(a+3)(2)}$

$= \dfrac{3a}{2(a+3)} - \dfrac{2a-2}{2(a+3)}$

$= \dfrac{3a-(2a-2)}{2(a+3)} = \dfrac{3a-2a+2}{2(a+3)} = \dfrac{a+2}{2(a+3)}$

43. $\dfrac{5}{2-x} + \dfrac{x}{2x-4} = \dfrac{5}{-(x-2)} + \dfrac{x}{2(x-2)}$

$= \dfrac{-5}{x-2} + \dfrac{x}{2(x-2)} = \dfrac{-5(2)}{(x-2)(2)} + \dfrac{x}{2(x-2)}$

$= \dfrac{-10}{2(x-2)} + \dfrac{x}{2(x-2)} = \dfrac{x-10}{2(x-2)}$

45. $\dfrac{-7}{y^2-3y+2} - \dfrac{2}{y-1} = \dfrac{-7}{(y-1)(y-2)} - \dfrac{2}{y-1}$

$= \dfrac{-7}{(y-1)(y-2)} - \dfrac{2(y-2)}{(y-1)(y-2)}$

$= \dfrac{-7-(2y-4)}{(y-1)(y-2)} = \dfrac{-7-2y+4}{(y-1)(y-2)}$

$= \dfrac{-3-2y}{(y-2)(y-1)}$

47. $\dfrac{13}{x^2-5x+6} - \dfrac{5}{x-3} = \dfrac{13}{(x-3)(x-2)} - \dfrac{5}{x-3}$

$= \dfrac{13}{(x-3)(x-2)} - \dfrac{5(x-2)}{(x-3)(x-2)}$

$= \dfrac{13-(5x-10)}{(x-3)(x-2)} = \dfrac{13-5x+10}{(x-3)(x-2)}$

$= \dfrac{-5x+23}{(x-3)(x-2)}$

49. $\dfrac{x+8}{x^2-5x-6}+\dfrac{x+1}{x^2-4x-5}$

$=\dfrac{x+8}{(x-6)(x+1)}+\dfrac{x+1}{(x-5)(x+1)}$

$=\dfrac{(x+8)(x-5)}{(x-6)(x+1)(x-5)}+\dfrac{(x+1)(x-6)}{(x-5)(x+1)(x-6)}$

$=\dfrac{x^2+3x-40+x^2-5x-6}{(x-6)(x+1)(x-5)}$

$=\dfrac{2x^2-2x-46}{(x-6)(x+1)(x-5)}$

51. $\quad 3x+5=7$

$3x+5-5=7-5$

$\qquad 3x=2$

$\qquad \dfrac{3x}{3}=\dfrac{2}{3}$

$\qquad x=\dfrac{2}{3}$

53. $\quad 2x^2-x-1=0$

$(2x+1)(x-1)=0$

$2x+1=0 \qquad$ or $\quad x-1=0$

$2x=-1 \qquad\qquad\qquad x=1$

$x=-\dfrac{1}{2}$

The solutions are $x=-\dfrac{1}{2}$ and $x=1$.

55. $\quad 4(x+6)+3=-3$

$4x+24+3=-3$

$4x+27=-3$

$4x+27-27=-3-27$

$\qquad 4x=-30$

$\qquad \dfrac{4x}{4}=-\dfrac{30}{4}$

$\qquad x=-\dfrac{30}{4}=\dfrac{-15\cdot 2}{2\cdot 2}=-\dfrac{15}{2}$

57. $\dfrac{3}{x+4}-\dfrac{1}{x-4}$

$=\dfrac{3(x-4)}{(x+4)(x-4)}-\dfrac{1(x+4)}{(x-4)(x+4)}$

$=\dfrac{3x-12}{(x+4)(x-4)}-\dfrac{x+4}{(x+4)(x-4)}$

$=\dfrac{3x-12-(x+4)}{(x+4)(x-4)}=\dfrac{3x-12-x-4}{(x+4)(x-4)}$

$=\dfrac{2x-16}{(x+4)(x-4)}$

The length of the other board is

$\dfrac{2x-16}{(x+4)(x-4)}$ inches.

59. $\dfrac{DA}{A+12}-\dfrac{D(A+1)}{24}$

$=\dfrac{DA(24)}{(A+12)(24)}-\dfrac{D(A+1)(A+12)}{24(A+12)}$

$=\dfrac{24DA}{24(A+12)}-\dfrac{D(A^2+13A+12)}{24(A+12)}$

$=\dfrac{24DA}{24(A+12)}-\dfrac{DA^2+13DA+12D}{24(A+12)}$

$=\dfrac{24DA-(DA^2+13DA+12D)}{24(A+12)}$

$=\dfrac{-DA^2+11DA-12D}{24(A+12)}$

The difference in the doses is

$\dfrac{-DA^2+11DA-12D}{24(A+12)}$.

61. Answers may vary.

63. $\dfrac{8}{x^2+6x+5}-\dfrac{3x}{x^2+4x-5}+\dfrac{2}{x^2-1}=\dfrac{8}{(x+5)(x+1)}-\dfrac{3x}{(x+5)(x-1)}+\dfrac{2}{(x-1)(x+1)}$

$=\dfrac{8(x-1)}{(x+5)(x+1)(x-1)}-\dfrac{3x(x+1)}{(x+5)(x-1)(x+1)}+\dfrac{2(x+5)}{(x-1)(x+1)(x+5)}$

$$= \frac{8(x-1) - 3x(x+1) + 2(x+5)}{(x+5)(x+1)(x-1)} = \frac{8x - 8 - 3x^2 - 3x + 2x + 10}{(x+5)(x+1)(x-1)}$$

$$= \frac{-3x^2 + 7x + 2}{(x+5)(x+1)(x-1)}$$

65. $\dfrac{10}{x^2 - 3x - 4} - \dfrac{8}{x^2 + 6x + 5} - \dfrac{9}{x^2 + x - 20} = \dfrac{10}{(x-4)(x+1)} - \dfrac{8}{(x+5)(x+1)} - \dfrac{9}{(x+5)(x-4)}$

$$= \frac{10(x+5)}{(x-4)(x+1)(x+5)} - \frac{8(x-4)}{(x+5)(x+1)(x-4)} - \frac{9(x+1)}{(x+5)(x-4)(x+1)}$$

$$= \frac{10(x+5) - 8(x-4) - 9(x+1)}{(x-4)(x+1)(x+5)} = \frac{10x + 50 - 8x + 32 - 9x - 9}{(x-4)(x+1)(x+5)}$$

$$= \frac{73 - 7x}{(x-4)(x+1)(x+5)}$$

Section 5.5

Practice Problems

1. $\dfrac{x}{4} + \dfrac{4}{5} = \dfrac{1}{20}$

$20\left(\dfrac{x}{4} + \dfrac{4}{5}\right) = 20\left(\dfrac{1}{20}\right)$

$20\left(\dfrac{x}{4}\right) + 20\left(\dfrac{4}{5}\right) = 20\left(\dfrac{1}{20}\right)$

$5x + 16 = 1$

$5x = -15$

$x = -3$

Check:

$\dfrac{x}{4} + \dfrac{4}{5} = \dfrac{1}{20}$

$\dfrac{-3}{4} + \dfrac{4}{5} \stackrel{?}{=} \dfrac{1}{20}$

$\dfrac{-15}{20} + \dfrac{16}{20} \stackrel{?}{=} \dfrac{1}{20}$

$\dfrac{1}{20} = \dfrac{1}{20}$ True

The solution is -3.

2. $\dfrac{x+2}{3} - \dfrac{x-1}{5} = \dfrac{1}{15}$

$15\left(\dfrac{x+2}{3} - \dfrac{x-1}{5}\right) = 15\left(\dfrac{1}{15}\right)$

$15\left(\dfrac{x+2}{3}\right) - 15\left(\dfrac{x-1}{5}\right) = 15\left(\dfrac{1}{15}\right)$

$5(x+2) - 3(x-1) = 1$

$5x + 10 - 3x + 3 = 1$

$2x + 13 = 1$

$2x = -12$

$x = -6$

Check:

$\dfrac{x+2}{3} - \dfrac{x-1}{5} = \dfrac{1}{15}$

$\dfrac{-6+2}{3} - \dfrac{-6-1}{5} \stackrel{?}{=} \dfrac{1}{15}$

$\dfrac{-4}{3} - \dfrac{-7}{5} \stackrel{?}{=} \dfrac{1}{15}$

$\dfrac{-20}{15} - \dfrac{-21}{15} \stackrel{?}{=} \dfrac{1}{15}$

$\dfrac{-20 + 21}{15} \stackrel{?}{=} \dfrac{1}{15}$

$\dfrac{1}{15} = \dfrac{1}{15}$ True

The solution is -6.

3. $2 + \dfrac{6}{x} = x + 7$

$x\left(2 + \dfrac{6}{x}\right) = x(x + 7)$

$x(2) + x\left(\dfrac{6}{x}\right) = x^2 + 7x$

$2x + 6 = x^2 + 7x$

$0 = x^2 + 5x - 6$

$0 = (x + 6)(x - 1)$

$x + 6 = 0 \quad \text{or} \quad x - 1 = 0$

$x = -6 \qquad\qquad x = 1$

Check:

$x = -6 \qquad\qquad\qquad x = 1$

$2 + \dfrac{6}{x} = x + 7 \qquad\qquad 2 + \dfrac{6}{x} = x + 7$

$2 + \dfrac{6}{-6} \overset{?}{=} -6 + 7 \qquad\qquad 2 + \dfrac{6}{1} \overset{?}{=} 1 + 7$

$2 + (-1) \overset{?}{=} 1 \qquad\qquad\qquad 2 + 6 \overset{?}{=} 8$

$1 = 1 \quad \text{True} \qquad\qquad\qquad 8 = 8 \quad \text{True}$

Both −6 and 1 are solutions.

4. $\dfrac{2}{x + 3} + \dfrac{3}{x - 3} = \dfrac{-2}{x^2 - 9}$

$(x + 3)(x - 3)\left(\dfrac{2}{x + 3} + \dfrac{3}{x - 3}\right) = (x + 3)(x - 3)\left(\dfrac{-2}{x^2 - 9}\right)$

$(x + 3)(x - 3)\left(\dfrac{2}{x + 3}\right) + (x + 3)(x - 3)\left(\dfrac{3}{x - 3}\right) = (x + 3)(x - 3)\left(\dfrac{-2}{x^2 - 9}\right)$

$2(x - 3) + 3(x + 3) = -2$

$2x - 6 + 3x + 9 = -2$

$5x + 3 = -2$

$5x = -5$

$x = -1$

Check:

$\dfrac{2}{x + 3} + \dfrac{3}{x - 3} = \dfrac{-2}{x^2 - 9}$

$\dfrac{2}{-1 + 3} + \dfrac{3}{-1 - 3} \overset{?}{=} \dfrac{-2}{(-1)^2 - 9}$

$\dfrac{2}{2} + \dfrac{3}{-4} \overset{?}{=} \dfrac{-2}{1 - 9}$

$1 - \dfrac{3}{4} \overset{?}{=} \dfrac{-2}{-8}$

$\dfrac{1}{4} = \dfrac{1}{4} \quad \text{True}$

The solution is −1.

5. $\dfrac{5x}{x-1} = \dfrac{5}{x-1} + 3$

$(x-1)\left(\dfrac{5x}{x-1}\right) = (x-1)\left(\dfrac{5}{x-1} + 3\right)$

$(x-1)\left(\dfrac{5x}{x-1}\right) = (x-1)\left(\dfrac{5}{x-1}\right) + (x-1)(3)$

$5x = 5 + 3x - 3$

$5x = 3x + 2$

$2x = 2$

$x = 1$

Notice that 1 makes the denominator 0 in the original equation. This equation has no solution.

6. $x - \dfrac{6}{x+3} = \dfrac{2x}{x+3} + 2$

$(x+3)\left(x - \dfrac{6}{x+3}\right) = (x+3)\left(\dfrac{2x}{x+3} + 2\right)$

$(x+3)(x) - (x+3)\left(\dfrac{6}{x+3}\right) = (x+3)\left(\dfrac{2x}{x+3}\right) + (x+3)(2)$

$x^2 + 3x - 6 = 2x + 2x + 6$

$x^2 + 3x - 6 = 4x + 6$

$x^2 - x - 12 = 0$

$(x-4)(x+3) = 0$

$x - 4 = 0 \quad \text{or} \quad x + 3 = 0$

$x = 4 \qquad\qquad x = -3$

Since –3 would make a denominator 0, –3 cannot be a solution. The only solution is 4.

7. $\dfrac{1}{a} + \dfrac{1}{b} = \dfrac{1}{x}$

$abx\left(\dfrac{1}{a} + \dfrac{1}{b}\right) = abx\left(\dfrac{1}{x}\right)$

$abx\left(\dfrac{1}{a}\right) + abx\left(\dfrac{1}{b}\right) = abx\left(\dfrac{1}{x}\right)$

$bx + ax = ab$

$bx = ab - ax$

$bx = a(b - x)$

$\dfrac{bx}{b-x} = a$

2. $\dfrac{x}{8} = 4$

$x = 32$

3. $\dfrac{z}{6} = 6$

$z = 36$

4. $\dfrac{y}{7} = 8$

$y = 56$

Mental Math

1. $\dfrac{x}{5} = 2$

$x = 10$

Exercise Set 5.5

1. $\dfrac{x}{5} + 3 = 9$

$5\left(\dfrac{x}{5} + 3\right) = 5(9)$

$5\left(\dfrac{x}{5}\right) + 5(3) = 5(9)$

$x + 15 = 45$

$x = 30$

Check:

$\dfrac{x}{5} + 3 = 9$

$\dfrac{30}{5} + 3 \overset{?}{=} 9$

$6 + 3 \overset{?}{=} 9$

$9 = 9$ True

The solution is 30.

3. $\dfrac{x}{2} + \dfrac{5x}{4} = \dfrac{x}{12}$

$12\left(\dfrac{x}{2} + \dfrac{5x}{4}\right) = 12\left(\dfrac{x}{12}\right)$

$12\left(\dfrac{x}{2}\right) + 12\left(\dfrac{5x}{4}\right) = 12\left(\dfrac{x}{12}\right)$

$6x + 15x = x$

$21x = x$

$20x = 0$

$x = 0$

Check:

$\dfrac{x}{2} + \dfrac{5x}{4} = \dfrac{x}{12}$

$\dfrac{0}{2} + \dfrac{5 \cdot 0}{4} \overset{?}{=} \dfrac{0}{12}$

$0 + \dfrac{0}{4} \overset{?}{=} 0$

$0 = 0$ True

The solution is 0.

5. $2 - \dfrac{8}{x} = 6$

$x\left(2 - \dfrac{8}{x}\right) = x(6)$

$x(2) - x\left(\dfrac{8}{x}\right) = x(6)$

$2x - 8 = 6x$

$-8 = 4x$

$-2 = x$

Check:

$2 - \dfrac{8}{x} = 6$

$2 - \dfrac{8}{-2} \overset{?}{=} 6$

$2 - (-4) \overset{?}{=} 6$

$2 + 4 \overset{?}{=} 6$

$6 = 6$ True

The solution is -2.

7. $2 + \dfrac{10}{x} = x + 5$

$x\left(2 + \dfrac{10}{x}\right) = x(x + 5)$

$x(2) + x\left(\dfrac{10}{x}\right) = x(x + 5)$

$2x + 10 = x^2 + 5x$

$0 = x^2 + 3x - 10$

$0 = (x + 5)(x - 2)$

$x + 5 = 0$ or $x - 2 = 0$

$x = -5$ $x = 2$

Check:

$x = -5$:

$2 + \dfrac{10}{x} = x + 5$

$2 + \dfrac{10}{-5} \overset{?}{=} -5 + 5$

$2 + (-2) \overset{?}{=} -5 + 5$

$0 = 0$ True

$x = 2$:

$2 + \dfrac{10}{x} = x + 5$

$2 + \dfrac{10}{2} \overset{?}{=} 2 + 5$

$2 + 5 \overset{?}{=} 2 + 5$

$7 = 7$ True

Both -5 and 2 are solutions.

9. $\dfrac{a}{5} = \dfrac{a - 3}{2}$

$10\left(\dfrac{a}{5}\right) = 10\left(\dfrac{a - 3}{2}\right)$

$2a = 5(a - 3)$

$2a = 5a - 15$

$-3a = -15$

$a = 5$

Check:

$$\frac{a}{5} = \frac{a-3}{2}$$

$$\frac{5}{5} \stackrel{?}{=} \frac{5-3}{2}$$

$$\frac{5}{5} \stackrel{?}{=} \frac{2}{2}$$

$1 = 1$ True

The solution is 5.

11. $\dfrac{x-3}{5} + \dfrac{x-2}{2} = \dfrac{1}{2}$

$$10\left(\frac{x-3}{5} + \frac{x-2}{2}\right) = 10\left(\frac{1}{2}\right)$$

$$10\left(\frac{x-3}{5}\right) + 10\left(\frac{x-2}{2}\right) = 10\left(\frac{1}{2}\right)$$

$$2(x-3) + 5(x-2) = 5$$

$$2x - 6 + 5x - 10 = 5$$

$$7x - 16 = 5$$

$$7x = 21$$

$$x = 3$$

Check:

$$\frac{x-3}{5} + \frac{x-2}{2} = \frac{1}{2}$$

$$\frac{3-3}{5} + \frac{3-2}{2} \stackrel{?}{=} \frac{1}{2}$$

$$\frac{0}{5} + \frac{1}{2} \stackrel{?}{=} \frac{1}{2}$$

$$0 + \frac{1}{2} \stackrel{?}{=} \frac{1}{2}$$

$$\frac{1}{2} = \frac{1}{2}$$ True

The solution is 3.

13. $\dfrac{2}{y} + \dfrac{1}{2} = \dfrac{5}{2y}$

$$2y\left(\frac{2}{y} + \frac{1}{2}\right) = 2y\left(\frac{5}{2y}\right)$$

$$2y\left(\frac{2}{y}\right) + 2y\left(\frac{1}{2}\right) = 2y\left(\frac{5}{2y}\right)$$

$$4 + y = 5$$

$$y = 1$$

Check:

$$\frac{2}{y} + \frac{1}{2} = \frac{5}{2y}$$

$$\frac{2}{1} + \frac{1}{2} \stackrel{?}{=} \frac{5}{2(1)}$$

$$\frac{4}{2} + \frac{1}{2} \stackrel{?}{=} \frac{5}{2}$$

$$\frac{5}{2} = \frac{5}{2}$$ True

The solution is 1.

15. $\dfrac{11}{2x} + \dfrac{2}{3} = \dfrac{7}{2x}$

$$6x\left(\frac{11}{2x} + \frac{2}{3}\right) = 6x\left(\frac{7}{2x}\right)$$

$$6x\left(\frac{11}{2x}\right) + 6x\left(\frac{2}{3}\right) = 6x\left(\frac{7}{2x}\right)$$

$$33 + 4x = 21$$

$$4x = -12$$

$$x = -3$$

Check:

$$\frac{11}{2x} + \frac{2}{3} = \frac{7}{2x}$$

$$\frac{11}{2(-3)} + \frac{2}{3} \stackrel{?}{=} \frac{7}{2(-3)}$$

$$\frac{11}{-6} + \frac{2}{3} \stackrel{?}{=} \frac{7}{-6}$$

$$\frac{-11}{6} + \frac{4}{6} \stackrel{?}{=} -\frac{7}{6}$$

$$\frac{-11+4}{6} \stackrel{?}{=} -\frac{7}{6}$$

$$-\frac{7}{6} = -\frac{7}{6}$$ True

The solution is –3.

17. $2 + \dfrac{3}{a-3} = \dfrac{a}{a-3}$

$$(a-3)\left(2 + \frac{3}{a-3}\right) = (a-3)\left(\frac{a}{a-3}\right)$$

$$(a-3)(2) + (a-3)\left(\frac{3}{a-3}\right) = (a-3)\left(\frac{a}{a-3}\right)$$

$$2a - 6 + 3 = a$$

$$2a - 3 = a$$

$$-3 = -a$$

$$3 = a$$

In the original equation, 3 makes a denominator 0. This equation has no solution.

19. $\dfrac{3}{2a-5} = -1$

$(2a-5)\left(\dfrac{3}{2a-5}\right) = (2a-5)(-1)$

$3 = -2a + 5$
$-2 = -2a$
$1 = a$
Check:

$\dfrac{3}{2a-5} = -1$

$\dfrac{3}{2(1)-5} \stackrel{?}{=} -1$

$\dfrac{3}{2-5} \stackrel{?}{=} -1$

$\dfrac{3}{-3} \stackrel{?}{=} -1$

$-1 = -1$ True
The solution is 1.

21. $\dfrac{y}{y+4} + \dfrac{4}{y+4} = 3$

$(y+4)\left(\dfrac{y}{y+4} + \dfrac{4}{y+4}\right) = (y+4)(3)$

$(y+4)\left(\dfrac{y}{y+4}\right) + (y+4)\left(\dfrac{4}{y+4}\right) = (y+4)(3)$

$y + 4 = 3y + 12$
$4 = 2y + 12$
$-8 = 2y$
$-\dfrac{8}{2} = y$
$-4 = y$
In the original equation, -4 makes a denominator zero. This equation has no solution.

23. $\dfrac{a}{a-6} = \dfrac{-2}{a-1}$

$(a-6)(a-1)\left(\dfrac{a}{a-6}\right) = (a-6)(a-1)\left(\dfrac{-2}{a-1}\right)$

$a(a-1) = -2(a-6)$
$a^2 - a = -2a + 12$
$a^2 + a - 12 = 0$
$(a+4)(a-3) = 0$
$a + 4 = 0$ or $a - 3 = 0$
$a = -4$ $a = 3$
Check:
$a = -4$:

$\dfrac{a}{a-6} = \dfrac{-2}{a-1}$

$\dfrac{-4}{-4-6} \stackrel{?}{=} \dfrac{-2}{-4-1}$

$\dfrac{-4}{-10} \stackrel{?}{=} \dfrac{-2}{-5}$

$\dfrac{2}{5} = \dfrac{2}{5}$ True

$a = 3$:

$\dfrac{a}{a-6} = \dfrac{-2}{a-1}$

$\dfrac{3}{3-6} \stackrel{?}{=} \dfrac{-2}{3-1}$

$\dfrac{3}{-3} \stackrel{?}{=} \dfrac{-2}{2}$

$-1 = -1$ True
The solutions are -4 and 3.

25. $\dfrac{2x}{x+2} - 2 = \dfrac{x-8}{x-2}$

$(x+2)(x-2)\left(\dfrac{2x}{x+2} - 2\right) = (x+2)(x-2)\left(\dfrac{x-8}{x-2}\right)$

$(x+2)(x-2)\left(\dfrac{2x}{x+2}\right) - (x+2)(x-2)(2) = (x+2)(x-2)\left(\dfrac{x-8}{x-2}\right)$

$2x(x-2) - 2(x^2-4) = (x+2)(x-8)$

$2x^2 - 4x - 2x^2 + 8 = x^2 - 6x - 16$

$-4x + 8 = x^2 - 6x - 16$

$0 = x^2 - 2x - 24$

$0 = (x-6)(x+4)$

$x - 6 = 0 \quad \text{or} \quad x + 4 = 0$

$x = 6 \qquad\qquad x = -4$

Check:

$x = 6$:

$\dfrac{2x}{x+2} - 2 = \dfrac{x-8}{x-2}$

$\dfrac{2(6)}{6+2} - 2 \overset{?}{=} \dfrac{6-8}{6-2}$

$\dfrac{12}{8} - 2 \overset{?}{=} -\dfrac{2}{4}$

$\dfrac{3}{2} - \dfrac{4}{2} \overset{?}{=} -\dfrac{1}{2}$

$\dfrac{3-4}{2} \overset{?}{=} -\dfrac{1}{2}$

$-\dfrac{1}{2} = -\dfrac{1}{2} \quad \text{True}$

$x = -4$:

$\dfrac{2x}{x+2} - 2 = \dfrac{x-8}{x-2}$

$\dfrac{2(-4)}{-4+2} - 2 \overset{?}{=} \dfrac{-4-8}{-4-2}$

$\dfrac{-8}{-2} - 2 \overset{?}{=} \dfrac{-12}{-6}$

$4 - 2 \overset{?}{=} 2$

$2 = 2 \quad \text{True}$

The solutions are 6 and -4.

27. $\dfrac{4y}{y-4} + 5 = \dfrac{5y}{y-4}$

$(y-4)\left(\dfrac{4y}{y-4} + 5\right) = (y-4)\left(\dfrac{5y}{y-4}\right)$

$(y-4)\left(\dfrac{4y}{y-4}\right) + (y-4)(5) = (y-4)\left(\dfrac{5y}{y-4}\right)$

$4y + 5y - 20 = 5y$

$9y - 20 = 5y$

$4y - 20 = 0$

$4y = 20$

$y = 5$

Check:
$$\frac{4y}{y-4}+5=\frac{5y}{y-4}$$
$$\frac{4(5)}{5-4}+5\overset{?}{=}\frac{5(5)}{5-4}$$
$$\frac{20}{1}+5\overset{?}{=}\frac{25}{1}$$
$$25=25 \quad \text{True}$$
The solution is 5.

29. $\dfrac{2}{x-2}+1=\dfrac{x}{x+2}$

$$(x-2)(x+2)\left(\frac{2}{x-2}+1\right)=(x-2)(x+2)\left(\frac{x}{x+2}\right)$$

$$(x-2)(x+2)\left(\frac{2}{x-2}\right)+(x-2)(x+2)=(x-2)(x+2)\left(\frac{x}{x+2}\right)$$

$$2(x+2)+(x-2)(x+2)=x(x-2)$$
$$2x+4+x^2-4=x^2-2x$$
$$2x+x^2=x^2-2x$$
$$2x=-2x$$
$$4x=0$$
$$x=0$$

Check:
$$\frac{2}{x-2}+1=\frac{x}{x+2}$$
$$\frac{2}{0-2}+1\overset{?}{=}\frac{0}{0+2}$$
$$\frac{2}{-2}+1\overset{?}{=}0$$
$$-1+1\overset{?}{=}0$$
$$0=0 \quad \text{True}$$
The solution is 0.

31. $\dfrac{t}{t-4}=\dfrac{t+4}{6}$

$$6(t-4)\left(\frac{t}{t-4}\right)=6(t-4)\left(\frac{t+4}{6}\right)$$

$$6t=t^2-16$$
$$0=t^2-6t-16$$
$$0=(t-8)(t+2)$$
$$t-8=0 \quad \text{or} \quad t+2=0$$
$$t=8 \qquad\qquad t=-2$$

Check:

$t = 8$:

$$\frac{t}{t-4} = \frac{t+4}{6}$$

$$\frac{8}{8-4} \overset{?}{=} \frac{8+4}{6}$$

$$\frac{8}{4} \overset{?}{=} \frac{12}{6}$$

$2 = 2$ True

$t = -2$:

$$\frac{t}{t-4} = \frac{t+4}{6}$$

$$\frac{-2}{-2-4} \overset{?}{=} \frac{-2+4}{6}$$

$$\frac{-2}{-6} \overset{?}{=} \frac{2}{6}$$

$$\frac{1}{3} = \frac{1}{3}$$ True

The solutions are 8 and −2.

33. $\dfrac{x+1}{3} - \dfrac{x-1}{6} = \dfrac{1}{6}$

$$6\left(\frac{x+1}{3} - \frac{x-1}{6}\right) = 6\left(\frac{1}{6}\right)$$

$$6\left(\frac{x+1}{3}\right) - 6\left(\frac{x-1}{6}\right) = 6\left(\frac{1}{6}\right)$$

$2(x+1) - (x-1) = 1$

$2x + 2 - x + 1 = 1$

$x + 3 = 1$

$x = -2$

Check:

$$\frac{x+1}{3} - \frac{x-1}{6} = \frac{1}{6}$$

$$\frac{-2+1}{3} - \frac{-2-1}{6} \overset{?}{=} \frac{1}{6}$$

$$-\frac{1}{3} - \frac{-3}{6} \overset{?}{=} \frac{1}{6}$$

$$-\frac{2}{6} - \frac{-3}{6} \overset{?}{=} \frac{1}{6}$$

$$\frac{-2-(-3)}{6} \overset{?}{=} \frac{1}{6}$$

$$\frac{-2+3}{6} \overset{?}{=} \frac{1}{6}$$

$$\frac{1}{6} = \frac{1}{6}$$ True

The solution is −2.

35. $\dfrac{y}{2y+2} + \dfrac{2y-16}{4y+4} = \dfrac{2y-3}{y+1}$

$\dfrac{y}{2(y+1)} + \dfrac{2y-16}{4(y+1)} = \dfrac{2y-3}{y+1}$

$4(y+1)\left(\dfrac{y}{2(y+1)} + \dfrac{2y-16}{4(y+1)}\right) = 4(y+1)\left(\dfrac{2y-3}{y+1}\right)$

$4(y+1)\left(\dfrac{y}{2(y+1)}\right) + 4(y+1)\left(\dfrac{2y-16}{4(y+1)}\right) = 4(y+1)\left(\dfrac{2y-3}{y+1}\right)$

$2y + 2y - 16 = 4(2y - 3)$
$4y - 16 = 8y - 12$
$-4y = 4$
$y = -1$
In the original equation, -1 makes a denominator 0. This equation has no solution.

37. $\dfrac{4r-4}{r^2+5r-14} + \dfrac{2}{r+7} = \dfrac{1}{r-2}$

$\dfrac{4r-4}{(r+7)(r-2)} + \dfrac{2}{r+7} = \dfrac{1}{r-2}$

$(r+7)(r-2)\left(\dfrac{4r-4}{(r+7)(r-2)} + \dfrac{2}{r+7}\right) = (r+7)(r-2)\left(\dfrac{1}{r-2}\right)$

$(r+7)(r-2)\left(\dfrac{4r-4}{(r+7)(r-2)}\right) + (r+7)(r-2)\left(\dfrac{2}{r+7}\right) = (r+7)(r-2)\left(\dfrac{1}{r-2}\right)$

$4r - 4 + 2(r-2) = (r+7)(1)$
$4r - 4 + 2r - 4 = r + 7$
$6r - 8 = r + 7$
$5r = 15$
$r = 3$
Check:

$\dfrac{4r-4}{r^2+5r-14} + \dfrac{2}{r+7} = \dfrac{1}{r-2}$

$\dfrac{4(3)-4}{3^2+5(3)-14} + \dfrac{2}{3+7} \overset{?}{=} \dfrac{1}{3-2}$

$\dfrac{12-4}{9+15-14} + \dfrac{2}{10} \overset{?}{=} \dfrac{1}{1}$

$\dfrac{8}{10} + \dfrac{2}{10} \overset{?}{=} 1$

$\dfrac{8+2}{10} \overset{?}{=} 1$

$\dfrac{10}{10} \overset{?}{=} 1$

$1 = 1$ True
The solution is 3.

39. $\dfrac{x+1}{x+3} = \dfrac{x^2-11x}{x^2+x-6} - \dfrac{x-3}{x-2}$

$\dfrac{x+1}{x+3} = \dfrac{x^2-11x}{(x+3)(x-2)} - \dfrac{x-3}{x-2}$

$(x+3)(x-2)\left(\dfrac{x+1}{x+3}\right) = (x+3)(x-2)\left(\dfrac{x^2-11x}{(x+3)(x-2)} - \dfrac{x-3}{x-2}\right)$

$(x+3)(x-2)\left(\dfrac{x+1}{x+3}\right) = (x+3)(x-2)\left(\dfrac{x^2-11x}{(x+3)(x-2)}\right) - (x+3)(x-2)\left(\dfrac{x-3}{x-2}\right)$

$(x-2)(x+1) = x^2-11x - (x+3)(x-3)$

$x^2 - x - 2 = x^2 - 11x - (x^2-9)$

$x^2 - x - 2 = x^2 - 11x - x^2 + 9$

$x^2 - x - 2 = -11x + 9$

$x^2 + 10x - 11 = 0$

$(x+11)(x-1) = 0$

$x+11=0 \quad\text{or}\quad x-1=0$

$x = -11 \qquad\qquad x = 1$

Check:

$x = -11$:

$\dfrac{x+1}{x+3} = \dfrac{x^2-11x}{x^2+x-6} - \dfrac{x-3}{x-2}$

$\dfrac{-11+1}{-11+3} \overset{?}{=} \dfrac{(-11)^2-11(-11)}{(-11)^2+(-11)-6} - \dfrac{-11-3}{-11-2}$

$\dfrac{-10}{-8} \overset{?}{=} \dfrac{121+121}{121-17} - \dfrac{-14}{-13}$

$\dfrac{5}{4} \overset{?}{=} \dfrac{242}{104} - \dfrac{14}{13}$

$\dfrac{5}{4} \overset{?}{=} \dfrac{121}{52} - \dfrac{14}{13}$

$\dfrac{65}{52} \overset{?}{=} \dfrac{121}{52} - \dfrac{56}{52}$

$\dfrac{65}{52} \overset{?}{=} \dfrac{121-56}{52}$

$\dfrac{65}{52} = \dfrac{65}{52}$ True

$x = 1$:

$\dfrac{x+1}{x+3} = \dfrac{x^2-11x}{x^2+x-6} - \dfrac{x-3}{x-2}$

$\dfrac{1+1}{1+3} \overset{?}{=} \dfrac{(1)^2-11(1)}{(1)^2+1-6} - \dfrac{1-3}{1-2}$

$\dfrac{2}{4} \overset{?}{=} \dfrac{1-11}{1+1-6} - \dfrac{-2}{-1}$

$\dfrac{1}{2} \overset{?}{=} \dfrac{-10}{-4} - 2$

$\dfrac{1}{2} \overset{?}{=} \dfrac{5}{2} - \dfrac{4}{2}$

$\dfrac{1}{2} = \dfrac{1}{2}$ True

The solutions are -11 and 1.

41. $\dfrac{d}{r} = t$

$r\left(\dfrac{d}{r}\right) = r(t)$

$d = rt$

$\dfrac{d}{t} = \dfrac{rt}{t}$

$\dfrac{d}{t} = r$

43. $T = \dfrac{V}{Q}$

$Q(T) = Q\left(\dfrac{V}{Q}\right)$

$QT = V$

$\dfrac{QT}{T} = \dfrac{V}{T}$

$Q = \dfrac{V}{T}$

45. $i = \dfrac{A}{t + B}$

$(t + B)(i) = (t + B)\left(\dfrac{A}{t + B}\right)$

$ti + Bi = A$

$ti = A - Bi$

$\dfrac{ti}{i} = \dfrac{A - Bi}{i}$

$t = \dfrac{A - Bi}{i}$

47. $N = R + \dfrac{V}{G}$

$N - R = \dfrac{V}{G}$

$G(N - R) = G\left(\dfrac{V}{G}\right)$

$G(N - R) = V$

$\dfrac{G(N - R)}{N - R} = \dfrac{V}{N - R}$

$G = \dfrac{V}{N - R}$

49. $\dfrac{C}{\pi r} = 2$

$\pi r\left(\dfrac{C}{\pi r}\right) = \pi r(2)$

$C = 2\pi r$

$\dfrac{C}{2\pi} = \dfrac{2\pi r}{2\pi}$

$\dfrac{C}{2\pi} = r$

51. The reciprocal of x is $\dfrac{1}{x}$.

53. The reciprocal of x added to the reciprocal of 2 is $\dfrac{1}{x} + \dfrac{1}{2}$.

55. If a tank is filled in 3 hours, the part of the tank filled in 1 hour is $\dfrac{1}{3}$.

57. $\dfrac{20x}{3} + \dfrac{32x}{6} = 180$

$6\left(\dfrac{20x}{3} + \dfrac{32x}{6}\right) = 6(180)$

$6\left(\dfrac{20x}{3}\right) + 6\left(\dfrac{32x}{6}\right) = 6(180)$

$40x + 32x = 1080$

$72x = 1080$

$\dfrac{72x}{72} = \dfrac{1080}{72}$

$x = 15$

$\dfrac{20x}{3} = \dfrac{20(15)}{3} = 100$

$\dfrac{32x}{6} = \dfrac{32(15)}{6} = 80$

The supplementary angles are $100°$ and $80°$.

59. $\dfrac{150}{x} + \dfrac{450}{x} = 90$

$x\left(\dfrac{150}{x} + \dfrac{450}{x}\right) = x(90)$

$x\left(\dfrac{150}{x}\right) + x\left(\dfrac{450}{x}\right) = x(90)$

$150 + 450 = 90x$

$600 = 90x$

$\dfrac{600}{90} = x$

$$\frac{20}{3} = x$$

$$\frac{150}{x} = \frac{150}{\frac{20}{3}} = 150\left(\frac{3}{20}\right) = \frac{45}{2} = 22.5$$

$$\frac{450}{x} = \frac{450}{\frac{20}{3}} = 450\left(\frac{3}{20}\right) = \frac{135}{2} = 67.5$$

The complementary angles are 22.5° and 67.5°.

61. $\dfrac{4}{a^2 + 4a + 3} + \dfrac{2}{a^2 + a - 6} - \dfrac{3}{a^2 - a - 2} = 0$

$$\frac{4}{(a+3)(a+1)} + \frac{2}{(a+3)(a-2)} - \frac{3}{(a-2)(a+1)} = 0$$

$$(a+3)(a+1)(a-2)\left(\frac{4}{(a+3)(a+1)} + \frac{2}{(a+3)(a-2)} - \frac{3}{(a-2)(a+1)}\right) = (a+3)(a+1)(a-2)(0)$$

$$(a+3)(a+1)(a-2)\left(\frac{4}{(a+3)(a+1)}\right) + (a+3)(a+1)(a-2)\left(\frac{2}{(a+3)(a-2)}\right)$$

$$- (a+3)(a+1)(a-2)\left(\frac{3}{(a-2)(a+1)}\right) = 0$$

$4(a-2) + 2(a+1) - 3(a+3) = 0$

$4a - 8 + 2a + 2 - 3a - 9 = 0$

$3a - 15 = 0$

$3a = 15$

$a = 5$

Integrated Review

1. expression

$$\frac{1}{x} + \frac{2}{3} = \frac{1(3)}{x(3)} + \frac{2(x)}{3(x)} = \frac{3}{3x} + \frac{2x}{3x} = \frac{3 + 2x}{3x}$$

2. expression

$$\frac{3}{a} + \frac{5}{6} = \frac{3(6)}{a(6)} + \frac{5(a)}{6(a)} = \frac{18}{6a} + \frac{5a}{6a} = \frac{18 + 5a}{6a}$$

3. equation

$$\frac{1}{x} + \frac{2}{3} = \frac{3}{x}$$

$$3x\left(\frac{1}{x} + \frac{2}{3}\right) = 3x\left(\frac{3}{x}\right)$$

$$3x\left(\frac{1}{x}\right) + 3x\left(\frac{2}{3}\right) = 3x\left(\frac{3}{x}\right)$$

$3 + 2x = 9$

$2x = 6$

$x = 3$

The solution is 3.

4. equation

$$\frac{3}{a} + \frac{5}{6} = 1$$

$$6a\left(\frac{3}{a} + \frac{5}{6}\right) = 6a(1)$$

$$6a\left(\frac{3}{a}\right) + 6a\left(\frac{5}{6}\right) = 6a$$

$18 + 5a = 6a$

$18 = a$

The solution is 18.

5. expression

$$\frac{2}{x+1} - \frac{1}{x} = \frac{2(x)}{(x+1)(x)} - \frac{1(x+1)}{x(x+1)}$$

$$= \frac{2x - (x+1)}{x(x+1)}$$

$$= \frac{2x - x - 1}{x(x+1)} = \frac{x-1}{x(x+1)}$$

6. expression

$$\frac{4}{x-3} - \frac{1}{x} = \frac{4(x)}{(x-3)(x)} - \frac{1(x-3)}{x(x-3)}$$

$$= \frac{4x-(x-3)}{x(x-3)} = \frac{4x-x+3}{x(x-3)} = \frac{3x+3}{x(x-3)}$$

$$= \frac{3(x+1)}{x(x-3)}$$

7. equation

$$\frac{2}{x+1} - \frac{1}{x} = 1$$

$$x(x+1)\left(\frac{2}{x+1} - \frac{1}{x}\right) = x(x+1)(1)$$

$$x(x+1)\left(\frac{2}{x+1}\right) - x(x+1)\left(\frac{1}{x}\right) = x(x+1)$$

$$2x - (x+1) = x(x+1)$$

$$2x - x - 1 = x^2 + x$$

$$x - 1 = x^2 + x$$

$$-1 = x^2$$

There is no real number solution.

8. equation

$$\frac{4}{x-3} - \frac{1}{x} = \frac{6}{x(x-3)}$$

$$x(x-3)\left(\frac{4}{x-3} - \frac{1}{x}\right) = x(x-3)\left(\frac{6}{x(x-3)}\right)$$

$$x(x-3)\left(\frac{4}{x-3}\right) - x(x-3)\left(\frac{1}{x}\right) = 6$$

$$4x - (x-3) = 6$$

$$4x - x + 3 = 6$$

$$3x + 3 = 6$$

$$3x = 3$$

$$x = 1$$

The solution is 1.

9. expression

$$\frac{15x}{x+8} \cdot \frac{2x+16}{3x} = \frac{15x \cdot (2x+16)}{(x+8) \cdot 3x}$$

$$= \frac{3 \cdot 5 \cdot x \cdot 2 \cdot (x+8)}{(x+8) \cdot 3 \cdot x} = 5 \cdot 2 = 10$$

10. expression

$$\frac{9z+5}{15} \cdot \frac{5z}{81z^2-25} = \frac{(9z+5) \cdot 5z}{15 \cdot (81z^2-25)}$$

$$= \frac{(9z+5) \cdot 5 \cdot z}{5 \cdot 3 \cdot (9z+5)(9z-5)} = \frac{z}{3(9z-5)}$$

11. expression

$$\frac{2x+1}{x-3} + \frac{3x+6}{x-3} = \frac{2x+1+3x+6}{x-3}$$

$$= \frac{5x+7}{x-3}$$

12. expression

$$\frac{4p-3}{2p+7} + \frac{3p+8}{2p+7} = \frac{4p-3+3p+8}{2p+7}$$

$$= \frac{7p+5}{2p+7}$$

13. equation

$$\frac{x+5}{7} = \frac{8}{2}$$

$$14\left(\frac{x+5}{7}\right) = 14\left(\frac{8}{2}\right)$$

$$2(x+5) = 56$$

$$2x + 10 = 56$$

$$2x = 46$$

$$x = 23$$

The solution is 23.

14. equation

$$\frac{1}{2} = \frac{x+1}{8}$$

$$8\left(\frac{1}{2}\right) = 8\left(\frac{x+1}{8}\right)$$

$$4 = x + 1$$

$$3 = x$$

The solution is 3.

15. expression

$$\frac{5a+10}{18} \div \frac{a^2-4}{10a} = \frac{5a+10}{18} \cdot \frac{10a}{a^2-4}$$

$$= \frac{5(a+2) \cdot 2 \cdot 5 \cdot a}{2 \cdot 9(a+2)(a-2)}$$

$$= \frac{5 \cdot 5 \cdot a}{9(a-2)} = \frac{25a}{9(a-2)}$$

16. expression

$$\frac{9}{x^2-1} \div \frac{12}{3x+3} = \frac{9}{x^2-1} \cdot \frac{3x+3}{12}$$

$$= \frac{3 \cdot 3 \cdot 3(x+1)}{(x-1)(x+1) \cdot 3 \cdot 4} = \frac{3 \cdot 3}{(x-1) \cdot 4} = \frac{9}{4(x-1)}$$

17. Answers may vary.

18. Answers may vary.

Section 5.6

Practice Problems

1. Let x = the unknown number.

$$\frac{x}{2} - \frac{1}{3} = \frac{x}{6}$$

$$6\left(\frac{x}{2} - \frac{1}{3}\right) = 6\left(\frac{x}{6}\right)$$

$$6\left(\frac{x}{2}\right) - 6\left(\frac{1}{3}\right) = 6\left(\frac{x}{6}\right)$$

$$3x - 2 = x$$
$$-2 = -2x$$
$$x = 1$$

The unknown number is 1.

2.

	Hours to Complete Total Job	Part of Job Completed in 1 Hour
Andrew	2	$\frac{1}{2}$
Timothy	3	$\frac{1}{3}$
Together	x	$\frac{1}{x}$

$$\frac{1}{2} + \frac{1}{3} = \frac{1}{x}$$

$$6x\left(\frac{1}{2}\right) + 6x\left(\frac{1}{3}\right) = 6x\left(\frac{1}{x}\right)$$

$$3x + 2x = 6$$
$$5x = 6$$

$$x = \frac{6}{5} \text{ or } 1\frac{1}{5}$$

Andrew and Timothy can recycle a batch in $1\frac{1}{5}$ hour.

3. Let r = the motorcycle's speed.

	distance	=	rate	·	time
car	280		$r + 10$		$\frac{280}{r+10}$
motorcycle	240		r		$\frac{240}{r}$

$$\frac{280}{r+10} = \frac{240}{r}$$
$$280r = 240(r+10)$$
$$280r = 240r + 2400$$
$$40r = 2400$$
$$r = 60$$
$$r + 10 = 70$$

The speed of the car is 70 miles per hour and the speed of the motorcycle is 60 miles per hour.

4. Since the triangles are similar, their corresponding sides are in proportion.

$$\frac{x}{12} = \frac{15}{9}$$
$$9x = 180$$
$$x = 20$$

The side has length 20 units.

Exercise Set 5.6

1. $3 \cdot \frac{1}{x} = 9 \cdot \frac{1}{6}$

$$\frac{3}{x} = \frac{9}{6}$$
$$6x\left(\frac{3}{x}\right) = 6x\left(\frac{9}{6}\right)$$
$$18 = 9x$$
$$x = 2$$

The unknown number is 2.

3. $\frac{3+2x}{x+1} = \frac{3}{2}$

$$2(x+1)\left(\frac{3+2x}{x+1}\right) = 2(x+1)\left(\frac{3}{2}\right)$$
$$2(3+2x) = 3(x+1)$$
$$6 + 4x = 3x + 3$$
$$x = -3$$

The unknown number is −3.

5. $\frac{2}{x-3} - \frac{4}{x+3} = 8 \cdot \frac{1}{x^2-9}$

$$(x-3)(x+3)\left(\frac{2}{x-3} - \frac{4}{x+3}\right)$$
$$= (x-3)(x+3)\left(\frac{8}{x^2-9}\right)$$
$$(x-3)(x+3)\left(\frac{2}{x-3}\right) - (x-3)(x+3)\left(\frac{4}{x+3}\right)$$
$$= 8$$
$$2(x+3) - 4(x-3) = 8$$
$$2x + 6 - 4x + 12 = 8$$
$$-2x + 18 = 8$$
$$-2x = -10$$
$$x = 5$$

The unknown number is 5.

7. $\frac{1}{4} = \frac{x}{8}$

$$8\left(\frac{1}{4}\right) = 8\left(\frac{x}{8}\right)$$
$$2 = x$$

The unknown number is 2.

9.

	Hours to Complete Total Job	Part of Job Completed in 1 Hour
Experienced	4	$\frac{1}{4}$
Apprentice	5	$\frac{1}{5}$
Together	x	$\frac{1}{x}$

$$\frac{1}{4} + \frac{1}{5} = \frac{1}{x}$$
$$20x\left(\frac{1}{4}\right) + 20x\left(\frac{1}{5}\right) = 20x\left(\frac{1}{x}\right)$$
$$5x + 4x = 20$$
$$9x = 20$$
$$x = \frac{20}{9} \text{ or } 2\frac{2}{9}$$

The experienced surveyor and apprentice surveyor, working together, can survey the road bed in $2\frac{2}{9}$ hours.

11.

	Minutes to Complete Total Job	Part of Job Completed in 1 Minute
Larger belt	2	$\frac{1}{2}$
Smaller belt	6	$\frac{1}{6}$
Both belts	x	$\frac{1}{x}$

$$\frac{1}{2} + \frac{1}{6} = \frac{1}{x}$$
$$6x\left(\frac{1}{2}\right) + 6x\left(\frac{1}{6}\right) = 6x\left(\frac{1}{x}\right)$$
$$3x + x = 6$$
$$4x = 6$$
$$x = \frac{6}{4} = \frac{3}{2} = 1\frac{1}{2}$$

Both belts together can move the cans to the storage area in $1\frac{1}{2}$ minute.

13.

	Hours to Complete Total Job	Part of Job Completed in 1 Hour
Marcus	6	$\frac{1}{6}$
Tony	4	$\frac{1}{4}$
Together	x	$\frac{1}{x}$

$$\frac{1}{6} + \frac{1}{4} = \frac{1}{x}$$

$$12x\left(\frac{1}{6}\right) + 12x\left(\frac{1}{4}\right) = 12x\left(\frac{1}{x}\right)$$

$$2x + 3x = 12$$

$$5x = 12$$

$$x = \frac{12}{5} = 2\frac{2}{5}$$

$$45\left(\frac{12}{5}\right) = 108$$

Together, Marcus and Tony work for $2\frac{2}{5}$ hours at \$45 per hour. The labor estimate should be \$108.

15.

	Hours to Complete Total Job	Part of Job Completed in 1 Hour
Custodian	3	$\frac{1}{3}$
Second Worker	x	$\frac{1}{x}$
Together	$1\frac{1}{2}$ or $\frac{3}{2}$	$\frac{2}{3}$

$$\frac{1}{3} + \frac{1}{x} = \frac{2}{3}$$

$$3x\left(\frac{1}{3}\right) + 3x\left(\frac{1}{x}\right) = 3x\left(\frac{2}{3}\right)$$

$$x + 3 = 2x$$

$$3 = x$$

It takes the second worker 3 hours to do the job alone.

17.

	Hours to Complete Total Job	Part of Job Completed in 1 Hour
First Pipe	20	$\frac{1}{20}$
Second Pipe	15	$\frac{1}{15}$
Third Pipe	x	$\frac{1}{x}$
3 Pipes Together	6	$\frac{1}{6}$

$\frac{1}{20} + \frac{1}{15} + \frac{1}{x} = \frac{1}{6}$

$60x\left(\frac{1}{20}\right) + 60x\left(\frac{1}{15}\right) + 60x\left(\frac{1}{x}\right) = 60x\left(\frac{1}{6}\right)$

$3x + 4x + 60 = 10x$

$7x + 60 = 10x$

$60 = 3x$

$20 = x$

It takes the third pipe 20 hours to fill the pond.

19.

	distance	=	rate	·	time
Trip to Park	3		$\frac{3}{x}$		x
Return Trip	9		$\frac{9}{x+1}$		$x+1$

$\frac{3}{x} = \frac{9}{x+1}$

$3(x + 1) = 9x$

$3x + 3 = 9x$

$3 = 6x$

$\frac{1}{2} = x$

The jogger spends $\frac{1}{2}$ hour on her trip to the park, so her rate is $\dfrac{3}{\frac{1}{2}} = \dfrac{3}{1} \cdot \dfrac{2}{1} = 6$ miles per hour.

21.

	distance	=	rate	·	time
First portion	20		r		$\frac{20}{r}$
Cooldown Portion	16		$r - 2$		$\frac{16}{r-2}$

$$\frac{20}{r} = \frac{16}{r-2}$$
$$20(r-2) = 16r$$
$$20r - 40 = 16r$$
$$-40 = -4r$$
$$r = 10 \text{ and } r - 2 = 10 - 2 = 8$$

His speed was 10 miles per hour during the first portion and 8 miles per hour during the cooldown portion.

23.

	distance	=	rate	·	time
Upstream	9		$r - 3$		$\frac{9}{r-3}$
Downstream	11		$r + 3$		$\frac{11}{r+3}$

$$\frac{9}{r-3} = \frac{11}{r+3}$$
$$9(r+3) = 11(r-3)$$
$$9r + 27 = 11r - 33$$
$$60 = 2r$$
$$r = 30$$

The speed of the boat in still water is 30 miles per hour.

25. Let w = the rate of the wind.

	distance	=	rate	·	time
With the wind	48		$16 + w$		$\frac{48}{16+w}$
Into the wind	16		$16 - w$		$\frac{16}{16-w}$

$$\frac{48}{16+w} = \frac{16}{16-w}$$
$$48(16 - w) = 16(16 + w)$$
$$768 - 48w = 256 + 16w$$
$$512 = 64w$$
$$w = 8$$

The rate of the wind is 8 miles per hour.

27. Let r = the speed of the car in still air.

	distance	=	rate	·	time
Into the wind	10		$r - 3$		$\frac{10}{r-3}$
With the wind	11		$r + 3$		$\frac{11}{r+3}$

$\dfrac{10}{r-3}=\dfrac{11}{r+3}$

$10(r+3)=11(r-3)$

$10r+30=11r-33$

$63=r$

The speed of the car in still air is 63 miles per hour.

29. $\dfrac{12}{4}=\dfrac{18}{x}$

$12x=72$

$x=6$

31. $\dfrac{x}{3.75}=\dfrac{12}{9}$

$9x=45$

$x=5$

33. $\dfrac{16}{10}=\dfrac{34}{y}$

$16y=340$

$y=21.25$

35. $\dfrac{11}{x}=\dfrac{5}{2}$ and $\dfrac{14}{y}=\dfrac{5}{2}$

$22=5x$ $28=5y$

$x=4.4$ $5.6=y$

The missing dimensions are $x=4.4$ feet and $y=5.6$ feet.

37. $\dfrac{\frac{3}{4}+\frac{1}{4}}{\frac{3}{8}+\frac{13}{8}}=\dfrac{\frac{3+1}{4}}{\frac{3+13}{8}}=\dfrac{\frac{4}{4}}{\frac{16}{8}}=\dfrac{1}{2}$

39. $\dfrac{\frac{2}{5}+\frac{1}{5}}{\frac{7}{10}+\frac{7}{10}}=\dfrac{\frac{2+1}{5}}{\frac{7+7}{10}}=\dfrac{\frac{3}{5}}{\frac{14}{10}}=\dfrac{3}{5}\div\dfrac{14}{10}=\dfrac{3}{5}\cdot\dfrac{10}{14}$

$=\dfrac{3\cdot2\cdot5}{5\cdot2\cdot7}=\dfrac{3}{7}$

41. Let $r=$ Zanardi's speed.

distance = rate · time

	distance	rate	time
Zanardi	1.952	r	$\dfrac{1.952}{r}$
Gugelmin	2	$r+5.6$	$\dfrac{2}{r+5.6}$

$\dfrac{1.952}{r}=\dfrac{2}{r+5.6}$

$1.952(r+5.6)=2r$

$1.952r+10.9312=2r$

$10.9312=0.048r$

$r\approx227.7$

$r+5.6\approx233.3$

Zanardi's fastest lap speed was approximately 227.7 miles per hour, and Gugelmin's fastest lap speed was approximately 233.3 miles per hour.

Section 5.7

Practice Problems

1. $\dfrac{\frac{3}{7}}{\frac{5}{9}}=\dfrac{3}{7}\cdot\dfrac{9}{5}=\dfrac{27}{35}$

2. $\dfrac{\frac{3}{4}-\frac{2}{3}}{\frac{1}{2}+\frac{3}{8}}=\dfrac{\frac{3(3)}{4(3)}-\frac{2(4)}{3(4)}}{\frac{1(4)}{2(4)}+\frac{3}{8}}=\dfrac{\frac{9}{12}-\frac{8}{12}}{\frac{4}{8}+\frac{3}{8}}=\dfrac{\frac{1}{12}}{\frac{7}{8}}$

$=\dfrac{1}{12}\cdot\dfrac{8}{7}=\dfrac{1\cdot4\cdot2}{3\cdot4\cdot7}=\dfrac{2}{21}$

3. $\dfrac{\frac{2}{5}-\frac{1}{x}}{\frac{x}{10}-\frac{1}{3}}=\dfrac{\frac{2x}{5x}-\frac{5}{5x}}{\frac{3x}{30}-\frac{10}{30}}=\dfrac{\frac{2x-5}{5x}}{\frac{3x-10}{30}}$

$=\dfrac{2x-5}{5x}\cdot\dfrac{30}{3x-10}=\dfrac{5\cdot6(2x-5)}{5\cdot x(3x-10)}$

$=\dfrac{6(2x-5)}{x(3x-10)}$

4. $\dfrac{\frac{3}{4}-\frac{2}{3}}{\frac{1}{2}+\frac{3}{8}}=\dfrac{24\left(\frac{3}{4}-\frac{2}{3}\right)}{24\left(\frac{1}{2}+\frac{3}{8}\right)}=\dfrac{24\left(\frac{3}{4}\right)-24\left(\frac{2}{3}\right)}{24\left(\frac{1}{2}\right)+24\left(\frac{3}{8}\right)}$

$=\dfrac{18-16}{12+9}=\dfrac{2}{21}$

5. $\dfrac{1+\frac{x}{y}}{\frac{2x+1}{y}}=\dfrac{y\left(1+\frac{x}{y}\right)}{y\left(\frac{2x+1}{y}\right)}=\dfrac{y(1)+y\left(\frac{x}{y}\right)}{y\left(\frac{2x+1}{y}\right)}=\dfrac{y+x}{2x+1}$

6. $\dfrac{\frac{5}{6y}+\frac{y}{x}}{\frac{y}{3}-x}=\dfrac{6xy\left(\frac{5}{6y}+\frac{y}{x}\right)}{6xy\left(\frac{y}{3}-x\right)}=\dfrac{6xy\left(\frac{5}{6y}\right)+6xy\left(\frac{y}{x}\right)}{6xy\left(\frac{y}{3}\right)-6xy(x)}$

$=\dfrac{5x+6y^2}{2xy^2-6x^2y}=\dfrac{5x+6y^2}{2xy(y-3x)}$

Exercise Set 5.7

1. $\dfrac{\frac{1}{2}}{\frac{3}{4}}=\dfrac{1}{2}\cdot\dfrac{4}{3}=\dfrac{1\cdot2\cdot2}{2\cdot3}=\dfrac{2}{3}$

3. $\dfrac{-\frac{4x}{9}}{-\frac{2x}{3}}=-\dfrac{4x}{9}\cdot-\dfrac{3}{2x}=\dfrac{2\cdot2\cdot3\cdot x}{3\cdot3\cdot2\cdot x}=\dfrac{2}{3}$

5. $\dfrac{\frac{-5}{12x^2}}{\frac{25}{16x^3}}=-\dfrac{5}{12x^2}\cdot\dfrac{16x^3}{25}=-\dfrac{5\cdot4\cdot4\cdot x^2\cdot x}{4\cdot3\cdot x^2\cdot5\cdot5}$

$=-\dfrac{4x}{15}$

7. $\dfrac{\frac{1}{3}}{\frac{1}{2}-\frac{1}{4}}=\dfrac{12\left(\frac{1}{3}\right)}{12\left(\frac{1}{2}-\frac{1}{4}\right)}=\dfrac{12\left(\frac{1}{3}\right)}{12\left(\frac{1}{2}\right)-12\left(\frac{1}{4}\right)}$

$=\dfrac{4}{6-3}=\dfrac{4}{3}$

9. $\dfrac{2+\frac{7}{10}}{1+\frac{3}{5}}=\dfrac{10\left(2+\frac{7}{10}\right)}{10\left(1+\frac{3}{5}\right)}$

$=\dfrac{10(2)+10\left(\frac{7}{10}\right)}{10(1)+10\left(\frac{3}{5}\right)}=\dfrac{20+7}{10+6}=\dfrac{27}{16}$

11. $\dfrac{\frac{m}{n}-1}{\frac{m}{n}+1}=\dfrac{n\left(\frac{m}{n}-1\right)}{n\left(\frac{m}{n}+1\right)}=\dfrac{n\left(\frac{m}{n}\right)-n(1)}{n\left(\frac{m}{n}\right)+n(1)}=\dfrac{m-n}{m+n}$

13. $\dfrac{\frac{1}{5}-\frac{1}{x}}{\frac{7}{10}+\frac{1}{x^2}}=\dfrac{10x^2\left(\frac{1}{5}-\frac{1}{x}\right)}{10x^2\left(\frac{7}{10}+\frac{1}{x^2}\right)}$

$=\dfrac{10x^2\left(\frac{1}{5}\right)-10x^2\left(\frac{1}{x}\right)}{10x^2\left(\frac{7}{10}\right)+10x^2\left(\frac{1}{x^2}\right)}=\dfrac{2x^2-10x}{7x^2+10}$

$=\dfrac{2x(x-5)}{7x^2+10}$

15. $\dfrac{1+\frac{1}{y-2}}{y+\frac{1}{y-2}}=\dfrac{(y-2)\left(1+\frac{1}{y-2}\right)}{(y-2)\left(y+\frac{1}{y-2}\right)}$

$=\dfrac{(y-2)(1)+(y-2)\left(\frac{1}{y-2}\right)}{(y-2)(y)+(y-2)\left(\frac{1}{y-2}\right)}$

$=\dfrac{y-2+1}{y^2-2y+1}=\dfrac{y-1}{(y-1)^2}=\dfrac{1}{y-1}$

17. $\dfrac{\frac{4y-8}{16}}{\frac{6y-12}{4}}=\dfrac{4y-8}{16}\cdot\dfrac{4}{6y-12}=\dfrac{4(y-2)\cdot4}{4\cdot4\cdot6(y-2)}$

$=\dfrac{1}{6}$

19. $\dfrac{\frac{x}{y}+1}{\frac{x}{y}-1}=\dfrac{y\left(\frac{x}{y}+1\right)}{y\left(\frac{x}{y}-1\right)}=\dfrac{y\left(\frac{x}{y}\right)+y(1)}{y\left(\frac{x}{y}\right)-y(1)}=\dfrac{x+y}{x-y}$

21. $\dfrac{1}{2+\frac{1}{3}}=\dfrac{3(1)}{3\left(2+\frac{1}{3}\right)}=\dfrac{3(1)}{3(2)+3\left(\frac{1}{3}\right)}$

$=\dfrac{3}{6+1}=\dfrac{3}{7}$

23. $\dfrac{\frac{ax+ab}{x^2-b^2}}{\frac{x+b}{x-b}}=\dfrac{ax+ab}{x^2-b^2}\cdot\dfrac{x-b}{x+b}$

$=\dfrac{a(x+b)\cdot(x-b)}{(x+b)(x-b)\cdot(x+b)}=\dfrac{a}{x+b}$

25. $\dfrac{\frac{8}{x+4}+2}{\frac{12}{x+4}-2} = \dfrac{(x+4)\left(\frac{8}{x+4}+2\right)}{(x+4)\left(\frac{12}{x+4}-2\right)}$

$= \dfrac{(x+4)\left(\frac{8}{x+4}\right)+(x+4)(2)}{(x+4)\left(\frac{12}{x+4}\right)-(x+4)(2)}$

$= \dfrac{8+2x+8}{12-2x-8} = \dfrac{16+2x}{4-2x} = \dfrac{2(8+x)}{2(2-x)}$

$= \dfrac{8+x}{2-x}$

27. $\dfrac{\frac{s}{r}+\frac{r}{s}}{\frac{s}{r}-\frac{r}{s}} = \dfrac{rs\left(\frac{s}{r}+\frac{r}{s}\right)}{rs\left(\frac{s}{r}-\frac{r}{s}\right)} = \dfrac{rs\left(\frac{s}{r}\right)+rs\left(\frac{r}{s}\right)}{rs\left(\frac{s}{r}\right)-rs\left(\frac{r}{s}\right)}$

$= \dfrac{s^2+r^2}{s^2-r^2}$

29. Answers may vary.

31. *Thriller* sold the most copies.

33. Tie between *Eagles Greatest Hits* and *Born in the U.S.A.* for the fewest copies sold.

35. $\dfrac{\frac{1}{3}+\frac{3}{4}}{2} = \dfrac{12\left(\frac{1}{3}+\frac{3}{4}\right)}{12(2)} = \dfrac{12\left(\frac{1}{3}\right)+12\left(\frac{3}{4}\right)}{12(2)}$

$= \dfrac{4+9}{24} = \dfrac{13}{24}$

37. $\dfrac{1}{\frac{1}{R_1}+\frac{1}{R_2}} = \dfrac{R_1 R_2 (1)}{R_1 R_2 \left(\frac{1}{R_1}+\frac{1}{R_2}\right)}$

$= \dfrac{R_1 R_2}{R_1 R_2 \left(\frac{1}{R_1}\right)+R_1 R_2 \left(\frac{1}{R_2}\right)} = \dfrac{R_1 R_2}{R_2 + R_1}$

39. $\dfrac{x^{-1}+2^{-1}}{x^{-2}-4^{-1}} = \dfrac{\frac{1}{x}+\frac{1}{2}}{\frac{1}{x^2}-\frac{1}{4}} = \dfrac{4x^2\left(\frac{1}{x}+\frac{1}{2}\right)}{4x^2\left(\frac{1}{x^2}-\frac{1}{4}\right)}$

$= \dfrac{4x^2\left(\frac{1}{x}\right)+4x^2\left(\frac{1}{2}\right)}{4x^2\left(\frac{1}{x^2}\right)-4x^2\left(\frac{1}{4}\right)} = \dfrac{4x+2x^2}{4-x^2}$

$= \dfrac{2x(2+x)}{(2-x)(2+x)} = \dfrac{2x}{2-x}$

41. $\dfrac{y^{-2}}{1-y^{-2}} = \dfrac{\frac{1}{y^2}}{1-\frac{1}{y^2}} = \dfrac{y^2\left(\frac{1}{y^2}\right)}{y^2\left(1-\frac{1}{y^2}\right)}$

$= \dfrac{y^2\left(\frac{1}{y^2}\right)}{y^2(1)-y^2\left(\frac{1}{y^2}\right)} = \dfrac{1}{y^2-1}$

Chapter 5 Review

1. The rational expression is undefined when
$x^2 - 4 = 0$
$(x-2)(x+2) = 0$
$x - 2 = 0$ or $x + 2 = 0$
$x = 2$ or $x = -2$

2. The rational expression is undefined when
$4x^2 - 4x - 15 = 0$
$(2x+3)(2x-5) = 0$
$2x + 3 = 0$ or $2x - 5 = 0$
$2x = -3$ or $2x = 5$
$x = -\dfrac{3}{2}$ or $x = \dfrac{5}{2}$

3. $\dfrac{2-z}{z+5} = \dfrac{2-(-2)}{-2+5} = \dfrac{2+2}{3} = \dfrac{4}{3}$

4. $\dfrac{x^2+xy-y^2}{x+y} = \dfrac{5^2+5\cdot 7-7^2}{5+7}$

$= \dfrac{25+35-49}{12} = \dfrac{11}{12}$

5. $\dfrac{2x+6}{x^2+3x} = \dfrac{2(x+3)}{x(x+3)} = \dfrac{2}{x}$

6. $\dfrac{3x-12}{x^2-4x} = \dfrac{3(x-4)}{x(x-4)} = \dfrac{3}{x}$

7. $\dfrac{x+2}{x^2-3x-10} = \dfrac{x+2}{(x-5)(x+2)} = \dfrac{1}{x-5}$

8. $\dfrac{x+4}{x^2+5x+4} = \dfrac{x+4}{(x+1)(x+4)} = \dfrac{1}{x+1}$

9. $\dfrac{x^3 - 4x}{x^2 + 3x + 2} = \dfrac{x(x^2 - 4)}{(x+2)(x+1)}$

$= \dfrac{x(x-2)(x+2)}{(x+2)(x+1)} = \dfrac{x(x-2)}{x+1}$

10. $\dfrac{5x^2 - 125}{x^2 + 2x - 15} = \dfrac{5(x^2 - 25)}{(x-3)(x+5)}$

$= \dfrac{5(x-5)(x+5)}{(x-3)(x+5)} = \dfrac{5(x-5)}{x-3}$

11. $\dfrac{x^2 - x - 6}{x^2 - 3x - 10} = \dfrac{(x-3)(x+2)}{(x-5)(x+2)} = \dfrac{x-3}{x-5}$

12. $\dfrac{x^2 - 2x}{x^2 + 2x - 8} = \dfrac{x(x-2)}{(x+4)(x-2)} = \dfrac{x}{x+4}$

13. $\dfrac{x^2 + xa + xb + ab}{x^2 - xc + bx - bc} = \dfrac{x(x+a) + b(x+a)}{x(x-c) + b(x-c)}$

$= \dfrac{(x+a)(x+b)}{(x-c)(x+b)} = \dfrac{x+a}{x-c}$

14. $\dfrac{x^2 + 5x - 2x - 10}{x^2 - 3x - 2x + 6} = \dfrac{x(x+5) - 2(x+5)}{x(x-3) - 2(x-3)}$

$= \dfrac{(x+5)(x-2)}{(x-3)(x-2)} = \dfrac{x+5}{x-3}$

15. $\dfrac{15x^3y^2}{z} \cdot \dfrac{z}{5xy^3} = \dfrac{15x^3y^2 \cdot z}{z \cdot 5xy^3}$

$= \dfrac{3 \cdot 5 \cdot x^2 \cdot x \cdot y^2 \cdot z}{z \cdot 5 \cdot x \cdot y^2 \cdot y} = \dfrac{3x^2}{y}$

16. $\dfrac{-y^3}{8} \cdot \dfrac{9x^2}{y^3} = -\dfrac{y^3 \cdot 9x^2}{8 \cdot y^3} = -\dfrac{9x^2}{8}$

17. $\dfrac{x^2 - 9}{x^2 - 4} \cdot \dfrac{x-2}{x+3} = \dfrac{(x^2 - 9) \cdot (x-2)}{(x^2 - 4) \cdot (x+3)}$

$= \dfrac{(x-3)(x+3)(x-2)}{(x+2)(x-2)(x+3)} = \dfrac{x-3}{x+2}$

18. $\dfrac{2x+5}{x-6} \cdot \dfrac{2x}{-x+6} = \dfrac{2x+5}{x-6} \cdot \dfrac{2x}{-(x-6)}$

$= \dfrac{2x+5}{x-6} \cdot \dfrac{-2x}{x-6} = \dfrac{(2x+5) \cdot (-2x)}{(x-6) \cdot (x-6)}$

$= \dfrac{-2x(2x+5)}{(x-6)^2}$

19. $\dfrac{x^2 - 5x - 24}{x^2 - x - 12} \div \dfrac{x^2 - 10x + 16}{x^2 + x - 6}$

$= \dfrac{x^2 - 5x - 24}{x^2 - x - 12} \cdot \dfrac{x^2 + x - 6}{x^2 - 10x + 16}$

$= \dfrac{(x-8)(x+3) \cdot (x+3)(x-2)}{(x-4)(x+3) \cdot (x-8)(x-2)}$

$= \dfrac{x+3}{x-4}$

20. $\dfrac{4x+4y}{xy^2} \div \dfrac{3x+3y}{x^2y} = \dfrac{4x+4y}{xy^2} \cdot \dfrac{x^2y}{3x+3y}$

$= \dfrac{4(x+y) \cdot x \cdot x \cdot y}{x \cdot y \cdot y \cdot 3(x+y)} = \dfrac{4x}{3y}$

21. $\dfrac{x^2 + x - 42}{x-3} \cdot \dfrac{(x-3)^2}{x+7}$

$= \dfrac{(x+7)(x-6) \cdot (x-3)(x-3)}{(x-3) \cdot (x+7)}$

$= (x-6)(x-3)$

22. $\dfrac{2a+2b}{3} \cdot \dfrac{a-b}{a^2 - b^2} = \dfrac{2(a+b) \cdot (a-b)}{3 \cdot (a+b)(a-b)} = \dfrac{2}{3}$

23. $\dfrac{x^2 - 9x + 14}{x^2 - 5x + 6} \cdot \dfrac{x+2}{x^2 - 5x - 14}$

$= \dfrac{(x-7)(x-2) \cdot (x+2)}{(x-3)(x-2) \cdot (x-7)(x+2)} = \dfrac{1}{x-3}$

24. $(x-3) \cdot \dfrac{x}{x^2 + 3x - 18}$

$= \dfrac{(x-3) \cdot x}{(x-3)(x+6)} = \dfrac{x}{x+6}$

25. $\dfrac{2x^2-9x+9}{8x-12} \div \dfrac{x^2-3x}{2x}$

$= \dfrac{2x^2-9x+9}{8x-12} \cdot \dfrac{2x}{x^2-3x}$

$= \dfrac{(2x-3)(x-3)\cdot 2x}{4(2x-3)\cdot x(x-3)}$

$= \dfrac{2}{4} = \dfrac{1}{2}$

26. $\dfrac{x^2-y^2}{x^2+xy} \div \dfrac{3x^2-2xy-y^2}{3x^2+6x}$

$= \dfrac{x^2-y^2}{x^2+xy} \cdot \dfrac{3x^2+6x}{3x^2-2xy-y^2}$

$= \dfrac{(x-y)(x+y)\cdot 3x(x+2)}{x(x+y)\cdot(3x+y)(x-y)}$

$= \dfrac{3(x+2)}{3x+y}$

27. $\dfrac{x}{x^2+9x+14} + \dfrac{7}{x^2+9x+14}$

$= \dfrac{x+7}{x^2+9x+14} = \dfrac{x+7}{(x+2)(x+7)} = \dfrac{1}{x+2}$

28. $\dfrac{x}{x^2+2x-15} + \dfrac{5}{x^2+2x-15} = \dfrac{x+5}{x^2+2x-15}$

$= \dfrac{x+5}{(x-3)(x+5)} = \dfrac{1}{x-3}$

29. $\dfrac{4x-5}{3x^2} - \dfrac{2x+5}{3x^2} = \dfrac{4x-5-(2x+5)}{3x^2}$

$= \dfrac{4x-5-2x-5}{3x^2} = \dfrac{2x-10}{3x^2}$

30. $\dfrac{9x+7}{6x^2} - \dfrac{3x+4}{6x^2} = \dfrac{9x+7-(3x+4)}{6x^2}$

$= \dfrac{9x+7-3x-4}{6x^2} = \dfrac{6x+3}{6x^2} = \dfrac{3(2x+1)}{3\cdot 2x^2}$

$= \dfrac{2x+1}{2x^2}$

31. $2x = 2 \cdot x$
$7x = 7 \cdot x$
$\text{LCD} = 2 \cdot 7 \cdot x = 14x$

32. $x^2-5x-24 = (x-8)(x+3)$
$x^2+11x+24 = (x+8)(x+3)$
$\text{LCD} = (x-8)(x+3)(x+8)$

33. $\dfrac{5}{7x} = \dfrac{5(2x^2y)}{7x(2x^2y)} = \dfrac{10x^2y}{14x^3y}$

34. $\dfrac{9}{4y} = \dfrac{9(4y^2x)}{4y(4y^2x)} = \dfrac{36y^2x}{16y^3x}$

35. $\dfrac{x+2}{x^2+11x+18} = \dfrac{x+2}{(x+9)(x+2)}$

$= \dfrac{(x+2)(x-5)}{(x+9)(x+2)(x-5)}$

$= \dfrac{x^2-3x-10}{(x+2)(x-5)(x+9)}$

36. $\dfrac{3x-5}{x^2+4x+4} = \dfrac{3x-5}{(x+2)^2}$

$= \dfrac{(3x-5)(x+3)}{(x+2)^2(x+3)} = \dfrac{3x^2+4x-15}{(x+2)^2(x+3)}$

37. $\dfrac{4}{5x^2} - \dfrac{6}{y} = \dfrac{4(y)}{5x^2(y)} - \dfrac{6(5x^2)}{y(5x^2)} = \dfrac{4y-30x^2}{5x^2y}$

38. $\dfrac{2}{x-3} - \dfrac{4}{x-1}$

$= \dfrac{2(x-1)}{(x-3)(x-1)} - \dfrac{4(x-3)}{(x-1)(x-3)}$

$= \dfrac{2(x-1)-4(x-3)}{(x-3)(x-1)} = \dfrac{2x-2-4x+12}{(x-3)(x-1)}$

$= \dfrac{-2x+10}{(x-3)(x-1)}$

39. $\dfrac{x+7}{x+3} - \dfrac{x-3}{x+7}$

$= \dfrac{(x+7)(x+7)}{(x+3)(x+7)} - \dfrac{(x-3)(x+3)}{(x+7)(x+3)}$

$= \dfrac{x^2+14x+49-(x^2-9)}{(x+3)(x+7)}$

$= \dfrac{x^2+14x+49-x^2+9}{(x+3)(x+7)} = \dfrac{14x+58}{(x+3)(x+7)}$

40. $\dfrac{4}{x+3} - 2 = \dfrac{4}{x+3} - \dfrac{2(x+3)}{x+3}$

$= \dfrac{4 - 2(x+3)}{x+3} = \dfrac{4 - 2x - 6}{x+3} = \dfrac{-2x - 2}{x+3}$

41. $\dfrac{3}{x^2 + 2x - 8} + \dfrac{2}{x^2 - 3x + 2}$

$= \dfrac{3}{(x+4)(x-2)} + \dfrac{2}{(x-1)(x-2)}$

$= \dfrac{3(x-1)}{(x+4)(x-2)(x-1)} + \dfrac{2(x+4)}{(x-1)(x-2)(x+4)}$

$= \dfrac{3(x-1) + 2(x+4)}{(x+4)(x-2)(x-1)}$

$= \dfrac{3x - 3 + 2x + 8}{(x+4)(x-2)(x-1)}$

$= \dfrac{5x + 5}{(x+4)(x-2)(x-1)}$

42. $\dfrac{2x - 5}{6x + 9} - \dfrac{4}{2x^2 + 3x}$

$= \dfrac{2x - 5}{3(2x+3)} - \dfrac{4}{x(2x+3)}$

$= \dfrac{(2x-5)(x)}{3(2x+3)(x)} - \dfrac{4(3)}{x(2x+3)(3)}$

$= \dfrac{2x^2 - 5x - 12}{3x(2x+3)} = \dfrac{(2x+3)(x-4)}{3x(2x+3)}$

$= \dfrac{x - 4}{3x}$

43. $\dfrac{x-1}{x^2 - 2x + 1} - \dfrac{x+1}{x-1} = \dfrac{x-1}{(x-1)^2} - \dfrac{x+1}{x-1}$

$= \dfrac{1}{x-1} - \dfrac{x+1}{x-1} = \dfrac{1 - (x+1)}{x-1}$

$= \dfrac{1 - x - 1}{x-1} = \dfrac{-x}{x-1} = -\dfrac{x}{x-1}$

44. $\dfrac{x-1}{x^2 + 4x + 4} + \dfrac{x-1}{x+2}$

$= \dfrac{x-1}{(x+2)^2} + \dfrac{(x-1)(x+2)}{(x+2)(x+2)}$

$= \dfrac{x - 1 + (x-1)(x+2)}{(x+2)^2}$

$= \dfrac{x - 1 + x^2 + x - 2}{(x+2)^2}$

$= \dfrac{x^2 + 2x - 3}{(x+2)^2}$

45. $P = 2l + 2w$

$P = 2\left(\dfrac{x}{8}\right) + 2\left(\dfrac{x+2}{4x}\right)$

$= \dfrac{x}{4} + \dfrac{2(x+2)}{4x}$

$= \dfrac{x \cdot x}{4 \cdot x} + \dfrac{2x + 4}{4x}$

$= \dfrac{x^2 + 2x + 4}{4x}$

$A = l \cdot w$

$A = \dfrac{x}{8} \cdot \dfrac{x+2}{4x} = \dfrac{x \cdot (x+2)}{8 \cdot 4x} = \dfrac{x+2}{32}$

The perimeter is $\dfrac{x^2 + 2x + 4}{4x}$ units and the

area is $\dfrac{x+2}{32}$ square units.

46. $P = \dfrac{3x}{4x - 4} + \dfrac{2x}{3x - 3} + \dfrac{x}{x - 1}$

$= \dfrac{3x}{4(x-1)} + \dfrac{2x}{3(x-1)} + \dfrac{x}{x-1}$

$= \dfrac{3x(3)}{4(x-1)(3)} + \dfrac{2x(4)}{3(x-1)(4)} + \dfrac{x(12)}{(x-1)(12)}$

$= \dfrac{9x + 8x + 12x}{12(x-1)} = \dfrac{29x}{12(x-1)}$

$A = \dfrac{1}{2} \cdot b \cdot h$

$A = \dfrac{1}{2} \cdot \dfrac{x}{x-1} \cdot \dfrac{6y}{5}$

$= \dfrac{1 \cdot x \cdot 2 \cdot 3y}{2 \cdot (x-1) \cdot 5}$

$= \dfrac{3xy}{5(x-1)}$

The perimeter is $\dfrac{29x}{12(x-1)}$ units and the area

is $\dfrac{3xy}{5(x-1)}$ square units.

47. $\dfrac{x+4}{9} = \dfrac{5}{9}$

$9\left(\dfrac{x+4}{9}\right) = 9\left(\dfrac{5}{9}\right)$

$x + 4 = 5$

$x = 1$

48. $\dfrac{n}{10} = 9 - \dfrac{n}{5}$

$10\left(\dfrac{n}{10}\right) = 10\left(9 - \dfrac{n}{5}\right)$

$10\left(\dfrac{n}{10}\right) = 10(9) - 10\left(\dfrac{n}{5}\right)$

$n = 90 - 2n$

$3n = 90$

$n = 30$

49. $\dfrac{5y-3}{7} = \dfrac{15y-2}{28}$

$28\left(\dfrac{5y-3}{7}\right) = 28\left(\dfrac{15y-2}{28}\right)$

$4(5y - 3) = 15y - 2$

$20y - 12 = 15y - 2$

$5y = 10$

$y = 2$

50. $\dfrac{2}{x+1} - \dfrac{1}{x-2} = -\dfrac{1}{2}$

$2(x+1)(x-2)\left(\dfrac{2}{x+1} - \dfrac{1}{x-2}\right) = 2(x+1)(x-2)\left(-\dfrac{1}{2}\right)$

$2(x+1)(x-2)\left(\dfrac{2}{x+1}\right) - 2(x+1)(x-2)\left(\dfrac{1}{x-2}\right) = 2(x+1)(x-2)\left(-\dfrac{1}{2}\right)$

$4(x - 2) - 2(x + 1) = -(x + 1)(x - 2)$

$4x - 8 - 2x - 2 = -(x^2 - x - 2)$

$2x - 10 = -x^2 + x + 2$

$x^2 + x - 12 = 0$

$(x + 4)(x - 3) = 0$

$x + 4 = 0$ or $x - 3 = 0$

$x = -4$ or $x = 3$

51. $\dfrac{1}{a+3} + \dfrac{1}{a-3} = -\dfrac{5}{a^2-9}$

$(a-3)(a+3)\left(\dfrac{1}{a+3} + \dfrac{1}{a-3}\right) = (a-3)(a+3)\left(-\dfrac{5}{(a-3)(a+3)}\right)$

$(a-3)(a+3)\left(\dfrac{1}{a+3}\right) + (a-3)(a+3)\left(\dfrac{1}{a-3}\right) = -5$

$a - 3 + a + 3 = -5$

$2a = -5$

$a = -\dfrac{5}{2}$

52. $\dfrac{y}{2y+2} + \dfrac{2y-16}{4y+4} = \dfrac{y-3}{y+1}$

$\dfrac{y}{2(y+1)} + \dfrac{2y-16}{4(y+1)} = \dfrac{y-3}{y+1}$

$4(y+1)\left(\dfrac{y}{2(y+1)} + \dfrac{2y-16}{4(y+1)}\right) = 4(y+1)\left(\dfrac{y-3}{y+1}\right)$

$4(y+1)\left(\dfrac{y}{2(y+1)}\right) + 4(y+1)\left(\dfrac{2y-16}{4(y+1)}\right) = 4(y+1)\left(\dfrac{y-3}{y+1}\right)$

$2y + 2y - 16 = 4(y-3)$

$4y - 16 = 4y - 12$

$-16 = -12$ False

The equation has no solution.

53. $\dfrac{4}{x+3} + \dfrac{8}{x^2-9} = 0$

$(x-3)(x+3)\left(\dfrac{4}{x+3} + \dfrac{8}{(x-3)(x+3)}\right) = (x-3)(x+3)(0)$

$(x-3)(x+3)\left(\dfrac{4}{x+3}\right) + (x-3)(x+3)\left(\dfrac{8}{(x-3)(x+3)}\right) = 0$

$4(x-3) + 8 = 0$

$4x - 12 + 8 = 0$

$4x - 4 = 0$

$4x = 4$

$x = 1$

54. $\dfrac{2}{x-3} - \dfrac{4}{x+3} = \dfrac{8}{x^2-9}$

$(x-3)(x+3)\left(\dfrac{2}{x-3} - \dfrac{4}{x+3}\right) = (x-3)(x+3)\left(\dfrac{8}{(x-3)(x+3)}\right)$

$(x-3)(x+3)\left(\dfrac{2}{x-3}\right) - (x-3)(x+3)\left(\dfrac{4}{x+3}\right) = 8$

$2(x+3) - 4(x-3) = 8$

$2x + 6 - 4x + 12 = 8$

$-2x + 18 = 8$

$-2x = -10$

$x = 5$

55. $\dfrac{x-3}{x+1} - \dfrac{x-6}{x+5} = 0$

$(x+1)(x+5)\left(\dfrac{x-3}{x+1} - \dfrac{x-6}{x+5}\right) = (x+1)(x+5)(0)$

$(x+1)(x+5)\left(\dfrac{x-3}{x+1}\right) - (x+1)(x+5)\left(\dfrac{x-6}{x+5}\right) = 0$

$(x+5)(x-3) - (x+1)(x-6) = 0$

$x^2 + 2x - 15 - (x^2 - 5x - 6) = 0$

$x^2 + 2x - 15 - x^2 + 5x + 6 = 0$

$7x - 9 = 0$

$7x = 9$

$x = \dfrac{9}{7}$

56. $x + 5 = \dfrac{6}{x}$

$x(x+5) = x\left(\dfrac{6}{x}\right)$

$x^2 + 5x = 6$

$x^2 + 5x - 6 = 0$

$(x+6)(x-1) = 0$

$x + 6 = 0$ or $x - 1 = 0$

$x = -6$ or $x = 1$

57. $\dfrac{4A}{5b} = x^2$

$4A = 5bx^2$

$\dfrac{4A}{5x^2} = \dfrac{5bx^2}{5x^2}$

$\dfrac{4A}{5x^2} = b$

58. $\dfrac{x}{7} + \dfrac{y}{8} = 10$

$56\left(\dfrac{x}{7}\right) + 56\left(\dfrac{y}{8}\right) = 56(10)$

$8x + 7y = 560$

$7y = 560 - 8x$

$y = \dfrac{560 - 8x}{7}$

59. $5 \cdot \dfrac{1}{x} = \dfrac{3}{2} \cdot \dfrac{1}{x} + \dfrac{7}{6}$

$\dfrac{5}{x} = \dfrac{3}{2x} + \dfrac{7}{6}$

$6x\left(\dfrac{5}{x}\right) = 6x\left(\dfrac{3}{2x}\right) + 6x\left(\dfrac{7}{6}\right)$

$30 = 9 + 7x$

$21 = 7x$

$x = 3$

The unknown number is 3.

60. $\dfrac{1}{x} = \dfrac{1}{4-x}$

$4 - x = x$

$4 = 2x$

$2 = x$

The unknown number is 2.

61. Let r = the speed of the faster car.

	distance	=	rate	·	time
First car	90		r		$\frac{90}{r}$
Second car	60		$r - 10$		$\frac{60}{r-10}$

$$\frac{90}{r} = \frac{60}{r-10}$$
$$90(r - 10) = 60r$$
$$90r - 900 = 60r$$
$$-900 = -30r$$
$$30 = r$$
$$r - 10 = 30 - 10 = 20$$

The rate of the first car is 30 miles per hour and the rate of the second car is 20 miles per hour.

62. Let r be the speed of the boat in still water.

	distance	=	rate	·	time
Upstream	48		$r - 4$		$\frac{48}{r-4}$
Downstream	72		$r + 4$		$\frac{72}{r+4}$

$$\frac{48}{r-4} = \frac{72}{r+4}$$
$$48(r + 4) = 72(r - 4)$$
$$48r + 192 = 72r - 288$$
$$480 = 24r$$
$$20 = r$$

The speed of the boat in still water is 20 miles per hour.

63.

	Hours to Complete Total Job	Part of Job Completed in 1 Hour
Mark	7	$\frac{1}{7}$
Maria	x	$\frac{1}{x}$
Together	5	$\frac{1}{5}$

$$\frac{1}{7} + \frac{1}{x} = \frac{1}{5}$$
$$35x\left(\frac{1}{7}\right) + 35x\left(\frac{1}{x}\right) = 35x\left(\frac{1}{5}\right)$$
$$5x + 35 = 7x$$
$$35 = 2x$$
$$x = \frac{35}{2} = 17\frac{1}{2}$$

It takes Maria $17\frac{1}{2}$ hours to complete the job alone.

64.

	Days to Complete Total Job	Part of Job Completed in 1 Day
Pipe A	20	$\frac{1}{20}$
Pipe B	15	$\frac{1}{15}$
Together	x	$\frac{1}{x}$

$$\frac{1}{20} + \frac{1}{15} = \frac{1}{x}$$

$$60x\left(\frac{1}{20}\right) + 60x\left(\frac{1}{15}\right) = 60x\left(\frac{1}{x}\right)$$

$3x + 4x = 60$

$7x = 60$

$x = \frac{60}{7} = 8\frac{4}{7}$

Both pipes fill the pond in $8\frac{4}{7}$ days.

65. $\frac{2}{3} = \frac{10}{x}$

$2x = 30$

$x = 15$

The missing length is 15.

66. $\frac{12}{4} = \frac{18}{x}$

$12x = 72$

$x = 6$

The missing length is 6.

67. $\frac{9}{7\frac{1}{5}} = \frac{x}{12}$

$108 = 7\frac{1}{5}x$

$108 = \frac{36}{5}x$

$540 = 36x$

$15 = x$

The missing length is 15.

68. $\frac{x}{5} = \frac{30}{2.5}$

$2.5x = 150$

$x = 60$

The missing length is 60.

69. $\dfrac{\frac{5x}{27}}{-\frac{10xy}{21}} = \frac{5x}{27} \cdot -\frac{21}{10xy} = -\frac{5x \cdot 3 \cdot 7}{3 \cdot 9 \cdot 5 \cdot 2 \cdot x \cdot y}$

$= -\frac{7}{18y}$

70. $\dfrac{\frac{8x}{x^2-9}}{\frac{4}{x+3}} = \frac{8x}{x^2-9} \cdot \frac{x+3}{4}$

$= \frac{2 \cdot 4 \cdot x \cdot (x+3)}{(x-3)(x+3) \cdot 4} = \frac{2x}{x-3}$

71. $\dfrac{\frac{3}{5} + \frac{2}{7}}{\frac{1}{5} + \frac{5}{6}} = \dfrac{\frac{21}{35} + \frac{10}{35}}{\frac{6}{30} + \frac{25}{30}} = \dfrac{\frac{31}{35}}{\frac{31}{30}} = \frac{31}{35} \cdot \frac{30}{31}$

$= \frac{31 \cdot 5 \cdot 6}{5 \cdot 7 \cdot 31} = \frac{6}{7}$

72. $\dfrac{2 + \frac{1}{x^2}}{\frac{1}{x} + \frac{2}{x^2}} = \dfrac{x^2\left(2 + \frac{1}{x^2}\right)}{x^2\left(\frac{1}{x} + \frac{2}{x^2}\right)} = \dfrac{x^2(2) + x^2\left(\frac{1}{x^2}\right)}{x^2\left(\frac{1}{x}\right) + x^2\left(\frac{2}{x^2}\right)}$

$= \frac{2x^2 + 1}{x + 2}$

73. $\dfrac{3 - \frac{1}{y}}{2 - \frac{1}{y}} = \dfrac{y\left(3 - \frac{1}{y}\right)}{y\left(2 - \frac{1}{y}\right)} = \dfrac{y(3) - y\left(\frac{1}{y}\right)}{y(2) - y\left(\frac{1}{y}\right)}$

$\ = \dfrac{3y - 1}{2y - 1}$

74. $\dfrac{\frac{6}{x+2} + 4}{\frac{8}{x+2} - 4} = \dfrac{(x+2)\left(\frac{6}{x+2} + 4\right)}{(x+2)\left(\frac{8}{x+2} - 4\right)}$

$\ = \dfrac{(x+2)\left(\frac{6}{x+2}\right) + (x+2)(4)}{(x+2)\left(\frac{8}{x+2}\right) - (x+2)(4)}$

$\ = \dfrac{6 + 4x + 8}{8 - 4x - 8} = \dfrac{4x + 14}{-4x} = -\dfrac{2(2x + 7)}{2 \cdot 2x}$

$\ = -\dfrac{2x + 7}{2x}$

Chapter 5 Test

1. The rational expression is undefined when
$x^2 + 4x + 3 = 0$
$(x + 3)(x + 1) = 0$
$x + 3 = 0 \ \text{or} \ x + 1 = 0$
$x = -3 \ \text{or} \ x = -1$

2. a. $C = \dfrac{100x + 3000}{x}$

$\ = \dfrac{100(200) + 3000}{200}$

$\ = \dfrac{20,000 + 3000}{200}$

$\ = \dfrac{23,000}{200} = 115$

The average cost per desk is \$115.

b. $C = \dfrac{100x + 3000}{x}$

$\ = \dfrac{100(1000) + 3000}{1000}$

$\ = \dfrac{100,000 + 3000}{1000}$

$\ = \dfrac{103,000}{1000} = 103$

The average cost per desk is \$103.

3. $\dfrac{3x - 6}{5x - 10} = \dfrac{3(x - 2)}{5(x - 2)} = \dfrac{3}{5}$

4. $\dfrac{x + 10}{x^2 - 100} = \dfrac{x + 10}{(x - 10)(x + 10)} = \dfrac{1}{x - 10}$

5. $\dfrac{x + 6}{x^2 + 12x + 36} = \dfrac{x + 6}{(x + 6)^2} = \dfrac{1}{x + 6}$

6. $\dfrac{7 - x}{x - 7} = \dfrac{-(x - 7)}{x - 7} = -1$

7. $\dfrac{2m^3 - 2m^2 - 12m}{m^2 - 5m + 6} = \dfrac{2m(m^2 - m - 6)}{(m - 3)(m - 2)}$

$\ = \dfrac{2m(m - 3)(m + 2)}{(m - 3)(m - 2)} = \dfrac{2m(m + 2)}{m - 2}$

8. $\dfrac{y - x}{x^2 - y^2} = \dfrac{-(x - y)}{(x - y)(x + y)} = -\dfrac{1}{x + y}$

9. $\dfrac{x^2 - 13x + 42}{x^2 + 10x + 21} \div \dfrac{x^2 - 4}{x^2 + x - 6}$

$\ = \dfrac{x^2 - 13x + 42}{x^2 + 10x + 21} \cdot \dfrac{x^2 + x - 6}{x^2 - 4}$

$\ = \dfrac{(x - 6)(x - 7) \cdot (x + 3)(x - 2)}{(x + 7)(x + 3) \cdot (x + 2)(x - 2)}$

$\ = \dfrac{(x - 6)(x - 7)}{(x + 7)(x + 2)}$

10. $\dfrac{3}{x - 1} \cdot (5x - 5) = \dfrac{3}{x - 1} \cdot 5(x - 1)$

$\ = \dfrac{3 \cdot 5(x - 1)}{x - 1} = 15$

11. $\dfrac{y^2 - 5y + 6}{2y + 4} \cdot \dfrac{y + 2}{2y - 6}$

$\ = \dfrac{(y - 3)(y - 2) \cdot (y + 2)}{2(y + 2) \cdot 2(y - 3)} = \dfrac{y - 2}{4}$

12. $\dfrac{5}{2x + 5} - \dfrac{6}{2x + 5} = \dfrac{5 - 6}{2x + 5} = \dfrac{-1}{2x + 5}$

13. $\dfrac{5a}{a^2-a-6}-\dfrac{2}{a-3}$

$=\dfrac{5a}{(a-3)(a+2)}-\dfrac{2(a+2)}{(a-3)(a+2)}$

$=\dfrac{5a-2(a+2)}{(a-3)(a+2)}=\dfrac{5a-2a-4}{(a-3)(a+2)}$

$=\dfrac{3a-4}{(a-3)(a+2)}$

14. $\dfrac{6}{x^2-1}+\dfrac{3}{x+1}$

$=\dfrac{6}{(x+1)(x-1)}+\dfrac{3(x-1)}{(x+1)(x-1)}$

$=\dfrac{6+3x-3}{(x+1)(x-1)}=\dfrac{3x+3}{(x+1)(x-1)}$

$=\dfrac{3(x+1)}{(x+1)(x-1)}=\dfrac{3}{x-1}$

15. $\dfrac{x^2-9}{x^2-3x}\div\dfrac{x^2+4x+1}{2x+10}$

$=\dfrac{x^2-9}{x^2-3x}\cdot\dfrac{2x+10}{x^2+4x+1}$

$=\dfrac{(x-3)(x+3)\cdot 2(x+5)}{x(x-3)\cdot(x^2+4x+1)}$

$=\dfrac{2(x+3)(x+5)}{x(x^2+4x+1)}$

16. $\dfrac{x+2}{x^2+11x+18}+\dfrac{5}{x^2-3x-10}$

$=\dfrac{x+2}{(x+9)(x+2)}+\dfrac{5}{(x-5)(x+2)}$

$=\dfrac{(x+2)(x-5)}{(x+9)(x+2)(x-5)}+\dfrac{5(x+9)}{(x-5)(x+2)(x+9)}$

$=\dfrac{(x+2)(x-5)+5(x+9)}{(x+9)(x+2)(x-5)}$

$=\dfrac{x^2-3x-10+5x+45}{(x+9)(x+2)(x-5)}$

$=\dfrac{x^2+2x+35}{(x+9)(x+2)(x-5)}$

17. $\dfrac{4y}{y^2+6y+5}-\dfrac{3}{y^2+5y+4}$

$=\dfrac{4y}{(y+5)(y+1)}-\dfrac{3}{(y+4)(y+1)}$

$=\dfrac{4y(y+4)}{(y+5)(y+1)(y+4)}-\dfrac{3(y+5)}{(y+4)(y+1)(y+5)}$

$=\dfrac{4y(y+4)-3(y+5)}{(y+5)(y+1)(y+4)}$

$=\dfrac{4y^2+16y-3y-15}{(y+5)(y+1)(y+4)}$

$=\dfrac{4y^2+13y-15}{(y+5)(y+1)(y+4)}$

18. $\dfrac{4}{y}-\dfrac{5}{3}=-\dfrac{1}{5}$

$15y\left(\dfrac{4}{y}-\dfrac{5}{3}\right)=15y\left(-\dfrac{1}{5}\right)$

$15y\left(\dfrac{4}{y}\right)-15y\left(\dfrac{5}{3}\right)=15y\left(-\dfrac{1}{5}\right)$

$60-25y=-3y$

$60=22y$

$\dfrac{60}{22}=y$

$y=\dfrac{30}{11}$

19. $\dfrac{5}{y+1}=\dfrac{4}{y+2}$

$5(y+2)=4(y+1)$

$5y+10=4y+4$

$y=-6$

20. $\dfrac{a}{a-3}=\dfrac{3}{a-3}-\dfrac{3}{2}$

$2(a-3)\left(\dfrac{a}{a-3}\right)=2(a-3)\left(\dfrac{3}{a-3}-\dfrac{3}{2}\right)$

$2a=2(a-3)\left(\dfrac{3}{a-3}\right)-2(a-3)\left(\dfrac{3}{2}\right)$

$2a=6-3(a-3)$

$2a=6-3a+9$

$2a=15-3a$

$5a=15$

$a=3$

In the original equation, 3 makes a denominator 0. This equation has no solution.

21. $\dfrac{10}{x^2-25}=\dfrac{3}{x+5}+\dfrac{1}{x-5}$

$(x+5)(x-5)\left(\dfrac{10}{(x+5)(x-5)}\right)=(x+5)(x-5)\left(\dfrac{3}{x+5}+\dfrac{1}{x-5}\right)$

$10=(x+5)(x-5)\left(\dfrac{3}{x+5}\right)+(x+5)(x-5)\left(\dfrac{1}{x-5}\right)$

$10=3(x-5)+x+5$

$10=3x-15+x+5$

$10=4x-10$

$20=4x$

$x=5$

In the original equation, 5 makes the denominator 0. This equation has no solution.

22. $\dfrac{\frac{5x^2}{yz^2}}{\frac{10x}{z^3}}=\dfrac{5x^2}{yz^2}\cdot\dfrac{z^3}{10x}=\dfrac{5\cdot x\cdot x\cdot z\cdot z^2}{y\cdot z^2\cdot 2\cdot 5\cdot x}$

$=\dfrac{xz}{2y}$

23. $\dfrac{\frac{b}{a}-\frac{a}{b}}{\frac{1}{b}+\frac{1}{a}}=\dfrac{ab\left(\frac{b}{a}-\frac{a}{b}\right)}{ab\left(\frac{1}{b}+\frac{1}{a}\right)}=\dfrac{ab\left(\frac{b}{a}\right)-ab\left(\frac{a}{b}\right)}{ab\left(\frac{1}{b}\right)+ab\left(\frac{1}{a}\right)}$

$=\dfrac{b^2-a^2}{a+b}=\dfrac{(b+a)(b-a)}{b+a}=b-a$

24. $\dfrac{5-\frac{1}{y^2}}{\frac{1}{y}+\frac{2}{y^2}}=\dfrac{y^2\left(5-\frac{1}{y^2}\right)}{y^2\left(\frac{1}{y}+\frac{2}{y^2}\right)}=\dfrac{y^2(5)-y^2\left(\frac{1}{y^2}\right)}{y^2\left(\frac{1}{y}\right)+y^2\left(\frac{2}{y^2}\right)}$

$=\dfrac{5y^2-1}{y+2}$

25. $\dfrac{8}{x}=\dfrac{10}{15}$

$8(15)=10x$

$120=10x$

$12=x$

26. $x+5\cdot\dfrac{1}{x}=6$

$x+\dfrac{5}{x}=6$

$x\left(x+\dfrac{5}{x}\right)=x(6)$

$x(x)+x\left(\dfrac{5}{x}\right)=x(6)$

$$x^2 + 5 = 6x$$
$$x^2 - 6x + 5 = 0$$
$$(x - 5)(x - 1) = 0$$
$$x - 5 = 0 \text{ or } x - 1 = 0$$
$$x = 5 \text{ or } x = 1$$

The unknown number is 5 or 1.

27. Let r = the speed of the boat in still water.

	distance	=	rate	·	time
Upstream	14		$r - 2$		$\frac{14}{r-2}$
Downstream	16		$r + 2$		$\frac{16}{r+2}$

$$\frac{14}{r - 2} = \frac{16}{r + 2}$$
$$14(r + 2) = 16(r - 2)$$
$$14r + 28 = 16r - 32$$
$$60 = 2r$$
$$r = 30$$

The speed of the boat in still water is 30 miles per hour.

28.

	Hours to Complete Total Job	Part of Job Completed in 1 Hour
First pipe	12	$\frac{1}{12}$
Second pipe	15	$\frac{1}{15}$
Both pipes	x	$\frac{1}{x}$

$$\frac{1}{12} + \frac{1}{15} = \frac{1}{x}$$
$$60x\left(\frac{1}{12}\right) + 60x\left(\frac{1}{15}\right) = 60x\left(\frac{1}{x}\right)$$
$$5x + 4x = 60$$
$$9x = 60$$
$$x = \frac{60}{9} = \frac{20}{3} = 6\frac{2}{3}$$

Together, the pipes can fill the tank in $6\frac{2}{3}$ hours.

Chapter 5 Cumulative Review

1. a. $\dfrac{15}{x} = 4$

   b. $12 - 3 = x$

   c. $4x + 17 = 21$

2. a. $3 + (-7) + (-8) = -4 + (-8) = -12$

   b. $[7 + (-10)] + [-2 + (-4)] = [-3] + [-6]$
   $= -9$

3. commutative property of multiplication

4. associative property of addition

5. $3 - x = 7$
   $3 - x - 3 = 7 - 3$
   $-x = 4$
   $\dfrac{-x}{-1} = \dfrac{4}{-1}$
   $x = -4$

6. Let x = the length of the shorter piece.
   Then $4x$ = the length of the longer piece.
   $x + 4x = 10$
   $5x = 10$
   $x = 2$
   The shorter piece is 2 feet long; the longer piece is $4 \cdot 2 = 8$ feet long.

7. $y = mx + b$
   $y - b = mx + b - b$
   $y - b = mx$
   $\dfrac{y - b}{m} = \dfrac{mx}{m}$
   $\dfrac{y - b}{m} = x$

8. $x + 4 \le -6$
   $x + 4 - 4 \le -6 - 4$
   $x \le -10$

9. $\dfrac{x^5}{x^2} = x^{5-2} = x^3$

10. $\dfrac{4^7}{4^3} = 4^{7-3} = 4^4 = 256$

11. $\dfrac{(-3)^5}{(-3)^2} = (-3)^{5-2} = (-3)^3 = -27$

12. $\dfrac{2x^5 y^2}{xy} = 2x^{5-1} y^{2-1} = 2x^4 y$

13. $2x^{-3} = \dfrac{2}{x^3}$

14. $(-2)^{-4} = \dfrac{1}{(-2)^4} = \dfrac{1}{16}$

15. $5x(2x^3 + 6) = 5x(2x^3) + 5x(6)$
    $= 10x^4 + 30x$

16. $-3x^2(5x^2 + 6x - 1)$
    $= -3x^2(5x^2) + (-3x^2)(6x) - (-3x^2)(1)$
    $= -15x^4 - 18x^3 + 3x^2$

17. Write $4x^2 + 7 + 8x^3$ as $8x^3 + 4x^2 + 0x + 7$
    before beginning long division.

    $$\begin{array}{r} 4x^2 - 4x + 6 \\ 2x+3\overline{)8x^3 + 4x^2 + 0x + 7} \\ \underline{8x^3 + 12x^2} \\ -8x^2 + 0x \\ \underline{-8x^2 - 12x} \\ 12x + 7 \\ \underline{12x + 18} \\ -11 \end{array}$$

    $\dfrac{4x^2 + 7 + 8x^3}{2x + 3} = 4x^2 - 4x + 6 + \dfrac{-11}{2x + 3}.$

18. $x^2 + 7x + 12 = (x + 3)(x + 4)$

19. $25x^2 + 20xy + 4y^2$
    $= (5x)^2 + 2 \cdot 5x \cdot 2y + (2y)^2$
    $= (5x + 2y)^2$

20. $x^2 - 9x - 22 = 0$
$(x - 11)(x + 2) = 0$
$x - 11 = 0$ or $x + 2 = 0$
$x = 11$ or $x = -2$
The solutions are 11 and –2.

21. $\dfrac{x^2 + x}{3x} \cdot \dfrac{6}{5x + 5} = \dfrac{x(x+1) \cdot 2 \cdot 3}{3x \cdot 5(x+1)} = \dfrac{2}{5}$

22. $\dfrac{3x^2 + 2x}{x - 1} - \dfrac{10x - 5}{x - 1}$

$= \dfrac{3x^2 + 2x - (10x - 5)}{x - 1}$

$= \dfrac{3x^2 + 2x - 10x + 5}{x - 1}$

$= \dfrac{3x^2 - 8x + 5}{x - 1} = \dfrac{(x-1)(3x-5)}{x-1}$

$= 3x - 5$

23. $\dfrac{6x}{x^2 - 4} - \dfrac{3}{x + 2} = \dfrac{6x}{(x+2)(x-2)} - \dfrac{3}{x+2}$

$= \dfrac{6x}{(x+2)(x-2)} - \dfrac{3(x-2)}{(x+2)(x-2)}$

$= \dfrac{6x - 3(x-2)}{(x+2)(x-2)} = \dfrac{6x - 3x + 6}{(x+2)(x-2)}$

$= \dfrac{3x + 6}{(x+2)(x-2)} = \dfrac{3(x+2)}{(x+2)(x-2)} = \dfrac{3}{x-2}$

24. $\dfrac{t - 4}{2} - \dfrac{t - 3}{9} = \dfrac{5}{18}$

$18\left(\dfrac{t-4}{2} - \dfrac{t-3}{9}\right) = 18\left(\dfrac{5}{18}\right)$

$9(t - 4) - 2(t - 3) = 5$
$9t - 36 - 2t + 6 = 5$
$7t - 30 = 5$
$7t = 35$
$t = 5$

25. Let $x =$ the time in hours it takes Sam and Frank to complete the job together. Then $\dfrac{1}{x} =$ the part of the job they complete in 1 hour. Since Sam completes $\dfrac{1}{3}$ of the job in 1 hour, and Frank completes $\dfrac{1}{7}$ of the job in 1 hour, we have

$\dfrac{1}{3} + \dfrac{1}{7} = \dfrac{1}{x}$

$21x\left(\dfrac{1}{3} + \dfrac{1}{7}\right) = 21x\left(\dfrac{1}{x}\right)$

$7x + 3x = 21$
$10x = 21$

$x = \dfrac{21}{10}$

$x = 2\dfrac{1}{10}$

Sam and Frank, working together, can complete the quality control tour in $2\dfrac{1}{10}$ hours.

26. $\dfrac{\dfrac{1}{z} - \dfrac{1}{2}}{\dfrac{1}{3} - \dfrac{z}{6}} = \dfrac{\dfrac{2 \cdot 1}{2 \cdot z} - \dfrac{z \cdot 1}{z \cdot 2}}{\dfrac{2 \cdot 1}{2 \cdot 3} - \dfrac{z}{6}} = \dfrac{\dfrac{2}{2z} - \dfrac{z}{2z}}{\dfrac{2}{6} - \dfrac{z}{6}} = \dfrac{\dfrac{2-z}{2z}}{\dfrac{2-z}{6}}$

$= \dfrac{2-z}{2z} \cdot \dfrac{6}{2-z} = \dfrac{2 \cdot 3 \cdot (2-z)}{2z \cdot (2-z)} = \dfrac{3}{z}$

Chapter 6

1. $(-4, 3)$, $(0, -2)$, and $(5, 0)$

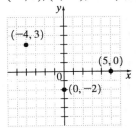

2. $8x - 3y = 2$
If $x = -2$,
$$8(-2) - 3y = 2$$
$$-16 - 3y = 2$$
$$-3y = 18$$
$$y = -6$$
$(-2, -6)$

3. $3x - y = 6$
If $x = 0$, $y = -6$,
If $y = 0$, $3x = 6$
$$x = 2$$
Plot using $(0, -6)$ and $(2, 0)$:

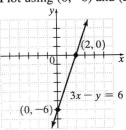

4. Plot using $(-1, 0)$ and $(0, 4)$:

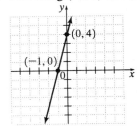

5. $3x + 2y \leq 6$
$3x + 2y = 6$
If $x = 0$, $2y = 6$
$$y = 3$$
If $y = 0$, $3x = 6$
$$x = 2$$
Plot using $(0, 3)$ and $(2, 0)$.

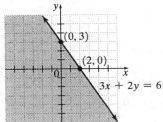

Test $(0, 0)$:
$3(0) + 6(0) \leq 6$?
$0 \leq 6$?
Yes; shade below the line.

6. $(-7, 8)$ and $(3, 5)$
$$m = \frac{y_2 - y_1}{x_2 - x_1} = \frac{5 - 8}{3 - (-7)} = -\frac{3}{10}$$

7. $4x - 5y = 20$
$$5y = 4x - 20$$
$$y = \frac{4}{5}x - 4$$
$$m = \frac{4}{5}$$

8. $x = 10$
Vertical line; has undefined slope

9. $m = -\frac{1}{3}$; $(3, -6)$
$$y - y_1 = m(x - x_1)$$
$$y - (-6) = -\frac{1}{3}(x - 3)$$
$$y + 6 = -\frac{1}{3}x + 1$$
$$y = -\frac{1}{3}x - 5$$
$$3y = -x - 15$$
$$x + 3y = -15$$

10. $(0, 0)$ and $(-8, 1)$

$$m = \frac{y_2 - y_1}{x_2 - x_1} = \frac{1 - 0}{-8 - 0} = -\frac{1}{8}$$

$$y - y_1 = m(x - x_1)$$

$$y - 0 = -\frac{1}{8}(x - 0)$$

$$y = -\frac{1}{8}x$$

$$8y = -x$$

$$x + 8y = 0$$

11. $m = \dfrac{2}{7}; \quad b = 14$

$$y = \frac{2}{7}x + 14$$

$$7y = 2x + 98$$

$$2x - 7y = -98$$

12. $\{(-3, 8), (7, -1), (0, 6), (2, -1)\}$
domain: $\{-3, 0, 2, 7\}$
range: $\{-1, 6, 8\}$

13. $\{(1, 7), (-8, 7), (6, 3), (9, 2)\}$
Every point has a unique x-value; it is a function.

14. $\{(0, 4), (1, 3), (2, -5), (1, 10), (-2, 8)\}$
Two points have the same x-value; it is not a function.

15. $f(x) = -3x + 8$

   a. $f(-1) = -3(-1) + 8 = 3 + 8 = 11$

   b. $f(0) = -3(0) + 8 = 0 + 8 = 8$

   c. $f(10) = -3(10) + 8 = -30 + 8 = -22$

Section 6.1

Practice Problems

1.

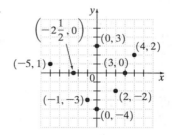

Point $(4, 2)$ lies in quadrant I.
Point $(-5, 1)$ lies in quadrant II.
Point $(-1, -3)$ lies in quadrant III.
Point $(2, -2)$ lies in quadrant IV.

Points $(0, 3)$, $\left(-2\dfrac{1}{2}, 0\right)$, $(3, 0)$, and $(0, -4)$

lie on axes, so they are not in any quadrant.

2. a. $(1986, 38), (1988, 44), (1990, 50),$
$(1992, 53), (1994, 57), (1996, 62)$

   b.

<!-- graph: Cable TV Subscribers (in millions) vs Year -->

   c. The number of cable TV subscribers has steadily increased.

3. $x + 2y = 8$

   a. $x = 0, \quad 0 + 2y = 8$
$$y = 4$$
$(0, 4)$

   b. $y = 3, \quad x + 2(3) = 8$
$$x + 6 = 8$$
$$x = 2$$
$(2, 3)$

c. $x = -4, \quad -4 + 2y = 8$
$$2y = 12$$
$$y = 6$$
$$(-4, 6)$$

4. $y = -2x$

a. $x = -3, \quad y = -2(-3)$
$$y = 6$$

b. $y = 0, \quad 0 = -2x$
$$x = 0$$

c. $y = 10, \quad 10 = -2x$
$$x = -5$$

x	y
-3	6
0	0
-5	10

5. $x = 5$

x	y
5	-2
5	0
5	4

6. $y = -50x + 400$

x	1	2	3	4	5	6	7
y	350	300	250	200	150	100	50

Mental Math

1. $x + y = 10$
Answers may vary; ex.: (5, 5), (7, 3)

2. $x + y = 6$
Answers may vary; ex.: (0, 6), (6, 0)

Exercise Set 6.1

1.

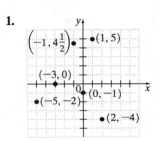

(1, 5) is in quadrant I

$\left(-1, 4\frac{1}{2}\right)$ is in quadrant II

(−5, −2) is in quadrant III
(2, −4) is in quadrant IV
(−3, 0) lies on the *x*-axis
(0, −1) lies on the *y*-axis

3. The graph of (a, b) is the same as the graph of (b, a) when $a = b$.

5. *A:* (0, 0)

7. *C:* (3, 2)

9. *E:* (−2, −2)

11. *G:* (2, −1)

13. *B:* (0, −3)

15. *D:* (1, 3)

17. *F:* (−3, −1)

19. a. (1991, 80), (1992, 79), (1993, 77), (1994, 74), (1995, 77), (1996, 85), (1997, 86)

b.

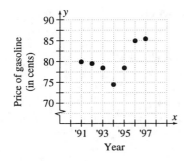

21. a. (1994, 578), (1995, 613), (1996, 654), (1997, 675), (1998, 717)

b.

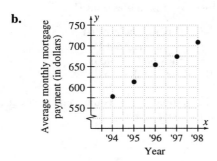

c. The scatter diagram shows a trend of increasing mortgage payments.

23. a. (0.50, 10), (0.75, 12), (1.00, 15), (1.25, 16), (1.50, 18), (1.50, 19), (1.75, 19), (2.00, 20)

b.

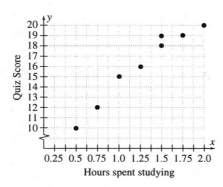

c. Minh might conclude that more time spent studying results in a better quiz score.

25. $x - 4y = 4$
Complete (, −2):
$$y = -2$$
$$x - 4(-2) = 4$$
$$x + 8 = 4$$
$$x = -4$$
(−4, −2)
Complete (4,):
$$x = 4$$
$$4 - 4y = 4$$
$$-4y = 0$$
$$y = 0$$
(4, 0)

27. $3x + y = 9$
Complete (0,):
$$x = 0$$
$$3(0) + y = 9$$
$$y = 9$$
(0, 9)
Complete (, 0):
$$y = 0$$
$$3x + 0 = 9$$
$$3x = 9$$
$$x = 3$$
(3, 0)

29. $y = -7$
Complete (11,):
$x = 11, y = -7$; (11, −7)
Complete (, −7):
$y = -7, x = $ any x

31. $x + 3y = 6$
Complete (0,):
$$x = 0$$
$$0 + 3y = 6$$
$$y = 2$$
(0, 2)
Complete (, 0):
$$y = 0$$
$$x + 3(0) = 6$$
$$x = 6$$
(6, 0)

Complete (, 1):
$$y = 1$$
$$x + 3(1) = 6$$
$$x + 3 = 6$$
$$x = 3$$
(3, 1)

x	y
0	2
6	0
3	1

33. $2x - y = 12$
Complete (0,):
$$x = 0$$
$$2(0) - y = 12$$
$$-y = 12$$
$$y = -12$$
(0, -12)
Complete (, -2):
$$y = -2$$
$$2x - (-2) = 12$$
$$2x + 2 = 12$$
$$2x = 10$$
$$x = 5$$
(5, -2)
Complete (3,):
$$x = 3$$
$$2(3) - y = 12$$
$$6 - y = 12$$
$$-y = 6$$
$$y = -6$$
(3, -6)

x	y
0	-12
5	-2
3	-6

35. $2x + 7y = 5$
Complete (0,):
$$x = 0$$
$$2(0) + 7y = 5$$
$$7y = 5$$
$$y = \frac{5}{7}$$
$\left(0, \dfrac{5}{7}\right)$
Complete (, 0):
$$y = 0$$
$$2x + 7(0) = 5$$
$$2x = 5$$
$$x = \frac{5}{2}$$
$\left(\dfrac{5}{2}, 0\right)$
Complete (, 1):
$$y = 1$$
$$2x + 7(1) = 5$$
$$2x + 7 = 5$$
$$2x = -2$$
$$x = -1$$
(-1, 1)

x	y
0	$\frac{5}{7}$
$\frac{5}{2}$	0
-1	1

37. $x = 3$
All x table values are 3.

x	y
3	0
3	-0.5
3	$\frac{1}{4}$

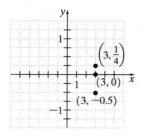

39. $x = -5y$
Complete (, 0):
$y = 0$
$x = 5(0)$
$x = 0$
$(0, 0)$
Complete (, 1):
$y = 1$
$x = -5(1)$
$x = -5$
$(-5, 1)$
Complete $(10,)$:
$x = 10$
$10 = -5y$
$y = -2$
$(10, -2)$

x	y
0	0
−5	1
10	−2

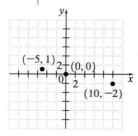

41. $y = 80x + 5000$

a. $x = 100$
$y = 80(100) + 5000$
$= 8000 + 5000$
$= 13,000$
$x = 200$

$y = 80(200) + 5000$
$= 16,000 + 5000$
$= 21,000$
$x = 300$
$y = 80(300) + 5000$
$= 24,000 + 5000$
$= 29,000$

x	100	200	300
y	13,000	21,000	29,000

b. $y = 8600$
$8600 = 80x + 5000$
$3600 = 80x$
$x = 45$ desks

43. $x + y = 5$
$y = 5 - x$

45. $2x + 4y = 5$
$4y = 5 - 2x$
$y = \dfrac{5 - 2x}{4}$

47. $10x = -5y$
$5y = -10x$
$y = -2x$

49. Plot the points:

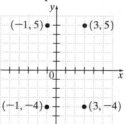

Rectangle is 9 units by 4 units, perimeter is
$9 + 9 + 4 + 4 = 26$ units.

51. years 0 to 1: $500 million
years 1 to 2: $1,500 million
years 2 to 3: $1,000 million
years 3 to 4: $1,500 million

53. answers may vary

55. $y = 0.364x + 21.939$

   a. $x = 20$

$$y = 0.364(20) + 21.939$$
$$= 7.28 + 21.939 = 29.219$$

$x = 65$

$$y = 0.364(65) + 21.939$$
$$= 23.66 + 21.939 = 45.599$$

$x = 90$

$$y = 0.364(90) + 21.939$$
$$= 32.76 + 21.939 = 54.699$$

x	20	65	90
y	29.219	45.599	54.699

   b. $y = 50$

$$50 = 0.364x + 21.939$$
$$0.364x = 28.061$$
$$x \approx 77.09$$

In 1977.

Section 6.2

Practice Problems

1. $x + 3y = 6$
Let $x = 0$
$0 + 3y = 6$
$3y = 6$
$y = 2$
Let $x = 3$
$3 + 3y = 6$
$3y = 3$
$y = 1$
Let $x = 6$
$6 + 3y = 6$
$3y = 0$
$y = 0$
Plot the ordered pairs $(0, 2)$, $(3, 1)$, and $(6, 0)$:

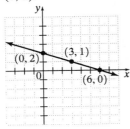

2. $-2x + 4y = 8$
Let $x = -2$
$-2(-2) + 4y = 8$
$4 + 4y = 8$
$4y = 4$
$y = 1$
Let $x = 0$
$-2(0) + 4y = 8$
$0 + 4y = 8$
$4y = 8$
$y = 2$
Let $x = 2$
$-2(2) + 4y = 8$
$-4 + 4y = 8$
$4y = 12$
$y = 3$
Plot the ordered pairs $(-2, 1)$, $(0, 2)$ and $(2, 3)$:

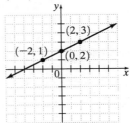

3. $y = 2x$

x	y
-2	-4
0	0
2	4

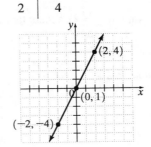

4. $y = -\dfrac{1}{2}x$

x	y
-4	2
0	0
4	-2

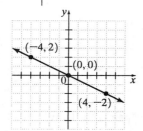

5. $y = 2x + 3$

x	y
-2	-1
0	3
2	7

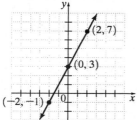

Same as the graph of $y = 2x$ except that the graph of $y = 2x + 3$ is moved 3 units upward.

6. $x = 3$
For any y-value, x is 3.

x	y
3	2
3	0
3	-4

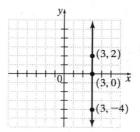

Graphing Calculator Explorations

1. $y = -3x + 7$

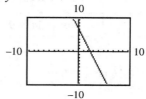

2. $y = -x + 5$

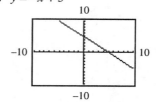

3. $y = 2.5x - 7.9$

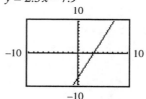

4. $y = -1.3x + 5.2$

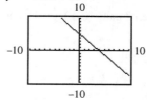

5. $y = -\dfrac{3}{10}x + \dfrac{32}{5}$

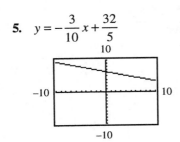

6. $y = \dfrac{2}{9}x - \dfrac{22}{3}$

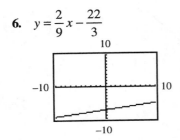

Exercise Set 6.2

1. $x - y = 6$
$y = 0,\ x = 6;\ (6, 0)$
$x = 4$
$4 - y = 6$
$\quad y = -2;\ (4, -2)$
$y = -1$
$x - (-1) = 6$
$\quad\quad x = 5;\ (5, -1)$

x	y
6	0
4	-2
5	-1

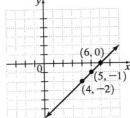

3. $y = -4x$
$x = 1,\ y = -4;\ (1, -4)$
$x = 0,\ y = 0;\ (0, 0)$
$x = -1,\ y = 4;\ (-1, 4)$

x	y
1	-4
0	0
-1	4

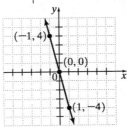

5. $y = \dfrac{1}{3}x$
$x = 0,\ y = 0;\ (0, 0)$
$x = 6,\ y = 2;\ (6, 2)$
$x = -3,\ y = -1;\ (-3, -1)$

x	y
0	0
6	2
-3	-1

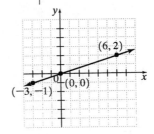

7. $y = -4x + 3$
$x = 0,\ y = 3;\ (0, 3)$
$x = 1$
$y = -4 + 3 = -1;\ (1, -1)$
$x = 2$
$y = -8 + 3 = -5;\ (2, -5)$

x	y
0	3
1	−1
2	−5

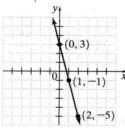

11. $x - y = -2$
Find 3 points:

x	y
0	2
−2	0
1	3

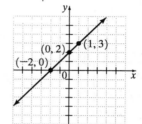

9. $x + y = 1$
Find 3 points:

x	y
0	1
1	0
−1	2

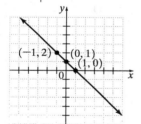

13. $x - 2y = 6$
Find 3 points:

x	y
0	−3
6	0
4	−1

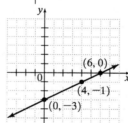

15. $y = 6x + 3$
Find 3 points:

x	y
0	3
−1	−3
1	9

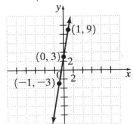

17. $x = -4$
   $y = $ any value

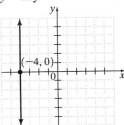

19. $y = 3$
   $x = $ any value

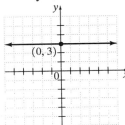

21. $y = x$
Find 3 points:

x	y
2	2
0	0
−2	−2

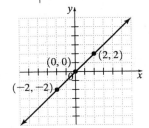

23. $y = 5x$
Find 3 points:

x	y
0	0
1	5
−1	−5

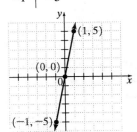

25. $x + 3y = 9$
Find 3 points:

x	y
0	3
3	2
−3	4

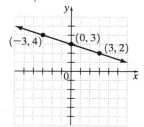

27. $y = \dfrac{1}{2}x - 1$
Find 3 points:

x	y
0	−1
2	0
4	1

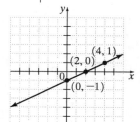

29. $3x - 2y = 12$
Find 3 points:

x	y
0	−6
4	0
2	−3

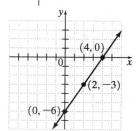

31. Find 3 points for each line:
$y = 5x$

x	y
0	0
1	5
−1	−5

$y = 5x + 4$

x	y
0	4
−1	−1
−2	−6

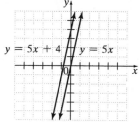

Answers vary.

33. Find 3 points for each line:
$y = -2x$

x	y
2	-4
0	0
-2	4

$y = -2x - 3$

x	y
-2	1
-1	-1
0	-3

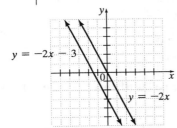

Answers vary.

35. Plot the points $(-2, 5)$, $(4, 5)$, and $(-2, -1)$.

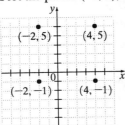

Complete the rectangle with point $(4, -1)$.

37. $x - y = -3$
$x = 0$
$0 - y = -3$
$\qquad y = 3; (0, 3)$
$y = 0$
$x - 0 = -3$
$\qquad x = -3; (-3, 0)$

x	y
0	3
-3	0

39. $y = 2x$
$x = 0$
$y = 0; (0, 0)$

x	y
0	0
0	0

41. $y = x^2$

x	y
0	0
1	1
-1	1
2	4
-2	4

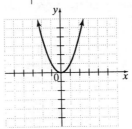

43. $x + y + 5 + 5 = 22$
$x + y = 12$
if $y = 3$, $x + 3 = 12$, $x = 9$ cm

45. $y = 43x + 2035$

   a. Find 3 points:

x	y
0	2035
4	2207
8	2379

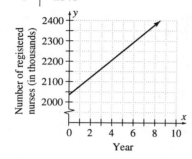

   b. Yes; 8 years after 1997 (or in 2005) there will be 2379 thousand registered nurses.

Section 6.3

Practice Problems

1. x-intercept: 2; y-intercept: -4; (2, 0) and (0, -4)

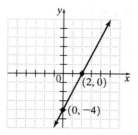

2. x-intercepts: -4, 2; y-intercept: 3; (-4, 0), (2, 0), and (0, 3)

3. x-intercept: none; y-intercept: 3; (0, 3)

4. $2x - y = 4$
If $x = 0$ then
$2(0) - y = 4$
$-y = 4$
$y = -4$
If $y = 0$ then
$2x - 0 = 4$
$2x = 4$
$x = 2$

5. $y = 3x$
If $x = 0$ then
$y = 3(0)$
$y = 0$
If $y = 0$ then
$0 = 3x$
$x = 0$
If $x = 1$ then
$y = 3(1)$
$y = 3$
If $x = -1$ then
$y = 3(-1)$
$y = -3$
Plot the points (0, 0), (1, 3), and (-1, -3):

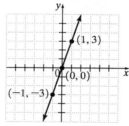

6. $x = -3$
For any y-value, x is -3.

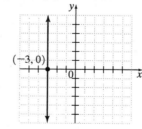

7. $y = 4$
For any x-value, y is 4.

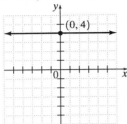

Graphing Calculator Explorations

1. $x = 3.78y$

$$y = \frac{x}{3.78}$$

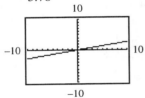

2. $-2.61y = x$

$$y = \frac{x}{-2.61}$$

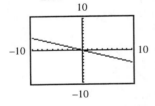

3. $-2.2x + 6.8y = 15.5$
$6.8y = 2.2x + 15.5$

$$y = \frac{2.2}{6.8}x + \frac{15.5}{6.8}$$

$$y = \frac{1.1}{3.4}x + \frac{15.5}{6.8}$$

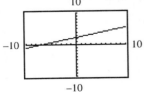

4. $5.9x - 0.8y = -10.4$
$0.8y = 5.9x + 10.4$

$$y = \frac{5.9}{0.8}x + \frac{10.4}{0.8}$$

$$y = 7.35x + 13$$

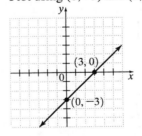

Exercise Set 6.3

1. $x = -1$; $y = 1$;
$(-1, 0)$; $(0, 1)$

3. $x = -2$; $x = 1$; $x = 3$; $y = 1$;
$(-2, 0)$; $(1, 0)$; $(3, 0)$; $(0, 1)$

5. infinite

7. 0

9. $x - y = 3$
If $x = 0$, then $y = -3$
If $y = 0$, then $x = 3$
Plot using $(0, -3)$ and $(3, 0)$:

11. $x = 5y$
If $x = 0$, then $y = 0$
Need another point:
If $y = 1$, then $x = 5$
Plot using (0, 0) and (5, 1):

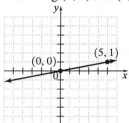

13. $-x + 2y = 6$
If $x = 0$, then $y = 3$
If $y = 0$, then $x = -6$
Plot using (0, 3) and (−6, 0):

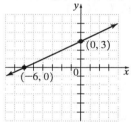

15. $2x - 4y = 8$
If $x = 0$, then $y = -2$
If $y = 0$, then $x = 4$
Plot using (0, −2) and (4, 0):

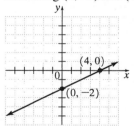

17. $x = 2y$
If $x = 0$, $y = 0$
Need another point:
If $y = 1$, $x = 2$

Plot using (0, 0) and (2, 1):

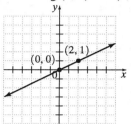

19. $y = 3x + 6$
If $x = 0$, $y = 6$
If $y = 0$, $x = -2$
Plot using (0, 6) and (−2, 0)

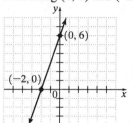

21. $x = y$
If $x = 0$, $y = 0$
Need another point:
If $x = 3$, $y = 3$
Plot using (0, 0) and (3, 3):

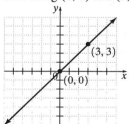

23. $x + 8y = 8$
If $x = 0$, $y = 1$
If $y = 0$, $x = 8$
Plot using (0, 1) and (8, 0):

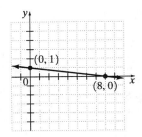

25. $5 = 6x - y$

If $x = 0$, $y = -5$

If $y = 0$, $x = \dfrac{5}{6}$

Plot using $(0, -5)$ and $\left(\dfrac{5}{6}, 0\right)$:

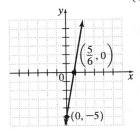

27. $-x + 10y = 11$

If $x = 0$, $y = \dfrac{11}{10}$

If $y = 0$, $x = -11$

Plot using $\left(0, \dfrac{11}{10}\right)$ and $(-11, 0)$:

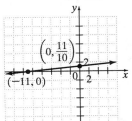

29. $x = -1$

For any y-value, x is -1.

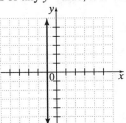

31. $y = 0$

For any x-value, y is 0.

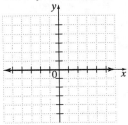

33. $y + 7 = 0$

$y = -7$

For any x-value, y is -7.

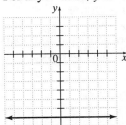

35. $x + 3 = 0$

$x = -3$

For any y-value, x is -3.

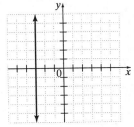

37. $\dfrac{-6-3}{2-8} = \dfrac{-9}{-6} = \dfrac{3}{2}$

39. $\dfrac{-8-(-2)}{-3-(-2)} = \dfrac{-6}{-1} = 6$

41. $\dfrac{0-6}{5-0} = \dfrac{-6}{5} = -\dfrac{6}{5}$

43. $y = 3$
For any x-value, y is 3.
C

45. $x = 3$
For any y-value, $x = 3$.
A

47. Answers vary.

49. $3x + 6y = 1200$

    a. If $x = 0,\ 6y = 1200$
                $y = 200$
    (0, 200) corresponds to no chairs and
    200 desks are manufactured.

    b. If $y = 0,\ 3x = 1200$
                $x = 400$
    (400, 0) corresponds to 400 chairs and
    no desks are manufactured.

    c. If $y = 50$,
        $3x + 6(50) = 1200$
          $3x + 300 = 1200$
              $3x = 900$
                $x = 300$
    300 chairs can be made.

51. Parallel to $y = -1$
y-intercept is (0, –4)

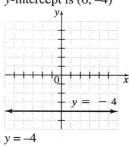

$y = -4$

53. $y = 1.2x + 23.6$

    a. If $x = 0$, $y = 23.6$

    b. (0, 23.6)

    c. In 1995, U.S. farm expenses for
    livestock feed were \$23.6 billion.

Section 6.4

Practice Problems

1. (–2, 3) and (4, –1)
$$m = \dfrac{y_2 - y_1}{x_2 - x_1}$$
$$= \dfrac{-1 - 3}{4 - (-2)}$$
$$= \dfrac{-4}{6} = -\dfrac{2}{3}$$

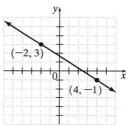

2. (–2, 1) and (3, 5)
$$m = \dfrac{y_2 - y_1}{x_2 - x_1}$$
$$= \dfrac{5 - 1}{3 - (-2)}$$
$$= \dfrac{4}{5}$$

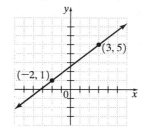

3. $5x + 4y = 10$
$4y = -5x + 10$
$$y = -\frac{5}{4}x + \frac{10}{4}$$
$$y = -\frac{5}{4}x + \frac{5}{2}$$
$$m = -\frac{5}{4}$$

4. $y = 3$
$y = 0x + 3$
$m = 0$

5. $x = -2$
Since this is a vertical line, its slope is undefined.

6. a. $x + y = 5$
$y = -x + 5$
The slope is -1.

$2x + y = 5$
$y = -2x + 5$
The slope is -2.
$-1(-2) = 2 \neq -1$
neither

b. $5y = 2x - 3$
$$y = \frac{2}{5}x - \frac{3}{5}$$
The slope is $\frac{2}{5}$.

$5x + 2y = 1$
$2y = -5x + 1$
$$y = -\frac{5}{2}x + \frac{1}{2}$$
The slope is $-\frac{5}{2}$.
$$\frac{2}{5}\left(-\frac{2}{5}\right) = -1$$
perpendicular

c. $y = 2x + 1$
The slope is 2.

$4x - 2y = 8$
$2y = 4x - 8$
$y = 2x - 4$
The slope is 2.
$2 = 2$
parallel

Graphing Calculator Explorations

1. $y_1 = 3.8x$
$y_2 = 3.8x - 3$
$y_3 = 3.8x + 6$

2. $y_1 = -4.9x$
$y_2 = -4.9x + 2$
$y_3 = -4.9x + 9$

3. $y_1 = \frac{1}{4}x$
$$y_2 = \frac{1}{4}x + 5$$
$$y_3 = \frac{1}{4}x - 8$$

4. $y_1 = -\dfrac{3}{4}x$

$y_2 = -\dfrac{3}{4}x - 5$

$y_3 = -\dfrac{3}{4}x + 6$

Mental Math

1. $m = \dfrac{7}{6}$

Positive slope means upward.

2. $m = -3$

Negative slope means downward.

3. $m = 0$

0 slope means horizontal

4. m is undefined.

Undefined slope means vertical.

Exercise Set 6.4

1. $p_1 = (-1, 2); \; p_2 = (2, -2)$

$m = \dfrac{y_2 - y_1}{x_2 - x_1}$

$= \dfrac{-2 - 2}{2 - (-1)}$

$= -\dfrac{4}{3}$

3. $p_1 = (-3, -2); \; p_2 = (-1, 3)$

$m = \dfrac{y_2 - y_1}{x_2 - x_1}$

$= \dfrac{3 - (-2)}{-1 - (-3)}$

$= \dfrac{5}{2}$

5. $(0, 0)$ and $(7, 8)$

$m = \dfrac{y_2 - y_1}{x_2 - x_1}$

$= \dfrac{8 - 0}{7 - 0}$

$= \dfrac{8}{7}$

7. $(-1, 5)$ and $(6, -2)$

$m = \dfrac{y_2 - y_1}{x_2 - x_1}$

$= \dfrac{-2 - 5}{6 - (-1)}$

$= \dfrac{-7}{7} = -1$

9. $(1, 4)$ and $(5, 3)$

$m = \dfrac{y_2 - y_1}{x_2 - x_1}$

$= \dfrac{3 - 4}{5 - 1}$

$= -\dfrac{1}{4}$

11. $(-2, 8)$ and $(1, 6)$

$m = \dfrac{y_2 - y_1}{x_2 - x_1}$

$= \dfrac{6 - 8}{1 - (-2)}$

$= -\dfrac{2}{3}$

13. $(5, 1)$ and $(-2, 1)$

$m = \dfrac{y_2 - y_1}{x_2 - x_1}$

$= \dfrac{1 - 1}{-2 - 5}$

$= \dfrac{0}{-7} = 0$

15. line 1 has a positive slope; line 2 has a negative slope; line 1

17. line 2 increases faster than line 1; line 2

19. $y = 5x - 2$

$m = 5$

21. $2x + y = 7$
$y = -2x + 7$
$m = -2$

23. $2x - 3y = 10$
$3y = 2x - 10$
$y = \dfrac{2}{3}x - \dfrac{10}{3}$
$m = \dfrac{2}{3}$

25. $x = 2y$
$y = \dfrac{1}{2}x$
$m = \dfrac{1}{2}$

27. $p_1 = (2, 3);\quad p_2 = (2, -1)$
$m = \dfrac{y_2 - y_1}{x_2 - x_1}$
$= \dfrac{-1 - 3}{2 - 2}$
$= -\dfrac{4}{0}$
undefined slope

29. $x = 1$
This is a vertical line, so it has an undefined slope.

31. $y = -3$
This is a horizontal line, so it has a slope of 0.

33. $x - 3y = -6$ $\qquad$ $3x - y = 0$
$\quad 3y = x + 6$ $\qquad\quad y = 3x$
$\quad y = \dfrac{1}{3}x + 2$ $\qquad\ m = 3$
$\quad m = \dfrac{1}{3}$
$\dfrac{1}{3}(3) = 1 \neq -1$
neither

35. $10 + 3x = 5y$ $\qquad$ $5x + 3y = 1$
$\quad 5y = 3x + 10$ $\qquad\quad 3y = -5x + 1$
$\quad y = \dfrac{3}{5}x + 2$ $\qquad\quad y = -\dfrac{5}{3}x + \dfrac{1}{3}$
$\quad m = \dfrac{3}{5}$ $\qquad\qquad\quad m = -\dfrac{5}{3}$
$\dfrac{3}{5}\left(-\dfrac{5}{3}\right) = -1$
perpendicular

37. $6x = 5y + 1$ $\qquad$ $-12x + 10y = 1$
$\quad 5y = 6x - 1$ $\qquad\quad 10y = 12x + 1$
$\quad y = \dfrac{6}{5}x - \dfrac{1}{5}$ $\qquad\quad y = \dfrac{6}{5}x + \dfrac{1}{10}$
$\quad m = \dfrac{6}{5}$ $\qquad\qquad\quad m = \dfrac{6}{5}$
parallel

39. $(-3, -3)$ and $(0, 0)$
$m = \dfrac{y_2 - y_1}{x_2 - x_1}$
$= \dfrac{0 - (-3)}{0 - (-3)}$
$= \dfrac{3}{3} = 1$

a. $m = 1$

b. $m = -1$

41. $(-8, -4)$ and $(3, 5)$
$m = \dfrac{y_2 - y_1}{x_2 - x_1}$
$= \dfrac{5 - (-4)}{3 - (-8)}$
$= \dfrac{9}{11}$

a. $\dfrac{9}{11}$

b. $-\dfrac{11}{9}$

43. $y - (-6) = 2(x - 4)$
$y + 6 = 2x - 8$
$y = 2x - 14$

45. $y - 1 = -6(x - (-2))$
$y - 1 = -6(x + 2)$
$y - 1 = -6x - 12$
$y = -6x - 11$

47. $p_1 = (0, 0); \ p_2 = (1, 1)$

$m = \dfrac{y_2 - y_1}{x_2 - x_1}$

$\quad = \dfrac{1 - 0}{1 - 0}$

$\quad = 1$

$m = 1; D$

49. A vertical line has an undefined slope. B

51. $p_1 = (2, 0); \ p_2 = (4, -1)$

$m = \dfrac{y_2 - y_1}{x_2 - x_1}$

$\quad = \dfrac{-1 - 0}{4 - 2}$

$\quad = -\dfrac{1}{2}$

E

53. $m = \dfrac{\Delta y}{\Delta x} = \dfrac{6}{10} = \dfrac{3}{5}$

55. $m = \dfrac{\Delta y}{\Delta x} = \dfrac{16}{100} = 16 = 16\%$

57. 20 mpg

59. $21.5 - 18.2 = 3.3$ mpg

61. 86 to 87

63. $y = -\dfrac{1}{3}x + 2$
$y = -2x + 2$
$y = -4x + 2$

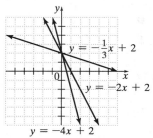

The line becomes steeper.

65. $y = -1800x + 18,000$

a. If $x = 0$,
$y = -1800(0) + 18,000$
$y = 18,000$
If $y = 0$,
$\quad 0 = -1800x + 18,000$
$1800x = 18,000$
$\quad\quad x = 10$
Intercepts: (0, 18,000) and (10, 0)
In 1991, 18,000 Hepatitis B cases were reported in the United States. In 1991 + 10 = 2001 the number of Hepatitis B cases reported in the United States will drop to zero.

b. $m = -1800$

c. The number of cases is decreasing by an average of 1800 per year.

Section 6.5

Practice Problems

1. $m = \dfrac{3}{5}; \ b = -2$

$y = mx + b$

$y = \dfrac{3}{5}x + (-2)$

$y = \dfrac{3}{5}x - 2$

2. $y = \dfrac{2}{3}x - 4$

$m = \dfrac{2}{3}$

y-intercept is -4.

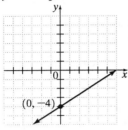

3. $3x + y = 2$
$y = -3x + 2$
$m = -3$
y-intercept is 2.

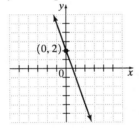

4. $(2, -4); \ m = -3$
$$y - y_1 = m(x - x_1)$$
$$y - (-4) = -3(x - 2)$$
$$y + 4 = -3x + 6$$
$$3x + y = 2$$

5. $(1, 3)$ and $(5, -2)$
$$m = \frac{y_2 - y_1}{x_2 - x_1} = \frac{-2 - 3}{5 - 1} = -\frac{5}{4}$$
$$y - y_1 = m(x - x_1)$$
$$y - 3 = -\frac{5}{4}(x - 1)$$
$$y - 3 = -\frac{5}{4}x + \frac{5}{4}$$
$$\frac{5}{4}x + y = \frac{17}{4}$$
$$5x + 4y = 17$$

6. a. $(10, 200)$ and $(9, 250)$
$$m = \frac{y_2 - y_1}{x_2 - x_1} = \frac{250 - 200}{9 - 10} = -50$$
$$y - y_1 = m(x - x_1)$$
$$y - 200 = -50(x - 10)$$
$$y - 200 = -50x + 500$$
$$y = -50x + 700$$

   b. $y = -50(7.50) + 700$
$$= -375 + 700$$
$$= 325$$

Graphing Calculator Explorations

1. $y_1 = x$
$y_2 = 6x$
$y_3 = -6x$

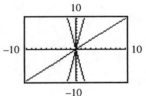

2. $y_1 = -x$
$y_2 = -5x$
$y_3 = -10x$

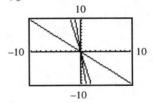

3. $y_1 = \dfrac{1}{2}x + 2$

$y_2 = \dfrac{3}{4}x + 2$

$y_3 = x + 2$

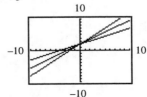

4. $y_1 = x + 1$

$y_2 = \dfrac{5}{4}x + 1$

$y_3 = \dfrac{5}{2}x + 1$

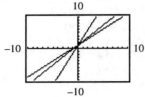

Mental Math

1. $y = 2x - 1$
$m = 2;\ (0, -1)$

2. $y = -7x + 3$
$m = -7;\ (0, 3)$

3. $y = x + \dfrac{1}{3}$
$m = 1;\ \left(0, \dfrac{1}{3}\right)$

4. $y = -x - \dfrac{2}{9}$
$m = -1;\ \left(0, \dfrac{2}{9}\right)$

5. $y = \dfrac{5}{7}x - 4$
$m = \dfrac{5}{7};\ (0, -4)$

6. $y = -\dfrac{1}{4}x + \dfrac{3}{5}$
$m = -\dfrac{1}{4};\ \left(0, \dfrac{3}{5}\right)$

Exercise Set 6.5

1. $m = 5,\ b = 3$
$y = 5x + 3$

3. $m = \dfrac{2}{3},\ b = 0$
$y = \dfrac{2}{3}x$

5. $m = -\dfrac{1}{5},\ b = \dfrac{1}{9}$
$y = -\dfrac{1}{5}x + \dfrac{1}{9}$

7. $y = 2x + 1$

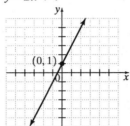

9. $y = \dfrac{2}{3}x + 5$

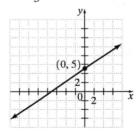

11. $y = -5x$

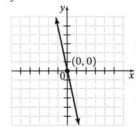

13. $4x + y = 6$
$y = -4x + 6$

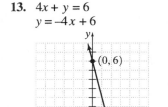

15. $4x - 7y = -14$
$-7y = -4x - 14$
$y = \dfrac{4}{7}x + 2$

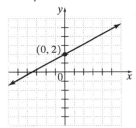

17. $m = 6; (2, 2)$
$y - y_1 = m(x - x_1)$
$y - 2 = 6(x - 2)$
$y - 2 = 6x - 12$
$-6x + y = -10$

19. $m = -8; (-1, -5)$
$y - y_1 = m(x - x_1)$
$y - (-5) = -8(x - (-1))$
$y + 5 = -8x - 8$
$8x + y = -13$

21. $m = \dfrac{1}{2}; (5, -6)$
$y - y_1 = m(x - x_1)$
$y - (-6) = \dfrac{1}{2}(x - 5)$
$y + 6 = \dfrac{1}{2}x - \dfrac{5}{2}$
$2y + 12 = x - 5$
$-x + 2y = -17$
$x - 2y = 17$

23. $m = -\dfrac{1}{2}; (-3, 0)$
$y - y_1 = m(x - x_1)$
$y - 0 = -\dfrac{1}{2}(x - (-3))$
$y = -\dfrac{1}{2}x - \dfrac{3}{2}$
$2y = -x - 3$
$x + 2y = -3$

25. $(3, 2)$ and $(5, 6)$
$m = \dfrac{y_2 - y_1}{x_2 - x_1} = \dfrac{6 - 2}{5 - 3} = \dfrac{4}{2} = 2$
$y - y_1 = m(x - x_1)$
$y - 2 = 2(x - 3)$
$y - 2 = 2x - 6$
$2x - y = 4$

27. $(-1, 3)$ and $(-2, -5)$
$m = \dfrac{y_2 - y_1}{x_2 - x_1} = \dfrac{-5 - 3}{-2 - (-1)} = \dfrac{-8}{-1} = 8$
$y - y_1 = m(x - x_1)$
$y - 3 = 8(x - (-1))$
$y - 3 = 8x + 8$
$8x - y = -11$

29. $(2, 3)$ and $(-1, -1)$
$m = \dfrac{y_2 - y_1}{x_2 - x_1} = \dfrac{-1 - 3}{-1 - 2} = \dfrac{-4}{-3} = \dfrac{4}{3}$
$y - y_1 = m(x - x_1)$
$y - (-1) = \dfrac{4}{3}(x - (-1))$
$y + 1 = \dfrac{4}{3}x + \dfrac{4}{3}$
$3y + 3 = 4x + 4$
$4x - 3y = -1$

31. $(10, 7)$ and $(7, 10)$
$m = \dfrac{y_2 - y_1}{x_2 - x_1} = \dfrac{10 - 7}{7 - 10} = \dfrac{3}{-3} = -1$
$y - y_1 = m(x - x_1)$
$y - 10 = -1(x - 7)$
$y - 10 = -x + 7$
$x + y = 17$

33. (1, 32) and (3, 96)

   a. $m = \dfrac{s_2 - s_1}{t_2 - t_1} = \dfrac{96 - 32}{3 - 1} = \dfrac{64}{2} = 32$

$s - s_1 = m(t - t_1)$
$s - 32 = 32(t - 1)$
$s - 32 = 32t - 32$
$s = 32t$

   b. If $t = 4$ then $s = 32(4) = 128$ ft/sec

35. **a.** (0, 27.6) and (3, 24.8)

   b. $m = \dfrac{y_2 - y_1}{x_2 - x_1} = \dfrac{24.8 - 27.6}{3 - 0} = -\dfrac{2.8}{3}$

$y - y_1 = m(x - x_1)$

$y - 27.6 = -\dfrac{2.8}{3}(x - 0)$

$y - 27.6 = -\dfrac{2.8}{3}x$

$y = -\dfrac{2.8}{3}x + 27.6$

   c. In 1996, $x = 4$

$y = -\dfrac{2.8}{3}(4) + 27.6$

≈ 23.87 heads / sq ft

For Exercises 37–39, evaluate $x^2 - 3x + 1$ for each given value of x.

37. 2

$(2)^2 - 3(2) + 1 = 4 - 6 + 1 = -1$

39. −1

$(-1)^2 - 3(-1) + 1 = 1 + 3 + 1 = 5$

41. No

43. Yes

45. $y = 2x + 1$
y-intercept = (0, 1)
slope = 2
B

47. $y = -3x - 2$
y-intercept = (0, –2)
slope = –2
D

49. $m = 3$; (–1, 2)
$y - y_1 = m(x - x_1)$
$y - 2 = 3(x - (-1))$
$y - 2 = 3x + 3$
$3x - y = -5$

51. Answers vary.

Integrated Review

1. (1, 2) and (2, 4)

$m = \dfrac{y_2 - y_1}{x_2 - x_1} = \dfrac{4 - 2}{2 - 1} = \dfrac{2}{1} = 2$

2. $m = 0$

3. (–3, 3) and (0, 1)

$m = \dfrac{y_2 - y_1}{x_2 - x_1} = \dfrac{1 - 3}{0 - (-3)} = -\dfrac{2}{3}$

4. m is undefined

5. $y = -2x$
$m = -2$
y-intercept = (0, 0)

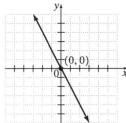

6. $x + y = 3$
$y = -x + 3$
$m = -1$
y-intercept $= (0, 3)$

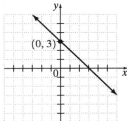

7. $x = -1$
For all y-values x is -1.

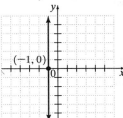

8. $y = 4$
For all x-values, y is 4.

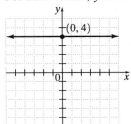

9. $x - 2y = 6$
$-2y = -x + 6$
$y = \dfrac{1}{2}x - 3$
$m = \dfrac{1}{2}$
y-intercept $= (0, -3)$

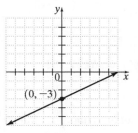

10. $y = 3x + 2$
$m = 3$
y-intercept $= (0, 2)$

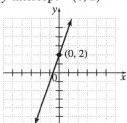

11. $m = 2; \ b = -\dfrac{1}{3}$
$y = mx + b$
$y = 2x - \dfrac{1}{3}$

12. $m = -4; \ (-1, 3)$
$y - y_1 = m(x - x_1)$
$y - 3 = -4(x - (-1))$
$y - 3 = -4x - 4$
$4x + y = -1$

13. $(2, 0)$ and $(-1, -3)$
$m = \dfrac{y_2 - y_1}{x_2 - x_1} = \dfrac{-3 - 0}{-1 - 2} = \dfrac{-3}{-3} = 1$
$y - y_1 = m(x - x_1)$
$y - 0 = 1(x - 2)$
$y = x - 2$
$-x + y = -2$

Section 6.6

Practice Problems

1. $\{(-3, 5), (-3, 1), (4, 6), (7, 0)\}$
   Domain: $\{-3, 4, 7\}$
   Range: $\{0, 1, 5, 6\}$

2. **a.** Every point has a unique x-value: it is a function.

   b. Two points have the same x-value: it is not a function.

3. **a.** This is the graph of the set of ordered pairs $\{(-3, -2), (-1, -1), (0, 0), (1, 1)\}$. Each x-coordinate has exactly one y-coordinate, so this graph is a function.

   b. This is the graph of the set of ordered pairs $\{(-1, -1), (-1, 2), (1, 0), (3, 1)\}$. The x-coordinate -1 is paired with two y-coordinates, -1 and -2, so this graph is not a function.

4. (a) and (b) pass the vertical line test—they are functions. (c) and (d) do not pass the vertical line test—they are not functions.

5. **a.** 6:30 a.m.

   b. end of March and middle of October

6. $f(x) = x^2 + 1$

   a. $f(1) = 1^2 + 1 = 2; (1, 2)$

   b. $f(-3) = (-3)^2 + 1 = 9 + 1 = 10; (-3, 10)$

   c. $f(0) = 0^2 + 1 = 1; (0, 1)$

Exercise Set 6.6

1. $\{(2, 4), (0, 0), (-7, 10), (10, -7)\}$
   Domain: $(-7, 0, 2, 10\}$
   Range: $\{-7, 0, 4, 10\}$

3. $\{(0, -2), (1, -2), (5, -2)\}$
   Domain: $\{0, 1, 5\}$
   Range: $\{-2\}$

5. Yes; each x-value is assigned to only one y-value.

7. No; the x-value -1 is paired with three y-values.

9. No

11. Yes

13. Yes

15. No

17. 5:20 A.M.

19. Answers may vary.

21. 9:00 P.M.

23. January 1st and December 1st

25. Yes; it passes the vertical line test.

For Exercises 27–33, find $f(-2), f(0)$ and $f(3)$.

27. $f(x) = 2x - 5$
    $f(-2) = 2(-2) - 5 = -4 - 5 = -9$
    $f(0) = 2(0) - 5 = -5$
    $f(3) = 2(3) - 5 = 6 - 5 = 1$

29. $f(x) = x^2 + 2$
    $f(-2) = (-2)^2 + 2 = 4 + 2 = 6$
    $f(0) = 0^2 + 2 = 2$
    $f(3) = 3^2 + 2 = 9 + 2 = 11$

31. $f(x) = 3x$
    $f(-2) = 3(-2) = -6$
    $f(0) = 3(0) = 0$
    $f(3) = 3(3) = 9$

33. $f(x) = |x|$
    $f(-2) = |-2| = 2$
    $f(0) = |0| = 0$
    $f(3) = |3| = 3$

For exercises 35–37, find $h(-1)$, $h(0)$ and $h(4)$.

35. $h(x) = -5x$
$h(-1) = -5(-1) = 5$
$h(0) = -5(0) = 0$
$h(4) = -5(4) = -20$

37. $h(x) = 2x^2 + 3$
$h(-1) = 2(-1)^2 + 3 = 2 + 3 = 5$
$h(0) = 2(0)^2 + 3 = 3$
$h(4) = 2(4)^2 + 3 = 32 + 3 = 35$

39. $2x + 5 < 7$
$2x < 2$
$x < 1$

41. $-x + 6 \leq 9$
$-x \leq 3$
$x \geq -3$

43. $P = \dfrac{3}{x} + \dfrac{3}{2x} + \dfrac{5}{x}$
$ = \dfrac{6}{2x} + \dfrac{3}{2x} + \dfrac{10}{2x}$
$ = \dfrac{19}{2x}$ meters

45. $f(x) = 2.59x + 47.24$

    a. $x = 46$
$ f(46) = 2.59(46) + 47.24 = 166.38$ cm

    b. $x = 39$
$ f(39) = 2.59(39) + 47.24 = 148.25$ cm

47. a. Answers vary.

    b. Answers vary.

    c. Answers vary.

49. $y = x + 7$
$f(x) = x + 7$

Section 6.7

Practice Problems

1. $x - 4y > 8$

    a. $(-3, 2)$
$ -3 - 4(2) > 8 \quad ?$
$ -3 - 8 > 8 \quad ?$
$ -11 > 8 \quad ?$
 No

    b. $(9, 0)$
$ 9 - 4(0) > 8 \quad ?$
$ 9 > 8 \quad ?$
 Yes

2. $x - y > 3$
$x - y = 3$
$y = x - 3$

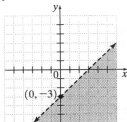

Test $(0, 0)$:
$0 - 0 > 3 \quad ?$
$ 0 > 3 \quad ?$
False
Shade below the line.

3. $x - 4y \leq 4$
$x - 4y = 4$
$-4y = -x + 4$
$y = \dfrac{1}{4}x - 1$

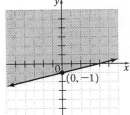

Test (0, 0):

$0 - 4(0) \le 4$?

$0 \le 4$?

True

Shade above the line.

4. $y < 3x$

$y = 3x$

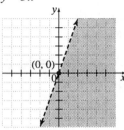

Test (0, 2):

$2 < 3(0)$?

$2 < 0$?

False

Shade below the line.

5. $3x + 2y \ge 12$

$3x + 2y = 12$

$2y = -3x + 12$

$y = -\dfrac{3}{2}x + 6$

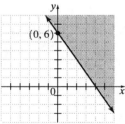

Test (0, 0):

$3(0) + 2(0) \ge 12$?

$0 \ge 12$?

False

Shade above the line.

6. $x < 2$

$x = 2$

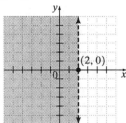

Test (0, 0):

$0 < 2$?

True

Shade to the left of the line.

Mental Math

1. Yes

2. No

3. Yes

4. No

5. $x + y > -5$; (0, 0)

$0 + 0 > -5$?

$0 > -5$?

Yes

6. $2x + 3y < 10$; (0, 0)

$2(0) + 3(0) < 10$?

$0 < 10$?

Yes

7. $x - y \le -1$; (0, 0)

$0 - 0 \le -1$?

$0 \le -1$?

No

8. $\dfrac{2}{3}x + \dfrac{5}{6}y > 4$; (0, 0)

$\dfrac{2}{3}(0) + \dfrac{5}{6}(0) > 4$?

$0 + 0 > 4$?

$0 > 4$?

No

Exercise Set 6.7

1. $x - y > 3$
   $(0, 3):\ 0 - 3 > 3\ \ ?$
   $\qquad\qquad -3 > 3\ \ ?$
   No
   $(2, -1):\ -2 - (-1) > 3\ \ ?$
   $\qquad\qquad\quad -1 > 3\ \ ?$
   No

3. $3x - 5y \le -4$
   $(2, 3);\ 3(2) - 5(3) \le -4\,?$
   $\qquad\qquad 6 - 15 \le -4\,?$
   $\qquad\qquad\quad -9 \le -4\,?$
   Yes
   $(-1, -1):\ 3(-1) - 5(-1) \le -4\,?$
   $\qquad\qquad\qquad -3 + 5 \le -4\,?$
   $\qquad\qquad\qquad\quad 2 \le -4\,?$
   No

5. $x < -y$
   $(0, 2):\ 0 < -2?$
   No
   $(-5, 1):\ -5 < -1?$
   Yes

7. $x + y \le 1$
   $x + y = 1$
   $\qquad y = -x + 1$

   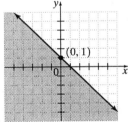

   Test $(0, 0)$:
   $0 + 0 \le 1?$
   $\quad 0 \le 1?$
   Yes; shade below the line.

9. $2x - y > -4$
   $2x - y = -4$
   $\qquad y = 2x + 4$

   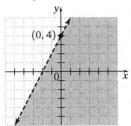

   Test $(0, 0)$:
   $2(0) - 0 > -4?$
   $\qquad 0 > -4?$
   Yes; shade below the line.

11. $y > 2x$
    $y = 2x$

    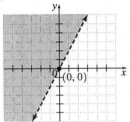

    Test $(0, 1)$:
    $1 > 2(0)?$
    $1 > 0?$
    Yes; shade above the line.

13. $x \le -3y$
    $x = -3y$
    $\qquad y = -\dfrac{1}{3}x$

    Test $(0, 1)$:
    $0 \le -3(1)?$
    $\quad 0 \le -3?$
    No; shade below the line.

15. $y \geq x + 5$
$y = x + 5$

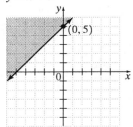

Test $(0, 0)$:
$0 \geq 0 + 5$?
$0 \geq 5$?
No; shade above the line.

17. $y < 4$
$y = 4$

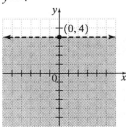

Test $(0, 0)$:
$0 < 4$?
Yes; shade below the line.

19. $x \geq -3$
$x = -3$

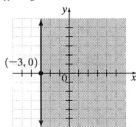

Test $(0, 0)$:
$0 \geq -3$?
Yes; shade to the right of the line.

21. $5x + 2y \leq 10$
$5x + 2y = 10$
$2y = -5x + 10$
$y = -\dfrac{5}{2}x + 5$

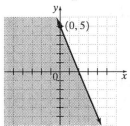

Test $(0, 0)$:
$5(0) + 2(0) \leq 10$?
$0 \leq 10$?
Yes; shade below the line.

23. $x > y$
$x = y$
$y = x$

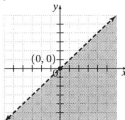

Test $(0, 1)$:
$0 > 1$?
No; shade below the line.

25. $x - y \leq 6$
$x - y = 6$
$y = x - 6$

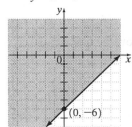

Test (0, 0):
$$0 - 0 \le 6?$$
$$0 \le 6?$$
Yes; shade above the line.

27. $x \ge 0$
$x = 0$

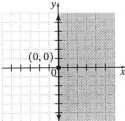

Test (1, 0):
$$1 \ge 0?$$
Yes; shade to the right of the line.

29. $2x + 7y > 5$
$2x + 7y = 5$
$$7y = -2x + 5$$
$$y = -\frac{2}{7}x + \frac{5}{7}$$

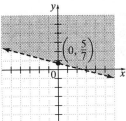

Test (0, 0):
$$2(0) + 7(0) > 5?$$
$$0 > 5?$$
No; shade above the line.

31. (–2, 1)

33. (–3, –1)

35. $x > 2$
a

37. $y \le 2$
b

39. Answers vary.

Chapter 6 Activity

1. Archer Daniels Midland:
(1995, 795.9); (1996, 695.9)

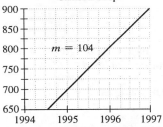
Archer Daniels Midland

Campbell Soup:
(1995, 698.0); (1996, 802.0)

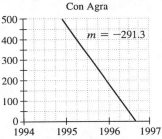
Cambell Soup

ConAgra: (1995, 471.6); (1996, 180.3)

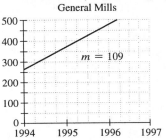
Con Agra

General Mills: (1995, 367.4); (1996, 476.4)

General Mills

H. J. Heinz: (1995, 591.0); (1996, 659.3)

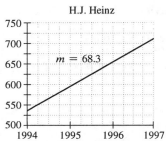

H.J. Heinz

$m = 68.3$

IBP: (1995, 257.9); (1996, 198.7)

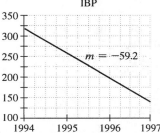

IBP

$m = -59.2$

RJR Nabisco Holdings
(1995, 501.0); (1996, 568.0)

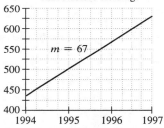

RJR Nabisco Holdings

$m = 67$

Sara Lee: (1995, 776.0); (1996, 889.0)

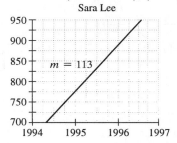

Sara Lee

$m = 113$

2. Archer Daniels Midland:

$$m = \frac{y_2 - y_1}{x_2 - x_1}$$
$$= \frac{695.9 - 795.9}{1996 - 1995}$$
$$= \frac{-100}{1} = -100$$

Campbell Soup:

$$m = \frac{y_2 - y_1}{x_2 - x_1}$$
$$= \frac{802.0 - 698.0}{1996 - 1995}$$
$$= \frac{104}{1} = 104$$

Con Agra:

$$m = \frac{y_2 - y_1}{x_2 - x_1}$$
$$= \frac{180.3 - 471.6}{1996 - 1995}$$
$$= \frac{-291.3}{1} = -291.3$$

General Mills:

$$m = \frac{y_2 - y_1}{x_2 - x_1}$$
$$= \frac{476.4 - 367.4}{1996 - 1995}$$
$$= \frac{109}{1} = 109$$

H.J. Heinz

$$m = \frac{y_2 - y_1}{x_2 - x_1}$$
$$= \frac{659.3 - 591.0}{1996 - 1995}$$
$$= \frac{68.3}{1} = 68.3$$

IBP:

$$m = \frac{y_2 - y_1}{x_2 - x_1}$$
$$= \frac{198.7 - 257.9}{1996 - 1995}$$
$$= \frac{-59.2}{1} = -59.2$$

RJR Nabisco Holdings:

$$m = \frac{y_2 - y_1}{x_2 - x_1}$$

$$= \frac{568.0 - 501.0}{1996 - 1995}$$

$$= \frac{67}{1} = 67$$

Sara Lee:

$$m = \frac{y_2 - y_1}{x_2 - x_1}$$

$$= \frac{889.0 - 776.0}{1996 - 1995}$$

$$= \frac{113}{1} = 113$$

3. Positive slopes: Campbell Soup, General Mills, H.J. Heinz, RJR Nabisco Holdings, Sara Lee; profit is increasing. Negative slopes: Archer Daniels Midland, Con Agra, IBP; profit is decreasing.

4. Answers vary.

5. Answers vary.

6. Answers vary.

Chapter 6 Review

1.–6.

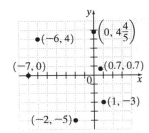

7. $-2 + y = 6x$
   $x = 7$:
   $-2 + y = 6(7)$
   $-2 + y = 42$
   $y = 44$
   $(7, 44)$

8. $y = 3x + 5$
   $y = -8$:
   $-8 = 3x + 5$
   $3x = -13$
   $x = -\frac{13}{3}$
   $\left(-\frac{13}{3}, -8\right)$

9. $9 = -3x + 4y$
   $y = 0$:
   $9 = -3x + 4(0)$
   $9 = -3x$
   $x = -3$
   $y = -3$:
   $9 = -3x + 4(3)$
   $9 = -3x + 12$
   $-3 = -3x$
   $x = 1$
   $x = 9$:
   $9 = -3(9) + 4y$
   $9 = -27 + 4y$
   $36 = 4y$
   $y = 9$

x	y
-3	0
1	3
9	9

10. $y = 5$

x	y
7	5
-7	5
0	5

11. $x = 2y$

x	y
0	0
10	5
−10	−5

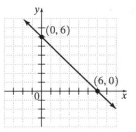

12. $y = 5x + 2000$

  a. $x = 1$:
$y = 5(1) + 2000 = 2005$
$x = 100$:
$y = 5(100) + 2000 = 2500$
$x = 1000$
$y = 5(1000) + 2000 = 7000$

x	1	100	1000
y	2005	2500	7000

  b. $y = 6430$
$6430 = 5x + 2000$
$5x = 4430$
$x = 886$ compact discs

13. $x - y = 1$
If $x = 0, y = -1$
If $y = 0, x = 1$
Plot using $(0, -1)$ and $(1, 0)$

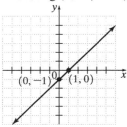

14. $x + y = 6$
If $x = 0, y = 6$
If $y = 0, x = 6$
Plot using $(0, 6)$ and $(6, 0)$

15. $x - 3y = 12$
If $x = 0, -3y = 12$
$y = -4$
If $y = 0, x = 12$
Plot using $(0, -4)$ and $(12, 0)$

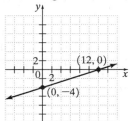

16. $5x - y = -8$
If $x = 0, y = 8$
If $y = -2, 5x - (-2) = -8$
$5x + 2 = -8$
$5x = -10$
$x = -2$
Plot using $(0, 8)$ and $(-2, -2)$

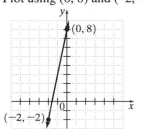

17. $x = 3y$
If $x = 0, y = 0$
If $y = 1, x = 3$
Plot using $(0, 0)$ and $(3, 1)$

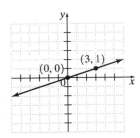

18. $y = -2x$
If $x = 0$, $y = 0$
If $x = 2$, $y = -4$
Plot using $(0, 0)$ and $(2, -4)$

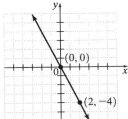

19. $x = 4$; $y = -2$;
$(4, 0)$; $(0, -2)$

20. $x = -2$; $x = 2$; $y = -2$; $y = 2$;
$(-2, 0)$; $(2, 0)$; $(0, -2)$; $(0, 2)$

21. $y = -3$
If $x = 0$, $y = -3$
Horizontal line, no x-intercept

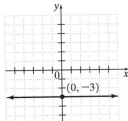

22. $x = 5$
If $y = 0$, $x = 5$
Vertical line, no y-intercept

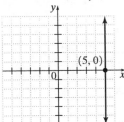

23. $x - 3y = 12$
If $x = 0$, $-3y = 12$
$\qquad\qquad y = -4$
If $y = 0$, $x = 12$
$(0, -4)$ and $(12, 0)$

24. $-4x + y = 8$
If $x = 0$, $y = 8$
If $y = 0$, $-4x = 8$
$\qquad\qquad x = -2$
$(0, 8)$ and $(-2, 0)$

25. $(-1, 2)$ and $(3, -1)$
$$m = \frac{y_2 - y_1}{x_2 - x_1} = \frac{-1 - 2}{3 - (-1)} = -\frac{3}{4}$$

26. $(-2, -2)$ and $(3, -1)$
$$m = \frac{y_2 - y_1}{x_2 - x_1} = \frac{-1 - (-2)}{3 - (-2)} = \frac{1}{5}$$

27. $m = 0$
horizontal line; d

28. $m = -1$
b

29. no slope
vertical line; c

30. $m = 4$
a

31. $(2, 5)$ and $(6, 8)$
$$m = \frac{y_2 - y_1}{x_2 - x_1} = \frac{8 - 5}{6 - 2} = \frac{3}{4}$$

32. $(4, 7)$ and $(1, 2)$

$$m = \frac{y_2 - y_1}{x_2 - x_1} = \frac{2 - 7}{1 - 4} = \frac{-5}{-3} = \frac{5}{3}$$

33. $(1, 3)$ and $(-2, -9)$

$$m = \frac{y_2 - y_1}{x_2 - x_1} = \frac{-9 - 3}{-2 - 1} = \frac{-12}{-3} = 4$$

34. $(-4, 1)$ and $(3, -6)$

$$m = \frac{y_2 - y_1}{x_2 - x_1} = \frac{-6 - 1}{3 - (-4)} = \frac{-7}{7} = -1$$

35. $y = 3x + 7$
$m = 3$

36. $x - 2y = 4$
$\qquad -2y = -x + 4$
$\qquad\quad y = \frac{1}{2}x - 2$
$m = \frac{1}{2}$

37. $y = -2$
$m = 0$

38. $x = 0$
undefined

39. $\quad x - y = -6 \qquad\quad x + y = 3$
$\qquad\quad y = x + 6 \qquad\quad y = -x + 3$
$\qquad\quad m = 1 \qquad\qquad m = -1$
$\quad 1(-1) = -1$; perpendicular

40. $\quad 3x + y = 7 \qquad\quad -3x - y = 10$
$\qquad\quad y = -3x + 7 \qquad\quad y = -3x - 10$
$\qquad\quad m = -3 \qquad\qquad m = -3$
$\quad$ parallel

41. $\quad y = 4x + \frac{1}{2} \qquad\quad 4x + 2y = 1$
$\qquad\quad m = 4 \qquad\qquad\quad 2y = -4x + 1$
$\qquad\qquad\qquad\qquad\qquad\quad y = -2x + \frac{1}{2}$
$\qquad\qquad\qquad\qquad\qquad\quad m = -2$

$\quad 4(-2) \neq -1$
$\quad$ neither

42. $3x + y = 7$
$\qquad y = -3x + 7$
$\qquad m = -3; (0, 7)$

43. $\quad x - 6y = -1$
$\qquad\quad 6y = x + 1$
$\qquad\qquad y = \frac{1}{6}x + \frac{1}{6}$
$\quad m = \frac{1}{6}; \left(0, \frac{1}{6}\right)$

44. $m = -5; \; b = \frac{1}{2}$

$\qquad y = -5x + \frac{1}{2}$

45. $m = \frac{2}{3}; \; b = 6$

$\qquad y = \frac{2}{3}x + 6$

46. $y = 2x + 1$
$\qquad m = 2; b = 1$
$\qquad$ D

47. $y = -4x$
$\qquad m = -4; b = 0$
$\qquad$ C

48. $y = 2x$
$\qquad m = 2; b = 0$
$\qquad$ A

49. $y = 2x - 1$
$\qquad m = 2; b = -1$
$\qquad$ B

50. $m = 4; (2, 0)$
$\qquad y - y_1 = m(x - x_1)$
$\qquad y - 0 = 4(x - 2)$
$\qquad\qquad y = 4x - 8$
$\qquad -4x + y = -8$

51. $m = -3; (0, -5)$
$\qquad y - y_1 = m(x - x_1)$
$\qquad y - (-5) = -3(x - 0)$
$\qquad\quad y + 5 = -3x$
$\qquad\quad 3x + y = -5$

52. $m = \dfrac{3}{5}; (1, 4)$

$$y - y_1 = m(x - x_1)$$
$$y - 4 = \frac{3}{5}(x - 1)$$
$$y - 4 = \frac{3}{5}x - \frac{3}{5}$$
$$5y - 20 = 3x - 3$$
$$3x - 5y = -17$$

53. $m = -\dfrac{1}{3}; (-3, 3)$

$$y - y_1 = m(x - x_1)$$
$$y - 3 = -\frac{1}{3}(x - (-3))$$
$$y - 3 = -\frac{1}{3}x - 1$$
$$y = -\frac{1}{3}x + 2$$
$$3y = -x + 6$$
$$x + 3y = 6$$

54. $(1, 7)$ and $(2, -7)$

$$m = \frac{y_2 - y_1}{x_2 - x_1} = \frac{-7 - 7}{2 - 1} = \frac{-14}{1} = -14$$
$$y - y_1 = m(x - x_1)$$
$$y - 7 = -14(x - 1)$$
$$y - 7 = -14x + 14$$
$$14x + y = 21$$

55. $(-2, 5)$ and $(-4, 6)$

$$m = \frac{y_2 - y_1}{x_2 - x_1} = \frac{6 - 5}{-4 - (-2)} = \frac{1}{-2} = -\frac{1}{2}$$
$$y - y_1 = m(x - x_1)$$
$$y - 5 = -\frac{1}{2}(x - (-2))$$
$$y - 5 = -\frac{1}{2}x - 1$$
$$2y - 10 = -x - 2$$
$$x + 2y = 8$$

56. $\{(7, 1), (7, 5), (2, 6)\}$
Two points have the same x-value; no

57. $\{(0, -1), (5, -1), (2, 2)\}$
Every point has a unique x-value; yes

58. yes

59. yes

60. no

61. yes

62. $f(x) = -2(x) + 6$

  a. $f(0) = -2(0) + 6 = 6$

  b. $f(-2) = -2(-2) + 6 = 4 + 6 = 10$

  c. $f\left(\dfrac{1}{2}\right) = -2\left(\dfrac{1}{2}\right) + 6 = -1 + 6 = 5$

63. $x + 6y < 6$
$x + 6y = 6$
If $x = 0$, $6y = 6$
$y = 1$
If $y = 0$, $x = 6$
Plot using $(0, 1)$ and $(6, 0)$

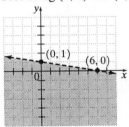

Test $(0, 0)$:
$0 + 6(0) < 6$?
$0 < 6$?
Yes; shade below the line.

64. $x + y > -2$
$x + y = -2$
If $x = 0$, $y = -2$
If $y = 0$, $x = -2$
Plot using $(0, -2)$ and $(-2, 0)$

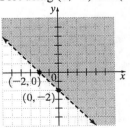

Test (0, 0):
0 + 0 > −2?
Yes; shade above the line.

65. $y \geq -7$
$y = -7$

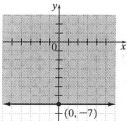

Test (0, 0):
$0 \geq -7$?
Yes; shade above the line.

66. $y \leq -4$
$y = -4$

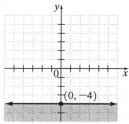

Test (0, 0):
$0 \leq -4$?
No; shade below the line.

67. $-x \leq y$
$-x = y$
If $x = 0$, $y = 0$
If $x = 3$, $y = -3$
Plot using (0, 0) and (3, −3)

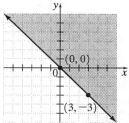

Test (0, 1):
$-0 \leq 1$?
Yes; shade above the line.

68. $x \geq -y$
$x = -y$
If $x = 0$, $y = 0$
If $x = 3$, $y = -3$
Plot using (0, 0) and (3, −3)

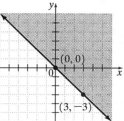

Test (0, 1)
$0 \geq -1$?
Yes; shade above the line.

Chapter 6 Test

1. $12y - 7y = 5$
If $x = 1$, $12y - 7(1) = 5$
$\qquad\qquad\qquad 12y = 12$
$\qquad\qquad\qquad\quad y = 1$

(1, 1)

2. $y = 17$
If $x = -4$, $y = 17$
(−4, 17)

3. (−1, −1) and (4, 1)
$$m = \frac{y_2 - y_1}{x_2 - x_1} = \frac{1 - (-1)}{4 - (-1)} = \frac{2}{5}$$

4. $m = 0$

5. (6, −5) and (−1, 2)
$$m = \frac{y_2 - y_1}{x_2 - x_1} = \frac{2 - (-5)}{-1 - 6} = \frac{7}{-7} = -1$$

6. (0, −8) and (−1, −1)
$$m = \frac{y_2 - y_1}{x_2 - x_1} = \frac{-1 - (-8)}{-1 - 0} = \frac{7}{-1} = -7$$

7. $-3x + y = 5$
$\qquad\quad y = 3x + 5$
$\qquad\quad m = 3$

8. $x = 6$
Vertical line; m is undefined.

9. $2x + y = 8$
If $x = 0$, $y = 8$
If $y = 0$, $2x = 8$
         $x = 4$
Plot using $(0, 8)$ and $(4, 0)$

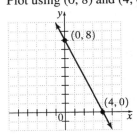

10. $-x + 4y = 5$
If $y = 0$, $x = -5$
If $y = 1$, $-x + 4 = 5$
           $-x = 1$
            $x = -1$
Plot using $(-5, 0)$ and $(-1, 1)$

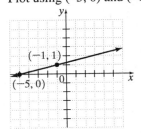

11. $x - y \geq -2$
$x - y = -2$
If $x = 0$, $y = 2$
If $y = 0$, $x = -2$
Plot using $(0, 2)$ and $(-2, 0)$

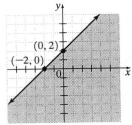

Test $(0, 0)$:
$0 - 0 \geq -2$?
    $0 \geq -2$?
Yes; shade below the line.

12. $y \geq -4x$
$y = -4x$
If $x = 0$, $y = 0$
If $x = 1$, $y = -4$
Plot using $(0, 0)$ and $(1, -4)$

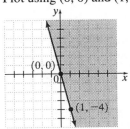

Test $(0, 1)$:
$1 \geq -4(0)$?
$1 \geq 0$?
Yes; shade above the line.

13. $5x - 7y = 10$
If $y = 0$, $5x = 10$
            $x = 2$
If $y = -5$, $5x - 7(-5) = 10$
              $5x + 35 = 10$
                  $5x = -25$
                    $x = -5$
Plot using $(2, 0)$ and $(-5, -5)$

14. $2x - 3y > -6$
$2x - 3y = -6$
If $x = 0$, $y = 2$
If $y = 0$, $x = -3$
Plot using $(0, 2)$ and $(-3, 0)$

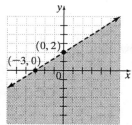

Test (0, 0):
$0 - 0 > -6$?
$0 > -6$?
Yes; shade below the line.

15. $6x + y > -1$
$6x + y = -1$
If $x = 0$, $y = -1$
If $x = -1$, $6(-1) + y = -1$
$-6 + y = -1$
$y = 5$
Plot using $(0, -1)$ and $(-1, 5)$

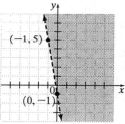

Test (0, 0):
$0 + 0 > -1$?
$0 > -1$?
Yes; shade above the line.

16. $y = -1$

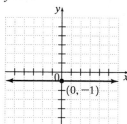

17. $y = 2x - 6$ $-4x = 2y$
$m = 2$ $y = -2x$
                              $m = -2$

neither

18. $m = -\dfrac{1}{4}$; $(2, 2)$
$y - y_1 = m(x - x_1)$
$y - 2 = -\dfrac{1}{4}(x - 2)$
$y - 2 = -\dfrac{1}{4}x + \dfrac{1}{2}$
$4y - 8 = -x + 2$
$x + 4y = 10$

19. $(0, 0)$ and $(6, -7)$
$m = \dfrac{y_2 - y_1}{x_2 - x_1} = \dfrac{-7 - 0}{6 - 0} = -\dfrac{7}{6}$
$y - y_1 = m(x - x_1)$
$y - 0 = -\dfrac{7}{6}(x - 0)$
$y = -\dfrac{7}{6}x$
$6y = -7x$
$7x + 6y = 0$

20. $(2, -5)$ and $(1, 3)$
$m = \dfrac{y_2 - y_1}{x_2 - x_1} = \dfrac{3 - (-5)}{1 - 2} = -8$
$y - y_1 = m(x - x_1)$
$y - 3 = -8(x - 1)$
$y - 3 = -8x + 8$
$8x + y = 11$

21. $m = \dfrac{1}{8}$; $b = 12$
$y = \dfrac{1}{8}x + 12$
$8y = x + 96$
$x - 8y = -96$

22. $\{(-1, 2), (-2, 4), (-3, 6), (-4, 8)\}$
Every point has a unique x-value; yes

23. $\{(-3, -3), (0, 5), (-3, 2), (0, 0)\}$
Two points have the same x-value; no

24. yes

25. yes

26. $f(x) = 2x - 4$

    a. $f(-2) = 2(-2) - 4 = -4 - 4 = -8$

    b. $f(0.2) = 2(0.2) - 4 = 0.4 - 4 = -3.6$

    c. $f(0) = 2(0) - 4 = 0 - 4 = -4$

27. $f(x) = x^3 - x$

    a. $f(-1) = (-1)^3 - (-1) = -1 + 1 = 0$

    b. $f(0) = 0^3 - 0 = 0$

    c. $f(4) = 4^3 - 4 = 64 - 4 = 60$

Chapter 6 Cumulative Review

1. $6 \div 3 + 5^2 = 2 + 25 = 27$

2. $3[4(5 + 2) - 10] = 3[4(7) - 10)]$
$$= 3[28 - 10]$$
$$= 3(18)$$
$$= 54$$

3. a. The Lion King; $315 million

    b. $215 - 145 = \$70$ million

4. $2x + 6$

5. $(x - 4) \div 7$

6. $5 + 3(x + 1) = 5 + 3x + 3$
$$= 8 + 3x$$

7. $\dfrac{5}{2}x = 15$
$$x = \dfrac{2}{5} \cdot 15$$
$$x = 6$$

8. $2x < -4$

$x < -2$

9. a. $-2t^2 + 3t + 6$
degree = 2; trinomial

    b. $15x - 10$
degree = 1; binomial

    c. $7x + 3x^3 + 2x^2 - 1$
degree = 3; none of these

10. $(-2x^2 + 5x - 1) + (-2x^2 + x + 3)$
$$= -4x^2 + 6x + 2$$

11. $(3y + 1)^2 = (3y)^2 + 2(3y) + 1^2$
$$= 9y^2 + 6y + 1$$

12. $-9a^5 + 18a^2 - 3a = 3a(-3a^4 + 6a - 1)$

13. $x^2 + 4x - 12 = (x - 2)(x + 6)$

14. $8x^2 - 22x + 5 = (4x - 1)(2x - 5)$

15. $x^2 - 9x = -20$
$$x^2 - 9x + 20 = 0$$
$$(x - 4)(x - 5) = 0$$
$$x - 4 = 0 \quad \text{or} \quad x - 5 = 0$$
$$x = 4 \quad \text{or} \quad x = 5$$

16. $\dfrac{2x^2 - 11x + 5}{5x - 25} \div \dfrac{4x - 2}{10}$

$$= \dfrac{2x^2 - 11x + 5}{5(x - 5)} \cdot \dfrac{10}{4x - 2}$$

$$= \dfrac{2(2x^2 - 11x + 5)}{(x - 5)(4x - 2)}$$

$$= \dfrac{2(x - 5)(2x - 1)}{2(x - 5)(2x - 1)}$$

$$= 1$$

17. $\dfrac{4b}{9a} \cdot \dfrac{3ab}{3ab} = \dfrac{12ab^2}{27a^2 b}$

18. $1 + \dfrac{m}{m + 1} = \dfrac{m + 1}{m + 1} + \dfrac{m}{m + 1} = \dfrac{2m + 1}{m + 1}$

19.
$$3 - \frac{6}{x} = x + 8$$
$$3x - 6 = x^2 + 8x$$
$$x^2 + 5x + 6 = 0$$
$$(x + 3)(x + 2) = 0$$
$$x + 3 = 0 \quad \text{or} \quad x + 2 = 0$$
$$x = -3 \quad \text{or} \quad x = -2$$

20. $\dfrac{\frac{x+1}{y}}{\frac{x}{y}+2} = \dfrac{\frac{x+1}{y}}{\frac{x+2y}{y}} = \dfrac{x+1}{y} \cdot \dfrac{y}{x+2y} = \dfrac{x+1}{x+2y}$

21. $3x + y = 12$

    a. $x = 0; \ 3(0) + y = 12$
$$y = 12$$
       $(0, 12)$

    b. $y = 6; \ 3x + 6 = 12$
$$3x = 6$$
$$x = 2$$
       $(2, 6)$

    c. $x = -1; \ 3(-1) + y = 12$
$$-3 + y = 12$$
$$y = 15$$
       $(-1, 15)$

22. $2x + y = 5$
If $x = 0$, then $y = 5$
If $x = 1$, then $y = 3$
Plot using $(0, 5)$ and $(1, 3)$

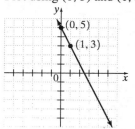

23. $-2x + 3y = 11$
$$3y = 2x + 11$$
$$y = \frac{2}{3}x + \frac{11}{3}$$
$$m = \frac{2}{3}$$

24. $m = -2; \ (-1, 5)$
$$y - y_1 = m(x - x_1)$$
$$y - 5 = -2(x - (-1))$$
$$y - 5 = -2x - 2$$
$$2x + y = 3$$

25. $f(x) = x^2 - 3$

    a. $f(2) = 2^2 - 3 = 4 - 3 = 1$
       $(2, 1)$

    b. $f(-2) = (-2)^2 - 3 = 4 - 3 = 1$
       $(-2, 1)$

    c. $f(0) = 0^2 - 3 = -3$
       $(0, -3)$

Chapter 7

1.

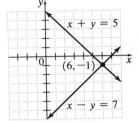

The solution of the system is the point of intersection, $(6, -1)$.

2.

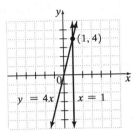

The solution of the system is the point of intersection, $(1, 4)$.

3. $\begin{cases} x + y = 6 \\ x = 3y - 2 \end{cases}$

Substitute $3y - 2$ for x in the first equation.
$x + y = 6$
$(3y - 2) + y = 6$
$4y - 2 = 6$
$4y = 8$
$y = 2$
To find the x-value, let $y = 2$ in the second equation.
$x = 3y - 2$
$x = 3(2) - 2$
$x = 6 - 2$
$x = 4$
The solution is $(4, 2)$.

4. $\begin{cases} 5x + y = 13 \\ 4x - 5y = 22 \end{cases}$

Solve the first equation for y.
$5x + y = 13$
$y = 13 - 5x$

Substitute $13 - 5x$ for y in the second equation.
$4x - 5y = 22$
$4x - 5(13 - 5x) = 22$
$4x - 65 + 25x = 22$
$29x - 65 = 22$
$29x = 87$
$x = 3$
To find the y-value, let $x = 3$ in the first equation.
$5x + y = 13$
$5(3) + y = 13$
$15 + y = 13$
$y = -2$
The solution is $(3, -2)$.

5. $\begin{cases} 7y = x - 6 \\ 2x + 3y = -5 \end{cases}$

Solve the first equation for x.
$7y = x - 6$
$7y + 6 = x$
Substitute $7y + 6$ for x in the second equation.
$2x + 3y = -5$
$2(7y + 6) + 3y = -5$
$14y + 12 + 3y = -5$
$17y + 12 = -5$
$17y = -17$
$y = -1$
To find the x-value, let $y = -1$ in the first equation.
$7y = x - 6$
$7(-1) = x - 6$
$-7 = x - 6$
$-1 = x$
The solution is $(-1, -1)$.

6. $\begin{cases} 4x = y + 6 \\ 8x - 2y = 12 \end{cases}$

Solve the first equation for y.
$4x = y + 6$
$4x - 6 = y$
Substitute $4x - 6$ for y in the second equation.

241

$8x - 2y = 12$
$8x - 2(4x - 6) = 12$
$8x - 8x + 12 = 12$
$12 = 12$
Thus, the linear equations in the original
system $\begin{cases} 4x = y + 6 \\ 8x - 2y = 12 \end{cases}$
are equivalent.
The system has an infinite number of solutions.

7. $\begin{cases} x - 5 = 3y \\ 6y - 2x = 10 \end{cases}$

Solve the first equation for x.
$x - 5 = 3y$
$x = 3y + 5$
Substitute $3y + 5$ for x in the second equation.
$6y - 2x = 10$
$6y - 2(3y + 5) = 10$
$6y - 6y - 10 = 10$
$-10 = 10$
Since this is a false statement, the system has no solution.

8. $\begin{cases} 4x + 6y = -14 \\ 6x + y = -1 \end{cases}$

Solve the second equation for y.
$6x + y = -1$
$y = -1 - 6x$
Substitute $-1 - 6x$ for y in the first equation.
$4x + 6y = -14$
$4x + 6(-1 - 6x) = -14$
$4x - 6 - 36x = -14$
$-6 - 32x = -14$
$-32x = -8$
$x = \dfrac{1}{4}$

To find the y-value, let $x = \dfrac{1}{4}$ in the second equation.
$6x + y = -1$
$6\left(\dfrac{1}{4}\right) + y = -1$
$\dfrac{6}{4} + y = -1$
$y = -1 - \dfrac{6}{4}$

$y = -\dfrac{4}{4} - \dfrac{6}{4}$
$y = -\dfrac{10}{4}$
$y = -\dfrac{5}{2}$
The solution is $\left(\dfrac{1}{4}, -\dfrac{5}{2}\right)$.

9. $\begin{cases} \dfrac{1}{5}x - y = 3 \\ x - 5y = 15 \end{cases}$

Solve the second equation for x.
$x - 5y = 15$
$x = 15 + 5y$
Substitute $15 + 5y$ for x in the first equation.
$\dfrac{1}{5}x - y = 3$

$\dfrac{1}{5}(15 + 5y) - y = 3$

$3 + y - y = 3$
$3 = 3$
Thus, the linear equations in the original system

$\begin{cases} \dfrac{1}{5}x - y = 3 \\ x - 5y = 15 \end{cases}$

are equivalent. This system has an infinite number of solutions.

10. $\begin{cases} y = 3x + 7 \\ y = 10x + 21 \end{cases}$

Substitute $3x + 7$ for y in the second equation.
$y = 10x + 21$
$3x + 7 = 10x + 21$
$7 = 7x + 21$
$-14 = 7x$
$-2 = x$
To find the y-value, let $x = -2$ in the first equation.
$y = 3x + 7$
$y = 3(-2) + 7$
$y = -6 + 7$
$y = 1$
The solution is $(-2, 1)$.

11. $\begin{cases} 2x + y = 11 \\ 3x - y = 29 \end{cases}$

$$2x + y = 11$$
$$\underline{3x - y = 29}$$
$$5x \quad\;\; = 40$$
$$x \quad = 8$$

To find the y-value, let $x = 8$ in the first equation.
$$2x + y = 11$$
$$2(8) + y = 11$$
$$16 + y = 11$$
$$y = -5$$
The solution is $(8, -5)$.

12. $\begin{cases} 4x - 3y = 13 \\ 5x - 9y = 53 \end{cases}$

Multiply the first equation by -3 and add it to the second equation.
$$-12x + 9y = -39$$
$$\underline{5x - 9y = \;\; 53}$$
$$-7x \quad\;\; = \;\; 14$$
$$x \quad = -2$$

To find the y-value, let $x = -2$ in the first equation.
$$4x - 3y = 13$$
$$4(-2) - 3y = 13$$
$$-8 - 3y = 13$$
$$-3y = 21$$
$$y = -7$$
The solution is $(-2, -7)$.

13. $\begin{cases} 6x + 8y = 92 \\ 5x - 3y = 9 \end{cases}$

Multiply the first equation by 3 and the second equation by 8. Then add the equations.
$$18x + 24y = 276$$
$$\underline{40x - 24y = \;\; 72}$$
$$58x \quad\quad\;\; = 348$$
$$x \quad\quad = \;\; 6$$

To find the y-value, let $x = 6$ in the first equation.
$$6x + 8y = 92$$
$$6(6) + 8y = 92$$
$$36 + 8y = 92$$
$$8y = 56$$
$$y = 7$$
The solution is $(6, 7)$.

14. $\begin{cases} 3x - 4y = 7 \\ -9x + 12y = 21 \end{cases}$

Multiply the first equation by 3 and add it to the second equation.
$$9x - 12y = 21$$
$$\underline{-9x + 12y = 21}$$
$$0 = 42$$

Since this is a false statement, there is no solution.

15. $\begin{cases} \dfrac{x}{2} + \dfrac{y}{3} = 2 \\ \dfrac{x}{6} - \dfrac{y}{4} = 5 \end{cases}$

Multiply the first equation by 18 and the second equation by 24. Then add the equations.
$$9x + 6y = \;\; 36$$
$$\underline{4x - 6y = 120}$$
$$13x \quad\quad = 156$$
$$x \quad\quad = \;\; 12$$

To find the y-value, let $x = 12$ in the first equation.
$$\frac{x}{2} + \frac{y}{3} = 2$$
$$\frac{12}{2} + \frac{y}{3} = 2$$
$$6 + \frac{y}{3} = 2$$
$$\frac{y}{3} = -4$$
$$y = -12$$
The solution is $(12, -12)$.

16. $\begin{cases} 6x + 10y = -4 \\ -x + y = -1 \end{cases}$

Multiply the second equation by 6, then add it to the first equation.
$$6x + 10y = -4$$
$$\underline{-6x + 6y = -6}$$
$$16y = -10$$
$$y = -\frac{10}{16}$$
$$y = -\frac{5}{8}$$

To find the x-value, let $y = -\dfrac{5}{8}$ in the second equation.

$-x + y = -1$

$-x + \left(-\dfrac{5}{8}\right) = -1$

$-x = -1 + \dfrac{5}{8}$

$-x = -\dfrac{3}{8}$

$x = \dfrac{3}{8}$

The solution is $\left(\dfrac{3}{8}, -\dfrac{5}{8}\right)$.

17. $\begin{cases} 2x = 8 - 3y \\ 9y = 24 - 6x \end{cases}$

Rewrite each equation in the standard form $Ax + By = C$.

$\begin{cases} 2x + 3y = 8 \\ 6x + 9y = 24 \end{cases}$

Multiply the first equation by -3, then add it to the second equation.

$\begin{array}{r} -6x - 9y = -24 \\ 6x + 9y = 24 \\ \hline 0 = 0 \end{array}$

Since this is a true statement, the system has an infinite number of solutions.

18. $\begin{cases} 11x = 5y + 30 \\ 3x + 4y = -24 \end{cases}$

Rewrite the first equation in the form $Ax + By = C$.

$\begin{cases} 11x - 5y = 30 \\ 3x + 4y = -24 \end{cases}$

Multiply the first equation by 4 and the second equation by 5, then add the equations.

$\begin{array}{r} 44x - 20y = 120 \\ 15x + 20y = -120 \\ \hline 59x = 0 \\ x = 0 \end{array}$

To find the y-value, let $x = 0$ in the first equation.

$11x = 5y + 30$

$0 = 5y + 30$

$-5y = 30$

$y = -6$

The solution is $(0, -6)$.

19. Let $x =$ the first number

$\quad\quad y =$ the second number

$\begin{cases} x + y = 97 \\ x - y = 65 \end{cases}$

Add the equations.

$\begin{array}{r} x + y = 97 \\ x - y = 65 \\ \hline 2x = 162 \\ x = 81 \end{array}$

Let $x = 81$ in the first equation.

$x + y = 97$

$81 + y = 97$

$y = 16$

The numbers are 81 and 16.

20. Two angles are complementary if the sum of their measures is $90°$.

Let $x =$ the measure of the first angle

$\quad\quad y =$ the measure of the second angle

$\begin{cases} x + y = 90 \\ x = 2y - 6 \end{cases}$

Rewrite the second equation in the form $Ax + By = C$.

$\begin{cases} x + y = 90 \\ x - 2y = -6 \end{cases}$

Multiply the first equation by -1, then add it to the second equation.

$\begin{array}{r} -x - y = -90 \\ x - 2y = -6 \\ \hline -3y = -96 \\ y = 32 \end{array}$

Let $y = 32$ in the first equation.

$x + y = 90$

$x + 32 = 90$

$x = 58$

The measures of the angles are $32°$ and $58°$.

Section 7.1

Practice Problems

1.

$5x - 2y = -3$	$y = 3x$
$5(3) - 2(9) \overset{?}{=} -3$	$9 \overset{?}{=} 3(3)$
$15 - 18 \overset{?}{=} -3$	$9 = 9$
$-3 = -3$	True
True	

Since $(3, 9)$ is a solution of both equations, it is a solution of the system.

2. $2x - y = 8$ $\qquad$ $x + 3y = 4$
$2(3) - (-2) \overset{?}{=} 8$ $\qquad$ $3 + 3(-2) \overset{?}{=} 4$
$6 + 2 \overset{?}{=} 8$ $\qquad$ $3 - 6 \overset{?}{=} 4$
$8 = 8$ $\qquad$ $-3 = 4$
True $\qquad\qquad$ False

Since $(3, -2)$ is not a solution of the second equation, it is not a solution of the system.

3. $\begin{cases} -3x + y = -10 \\ x - y = 6 \end{cases}$

Graph each linear equation on a single set of axes.

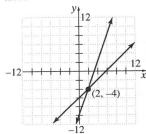

The solution is $(2, -4)$, the intersection of the two lines.

4. $\begin{cases} x + 3y = -1 \\ y = 1 \end{cases}$

Graph each linear equation on a single set of axes.

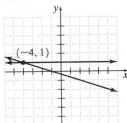

The solution is $(-4, 1)$, the intersection of the two lines.

5. $\begin{cases} 3x - y = 6 \\ 6x = 2y \end{cases}$

Graph each linear equation on a single set of axes.

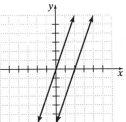

There is no solution because the lines are parallel and do not intersect.

6. $\begin{cases} 3x + 4y = 12 \\ 9x + 12y = 36 \end{cases}$

Graph each linear equation on a single set of axes.

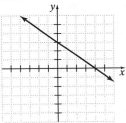

There are an infinite number of solutions because the graphs of the equations are the same line.

Mental Math

1. 1 solution, $(-1, 3)$

2. no solution

3. Infinite number of solutions

4. 1 solution, $(3, 4)$

5. no solution

6. Infinite number of solutions

7. 1 solution, $(3, 2)$

8. 1 solution, $(0, -3)$

Exercise Set 7.1

1. a. $x + y = 8$ $3x + 2y = 21$
     $2 + 4 \stackrel{?}{=} 8$ $3(2) + 2(4) \stackrel{?}{=} 21$
     $6 = 8$ $6 + 8 \stackrel{?}{=} 21$
     False $14 = 21$
                     False

     No, (2, 4) is not a solution of the system.

b. $x + y = 8$ $3x + 2y = 21$
     $5 + 3 \stackrel{?}{=} 8$ $3(5) + 2(3) \stackrel{?}{=} 21$
     $8 = 8$ $15 + 6 \stackrel{?}{=} 21$
     True $21 = 21$

     Yes, (5, 3) is a solution of the system.

3. a. $3x - y = 5$ $x + 2y = 11$
     $3(3) - 4 \stackrel{?}{=} 5$ $3 + 2(4) \stackrel{?}{=} 11$
     $9 - 4 \stackrel{?}{=} 5$ $3 + 8 \stackrel{?}{=} 11$
     $5 = 5$ $11 = 11$
     True   True

     Yes, (3, 4) is a solution of the system.

b. $3x - y = 5$ $x + 2y = 11$
     $3(0) - (-5) \stackrel{?}{=} 5$ $0 + 2(-5) \stackrel{?}{=} 11$
     $0 + 5 \stackrel{?}{=} 5$ $0 - 10 \stackrel{?}{=} 11$
     $5 = 5$ $-10 = 11$
     True   False

     No, (0, −5) is not a solution of the system.

5. a. $2y = 4x$ $2x - y = 0$
     $2(-6) \stackrel{?}{=} 4(-3)$ $2(-3) - (-6) \stackrel{?}{=} 0$
     $-12 = -12$ $-6 + 6 \stackrel{?}{=} 0$
     True $0 = 0$
                    True

     Yes, (−3, −6) is a solution of the system.

b. $2y = 4x$ $2x - y = 0$
     $2(0) \stackrel{?}{=} 4(0)$ $2(0) - 0 \stackrel{?}{=} 0$
     $0 = 0$ $0 = 0$
     True   True

     Yes, (0, 0) is a solution of the system.

7. $\begin{cases} x + y = -1 \\ x - y = 2 \end{cases}$

Graph each linear equation on a single set of axes.

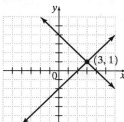

The solution is the intersection point of the two lines, (3, 1).

9. $\begin{cases} x + y = 6 \\ -x + y = -6 \end{cases}$

Graph each linear equation on a single set of axes.

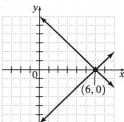

The solution is the intersection point of the two lines, (6, 0).

11. $\begin{cases} y = 2x \\ 3x - y = -2 \end{cases}$

Graph each linear equation on a single set of axes.

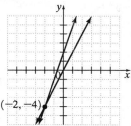

The solution is the intersection point of the two lines, (−2, −4).

13. $\begin{cases} y = x + 1 \\ y = 2x - 1 \end{cases}$

Graph each linear equation on a single set of axes.

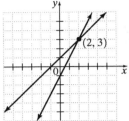

The solution is the intersection point of the two lines, (2, 3).

15. $\begin{cases} 2x + y = 0 \\ 3x + y = 1 \end{cases}$

Graph each linear equation on a single set of axes.

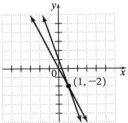

The solution is the intersection point of the two lines, (1, −2).

17. $\begin{cases} y = -x - 1 \\ y = 2x + 5 \end{cases}$

Graph each linear equation on a single set of axes.

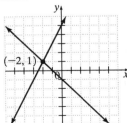

The solution is the intersection point of the two lines, (−2, 1).

19. $\begin{cases} 2x - y = 6 \\ y = 2 \end{cases}$

Graph each linear equation on a single set of axes.

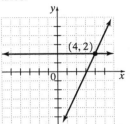

The solution is the intersection point of the two lines, (4, 2).

21. $\begin{cases} x + y = 5 \\ x + y = 6 \end{cases}$

Graph each linear equation on a single set of axes.

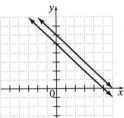

Since the lines are parallel, the system has no solution.

23. $\begin{cases} 2x + y = 4 \\ x + y = 2 \end{cases}$

Graph each linear equation on a single set of axes.

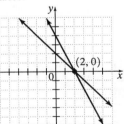

The solution is the intersection point of the two lines, (2, 0).

25. $\begin{cases} x - 2y = 2 \\ 3x + 2y = -2 \end{cases}$

Graph each linear equation on a single set of axes.

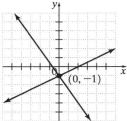

The solution is the intersection point of the two lines, $(0, -1)$.

27. $\begin{cases} y - 3x = -2 \\ 6x - 2y = 4 \end{cases}$

Graph each linear equation on a single set of axes.

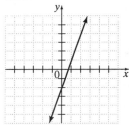

Since the graphs of the equations are the same line, there are infinite solutions.

29. $\begin{cases} x = 3 \\ y = -1 \end{cases}$

Graph each linear equation on a single set of axes.

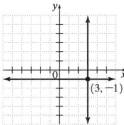

The solution is the intersection point of the two lines, $(3, -1)$.

31. $\begin{cases} y = x - 2 \\ y = 2x + 3 \end{cases}$

Graph each linear equation on a single set of axes.

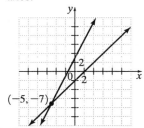

The solution is the intersection point of the two lines, $(-5, -7)$.

33. $\begin{cases} 2x - 3y = -2 \\ -3x + 5y = 5 \end{cases}$

Graph each linear equation on a single set of axes.

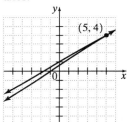

The solution is the intersection point of the two lines, $(5, 4)$.

35. Possible answer:

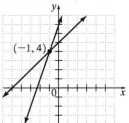

The two lines intersect at $(-1, 4)$.

37. Possible answer:

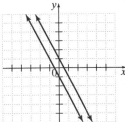

Any two parallel lines will meet the condition.

39. The two lines intersect when the x-coordinate is 1984 or 1988. Therefore, the years when the quantities are equal are 1984 and 1988.

41. a. Each of the tables includes the point $(4, 9)$. Therefore, $(4, 9)$ is a solution to the system.

b.

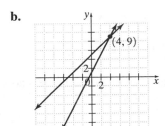

Yes, the lines intersect at the point $(4, 9)$.

43. $-2x + 3(x + 6) = 17$
$-2x + 3x + 18 = 17$
$x + 18 = 17$
$x = -1$
The solution is $x = -1$.

45. $-y + 12\left(\dfrac{y-1}{4}\right) = 3$

$-y + 3(y - 1) = 3$
$-y + 3y - 3 = 3$
$2y - 3 = 3$
$2y = 6$
$y = 3$
The solution is $y = 3$.

47. $3z - (4z - 2) = 9$
$3z - 4z + 2 = 9$
$-z + 2 = 9$
$-z = 7$
$z = -7$
The solution is $z = -7$.

49. Answers may vary.
$\begin{cases} 3x - 2y = -2 \\ 6x + 4y = 4 \end{cases}$

51. Answers may vary.

Section 7.2

Practice Problems

1. $\begin{cases} 2x + 3y = 13 \\ x = y + 4 \end{cases}$
Substitute $y + 4$ for x in the first equation.
$2(y + 4) + 3y = 13$
$2y + 8 + 3y = 13$
$5y = 5$
$y = 1$
Replace y with 1 in the second equation.
$x = 1 + 4$
$x = 5$
The solution is $(5, 1)$.

2. $\begin{cases} 4x - y = 2 \\ y = 5x \end{cases}$
Substitute $5x$ for y in the first equation.
$4x - 5x = 2$
$-x = 2$
$x = -2$
Replace x with -2 in the second equation.
$y = 5(-2)$
$y = -10$
The solution is $(-2, -10)$.

3. $\begin{cases} 3x + y = 5 \\ 3x - 2y = -7 \end{cases}$
Solve the first equation for y.
$y = 5 - 3x$
Substitute $5 - 3x$ for y in the second equation.
$3x - 2(5 - 3x) = -7$
$3x - 10 + 6x = -7$
$9x = 3$
$x = \dfrac{1}{3}$

Substitute $\dfrac{1}{3}$ for x in $y = 5 - 3x$.

$y = 5 - 3\left(\dfrac{1}{3}\right)$

$y = 5 - 1$

$y = 4$

The solution is $\left(\dfrac{1}{3},\ 4\right)$.

4. $\begin{cases} 5x - 2y = 6 \\ -3x + y = -3 \end{cases}$

Solve the second equation for y.
$y = 3x - 3$
Substitute $3x - 3$ for y in the first equation.
$5x - 2(3x - 3) = 6$
$5x - 6x + 6 = 6$
$-x = 0$
$x = 0$
Substitute 0 for x in $y = 3x - 3$.
$y = 3(0) - 3$
$y = -3$
The solution is $(0, -3)$.

5. $\begin{cases} -x + 3y = 6 \\ y = \dfrac{1}{3}x + 2 \end{cases}$

Substitute $\dfrac{1}{3}x + 2$ for y in the first equation.

$-x + 3\left(\dfrac{1}{3}x + 2\right) = 6$

$-x + x + 6 = 6$
$6 = 6$
Since this is a true statement, the system has an infinite number of solutions.

6. $\begin{cases} 2x - 3y = 6 \\ -4x + 6y = -12 \end{cases}$

Solve the first equation for x.
$2x - 3y = 6$
$2x = 3y + 6$
$x = \dfrac{3y + 6}{2}$
Substitute $\dfrac{3y + 6}{2}$ for x in the second equation.

$-4\left(\dfrac{3y + 6}{2}\right) + 6y = -12$
$-2(3y + 6) + 6y = -12$
$-6y - 12 + 6y = -12$
$-12 = -12$
Since this is a true statement, the system has an infinite number of solutions.

Exercise Set 7.2

1. $\begin{cases} x + y = 3 \\ x = 2y \end{cases}$

Substitute $2y$ for x in the first equation. Then solve for y.
$2y + y = 3$
$3y = 3$
$y = 1$
Substitute 1 for y in the second equation. Then solve for x.
$x = 2(1)$
$x = 2$
The solution is $(2, 1)$.

3. $\begin{cases} x + y = 6 \\ y = -3x \end{cases}$

Substitute $-3x$ for y in the first equation. Then solve for x.
$x + (-3x) = 6$
$-2x = 6$
$x = -3$
Substitute -3 for x in the second equation. Then solve for y.
$y = -3(-3)$
$y = 9$
The solution is $(-3, 9)$.

5. $\begin{cases} 3x + 2y = 16 \\ x = 3y - 2 \end{cases}$

Substitute $3y - 2$ for x in the first equation. Then solve for y.
$3(3y - 2) + 2y = 16$
$9y - 6 + 2y = 16$
$11y = 22$
$y = 2$

Substitute 2 for y in the second equation.
Then solve for x.
$x = 3(2) - 2$
$x = 6 - 2$
$x = 4$
The solution is $(4, 2)$.

7. $\begin{cases} 3x - 4y = 10 \\ x = 2y \end{cases}$

Substitute $2y$ for x in the first equation. Then
solve for y.
$3(2y) - 4y = 10$
$6y - 4y = 10$
$2y = 10$
$y = 5$
Substitute 5 for y in the second equation.
Then solve for x.
$x = 2(5)$
$x = 10$
The solution is $(10, 5)$.

9. $\begin{cases} y = 3x + 1 \\ 4y - 8x = 12 \end{cases}$

Substitute $3x + 1$ for y in the second
equation. Then solve for x.
$4(3x + 1) - 8x = 12$
$12x + 4 - 8x = 12$
$4x = 8$
$x = 2$
Substitute 2 for x in the first equation. Then
solve for y.
$y = 3(2) + 1$
$y = 6 + 1$
$y = 7$
The solution is $(2, 7)$.

11. $\begin{cases} y = 2x + 9 \\ y = 7x + 10 \end{cases}$

Substitute $2x + 9$ for y in the second
equation. Then solve for x.
$2x + 9 = 7x + 10$
$-1 = 5x$
$-\frac{1}{5} = x$

Substitute $-\frac{1}{5}$ for x in the first equation.

Then solve for y.
$y = 2\left(-\frac{1}{5}\right) + 9$
$y = -\frac{2}{5} + \frac{45}{5}$
$y = \frac{43}{5}$
The solution is $\left(-\frac{1}{5}, \frac{43}{5}\right)$.

13. $\begin{cases} x + 2y = 6 \\ 2x + 3y = 8 \end{cases}$

Solve the first equation for x.
$x = 6 - 2y$
Substitute $6 - 2y$ for x in the second
equation. Then solve for y.
$2(6 - 2y) + 3y = 8$
$12 - 4y + 3y = 8$
$-y = -4$
$y = 4$
Substitute 4 for y in $x = 6 - 2y$. Then solve
for x.
$x = 6 - 2(4)$
$x = 6 - 8$
$x = -2$
The solution is $(-2, 4)$.

15. $\begin{cases} 2x - 5y = 1 \\ 3x + y = -7 \end{cases}$

Solve the second equation for y.
$y = -3x - 7$
Substitute $-3x - 7$ for y in the first equation.
Then solve for x.
$2x - 5(-3x - 7) = 1$
$2x + 15x + 35 = 1$
$17x = -34$
$x = -2$
Substitute -2 for x in $y = -3x - 7$. Then solve
for y.
$y = -3(-2) - 7$
$y = 6 - 7$
$y = -1$
The solution is $(-2, -1)$.

17. $\begin{cases} 2y = x + 2 \\ 6x - 12y = 0 \end{cases}$

Solve the first equation for x.
$x = 2y - 2$
Substitute $2y - 2$ for x in the second

equation.
$6(2y - 2) - 12y = 0$
$12y - 12 - 12y = 0$
$-12 = 0$
Since this is false, the system has no solution.

19. $\begin{cases} 4x + y = 11 \\ 2x + 5y = 1 \end{cases}$

Solve the first equation for y.
$y = 11 - 4x$
Substitute $11 - 4x$ for y in the second equation. Then solve for x.
$2x + 5(11 - 4x) = 1$
$2x + 55 - 20x = 1$
$-18x = -54$
$x = 3$
Substitute 3 for x in $y = 11 - 4x$. Then solve for y.
$y = 11 - 4(3)$
$y = 11 - 12$
$y = -1$
The solution is $(3, -1)$.

21. $\begin{cases} 2x - 3y = -9 \\ 3x = y + 4 \end{cases}$

Solve the second equation for y.
$y = 3x - 4$
Substitute $3x - 4$ for y in the first equation. Then solve for x.
$2x - 3(3x - 4) = -9$
$2x - 9x + 12 = -9$
$-7x = -21$
$x = 3$
Substitute 3 for x in $y = 3x - 4$. Then solve for y.
$y = 3(3) - 4$
$y = 9 - 4$
$y = 5$
The solution is $(3, 5)$.

23. $\begin{cases} 6x - 3y = 5 \\ x + 2y = 0 \end{cases}$

Solve the second equation for x.
$x = -2y$
Substitute $-2y$ for x in the first equation. Then solve for y.

$6(-2y) - 3y = 5$
$-12y - 3y = 5$
$-15y = 5$
$y = -\dfrac{1}{3}$
Substitute $-\dfrac{1}{3}$ for y in $x = -2y$. Then solve for x.
$x = -2\left(-\dfrac{1}{3}\right)$
$x = \dfrac{2}{3}$
The solution is $\left(\dfrac{2}{3}, -\dfrac{1}{3}\right)$.

25. $\begin{cases} 3x - y = 1 \\ 2x - 3y = 10 \end{cases}$

Solve the first equation for y.
$y = 3x - 1$
Substitute $3x - 1$ for y in the second equation. Then solve for x.
$2x - 3(3x - 1) = 10$
$2x - 9x + 3 = 10$
$-7x = 7$
$x = -1$
Substitute -1 for x in $y = 3x - 1$. Then solve for y.
$y = 3(-1) - 1$
$y = -3 - 1$
$y = -4$
The solution is $(-1, -4)$.

27. $\begin{cases} -x + 2y = 10 \\ -2x + 3y = 18 \end{cases}$

Solve the first equation for x.
$x = 2y - 10$
Substitute $2y - 10$ for x in the second equation. Then solve for y.
$-2(2y - 10) + 3y = 18$
$-4y + 20 + 3y = 18$
$-y = -2$
$y = 2$
Substitute 2 for y in $x = 2y - 10$. Then solve for x.
$x = 2(2) - 10$
$x = 4 - 10$
$x = -6$
The solution is $(-6, 2)$.

29. $\begin{cases} 5x + 10y = 20 \\ 2x + 6y = 10 \end{cases}$

Solve the first equation for x.
$x + 2y = 4$
$x = 4 - 2y$
Substitute $4 - 2y$ for x in the second equation. Then solve for y.
$2(4 - 2y) + 6y = 10$
$8 - 4y + 6y = 10$
$2y = 2$
$y = 1$
Substitute 1 for y in $x = 4 - 2y$. Then solve for x.
$x = 4 - 2(1)$
$x = 2$
The solution is $(2, 1)$.

31. $\begin{cases} 3x + 6y = 9 \\ 4x + 8y = 16 \end{cases}$

Solve the first equation for x.
$x + 2y = 3$
$x = 3 - 2y$
Substitute $3 - 2y$ for x in the second equation.
$4(3 - 2y) + 8y = 16$
$12 - 8y + 8y = 16$
$12 = 16$
Since this is false, the system has no solution.

33. $\begin{cases} \frac{1}{3}x - y = 2 \\ x - 3y = 6 \end{cases}$

Solve the second equation for x.
$x = 3y + 6$
Substitute $3y + 6$ for x in the first equation.
$\frac{1}{3}(3y + 6) - y = 2$
$y + 2 - y = 2$
$2 = 2$
Since this is always true, the system has an infinite number of solutions.

35. Answers may vary.

37. $3x + 2y = 6$
$-2(3x + 2y) = -2(6)$
$-6x - 4y = -12$

39. $-4x + y = 3$
$3(-4x + y) = 3(3)$
$-12x + 3y = 9$

41. $3n + 6m$
$\underline{2n - 6m}$
$5n$

43. $-5a - 7b$
$\underline{5a - 8b}$
$-15b$

45. Simplify the first equation.
$-5y + 6y = 3x + 2(x - 5) - 3x + 5$
$y = 3x + 2x - 10 - 3x + 5$
$y = 2x - 5$
Simplify the second equation.
$4(x + y) - x + y = -12$
$4x + 4y - x + y = -12$
$3x + 5y = -12$
Solve the system.
$\begin{cases} y = 2x - 5 \\ 3x + 5y = -12 \end{cases}$
Substitute $2x - 5$ for y in the second equation. Then solve for x.
$3x + 5(2x - 5) = -12$
$3x + 10x - 25 = -12$
$13x = 13$
$x = 1$
Substitute 1 for x in the first equation. Then solve for y.
$y = 2(1) - 5$
$y = 2 - 5$
$y = -3$
The solution is $(1, -3)$.

47. $\begin{cases} y = 5.1x + 14.56 \\ y = -2x - 3.9 \end{cases}$

Let $y_1 = 5.1x + 14.56$ and $y_2 = -2x - 3.9$ and find the intersection.

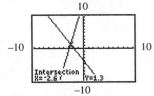

The solution is $(-2.6, 1.3)$.

49. $\begin{cases} 3x + 2y = 14.05 \\ 5x + y = 18.5 \end{cases}$

Let $y_1 = \dfrac{14.05 - 3x}{2}$ and $y_2 = 18.5 - 5x$ and

find the intersection.

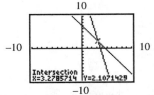

The approximate solution is (3.28, 2.11).

51. a. $\begin{cases} y = -0.65x + 32.02 \\ y = 0.78x + 1.32 \end{cases}$

Substitute $-0.65x + 32.02$ for y in the
second equation. Then solve for x.
$-0.65x + 32.02 = 0.78x + 1.32$
$-1.43x = -30.7$
$x \approx 21.47$
Substitute 21.47 for x in the first
equation. Then solve for y.
$y \approx -0.65(21.47) + 32.02$
$y \approx 18.06$
The solution is approximately (21, 18).

b. Answers may vary.

c.

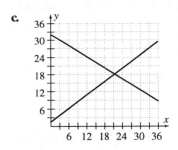

Answers may vary.

Section 7.3

Practice Problems

1. $\begin{cases} x + y = 13 \\ x - y = 5 \end{cases}$

$\begin{array}{r} x + y = 13 \\ x - y = 5 \\ \hline 2x = 18 \\ x = 9 \end{array}$

Let $x = 9$ in the first equation.
$9 + y = 13$
$y = 4$
The solution is (9, 4).

2. $\begin{cases} 2x - y = -6 \\ -x + 4y = 17 \end{cases}$

Multiply the second equation by 2 and add it
to the first equation.
$\begin{array}{r} 2x - y = -6 \\ -2x + 8y = 34 \\ \hline 7y = 28 \\ y = 4 \end{array}$

Let $y = 4$ in the first equation.
$2x - 4 = -6$
$2x = -2$
$x = -1$
The solution is (−1, 4).

3. $\begin{cases} x - 3y = -2 \\ -3x + 9y = 5 \end{cases}$

Multiply the first equation by 3 and add it to
the second equation.
$\begin{array}{r} 3x - 9y = -6 \\ -3x + 9y = 5 \\ \hline 0 = -1 \end{array}$

Since this is a false statement, the system
has no solution.

4. $\begin{cases} 2x + 5y = 1 \\ -4x - 10y = -2 \end{cases}$

Multiply the first equation by 2 and add it to
the second equation.
$\begin{array}{r} 4x + 10y = 2 \\ -4x - 10y = -2 \\ \hline 0 = 0 \end{array}$

Since this is a true statement, the system has
an infinite number of solutions.

5. $\begin{cases} 4x + 5y = 14 \\ 3x - 2y = -1 \end{cases}$

Multiply the first equation by 2 and the second equation by 5. Add the resulting systems.
$$8x + 10y = 28$$
$$\underline{15x - 10y = -5}$$
$$23x = 23$$
$$x = 1$$

Let $x = 1$ in the first equation.
$$4(1) + 5y = 14$$
$$4 + 5y = 14$$
$$5y = 10$$
$$y = 2$$
The solution is $(1, 2)$.

6. $\begin{cases} -\dfrac{x}{3} + y = \dfrac{4}{3} \\ \dfrac{x}{2} - \dfrac{5}{2}y = -\dfrac{1}{2} \end{cases}$

To clear fractions, multiply the first equation by 3 and the second equation by 2. The system simplifies to
$$\begin{cases} -x + 3y = 4 \\ x - 5y = -1 \end{cases}$$

Add the resulting equations.
$$-x + 3y = 4$$
$$\underline{x - 5y = -1}$$
$$-2y = 3$$
$$y = -\frac{3}{2}$$

To find x, go back to the simplified system. Multiply the first equation by 5 and the second equation by 3. Add the resulting equations.
$$-5x + 15y = 20$$
$$\underline{3x - 15y = -3}$$
$$-2x = 17$$
$$x = -\frac{17}{2}$$

The solution is $\left(-\dfrac{17}{2}, -\dfrac{3}{2} \right)$.

Exercise Set 7.3

1. $\begin{cases} 3x + y = 5 \\ 6x - y = 4 \end{cases}$
$$\underline{}$$
$$9x = 9$$
$$x = 1$$

Let $x = 1$ in the first equation.
$$3(1) + y = 5$$
$$y = 2$$
The solution is $(1, 2)$.

3. $\begin{cases} x - 2y = 8 \\ -x + 5y = -17 \end{cases}$
$$\underline{}$$
$$3y = -9$$
$$y = -3$$

Let $y = -3$ in the first equation.
$$x - 2(-3) = 8$$
$$x + 6 = 8$$
$$x = 2$$
The solution is $(2, -3)$.

5. $\begin{cases} 3x + 2y = 11 \\ 5x - 2y = 29 \end{cases}$
$$\underline{}$$
$$8x = 40$$
$$x = 5$$

Let $x = 5$ in the first equation.
$$3(5) + 2y = 11$$
$$15 + 2y = 11$$
$$2y = -4$$
$$y = -2$$
The solution is $(5, -2)$.

7. $\begin{cases} x + y = 6 \\ x - y = 6 \end{cases}$
$$\underline{}$$
$$2x = 12$$
$$x = 6$$

Let $x = 6$ in the first equation.
$$6 + y = 6$$
$$y = 0$$
The solution is $(6, 0)$.

9. $\begin{cases} 3x + y = -11 \\ 6x - 2y = -2 \end{cases}$

$\begin{cases} 2(3x + y) = 2(-11) \\ 6x - 2y = -2 \end{cases}$

$\begin{cases} 6x + 2y = -22 \\ 6x - 2y = -2 \end{cases}$
$$\overline{12x = -24}$$
$$x = -2$$

Let $x = -2$ in the first equation.
$3(-2) + y = -11$
$-6 + y = -11$
$y = -5$
The solution is $(-2, -5)$.

11. $\begin{cases} x + 5y = 18 \\ 3x + 2y = -11 \end{cases}$

$\begin{cases} -3(x + 5y) = -3(18) \\ 3x + 2y = -11 \end{cases}$

$\begin{cases} -3x - 15y = -54 \\ 3x + 2y = -11 \end{cases}$
$$\overline{-13y = -65}$$
$$y = 5$$

Let $y = 5$ in the first equation.
$x + 5(5) = 18$
$x + 25 = 18$
$x = -7$
The solution is $(-7, 5)$.

13. $\begin{cases} 2x - 5y = 4 \\ 3x - 2y = 4 \end{cases}$

$\begin{cases} -3(2x - 5y) = -3(4) \\ 2(3x - 2y) = 2(4) \end{cases}$

$\begin{cases} -6x + 15y = -12 \\ 6x - 4y = 8 \end{cases}$
$$\overline{11y = -4}$$
$$y = -\frac{4}{11}$$

$\begin{cases} -2(2x - 5y) = -2(4) \\ 5(3x - 2y) = 5(4) \end{cases}$

$\begin{cases} -4x + 10y = -8 \\ 15x - 10y = 20 \end{cases}$
$$\overline{11x = 12}$$
$$x = \frac{12}{11}$$

The solution is $\left(\frac{12}{11}, -\frac{4}{11} \right)$.

15. $\begin{cases} 2x + 3y = 0 \\ 4x + 6y = 3 \end{cases}$

$\begin{cases} -2(2x + 3y) = -2(0) \\ 4x + 6y = 3 \end{cases}$

$\begin{cases} -4x - 6y = 0 \\ 4x + 6y = 3 \end{cases}$
$$\overline{0 = 3}$$

Since this is false, the system has no solution.

17. $\begin{cases} 3x + y = 4 \\ 9x + 3y = 6 \end{cases}$

$\begin{cases} -3(3x + y) = -3(4) \\ 9x + 3y = 6 \end{cases}$

$\begin{cases} -9x - 3y = -12 \\ 9x + 3y = 6 \end{cases}$
$$\overline{0 = -6}$$

Since this is false, the system has no solution.

19. $\begin{cases} 3x - 2y = 7 \\ 5x + 4y = 8 \end{cases}$

$\begin{cases} 2(3x - 2y) = 2(7) \\ 5x + 4y = 8 \end{cases}$

$\begin{cases} 6x - 4y = 14 \\ 5x + 4y = 8 \end{cases}$
$$\overline{11x = 22}$$
$$x = 2$$

Let $x = 2$ in the first equation.
$3(2) - 2y = 7$
$6 - 2y = 7$
$-2y = 1$
$y = -\frac{1}{2}$

The solution is $\left(2, -\frac{1}{2} \right)$.

21. $\begin{cases} \frac{2}{3}x + 4y = -4 \\ 5x + 6y = 18 \end{cases}$

$\begin{cases} 3\left(\frac{2}{3}x + 4y\right) = 3(-4) \\ -2(5x + 6y) = -2(18) \end{cases}$

$\begin{cases} 2x + 12y = -12 \\ -10x - 12y = -36 \end{cases}$

$\begin{array}{l} -8x = -48 \\ x = 6 \end{array}$

Let $x = 6$ in the second equation.
$5(6) + 6y = 18$
$30 + 6y = 18$
$6y = -12$
$y = -2$
The solution is $(6, -2)$.

23. $\begin{cases} 4x - 6y = 8 \\ 6x - 9y = 12 \end{cases}$

$\begin{cases} -3(4x - 6y) = -3(8) \\ 2(6x - 9y) = 2(12) \end{cases}$

$\begin{cases} -12x + 18y = -24 \\ 12x - 18y = 24 \end{cases}$

$ 0 = 0$

Since this is always true, the system has an infinite number of solutions.

25. $\begin{cases} 8x = -11y - 16 \\ 2x + 3y = -4 \end{cases}$

$\begin{cases} 8x + 11y = -16 \\ -4(2x + 3y) = -4(-4) \end{cases}$

$\begin{cases} 8x + 11y = -16 \\ -8x - 12y = 16 \end{cases}$

$\begin{array}{l} -y = 0 \\ = 0 \\ y = 0 \end{array}$

Let $y = 0$ in the first equation.
$8x = -11(0) - 16$
$8x = -16$
$x = -2$
The solution is $(-2, 0)$.

27. Answers may vary.

29. $\begin{cases} \frac{x}{3} + \frac{y}{6} = 1 \\ \frac{x}{2} - \frac{y}{4} = 0 \end{cases}$

$\begin{cases} 6\left(\frac{x}{3} + \frac{y}{6}\right) = 6(1) \\ 4\left(\frac{x}{2} - \frac{y}{4}\right) = 4(0) \end{cases}$

$\begin{cases} 2x + y = 6 \\ 2x - y = 0 \end{cases}$

$\begin{array}{l} 4x = 6 \\ x = \frac{3}{2} \end{array}$

To find y, we may multiply the second equation of the simplified system above by -1.

$\begin{cases} 2x + y = 6 \\ -2x + y = 0 \end{cases}$

$2y = 6$

$y = 3$

The solution is $\left(\frac{3}{2}, 3\right)$.

31. $\begin{cases} x - \frac{y}{3} = -1 \\ -\frac{x}{2} + \frac{y}{8} = \frac{1}{4} \end{cases}$

$\begin{cases} 3\left(x - \frac{y}{3}\right) = 3(-1) \\ 8\left(-\frac{x}{2} + \frac{y}{8}\right) = 8\left(\frac{1}{4}\right) \end{cases}$

$\begin{cases} 3x - y = -3 \\ -4x + y = 2 \end{cases}$

$\begin{array}{l} -x = -1 \\ x = 1 \end{array}$

Let $x = 1$ in the first equation.
$1 - \frac{y}{3} = -1$

$-\frac{y}{3} = -2$

$y = 6$
The solution is $(1, 6)$.

33.
$$\begin{cases} \dfrac{x}{3} - y = 2 \\ -\dfrac{x}{2} + \dfrac{3y}{2} = -3 \end{cases}$$

$$\begin{cases} 3\left(\dfrac{x}{3} - y\right) = 3(2) \\ 2\left(-\dfrac{x}{2} + \dfrac{3y}{2}\right) = 2(-3) \end{cases}$$

$$\begin{array}{r} x - 3y = 6 \\ -x + 3y = -6 \\ \hline 0 = 0 \end{array}$$

Since this is always true, the system has an infinite number of solutions.

35. Let x = a number
$$2x + 6 = x - 3$$

37. Let x = a number
$$20 - 3x = 2$$

39. Let n = a number
$$4(n + 6) = 2n$$

41. a.
$$\begin{cases} x + y = 5 \\ 3x + 3y = b \end{cases}$$

The system will have an infinite number of solutions if the second equation is 3 times the first equation.
$$3(x + y) = 3(5)$$
$$3x + 3y = 15$$
Therefore, $b = 15$.

b. There are no solutions to the system if b is any real number except 15.

43.
$$\begin{cases} 2x + 3y = 14 \\ 3x - 4y = -69.1 \end{cases}$$

$$\begin{cases} -3(2x + 3y) = -3(14) \\ 2(3x - 4y) = 2(-69.1) \end{cases}$$

$$\begin{array}{r} -6x - 9y = -42 \\ 6x - 8y = -138.2 \\ \hline -17y = -180.2 \\ y = 10.6 \end{array}$$

Let $y = 10.6$ in the first equation.
$$2x + 3(10.6) = 14$$
$$2x + 31.8 = 14$$
$$2x = -17.8$$
$$x = -8.9$$
The solution is $(-8.9, 10.6)$.

45. a.
$$\begin{cases} 21x - 7y = -3779 \\ -445x + 28y = 14{,}017 \end{cases}$$

$$\begin{cases} 4(21x - 7y) = 4(-3779) \\ -445x + 28y = 14{,}017 \end{cases}$$

$$\begin{array}{r} 84x - 28y = -15{,}116 \\ -445x + 28y = 14{,}017 \\ \hline -361x = -1099 \\ x \approx 3.04 \end{array}$$

Substitute 3.04 for x in the first equation.
$$21(3.04) - 7y = -3779$$
$$63.84 - 7y = -3779$$
$$-7y = -3842.84$$
$$y \approx 548.98$$
The solution is approximately $(3, 549)$.

b. Answers may vary.

c. Since they were equal 3 years after 1988, then from 1988 + 4 = 1992 to 1994 there were more UHF stations than VHF stations.

Integrated Review

1.
$$\begin{cases} 2x - 3y = -11 \\ y = 4x - 3 \end{cases}$$

Substitute $4x - 3$ for y in the first equation. Then solve for x.
$$2x - 3(4x - 3) = -11$$
$$2x - 12x + 9 = -11$$
$$-10x = -20$$
$$x = 2$$
Substitute 2 for x in the second equation.
$$y = 4(2) - 3$$
$$y = 8 - 3$$
$$y = 5$$
The solution is $(2, 5)$.

2.
$$\begin{cases} 4x - 5y = 6 \\ y = 3x - 10 \end{cases}$$

Substitute $3x - 10$ for y in the first equation. Then solve for x.
$$4x - 5(3x - 10) = 6$$
$$4x - 15x + 50 = 6$$
$$-11x = -44$$
$$x = 4$$
Substitute 4 for x in the second equation.

$y = 3(4) - 10$
$y = 12 - 10$
$y = 2$
The solution is (4, 2).

3. $\begin{cases} x + y = 3 \\ x - y = 7 \end{cases}$
$\quad \overline{2x \quad\quad = 10}$
$\quad\quad\quad x = 5$

Let $x = 5$ in the first equation.
$5 + y = 3$
$y = -2$
The solution is (5, –2).

4. $\begin{cases} x - y = 20 \\ x + y = -8 \end{cases}$
$\quad \overline{2x \quad\quad = 12}$
$\quad\quad\quad x = 6$

Let $x = 6$ in the second equation.
$6 + y = -8$
$y = -14$
The solution is (6, –14).

5. $\begin{cases} x + 2y = 1 \\ 3x + 4y = -1 \end{cases}$

$\begin{cases} -2(x + 2y) = -2(1) \\ 3x + 4y = -1 \end{cases}$

$\begin{cases} -2x - 4y = -2 \\ 3x + 4y = -1 \end{cases}$
$\quad \overline{x \quad\quad = -3}$
Let $x = -3$ in the first equation.
$-3 + 2y = 1$
$2y = 4$
$y = 2$
The solution is (–3, 2).

6. $\begin{cases} x + 3y = 5 \\ 5x + 6y = -2 \end{cases}$

$\begin{cases} -5(x + 3y) = -5(5) \\ 5x + 6y = -2 \end{cases}$

$\begin{cases} -5x - 15y = -25 \\ 5x + 6y = -2 \end{cases}$
$\quad \overline{ -9y = -27}$
$\quad\quad\quad\quad y = 3$
Let $y = 3$ in the first equation.

$x + 3(3) = 5$
$x + 9 = 5$
$x = -4$
The solution is (–4, 3).

7. $\begin{cases} y = x + 3 \\ 3x - 2y = -6 \end{cases}$

Substitute $x + 3$ for y in the second equation.
Then solve for x.
$3x - 2(x + 3) = -6$
$3x - 2x - 6 = -6$
$x = 0$
Substitute 0 for x in the first equation.
$y = 0 + 3$
$y = 3$
The solution is (0, 3).

8. $\begin{cases} y = -2x \\ 2x - 3y = -16 \end{cases}$

Substitute $-2x$ for y in the second equation.
Then solve for x.
$2x - 3(-2x) = -16$
$2x + 6x = -16$
$8x = -16$
$x = -2$
Let $x = -2$ in the first equation
$y = -2(-2)$
$y = 4$
The solution is (–2, 4).

9. $\begin{cases} y = 2x - 3 \\ y = 5x - 18 \end{cases}$

Substitute $2x - 3$ for y in the second
equation. Then solve for x.
$2x - 3 = 5x - 18$
$15 = 3x$
$5 = x$
Let $x = 5$ in the first equation.
$y = 2(5) - 3$
$y = 10 - 3$
$y = 7$
The solution is (5, 7).

10. $\begin{cases} y = 6x - 5 \\ y = 4x - 11 \end{cases}$

Substitute $6x - 5$ for y in the second equation. Then solve for x.

$6x - 5 = 4x - 11$

$2x = -6$

$x = -3$

Let $x = -3$ in the first equation.

$y = 6(-3) - 5$

$y = -18 - 5$

$y = -23$

The solution is $(-3, -23)$.

11. $\begin{cases} x + \dfrac{1}{6}y = \dfrac{1}{2} \\ 3x + 2y = 3 \end{cases}$

$\begin{cases} -12\left(x + \dfrac{1}{6}y\right) = -12\left(\dfrac{1}{2}\right) \\ 3x + 2y = 3 \end{cases}$

$\begin{cases} -12x - 2y = -6 \\ 3x + 2y = 3 \end{cases}$

$\begin{array}{r} -9x = -3 \\ x = \dfrac{1}{3} \end{array}$

Let $x = \dfrac{1}{3}$ in the second equation.

$3\left(\dfrac{1}{3}\right) + 2y = 3$

$1 + 2y = 3$

$2y = 2$

$y = 1$

The solution is $\left(\dfrac{1}{3}, 1\right)$.

12. $\begin{cases} x + \dfrac{1}{3}y = \dfrac{5}{12} \\ 8x + 3y = 4 \end{cases}$

$\begin{cases} -24\left(x + \dfrac{1}{3}y\right) = -24\left(\dfrac{5}{12}\right) \\ 3(8x + 3y) = 3(4) \end{cases}$

$\begin{array}{r} -24x - 8y = -10 \\ 24x + 9y = 12 \\ \hline y = 2 \end{array}$

Let $y = 2$ in the second equation.

$8x + 3(2) = 4$

$8x + 6 = 4$

$8x = -2$

$x = -\dfrac{1}{4}$

The solution is $\left(-\dfrac{1}{4}, 2\right)$.

13. $\begin{cases} x - 5y = 1 \\ -2x + 10y = 3 \end{cases}$

$\begin{cases} 2(x - 5y) = 2(1) \\ -2x + 10y = 3 \end{cases}$

$\begin{array}{r} 2x - 10y = 2 \\ -2x + 10y = 3 \\ \hline 0 = 5 \end{array}$

Since this is false, the system has no solution.

14. $\begin{cases} -x + 2y = 3 \\ 3x - 6y = -9 \end{cases}$

$\begin{cases} 3(-x + 2y) = 3(3) \\ 3x - 6y = -9 \end{cases}$

$\begin{array}{r} -3x + 6y = 9 \\ 3x - 6y = -9 \\ \hline 0 = 0 \end{array}$

Since this is always true, the system has an infinite number of solutions.

15. Answers may vary.

16. Answers may vary.

Section 7.4

Practice Problems

1. Let $x =$ first number
$ y =$ second number

$\begin{cases} x + y = 50 \\ x - y = 22 \end{cases}$

Add the equations.

$\begin{array}{r} x + y = 50 \\ x - y = 22 \\ \hline 2x = 72 \\ x = 36 \end{array}$

Let $x = 36$ in the first equation.
$36 + y = 50$
$y = 14$
The numbers are 36 and 14.

2. Let $C =$ number of children
$A =$ number of adults
$$\begin{cases} 5C + 7A = 3379 \\ C + A = 587 \end{cases}$$
Multiply the second equation by –5 and add it to the first equation.
$$\begin{array}{r} 5C + 7A = 3379 \\ -5C - 5A = -2935 \\ \hline 2A = 444 \end{array}$$
$$A = 222$$
Let $A = 222$ in the second equation.
$C + 222 = 587$
$C = 365$
There were 365 children and 222 adults.

3. Let $x =$ speed of faster car in miles per hour
$y =$ speed of slower car in miles per hour

	r	$\cdot$	t	$=$	d
Faster car	x		3		$3x$
Slower car	y		3		$3y$

$$\begin{cases} x = y + 10 \\ 3x + 3y = 440 \end{cases}$$
Substitute $y + 10$ for x in the second equation.
$3(y + 10) + 3y = 440$
$3y + 30 + 3y = 440$
$6y = 410$
$y = \dfrac{410}{6}$
$y = 68\dfrac{1}{3}$

Let $y = 68\dfrac{1}{3}$ in the first equation.

$x = 68\dfrac{1}{3} + 10$

$x = 78\dfrac{1}{3}$

One car's speed is $68\dfrac{1}{3}$ mph and the other

car's speed is $78\dfrac{1}{3}$ mph.

4. Let $x =$ number of liters of 20% solution
$y =$ number of liters of 70% solution

	Concentration rate	Liters of solution	Liters of pure alcohol
First solution	20%	x	$0.2x$
Second solution	70%	y	$0.7y$
Mixture needed	60%	50	$0.6(50)$

$$\begin{cases} x + y = 50 \\ 0.2x + 0.7y = 0.6(50) \end{cases}$$
Multiply the first equation by –0.2 and add to the second equation.
$$\begin{array}{r} -0.2x - 0.2y = -10 \\ 0.2x + 0.7y = 30 \\ \hline 0.5y = 20 \\ y = 40 \end{array}$$

Substitute 40 for y in the first equation.
$x + 40 = 50$
$x = 10$
She needs 10 liters of a 20% alcohol solution and 40 liters of a 70% alcohol solution.

Mental Math

1. c

2. b

3. b

4. c

5. a

6. c

Exercise Set 7.4

1. Let $x =$ one number
$y =$ another number
$$\begin{cases} x + y = 15 \\ x - y = 7 \end{cases}$$

3. Let $x =$ amount invested in larger account
$y =$ amount invested in smaller account
$$\begin{cases} x + y = 6500 \\ x = y + 800 \end{cases}$$

5. Let $x =$ one number
$y =$ another number
$$\begin{cases} x + y = 83 \\ x - y = 17 \end{cases}$$
$$\begin{aligned} 2x &= 100 \\ x &= 50 \end{aligned}$$
Let $x = 50$ in the first equation.
$50 + y = 83$
$y = 33$
The two numbers are 33 and 50.

7. Let $x =$ first number
$y =$ second number
$$\begin{cases} x + 2y = 8 \\ 2x + y = 25 \end{cases}$$

$$\begin{cases} -2(x + 2y) = -2(8) \\ 2x + y = 25 \end{cases}$$
$$\begin{cases} -2x - 4y = -16 \\ 2x + y = 25 \end{cases}$$
$$\begin{aligned} -3y &= 9 \\ y &= -3 \end{aligned}$$
Let $y = -3$ in the first equation.
$x + 2(-3) = 8$
$x - 6 = 8$
$x = 14$
The numbers are 14 and –3.

9. Let $x =$ number of points Cooper scored
$y =$ number of points
Bolton - Holifield scored
$$\begin{cases} x = 174 + y \\ x + y = 1068 \end{cases}$$
Substitute $174 + y$ for x in the second equation.
$174 + y + y = 1068$
$2y = 894$
$y = 447$
Let $y = 447$ in the first equation.
$x = 174 + 447$
$x = 621$
Cooper scored 621 points and Bolton-Holifield scores 447 points.

11. Let $x =$ price of adult's ticket
$y =$ price of child's ticket
$$\begin{cases} 3x + 4y = 159 \\ 2x + 3y = 112 \end{cases}$$
$$\begin{cases} -2(3x + 4y) = -2(159) \\ 3(2x + 3y) = 3(112) \end{cases}$$
$$\begin{cases} -6x - 8y = -318 \\ 6x + 9y = 336 \end{cases}$$
$$y = 18$$
Let $y = 18$ in the first equation.
$3x + 4(18) = 159$
$3x + 72 = 159$
$3x = 87$
$x = 29$
An adult's ticket is $29 and a child's ticket is $18.

13. Let $x =$ number of quarters

$y =$ number of nickels

$$\begin{cases} x + y = 80 \\ 0.25x + 0.05y = 14.6 \end{cases}$$

Substitute $80 - x$ for y in the second equation.

$0.25x + 0.05(80 - x) = 14.6$

$0.25x + 4 - 0.05x = 14.6$

$0.2x = 10.6$

$x = 53$

Let $x = 53$ in the first equation.

$53 + y = 80$

$y = 27$

There are 53 quarters and 27 nickels.

15. Let $x =$ price of Procter and Gamble stock

$y =$ price of Microsoft stock

$$\begin{cases} 40x + 25y = 6058.75 \\ y = x + 64.25 \end{cases}$$

Substitute $x + 64.25$ for y in the first equation.

$40x + 25(x + 64.25) = 6058.75$

$40x + 25x + 1606.25 = 6058.75$

$65x = 4452.5$

$x = 68.5$

Let $x = 68.5$ in the second equation.

$y = 68.5 + 64.25$

$y = 132.75$

The stock of Procter and Gamble was $68.50 and the stock of Microsoft was $132.75.

17. Let $x =$ Pratap's rate in still water, in miles per hour

$y =$ rate of current in miles per hour

$$\begin{cases} 2(x + y) = 18 \\ 4.5(x - y) = 18 \end{cases}$$

$$\begin{cases} 2x + 2y = 18 \\ 4.5x - 4.5y = 18 \end{cases}$$

$$\begin{cases} 2.25(2x + 2y) = 2.25(18) \\ 4.5x - 4.5y = 18 \end{cases}$$

$$\begin{cases} 4.5x + 4.5y = 40.5 \\ 4.5x - 4.5y = 18 \end{cases}$$

$$\overline{\quad 9x \qquad\qquad = 58.5}$$

$$x = 6.5$$

Let $x = 6.5$ in the first equation.

$2(6.5) + 2y = 18$

$2y = 5$

$y = 2.5$

Pratap's rate in still water was 6.5 mph and the current's rate was 2.5 mph.

19. Let $x =$ speed of plane in still air, in miles per hour

$y =$ speed of wind, in miles per hour

$$\begin{cases} 780 = 2(x - y) \\ 780 = 1.5(x + y) \end{cases}$$

$$\begin{cases} 390 = x - y \\ 520 = x + y \end{cases}$$

$$\overline{\quad 910 = 2x}$$

$$455 = x$$

Let $x = 455$ in the equation $520 = x + y$.

$520 = 455 + y$

$65 = y$

The plane's speed is 455 mph in still air and the wind's speed is 65 mph.

21. Let $x =$ amount of 12% solution

$y =$ amount of 4% solution

$$\begin{cases} x + y = 12 \\ 0.12x + 0.04y = 0.09(12) \end{cases}$$

Substitute $12 - x$ for y in the second equation.

$0.12x + 0.04(12 - x) = 0.09(12)$

$0.12x + 0.48 - 0.04x = 1.08$

$0.08x = 0.6$

$x = 7.5$

Let $x = 7.5$ in the first equation.

$7.5 + y = 12$

$y = 4.5$

She needs $7\dfrac{1}{2}$ oz of 12% solution and

$4\dfrac{1}{2}$ oz of 4% solution.

23. Let $x =$ number of pounds of $4.95 per pound beans

$y =$ number of pounds of $2.65 per pound beans

$$\begin{cases} x + y = 200 \\ 4.95x + 2.65y = 200(3.95) \end{cases}$$

Substitute $200 - x$ for y in the second equation.

$4.95x + 2.65(200 - x) = 200(3.95)$
$4.95x + 530 - 2.65x = 790$
$2.3x = 260$
$x \approx 113.04$
Substitute 113.04 for x in the first equation.
$113 + y \approx 200$
$y \approx 86.96$
To the nearest pound, he needs 113 pounds of the $4.95 per pound beans and 87 pounds of the $2.65 per pound beans.

25. $4^2 = 16$

27. $(6x)^2 = 6^2 \cdot x^2 = 36x^2$

29. $\left(10y^3\right)^2 = (10)^2 \cdot y^{3 \cdot 2} = 100y^6$

31. Let $x =$ first angle
    $y =$ second angle
    $\begin{cases} x + y = 90 \\ x = 2y \end{cases}$
    Let $x = 2y$ in the first equation.
    $2y + y = 90$
    $3y = 90$
    $y = 30$
    Let $y = 30$ in the second equation.
    $x = 2(30)$
    $x = 60$
    The angles measure $60°$ and $30°$.

33. Let $x =$ first angle
    $y =$ second angle
    $\begin{cases} x + y = 90 \\ y = 3x + 10 \end{cases}$
    Let $y = 3x + 10$ in the first equation.
    $x + 3x + 10 = 90$
    $4x = 80$
    $x = 20$
    Let $x = 20$ in the second equation.
    $y = 3(20) + 10$
    $y = 70$
    The angles measure $20°$ and $70°$.

35. Let $x =$ number sold at $9.50
    $y =$ number sold at $7.50
    $\begin{cases} x + y = 90 \\ 9.5x + 7.5y = 721 \end{cases}$
    Substitute $90 - y$ for x in the second equation.
    $9.5(90 - y) + 7.5y = 721$
    $855 - 9.5y + 7.5y = 721$
    $-2y = -134$
    $y = 67$
    Let $y = 67$ in the first equation.
    $x + 67 = 90$
    $x = 23$
    They sold 23 at $9.50 each and 67 at $7.50 each.

37. Let $x =$ width
    $y =$ length
    $\begin{cases} 2x + y = 33 \\ y = 2x - 3 \end{cases}$
    Substitute $2x - 3$ for y in the first equation.
    $2x + 2x - 3 = 33$
    $4x = 36$
    $x = 9$
    Let $x = 9$ in the second equation.
    $y = 2(9) - 3$
    $y = 15$
    The width is 9 feet and the length is 15 feet.

39. Answers may vary.

Chapter 7 Review

1. **a.** $\begin{array}{ll} 2x - 3y = 12 & 3x + 4y = 1 \\ 2(12) - 3(4) \stackrel{?}{=} 12 & 3(12) + 4(4) \stackrel{?}{=} 1 \\ 24 - 12 \stackrel{?}{=} 12 & 36 + 16 \stackrel{?}{=} 1 \\ 12 = 12 & 52 = 1 \\ \text{True} & \text{False} \end{array}$

   No, $(12, 4)$ is not a solution of the system.

   b. $\begin{array}{ll} 2x - 3y = 12 & 3x + 4y = 1 \\ 2(3) - 3(-2) \stackrel{?}{=} 12 & 3(3) + 4(-2) \stackrel{?}{=} 1 \\ 6 + 6 \stackrel{?}{=} 12 & 9 - 8 \stackrel{?}{=} 1 \\ 12 = 12 & 1 = 1 \\ \text{True} & \text{True} \end{array}$

   Yes, $(3, -2)$ is a solution of the system.

2. a. $4x + y = 0$ $-8x - 5y = 9$

$4\left(\dfrac{3}{4}\right) + (-3) \stackrel{?}{=} 0$ $-8\left(\dfrac{3}{4}\right) - 5(-3) \stackrel{?}{=} 9$

$3 - 3 \stackrel{?}{=} 0$ $-6 + 15 \stackrel{?}{=} 9$

$0 = 0$ $9 = 9$

True   True

Yes, $\left(\dfrac{3}{4}, -3\right)$ is a solution of the system.

b. $4x + y = 0$ $-8x - 5y = 9$

$4(-2) + 8 \stackrel{?}{=} 0$ $-8(-2) - 5(8) \stackrel{?}{=} 9$

$-8 + 8 \stackrel{?}{=} 0$ $16 - 40 \stackrel{?}{=} 9$

$0 = 0$ $-24 = 9$

True   False

No, $(-2, 8)$ is not a solution of the system.

3. a. $5x - 6y = 18$ $2y - x = -4$

$5(-6) - 6(-8) \stackrel{?}{=} 18$ $2(-8) - (-6) \stackrel{?}{=} -4$

$-30 + 48 \stackrel{?}{=} 18$ $-16 + 6 \stackrel{?}{=} -4$

$18 = 18$ $-10 = 4$

True   False

No, $(-6, -8)$ is not a solution of the system.

b. $5x - 6y = 18$ $2y - x = -4$

$5(3) - 6\left(\dfrac{5}{2}\right) \stackrel{?}{=} 18$ $2\left(\dfrac{5}{2}\right) - 3 \stackrel{?}{=} -4$

$15 - 15 \stackrel{?}{=} 18$ $5 - 3 \stackrel{?}{=} -4$

$0 = 18$ $2 = -4$

False   False

No, $\left(3, \dfrac{5}{2}\right)$ is not a solution of the system.

4. a. $2x + 3y = 1$ $3y - x = 4$

$2(2) + 3(2) \stackrel{?}{=} 1$ $3(2) - 2 \stackrel{?}{=} 4$

$4 + 6 \stackrel{?}{=} 1$ $6 - 2 \stackrel{?}{=} 4$

$10 = 1$ $4 = 4$

False   True

No, $(2, 2)$ is not a solution of the system.

b. $2x + 3y = 1$ $3y - x = 4$

$2(-1) + 3(1) \stackrel{?}{=} 1$ $3(1) - (-1) \stackrel{?}{=} 4$

$-2 + 3 \stackrel{?}{=} 1$ $3 + 1 \stackrel{?}{=} 4$

$1 = 1$ $4 = 4$

True   True

Yes, $(-1, 1)$ is a solution of the system.

5. $\begin{cases} x + y = 5 \\ x - y = 1 \end{cases}$

Graph each linear equation on a single set of axes.

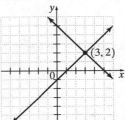

The solution is the intersection point of the two lines, $(3, 2)$.

6. $\begin{cases} x + y = 3 \\ x - y = -1 \end{cases}$

Graph each linear equation on a single set of axes.

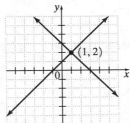

The solution is the intersection point of the two lines, $(1, 2)$.

7. $\begin{cases} x = 5 \\ y = -1 \end{cases}$

Graph each linear equation on a single set of axes.

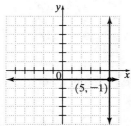

The solution is the intersection point of the two lines (5, –1).

8. $\begin{cases} x = -3 \\ y = 2 \end{cases}$

Graph each linear equation on a single set of axes.

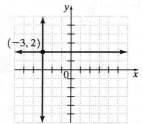

The solution is the intersection point of the two lines, (–3, 2).

9. $\begin{cases} 2x + y = 5 \\ x = -3y \end{cases}$

Graph each linear equation on a single set of axes.

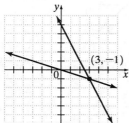

The solution is the intersection point of the two lines, (3, –1).

10. $\begin{cases} 3x + y = -2 \\ y = -5x \end{cases}$

Graph each linear equation on a single set of axes.

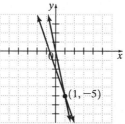

The solution is the intersection point of the two lines, (1, –5).

11. $\begin{cases} y = 2x + 4 \\ y = -x - 5 \end{cases}$

Graph each linear equation on a single set of axes.

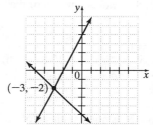

The solution is the intersection point of the two lines, (–3, –2).

12. $\begin{cases} y = x - 5 \\ y = -2x + 2 \end{cases}$

Graph each linear equation on a single set of axes.

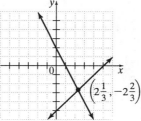

The solution is the intersection point of the

two lines, $\left(2\frac{1}{3}, -2\frac{2}{3} \right)$.

13. $\begin{cases} y = 3x \\ -6x + 2y = 6 \end{cases}$

Graph each linear equation on a single set of axes.

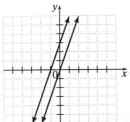

Since the lines are parallel, the system has no solution.

14. $\begin{cases} x - 2y = 2 \\ -2x + 4y = -4 \end{cases}$

Graph each linear equation on a single set of axes.

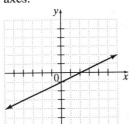

The lines are the same. The system has an infinite number of solutions.

15. $\begin{cases} x = 2y \\ 2x - 3y = 2 \end{cases}$

Substitute $2y$ for x in the second equation.
$2(2y) - 3y = 2$
$4y - 3y = 2$
$y = 2$
Let $y = 2$ in the first equation.
$x = 2(2)$
$x = 4$
The solution is $(4, 2)$.

16. $\begin{cases} x = 5y \\ x - 4y = 1 \end{cases}$

Substitute $5y$ for x in the second equation.
$5y - 4y = 1$
$y = 1$
Let $y = 1$ in the first equation.
$x = 5(1)$
$x = 5$
The solution is $(5, 1)$.

17. $\begin{cases} y = 2x + 6 \\ 3x - 2y = -11 \end{cases}$

Substitute $2x + 6$ for y in the second equation.
$3x - 2(2x + 6) = -11$
$3x - 4x - 12 = -11$
$-x = 1$
$x = -1$
Let $x = -1$ in the first equation.
$y = 2(-1) + 6$
$y = 4$
The solution is $(-1, 4)$.

18. $\begin{cases} y = 3x - 7 \\ 2x - 3y = 7 \end{cases}$

Substitute $3x - 7$ for y in the second equation.
$2x - 3(3x - 7) = 7$
$2x - 9x + 21 = 7$
$-7x = -14$
$x = 2$
Let $x = 2$ in the first equation.
$y = 3(2) - 7$
$y = -1$
The solution is $(2, -1)$.

19. $\begin{cases} x + 3y = -3 \\ 2x + y = 4 \end{cases}$

Solve the first equation for x.
$x = -3 - 3y$
Substitute $-3 - 3y$ for x in the second equation.
$2(-3 - 3y) + y = 4$
$-6 - 6y + y = 4$
$-5y = 10$
$y = -2$
Let $y = -2$ in $x = -3 - 3y$.
$x = -3 - 3(-2)$
$x = -3 + 6$
$x = 3$
The solution is $(3, -2)$.

20. $\begin{cases} 3x + y = 11 \\ x + 2y = 12 \end{cases}$

Solve the first equation for y.
$y = 11 - 3x$
Substitute $11 - 3x$ for y in the second equation.

$x + 2(11 - 3x) = 12$
$x + 22 - 6x = 12$
$-5x = -10$
$x = 2$
Let $x = 2$ in $y = 11 - 3x$.
$y = 11 - 3(2)$
$y = 5$
The solution is (2, 5).

21. $\begin{cases} 4y = 2x - 3 \\ x - 2y = 4 \end{cases}$

Solve the second equation for x.
$x = 2y + 4$
Substitute $2y + 4$ for x in the first equation.
$4y = 2(2y + 4) - 3$
$4y = 4y + 8 - 3$
$0 = 5$
Since this is false, the system has no solution.

22. $\begin{cases} 2x = 3y - 18 \\ x + 4y = 2 \end{cases}$

Solve the second equation for x.
$x = 2 - 4y$
Substitute $2 - 4y$ for x in the first equation.
$2(2 - 4y) = 3y - 18$
$4 - 8y = 3y - 18$
$-11y = -22$
$y = 2$
Let $y = 2$ in $x = 2 - 4y$.
$x = 2 - 4(2)$
$x = -6$
The solution is (-6, 2).

23. $\begin{cases} x + y = 6 \\ y = -x - 4 \end{cases}$

Substitute $-x - 4$ for y in the first equation.
$x + (-x - 4) = 6$
$x - x - 4 = 6$
$-4 = 6$
Since this is false, the system has no solution.

24. $\begin{cases} -3x + y = 6 \\ y = 3x + 2 \end{cases}$

Substitute $3x + 2$ for y in the first equation.
$-3x + 3x + 2 = 6$
$2 = 6$
Since this is false, the system has no solution.

25. $\begin{cases} x + y = 14 \\ x - y = 18 \end{cases}$
$\overline{2x = 32}$
$x = 16$

Let $x = 16$ in the first equation.
$16 + y = 14$
$y = -2$
The solution is (16, -2).

26. $\begin{cases} x + y = 9 \\ x - y = 13 \end{cases}$
$\overline{2x = 22}$
$x = 11$

Let $x = 11$ in the first equation.
$11 + y = 9$
$y = -2$
The solution is (11, -2).

27. $\begin{cases} 2x + 3y = -6 \\ x - 3y = -12 \end{cases}$
$\overline{3x = -18}$
$x = -6$

Let $x = -6$ in the second equation.
$-6 - 3y = -12$
$-3y = -6$
$y = 2$
The solution is (-6, 2).

28. $\begin{cases} 4x + y = 15 \\ -4x + 3y = -19 \end{cases}$
$\overline{4y = -4}$
$y = -1$

Let $y = -1$ in the first equation.
$4x - 1 = 15$
$4x = 16$
$x = 4$
The solution is (4, -1).

29. $\begin{cases} 2x - 3y = -15 \\ x + 4y = 31 \end{cases}$

$\begin{cases} 2x - 3y = -15 \\ -2x - 8y = -62 \end{cases}$
$\overline{-11y = -77}$
$y = 7$

Let $y = 7$ in the second equation.

$x + 4(7) = 31$
$x = 3$
The solution is (3, 7).

30. $\begin{cases} x - 5y = -22 \\ 4x + 3y = 4 \end{cases}$

$\begin{cases} -4x + 20y = 88 \\ 4x + 3y = 4 \end{cases}$
$23y = 92$
$y = 4$

Let $y = 4$ in the first equation.
$x - 5(4) = -22$
$x = -2$
The solution is (-2, 4).

31. $\begin{cases} 2x - 6y = -1 \\ -x + 3y = \frac{1}{2} \end{cases}$

$\begin{cases} 2x - 6y = -1 \\ -2x + 6y = 1 \end{cases}$
$0 = 0$

Since this is always true, the system has an infinite number of solutions.

32. $\begin{cases} -4x - 6y = 8 \\ 2x + 3y = -3 \end{cases}$

$\begin{cases} -4x - 6y = 8 \\ 4x + 6y = -6 \end{cases}$
$0 = 2$

Since this is false, the system has no solution.

33. $\begin{cases} \frac{3}{4}x + \frac{2}{3}y = 2 \\ 3x + y = 18 \end{cases}$

$\begin{cases} 9x + 8y = 24 \\ -9x - 3y = -54 \end{cases}$
$5y = -30$
$y = -6$

Let $y = -6$ in the second equation.
$3x - 6 = 18$
$3x = 24$
$x = 8$
The solution is (8, -6).

34. $\begin{cases} \frac{2}{5}x + \frac{3}{4}y = 1 \\ x + 3y = -2 \end{cases}$

$\begin{cases} 8x + 15y = 20 \\ -5x - 15y = 10 \end{cases}$
$3x = 30$
$x = 10$

Let $x = 10$ in the second equation.
$10 + 3y = -2$
$3y = -12$
$y = -4$
The solution is (10, -4).

35. $\begin{cases} 5x + 2y = 9 \\ 3x + 4y = 25 \end{cases}$

$\begin{cases} -10x - 4y = -18 \\ 3x + 4y = 25 \end{cases}$
$-7x = 7$
$x = -1$

Let $x = -1$ in the first equation.
$5(-1) + 2y = 9$
$2y = 14$
$y = 7$
The solution is (-1, 7).

36. $\begin{cases} 6x - 3y = -15 \\ 4x - 2y = -10 \end{cases}$

$\begin{cases} 12x - 6y = -30 \\ -12x + 6y = 30 \end{cases}$
$0 = 0$

Since this is always true, the system has an infinite number of solutions.

37. Let $x =$ smaller number
$ y =$ larger number

$\begin{cases} x + y = 16 \\ 3y - x = 72 \end{cases}$

Solve the first equation for x.
$x = 16 - y$
Let $x = 16 - y$ in the second equation.
$3y - (16 - y) = 72$
$3y - 16 + y = 72$
$4y = 88$
$y = 22$

Let $y = 22$ in $x = 16 - y$.
$x = 16 - 22$
$x = -6$
The numbers are –6 and 22.

38. Let $x =$ number of orchestra tickets
$y =$ number of balcony tickets

$\begin{cases} x + y = 360 \\ 45x + 35y = 15,150 \end{cases}$

Solve the first equation for x.
$x = 360 - y$
Let $x = 360 - y$ in the second equation.
$45(360 - y) + 35y = 15,150$
$16,200 - 45y + 35y = 15,150$
$-10y = -1050$
$y = 105$
Let $y = 105$ in $x = 360 - y$.
$x = 360 - 105$
$x = 255$
There are 255 orchestra seats and 105 balcony seats.

39. Let $x =$ speed of ship in still water, in miles per
hour
$y =$ current of river in miles per hour

$\begin{cases} 19(x - y) = 340 \\ 14(x + y) = 340 \end{cases}$

$\begin{cases} 19x - 19y = 340 \\ 14x + 14y = 340 \end{cases}$

$\begin{cases} 266x - 266y = 4760 \\ 266x + 266y = 6460 \end{cases}$

$\overline{\quad 532x \qquad = 11,220}$
$\qquad x \approx 21.1$

Substitute 21.1 for x in the first equation.
$19(21.1 - y) \approx 340$
$-19y \approx -60.9$
$y \approx 3.2$
The current of the river is 3.2 mph and the speed of the ship is 21.1 mph.

40. Let $x =$ amount invested at 6%
$y =$ amount invested at 10%

$\begin{cases} x + y = 9000 \\ 0.06x + 0.1y = 652.8 \end{cases}$

Solve the first equation for y.
$y = 9000 - x$
Let $y = 9000 - x$ in the second equation.

$0.06x + 0.1(9000 - x) = 652.8$
$0.06x + 900 - 0.1x = 652.8$
$-0.04x = -247.2$
$x = 6180$
Let $x = 6180$ in $y = 9000 - x$.
$y = 9000 - 6180$
$y = 2820$
They invested $6180 at 6% and $2820 at 10%.

41. Let $x =$ length
$y =$ width

$\begin{cases} 2x + 2y = 6 \\ x = 1.6y \end{cases}$

Let $x = 1.6y$ in the first equation.
$2(1.6y) + 2y = 6$
$5.2y = 6$
$y \approx 1.154$
Let $y = 1.154$ in the second equation.
$x \approx 1.6(1.154)$
$x \approx 1.846$
The length is about 1.85 feet and the width is 1.15 feet.

42. Let $x =$ amount of 6% solution
$y =$ amount of 14% solution

$\begin{cases} x + y = 50 \\ 0.06x + 0.14y = 0.12(50) \end{cases}$

Solve the first equation for y.
$y = 50 - x$
Let $y = 50 - x$ in the second equation.
$0.06x + 0.14(50 - x) = 0.12(50)$
$0.06x + 7 - 0.14x = 6$
$-0.08x = -1$
$x = 12.5$
Let $x = 12.5$ in $y = 50 - x$.
$y = 50 - 12.5$
$y = 37.5$

Pat should combine $12\frac{1}{2}$ cc of 6% solution

and $37\frac{1}{2}$ cc of 14% solution.

43. Let $x =$ cost of each egg
$y =$ cost of each strip of bacon

$$\begin{cases} 3x + 4y = 3.80 \\ 2x + 3y = 2.75 \end{cases}$$

$$\begin{cases} -6x - 8y = -7.60 \\ \underline{6x + 9y = 8.25} \\ y = 0.65 \end{cases}$$

Let $y = 0.65$ in the first equation.
$3x + 4(0.65) = 3.80$
$3x = 1.2$
$x = 0.4$
Each egg costs 40¢ and each strip of bacon costs 65¢.

44. Let $x =$ number of hours spent jogging
$y =$ number of hours spent walking

$$\begin{cases} x + y = 3 \\ 7.5x + 4y = 15 \end{cases}$$

$$\begin{cases} -4x - 4y = -12 \\ \underline{7.5x + 4y = 15} \\ 3.5x = 3 \\ x \approx 0.86 \end{cases}$$

Subsitute 0.86 for x in the first equation.
$0.86 + y \approx 3$
$y \approx 2.14$
He spent 0.86 hour jogging and 2.14 hours walking.

Chapter 7 Test

1. $\begin{cases} x - y = 2 \\ 3x - y = -2 \end{cases}$

Graph each linear equation on a single set of axes.

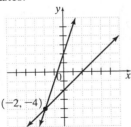

$(-2, -4)$

The solution is the intersection point of the two lines, $(-2, -4)$.

2. $\begin{cases} y = -3x \\ 3x + y = 6 \end{cases}$

Graph each linear equation on a single set of axes.

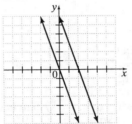

Since the two lines are parallel, there is no solution.

3. $\begin{cases} 3x - 2y = -14 \\ y = x + 5 \end{cases}$

Substitute $x + 5$ for y in the first equation.
$3x - 2(x + 5) = -14$
$3x - 2x - 10 = -14$
$x = -4$
Let $x = -4$ in the second equation.
$y = -4 + 5$
$y = 1$
The solution is $(-4, 1)$.

4. $\begin{cases} 3x + y = 7 \\ 4x + 3y = 1 \end{cases}$

Solve the first equation for y.
$y = 7 - 3x$
Let $y = 7 - 3x$ in the second equation.
$4x + 3(7 - 3x) = 1$
$4x + 21 - 9x = 1$
$-5x = -20$
$x = 4$
Let $x = 4$ in $y = 7 - 3x$.
$y = 7 - 3(4)$
$y = -5$
The solution is $(4, -5)$.

5. $\begin{cases} x - y = 4 \\ x - 2y = 11 \end{cases}$

Solve the first equation for x.
$x = y + 4$
Let $x = y + 4$ in the second equation.
$y + 4 - 2y = 11$
$-y = 7$
$y = -7$

Let $y = -7$ in $x = y + 4$.
$x = -7 + 4$
$x = -3$
The solution is $(-3, -7)$.

6. $\begin{cases} 8x - 4y = 12 \\ y = 2x - 3 \end{cases}$

Let $y = 2x - 3$ in the first equation.
$8x - 4(2x - 3) = 12$
$8x - 8x + 12 = 12$
$0 = 0$
Since this is always true, the system has an infinite number of solutions.

7. $\begin{cases} x + y = 28 \\ x - y = 12 \end{cases}$
$\overline{2x = 40}$
$x = 20$

Let $x = 20$ in the first equation.
$20 + y = 28$
$y = 8$
The solution is $(20, 8)$.

8. $\begin{cases} y - x = 6 \\ y + 2x = -6 \end{cases}$

$\begin{cases} 2y - 2x = 12 \\ y + 2x = -6 \end{cases}$
$\overline{3y = 6}$
$y = 2$

Let $y = 2$ in the first equation.
$2 - x = 6$
$x = -4$
The solution is $(-4, 2)$.

9. $\begin{cases} 5x - 6y = 7 \\ 7x - 4y = 12 \end{cases}$

$\begin{cases} -10x + 12y = -14 \\ 21x - 12y = 36 \end{cases}$
$\overline{11x = 22}$
$x = 2$

Let $x = 2$ in the first equation.
$5(2) - 6y = 7$
$-6y = -3$
$y = \dfrac{1}{2}$

The solution is $\left(2, \dfrac{1}{2}\right)$.

10. $\begin{cases} x - \dfrac{2}{3}y = 3 \\ -2x + 3y = 10 \end{cases}$

$\begin{cases} 9x - 6y = 27 \\ -4x + 6y = 20 \end{cases}$
$\overline{5x = 47}$
$x = \dfrac{47}{5}$
$x = 9\dfrac{2}{5}$

$\begin{cases} 6x - 4y = 18 \\ -6x + 9y = 30 \end{cases}$
$\overline{5y = 48}$
$y = \dfrac{48}{5}$
$y = 9\dfrac{3}{5}$

The solution is $\left(9\dfrac{2}{5}, 9\dfrac{3}{5}\right)$.

11. Let $x =$ larger number
    $y =$ small number

$\begin{cases} x + y = 124 \\ x - y = 32 \end{cases}$
$\overline{2x = 156}$
$x = 78$

Let $x = 78$ in the first equation.
$78 + y = 124$
$y = 46$
The numbers are 78 and 46.

12. Let $x =$ number of \$1 bills
    $y =$ number of \$5 bills

$\begin{cases} x + y = 62 \\ x + 5y = 230 \end{cases}$
Solve the first equation for x.
$x = 62 - y$
Let $x = 62 - y$ in the second equation.
$62 - y + 5y = 230$
$4y = 168$
$y = 42$
Let $y = 42$ in $x = 62 - y$.
$x = 62 - 42$
$x = 20$
She had 20 \$1 bills and 42 \$5 bills.

13. Let $x =$ amount of 30% solution
$y =$ amount of 70% solution
$$\begin{cases} x + y = 10 \\ 0.3x + 0.7y = 0.4(10) \end{cases}$$
Solve the first equation for x.
$x = 10 - y$
Let $x = 10 - y$ in the second equation.
$0.3(10 - y) + 0.7y = 0.4(10)$
$3 - 0.3y + 0.7y = 4$
$0.4y = 1$
$y = 2.5$
Let $y = 2.5$ in $x = 10 - y$.
$x = 10 - 2.5$
$x = 7.5$
There should be 7.5 liters of 30% solution and 2.5 liters of 70% solution.

14. Let $x =$ slow hiker's speed
$y =$ fast hiker's speed
$$\begin{cases} 4x + 4y = 36 \\ y = 2x \end{cases}$$
Let $y = 2x$ in the first equation.
$4x + 4(2x) = 36$
$12x = 36$
$x = 3$
Let $x = 3$ in $y = 2x$.
$y = 2(3)$
$y = 6$
The speeds are 3 mph and 6 mph.

15. The graphs intersect at point where x-value is 1992. Therefore, the earnings are equal in 1992.

16. The line representing Illinois is below that of Wisconsin for the years after 1992. Therefore, Illinois earnings was less for years 1993–1995.

Chapter 7 Cumulative Review

1. a. $-14 - 8 + 10 - (-6) = -14 + (-8) + 10 + 6$
$= -6$

   b. $1.6 - (-10.3) + (-5.6) = 1.6 + 10.3 + (-5.6) = 6.3$

2. The reciprocal of 22 is $\frac{1}{22}$ since
$22 \cdot \frac{1}{22} = 1.$

3. The reciprocal of $\frac{3}{16}$ is $\frac{16}{3}$ since
$\frac{3}{16} \cdot \frac{16}{3} = 1.$

4. The reciprocal of -10 is $-\frac{1}{10}$ since
$-10 \cdot -\frac{1}{10} = 1.$

5. The reciprocal of $-\frac{9}{13}$ is $-\frac{13}{9}$ since
$-\frac{9}{13} \cdot -\frac{13}{9} = 1.$

6. The reciprocal of 1.7 is $\frac{1}{1.7}$ since
$1.7 \cdot \frac{1}{1.7} = 1.$

7. a. If the sum of two numbers is 8 and one number is 3, the other number is $8 - 3 = 5$.

   b. If the sum of two numbers is 8 and one number is x, the other number is $8 - x$.

8. $-2(x - 5) + 10 = -3(x + 2) + x$
$-2x + 10 + 10 = -3x - 6 + x$
$-2x + 20 = -2x - 6$
$-2x + 20 + 2x = -2x - 6 + 2x$
$20 = -6$
Since there is no value for x that makes $20 = -6$ a true statement, there is no solution to the equation.

9. a. The ratio of 2 parts salt to 5 parts water is $\frac{2}{5}$.

b. 2 feet $= 2 \cdot 12$ inches $= 24$ inches
The ratio of 18 inches to 2 feet is then
$\dfrac{18}{24}$, which is $\dfrac{3}{4}$ in lowest terms.

10. $-5x + 7 < 2(x - 3)$
$-5x + 7 < 2x - 6$
$-5x + 7 - 2x < 2x - 6 - 2x$
$-7x + 7 < -6$
$-7x + 7 - 7 < -6 - 7$
$-7x < -13$
$\dfrac{-7x}{-7} > \dfrac{-13}{-7}$
$x > \dfrac{13}{7}$

11. $\left(\dfrac{m}{n}\right)^7 = \dfrac{m^7}{n^7}, \; n \neq 0$

12. $\left(\dfrac{2x^4}{3y^5}\right)^4 = \dfrac{2^4 \cdot (x^4)^4}{3^4 \cdot (y^5)^4} = \dfrac{16x^{16}}{81y^{20}}, \; n \neq 0$

13. $(2x^3 + 8x^2 - 6x) - (2x^3 - x^2 + 1)$
$= 2x^3 + 8x^2 - 6x - 2x^3 + x^2 - 1$
$= 2x^3 - 2x^3 + 8x^2 + x^2 - 6x - 1$
$= 9x^2 - 6x - 1$

14.
$$
\begin{array}{r}
2x + 4 \\
3x - 1 \overline{)6x^2 + 10x - 5} \\
\underline{6x^2 - 2x} \\
12x - 5 \\
\underline{12x - 4} \\
-1
\end{array}
$$

Thus, $\dfrac{6x^2 + 10x - 5}{3x - 1} = 2x + 4 - \dfrac{1}{3x - 1}$.

15. $x(2x - 7) = 4$
$2x^2 - 7x = 4$
$2x^2 - 7x - 4 = 0$
$(2x + 1)(x - 4) = 0$
$2x + 1 = 0 \qquad$ or $\quad x - 4 = 0$
$2x = -1 \qquad\qquad\qquad x = 4$
$x = -\dfrac{1}{2}$

The solutions are $x = -\dfrac{1}{2}$ and $x = 4$.

16. Let $x =$ the length of the shorter leg. Then
$x + 2 =$ the length of the longer leg, and
$x + 4 =$ the length of the hypotenuse.
By the Pythagorean theorem, we have
$(x + 4)^2 = x^2 + (x + 2)^2$
$x^2 + 8x + 16 = x^2 + x^2 + 4x + 4$
$x^2 + 8x + 16 = 2x^2 + 4x + 4$
$x^2 - 4x - 12 = 0$
$(x - 6)(x + 2) = 0$
$x - 6 = 0 \qquad$ or $\quad x + 2 = 0$
$x = 6 \qquad\qquad\qquad x = -2$
Since the length cannot be negative, we discard the result -2. The sides of the triangle have lengths 6 units, 8 units, and 10 units.

17. $\dfrac{2y}{2y - 7} - \dfrac{7}{2y - 7} = \dfrac{2y - 7}{2y - 7} = 1$

18. $\dfrac{\frac{x}{y} + \frac{3}{2x}}{\frac{x}{2} + y} = \dfrac{2xy\left(\frac{x}{y} + \frac{3}{2x}\right)}{2xy\left(\frac{x}{2} + y\right)} = \dfrac{2xy\left(\frac{x}{y}\right) + 2xy\left(\frac{3}{2x}\right)}{2xy\left(\frac{x}{2}\right) + 2xy(y)}$

$= \dfrac{2x^2 + 3y}{x^2 y + 2xy^2}$

19. To find the slope, select any two points on the line. Since $y = -1$ for every value of x, two such points are $(x_1, y_1) = (0, -1)$ and $(x_2, y_2) = (2, -1)$.

$$m = \frac{y_2 - y_1}{x_2 - x_1}$$

$$m = \frac{-1 - (-1)}{2 - 0}$$

$$m = \frac{-1 + 1}{2}$$

$$m = \frac{0}{2}$$

$$m = 0$$

Thus, the slope is 0.

20. Let $(x_1, y_1) = (2, 5)$ and $(x_2, y_2) = (-3, 4)$.

$$m = \frac{y_2 - y_1}{x_2 - x_1}$$

$$m = \frac{4 - 5}{-3 - 2}$$

$$m = \frac{-1}{-5}$$

$$m = \frac{1}{5}$$

Using point-slope form with $m = \frac{1}{5}$, we

have

$$y - y_1 = m(x - x_1)$$

$$y - 5 = \frac{1}{5}(x - 2)$$

$$y - 5 = \frac{1}{5}x - \frac{2}{5}$$

$$5(y - 5) = 5\left(\frac{1}{5}x - \frac{2}{5}\right)$$

$$5y - 25 = x - 2$$

$$-x + 5y = 23$$

21. The domain is $\{-1, 0, 3\}$.
The range is $\{-2, 0, 2, 3\}$.

22. Replace x with 12 and y with 6 in both equations. For the first equation we obtain,
$$2x - 3y = 6$$
$$2(12) - 3(6) \overset{?}{=} 6$$
$$24 - 18 \overset{?}{=} 6$$
$$6 = 6$$
True

$$x = 2y$$
$$12 \overset{?}{=} 2(6)$$
$$12 = 12$$
True
Since $(12, 6)$ is a solution of both equations, it is a solution of the system.

23. $\begin{cases} x + 2y = 7 \\ 2x + 2y = 13 \end{cases}$

Solve the first equation for x.
$$x + 2y = 7$$
$$x = 7 - 2y$$
Substitute $7 - 2y$ for x in the second equation and solve for y.
$$2x + 2y = 13$$
$$2(7 - 2y) + 2y = 13$$
$$14 - 4y + 2y = 13$$
$$14 - 2y = 13$$
$$-2y = -1$$
$$y = \frac{1}{2}$$

To find x, let $y = \frac{1}{2}$ in the first equation.

$$x + 2y = 7$$
$$x + 2\left(\frac{1}{2}\right) = 7$$
$$x + 1 = 7$$
$$x = 6$$

The solution is $\left(6, \frac{1}{2}\right)$.

24. $\begin{cases} -x - \dfrac{y}{2} = \dfrac{5}{2} \\ -\dfrac{x}{2} + \dfrac{y}{4} = 0 \end{cases}$

Multiply both sides of the first equation by 2, and multiply both sides of the second equation by 4 to obtain a simplified system without fractions. Then add the resulting equations.
$$\begin{array}{r} -2x - y = 5 \\ -2x + y = 0 \\ \hline -4x = 5 \end{array}$$
$$x = -\frac{5}{4}$$

To find y, multiply the first equation in the simplified system by -1. Then add the resulting equations.

$$\begin{cases} 2x + y = -5 \\ -2x + y = 0 \end{cases}$$
$$\overline{2y = -5}$$
$$y = -\frac{5}{2}$$

The solution is $\left(-\dfrac{5}{4}, -\dfrac{5}{2}\right)$.

25. Let $x =$ the first number and let $y =$ the second number.
$$\begin{cases} x + y = 37 \\ x - y = 21 \end{cases}$$
Add the equations.
$$x + y = 37$$
$$\underline{x - y = 21}$$
$$2x = 58$$
$$x = 29$$

To find y, let $x = 29$ in the first equation.
$$x + y = 37$$
$$29 + y = 37$$
$$y = 8$$
The numbers are 29 and 8.

Chapter 8

1. $-\sqrt{49} = -7$

2. $\sqrt{\dfrac{4}{25}} = \dfrac{2}{5}$

3. $\sqrt[3]{-64} = -4$

4. $\sqrt{120} = \sqrt{4 \cdot 30} = \sqrt{4} \cdot \sqrt{30} = 2\sqrt{30}$

5. $\sqrt{\dfrac{24}{y^6}} = \dfrac{\sqrt{24}}{\sqrt{y^6}} = \dfrac{\sqrt{4} \cdot \sqrt{6}}{\sqrt{(y^3)^2}} = \dfrac{2\sqrt{6}}{y^3}$

6. $\sqrt[3]{112} = \sqrt[3]{8 \cdot 14} = \sqrt[3]{8} \cdot \sqrt[3]{14} = 2\sqrt[3]{14}$

7. $\sqrt{15} + 2\sqrt{15} - 6\sqrt{15} = (1 + 2 - 6)\sqrt{15}$
$= -3\sqrt{15}$

8. $3\sqrt{12} - 2\sqrt{27} = 3\sqrt{4 \cdot 3} - 2\sqrt{9 \cdot 3}$
$= 3\sqrt{4} \cdot \sqrt{3} - 2\sqrt{9} \cdot \sqrt{3} = 6\sqrt{3} - 6\sqrt{3}$
$= (6 - 6)\sqrt{3} = 0$

9. $\sqrt{\dfrac{7}{4}} + \sqrt{\dfrac{7}{25}} = \dfrac{\sqrt{7}}{\sqrt{4}} + \dfrac{\sqrt{7}}{\sqrt{25}} = \dfrac{\sqrt{7}}{2} + \dfrac{\sqrt{7}}{5}$
$= \dfrac{5\sqrt{7}}{5 \cdot 2} + \dfrac{2\sqrt{7}}{2 \cdot 5} = \dfrac{5\sqrt{7}}{10} + \dfrac{2\sqrt{7}}{10}$
$= \dfrac{(5 + 2)\sqrt{7}}{10} = \dfrac{7\sqrt{7}}{10}$

10. $\sqrt{6} \cdot \sqrt{18} = \sqrt{6 \cdot 18} = \sqrt{108} = \sqrt{36 \cdot 3}$
$= \sqrt{36} \cdot \sqrt{3} = 6\sqrt{3}$

11. $\sqrt{2}\left(\sqrt{14} - \sqrt{5}\right) = \sqrt{2} \cdot \sqrt{14} - \sqrt{2} \cdot \sqrt{5}$
$= \sqrt{28} - \sqrt{10} = \sqrt{4} \cdot \sqrt{7} - \sqrt{10}$
$= 2\sqrt{7} - \sqrt{10}$

12. $\left(\sqrt{y} - 3\right)^2 = \left(\sqrt{y} - 3\right)\left(\sqrt{y} - 3\right)$
$= \sqrt{y} \cdot \sqrt{y} - 3\sqrt{y} - 3\sqrt{y} - (-3)(3)$
$= y - 6\sqrt{y} + 9$

13. $\dfrac{\sqrt{56x^5}}{\sqrt{2x^3}} = \sqrt{\dfrac{56x^5}{2x^3}} = \sqrt{28x^2} = \sqrt{4 \cdot 7 \cdot x^2}$
$= 2x\sqrt{7}$

14. $\sqrt{\dfrac{5}{11}} = \dfrac{\sqrt{5}}{\sqrt{11}} = \dfrac{\sqrt{5} \cdot \sqrt{11}}{\sqrt{11} \cdot \sqrt{11}} = \dfrac{\sqrt{55}}{11}$

15. $\dfrac{16}{\sqrt{2a}} = \dfrac{16 \cdot \sqrt{2a}}{\sqrt{2a} \cdot \sqrt{2a}} = \dfrac{16\sqrt{2a}}{2a} = \dfrac{8\sqrt{2a}}{a}$

16. $\dfrac{3}{2 - \sqrt{x}} = \dfrac{3\left(2 + \sqrt{x}\right)}{\left(2 - \sqrt{x}\right)\left(2 + \sqrt{x}\right)} = \dfrac{6 + 3\sqrt{x}}{4 - x}$

17. $\sqrt{x} + 9 = 16$
$\sqrt{x} = 7$
$\left(\sqrt{x}\right)^2 = 7^2$
$x = 49$

18. $\sqrt{x + 4} = \sqrt{x} + 1$
$\left(\sqrt{x + 4}\right)^2 = \left(\sqrt{x} + 1\right)^2$
$x + 4 = x + 2\sqrt{x} + 1$
$4 = 2\sqrt{x} + 1$
$3 = 2\sqrt{x}$
$\dfrac{3}{2} = \sqrt{x}$
$\left(\dfrac{3}{2}\right)^2 = \left(\sqrt{x}\right)^2$
$x = \dfrac{9}{4}$

19. Let b = the length of the unknown leg.
By the Pythagorean theorem we have
$$a^2 + b^2 = c^2$$
$$6^2 + b^2 = 14^2$$
$$36 + b^2 = 196$$
$$b^2 = 160$$
$$b = \sqrt{160}$$
$$b = \sqrt{16} \cdot \sqrt{10}$$
$$b = 4\sqrt{10}$$
The length of the unknown leg is $4\sqrt{10}$ cm.

20. Let $S = 80$.
$$r = \sqrt{\frac{S}{4\pi}}$$
$$r = \sqrt{\frac{80}{4\pi}}$$
$$r = \sqrt{\frac{20}{\pi}}$$
$$r = \frac{\sqrt{4} \cdot \sqrt{5}}{\sqrt{\pi}} \approx 2.52$$
The radius of the sphere is exactly $\dfrac{2\sqrt{5}}{\sqrt{\pi}}$ in.,
or approximately 2.52 in.

Section 8.1

Practice Problems

1. $\sqrt{100} = 10$ because $10^2 = 100$ and 10 is positive.

2. $\sqrt{9} = 3$ because $3^2 = 9$ and 3 is positive.

3. $-\sqrt{36} = -6$
The negative sign in front of the radical indicates the negative square root of 36.

4. $\sqrt{\dfrac{25}{81}} = \dfrac{5}{9}$ because $\left(\dfrac{5}{9}\right)^2 = \dfrac{25}{81}$ and $\dfrac{5}{9}$ is positive.

5. $\sqrt{1} = 1$ because $1^2 = 1$ and 1 is positive.

6. $\sqrt[3]{27} = 3$ because $3^3 = 27$.

7. $\sqrt[3]{-8} = -2$ because $(-2)^3 = -8$.

8. $\sqrt[3]{\dfrac{1}{64}} = \dfrac{1}{4}$ because $\left(\dfrac{1}{4}\right)^3 = \dfrac{1}{64}$.

9. $\sqrt[4]{-16}$ is not a real number since the index is even and the radicand is negative.

10. $\sqrt[5]{-1} = -1$ because $(-1)^5 = -1$.

11. $\sqrt[4]{81} = 3$ because $3^4 = 81$.

12. $\sqrt[6]{-64}$ is not a real number since the index is even and the radicand is negative.

13. $\sqrt{10} \approx 3.162$

14. $\sqrt{x^8} = x^4$ because $(x^4)^2 = x^8$.

15. $\sqrt{x^{20}} = x^{10}$ because $(x^{10})^2 = x^{20}$.

16. $\sqrt{4x^6} = 2x^3$ because $(2x^3)^2 = 4x^6$.

17. $\sqrt[3]{8y^{12}} = 2y^4$ because $(2y^4)^3 = 8y^{12}$.

Calculator Explorations

1. $\sqrt{7} \approx 2.646$

2. $\sqrt{14} \approx 3.742$

3. $\sqrt{11} \approx 3.317$

4. $\sqrt{200} \approx 14.142$

5. $\sqrt{82} \approx 9.055$

6. $\sqrt{46} \approx 6.782$

7. $\sqrt[3]{40} \approx 3.420$

8. $\sqrt[3]{71} \approx 4.141$

9. $\sqrt[4]{20} \approx 2.115$

10. $\sqrt[4]{15} \approx 1.968$

11. $\sqrt[5]{18} \approx 1.783$

12. $\sqrt[8]{2} \approx 1.122$

Exercise Set 8.1

1. $\sqrt{16} = 4$ because $4^2 = 16$ and 4 is positive.

3. $\sqrt{81} = 9$ because $9^2 = 81$ and 9 is positive.

5. $\sqrt{\frac{1}{25}} = \frac{1}{5}$ because $\left(\frac{1}{5}\right)^2 = \frac{1}{25}$ and $\frac{1}{5}$ is positive.

7. $-\sqrt{100} = -10$ because the negative sign in front of the radical indicates the negative square root of 100.

9. $\sqrt{-4}$ is not a real number because the index is even and the radicand is negative.

11. $-\sqrt{121} = -11$ because the negative sign in front of the radicals indicates the negative square root of 121.

13. $\sqrt{\frac{9}{25}} = \frac{3}{5}$ because $\left(\frac{3}{5}\right)^2 = \frac{9}{25}$ and $\frac{3}{5}$ is positive.

15. $\sqrt[3]{125} = 5$ because $5^3 = 125$.

17. $\sqrt[3]{64} = -4$ because $(-4)^3 = -64$

19. $-\sqrt[3]{8} = -2$. because $\sqrt[3]{8} = 2$.

21. $\sqrt[3]{\frac{1}{8}} = \frac{1}{2}$ because $\left(\frac{1}{2}\right)^3 = \frac{1}{8}$.

23. $\sqrt[3]{-125} = -5$ because $(-5)^3 = -125$.

25. Answers may vary.

27. $\sqrt[5]{32} = 2$ because $2^5 = 32$.

29. $\sqrt[4]{-16}$ is not a real number.

31. $-\sqrt[4]{625} = -5$ because $\sqrt[4]{625} = 5$.

33. $\sqrt[6]{1} = 1$ because $1^6 = 1$.

35. $\sqrt{37} \approx 6.083$

37. $\sqrt{136} \approx 11.662$

39. $\sqrt{2} \approx 1.41$
$90 \cdot \sqrt{2} \approx 90(1.41) = 126.90$ feet

41. $\sqrt{z^2} = z$ because $(z)^2 = z^2$.

43. $\sqrt{x^4} = x^2$ because $\left(x^2\right)^2 = x^4$.

45. $\sqrt{9x^8} = 3x^4$ because $\left(3x^4\right)^2 = 9x^8$.

47. $\sqrt{81x^2} = 9x$ because $(9x)^2 = 81x^2$.

49. $50 = 25 \cdot 2$, 25 is a perfect square.

51. $32 = 16 \cdot 2$, 16 is a perfect square or $32 = 4 \cdot 8$, 4 is a perfect square.

53. $28 = 4 \cdot 7$, 4 is a perfect square.

55. $27 = 9 \cdot 3$, 9 is a perfect square.

57. $\sqrt{\sqrt{81}} = \sqrt{9} = 3$

59.

x	$y = \sqrt{x}$
0	0
1	1
3	1.7(approx.)
4	2
9	3

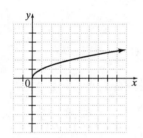

61. $y = \sqrt{x - 2}$

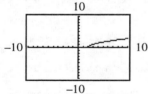

The graph starts at the point $(2, 0)$. $x - 2$ is greater than or equal to zero for $x \geq 2$.

63. $y = \sqrt{x + 4}$

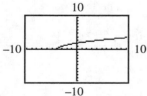

The graph starts at the point $(-4, 0)$. $x + 4$ is greater than or equal to zero for $x \geq -4$.

65. $\sqrt[3]{195,112} = 58$

because $(58)^3 = 195,112$. Each side would be 58 feet.

Section 8.2

Practice Problems

1. $\sqrt{40} = \sqrt{4 \cdot 10} = \sqrt{4} \cdot \sqrt{10} = 2\sqrt{10}$

2. $\sqrt{18} = \sqrt{9 \cdot 2} = \sqrt{9} \cdot \sqrt{2} = 3\sqrt{2}$

3. $\sqrt{700} = \sqrt{100 \cdot 7} = \sqrt{100} \cdot \sqrt{7} = 10\sqrt{7}$

4. $\sqrt{15}$ cannot be simplified.

5. $\sqrt{\dfrac{16}{81}} = \dfrac{\sqrt{16}}{\sqrt{81}} = \dfrac{4}{9}$

6. $\sqrt{\dfrac{2}{25}} = \dfrac{\sqrt{2}}{\sqrt{25}} = \dfrac{\sqrt{2}}{5}$

7. $\sqrt{\dfrac{45}{49}} = \dfrac{\sqrt{45}}{\sqrt{49}} = \dfrac{\sqrt{9} \cdot \sqrt{5}}{7} = \dfrac{3\sqrt{5}}{7}$

8. $\sqrt{x^{11}} = \sqrt{x^{10} \cdot x} = \sqrt{x^{10}} \cdot \sqrt{x} = x^5 \sqrt{x}$

9. $\sqrt{18x^4} = \sqrt{9 \cdot 2 \cdot x^4} = \sqrt{9x^4 \cdot 2} = \sqrt{9x^4} \cdot \sqrt{2}$
$= 3x^2\sqrt{2}$

10. $\sqrt{\dfrac{27}{x^8}} = \dfrac{\sqrt{27}}{\sqrt{x^8}} = \dfrac{\sqrt{9 \cdot 3}}{x^4} = \dfrac{\sqrt{9} \cdot \sqrt{3}}{x^4} = \dfrac{3\sqrt{3}}{x^4}$

11. $\sqrt[3]{40} = \sqrt[3]{8 \cdot 5} = \sqrt[3]{8} \cdot \sqrt[3]{5} = 2\sqrt[3]{5}$

12. $\sqrt[3]{50}$ cannot be simplified.

13. $\sqrt[3]{\dfrac{10}{27}} = \dfrac{\sqrt[3]{10}}{\sqrt[3]{27}} = \dfrac{\sqrt[3]{10}}{3}$

14. $\sqrt[3]{\dfrac{81}{8}} = \dfrac{\sqrt[3]{81}}{\sqrt[3]{8}} = \dfrac{\sqrt[3]{27 \cdot 3}}{2} = \dfrac{\sqrt[3]{27} \cdot \sqrt[3]{3}}{2}$
$= \dfrac{3\sqrt[3]{3}}{2}$

Mental Math

1. $\sqrt{4 \cdot 9} = 6$

2. $\sqrt{9 \cdot 36} = 18$

3. $\sqrt{x^2} = x$

4. $\sqrt{y^4} = y^2$

5. $\sqrt{0} = 0$

6. $\sqrt{1} = 1$

7. $\sqrt{25x^4} = 5x^2$

8. $\sqrt{49x^2} = 7x$

Exercise Set 8.2

1. $\sqrt{20} = \sqrt{4 \cdot 5} = \sqrt{4} \cdot \sqrt{5} = 2\sqrt{5}$

3. $\sqrt{18} = \sqrt{9 \cdot 2} = \sqrt{9} \cdot \sqrt{2} = 3\sqrt{2}$

5. $\sqrt{50} = \sqrt{25 \cdot 2} = \sqrt{25} \cdot \sqrt{2} = 5\sqrt{2}$

7. $\sqrt{33}$ cannot be simplified further.

9. $\sqrt{60} = \sqrt{4 \cdot 15} = \sqrt{4} \cdot \sqrt{15} = 2\sqrt{15}$

11. $\sqrt{180} = \sqrt{36 \cdot 5} = \sqrt{36} \cdot \sqrt{5} = 6\sqrt{5}$

13. $\sqrt{52} = \sqrt{4 \cdot 13} = \sqrt{4} \cdot \sqrt{13} = 2\sqrt{13}$

15. $\sqrt{\dfrac{8}{25}} = \dfrac{\sqrt{8}}{\sqrt{25}} = \dfrac{\sqrt{4} \cdot \sqrt{2}}{5} = \dfrac{2\sqrt{2}}{5}$

17. $\sqrt{\dfrac{27}{121}} = \dfrac{\sqrt{27}}{\sqrt{121}} = \dfrac{\sqrt{9} \cdot \sqrt{3}}{11} = \dfrac{3\sqrt{3}}{11}$

19. $\sqrt{\dfrac{9}{4}} = \dfrac{\sqrt{9}}{\sqrt{4}} = \dfrac{3}{2}$

21. $\sqrt{\dfrac{125}{9}} = \dfrac{\sqrt{125}}{\sqrt{9}} = \dfrac{\sqrt{25} \cdot \sqrt{5}}{3} = \dfrac{5\sqrt{5}}{3}$

23. $\sqrt{\dfrac{11}{36}} = \dfrac{\sqrt{11}}{\sqrt{36}} = \dfrac{\sqrt{11}}{6}$

25. $-\sqrt{\dfrac{27}{144}} = -\dfrac{\sqrt{27}}{\sqrt{144}} = -\dfrac{\sqrt{9} \cdot \sqrt{3}}{12} = -\dfrac{3\sqrt{3}}{12}$

$= -\dfrac{\sqrt{3}}{4}$

27. $\sqrt{x^7} = \sqrt{x^6 \cdot x} = \sqrt{x^6} \cdot \sqrt{x} = x^3\sqrt{x}$

29. $\sqrt{x^{13}} = \sqrt{x^{12} \cdot x} = \sqrt{x^{12}} \cdot \sqrt{x} = x^6\sqrt{x}$

31. $\sqrt{75x^2} = \sqrt{25x^2 \cdot 3} = \sqrt{25x^2} \cdot \sqrt{3} = 5x\sqrt{3}$

33. $\sqrt{96x^4} = \sqrt{16x^4 \cdot 6} = \sqrt{16x^4} \cdot \sqrt{6}$

$= 4x^2\sqrt{6}$

35. $\sqrt{\dfrac{12}{y^2}} = \dfrac{\sqrt{12}}{\sqrt{y^2}} = \dfrac{\sqrt{4} \cdot \sqrt{3}}{y} = \dfrac{2\sqrt{3}}{y}$

37. $\sqrt{\dfrac{9x}{y^2}} = \dfrac{\sqrt{9x}}{\sqrt{y^2}} = \dfrac{\sqrt{9} \cdot \sqrt{x}}{y} = \dfrac{3\sqrt{x}}{y}$

39. $\sqrt{\dfrac{88}{x^4}} = \dfrac{\sqrt{88}}{\sqrt{x^4}} = \dfrac{\sqrt{4} \cdot \sqrt{22}}{x^2} = \dfrac{2\sqrt{22}}{x^2}$

41. $\sqrt[3]{24} = \sqrt[3]{8 \cdot 3} = \sqrt[3]{8} \cdot \sqrt[3]{3} = 2\sqrt[3]{3}$

43. $\sqrt[3]{250} = \sqrt[3]{125 \cdot 2} = \sqrt[3]{125} \cdot \sqrt[3]{2} = 5\sqrt[3]{2}$

45. $\sqrt[3]{\dfrac{5}{64}} = \dfrac{\sqrt[3]{5}}{\sqrt[3]{64}} = \dfrac{\sqrt[3]{5}}{4}$

47. $\sqrt[3]{\dfrac{7}{8}} = \dfrac{\sqrt[3]{7}}{\sqrt[3]{8}} = \dfrac{\sqrt[3]{7}}{2}$

49. $\sqrt[3]{\dfrac{15}{64}} = \dfrac{\sqrt[3]{15}}{\sqrt[3]{64}} = \dfrac{\sqrt[3]{15}}{4}$

51. $\sqrt[3]{80} = \sqrt[3]{8 \cdot 10} = \sqrt[3]{8} \cdot \sqrt[3]{10} = 2\sqrt[3]{10}$

53. $6x + 8x = (6 + 8)x = 14x$

55. $(2x + 3)(x - 5) = 2x(x - 5) + 3(x - 5)$
$= 2x^2 - 10x + 3x - 15 = 2x^2 - 7x - 15$

57. $9y^2 - 9y^2 = 0$

59. $\sqrt{x^6 y^3} = \sqrt{x^6 y^2 \cdot y} = \sqrt{x^6 y^2} \cdot \sqrt{y}$
$= x^3 y \sqrt{y}$

61. $\sqrt{x^2 + 4x + 4} = \sqrt{(x + 2)^2} = x + 2$

63. $\sqrt[3]{80} = \sqrt[3]{8 \cdot 10} = \sqrt[3]{8} \cdot \sqrt[3]{10} = 2\sqrt[3]{10}$
Each side has length $2\sqrt[3]{10}$ in.

65. $C = 100\sqrt[3]{n} + 700$
$C = 100\sqrt[3]{1000} + 700$
$C = 100(10) + 700$
$C = 1000 + 700$
$C = 1700$
The cost is $1700.

67. Answers may vary.

Section 8.3

Practice Problems

1. $6\sqrt{11} + 9\sqrt{11} = (6 + 9)\sqrt{11} = 15\sqrt{11}$

2. $\sqrt{7} - 3\sqrt{7} = 1\sqrt{7} - 3\sqrt{7} = (1 - 3)\sqrt{7} = -2\sqrt{7}$

3. $\sqrt{2} + \sqrt{2} = 2\sqrt{2}$ cannot be simplified.

4. $3\sqrt{3} - 3\sqrt{2}$ cannot be simplified.

5. $\sqrt{27} + \sqrt{75} = \sqrt{9 \cdot 3} + \sqrt{25 \cdot 3}$
$= \sqrt{9} \cdot \sqrt{3} + \sqrt{25} \cdot \sqrt{3} = 3\sqrt{3} + 5\sqrt{3}$
$= 8\sqrt{3}$

6. $3\sqrt{20} - 7\sqrt{45} = 3\sqrt{4 \cdot 5} - 7\sqrt{9 \cdot 5}$
$= 3\sqrt{4} \cdot \sqrt{5} - 7\sqrt{9} \cdot \sqrt{5}$
$= 3(2)\sqrt{5} - 7(3)\sqrt{5} = 6\sqrt{5} - 21\sqrt{5}$
$= -15\sqrt{5}$

7. $\sqrt{36} - \sqrt{48} - 4\sqrt{3} - \sqrt{9}$
$= 6 - \sqrt{16 \cdot 3} - 4\sqrt{3} - 3$
$= 6 - \sqrt{16} \cdot \sqrt{3} - 4\sqrt{3} - 3$
$= 6 - 4\sqrt{3} - 4\sqrt{3} - 3$
$= 3 - 8\sqrt{3}$

8. $\sqrt{9x^4} - \sqrt{36x^3} + \sqrt{x^3}$
$= 3x^2 - \sqrt{36x^2 \cdot x} + \sqrt{x^2 \cdot x}$
$= 3x^2 - \sqrt{36x^2} \cdot \sqrt{x} + \sqrt{x^2} \cdot \sqrt{x}$
$= 3x^2 - 6x\sqrt{x} + x\sqrt{x} = 3x^2 - 5x\sqrt{x}$

Mental Math

1. $3\sqrt{2} + 5\sqrt{2} = 8\sqrt{2}$

2. $3\sqrt{5} + 7\sqrt{5} = 10\sqrt{5}$

3. $5\sqrt{x} + 2\sqrt{x} = 7\sqrt{x}$

4. $8\sqrt{x} + 3\sqrt{x} = 11\sqrt{x}$

5. $5\sqrt{7} - 2\sqrt{7} = 3\sqrt{7}$

6. $8\sqrt{6} - 5\sqrt{6} = 3\sqrt{6}$

Exercise Set 8.3

1. $4\sqrt{3} - 8\sqrt{3} = (4 - 8)\sqrt{3} = -4\sqrt{3}$

3. $3\sqrt{6} + 8\sqrt{6} - 2\sqrt{6} - 5 = (3 + 8 - 2)\sqrt{6} - 5$
$= 9\sqrt{6} - 5$

5. $6\sqrt{5} - 5\sqrt{5} + \sqrt{2} = (6 - 5)\sqrt{5} + \sqrt{2}$
$= \sqrt{5} + \sqrt{2}$

7. $2\sqrt{3} + 5\sqrt{3} - \sqrt{3} = 2\sqrt{3} + 5\sqrt{3} - 1\sqrt{3}$
$= (2 + 5 - 1)\sqrt{3} = 6\sqrt{3}$

9. $2\sqrt{2} - 7\sqrt{2} - 6 = (2 - 7)\sqrt{2} - 6$
$= -5\sqrt{2} - 6$

11. $12\sqrt{5} - \sqrt{5} - 4\sqrt{5} = 12\sqrt{5} - 1\sqrt{5} - 4\sqrt{5}$
$= (12 - 1 - 4)\sqrt{5} = 7\sqrt{5}$

13. $\sqrt{5} + \sqrt{5} = 1\sqrt{5} + 1\sqrt{5} = (1 + 1)\sqrt{5} = 2\sqrt{5}$

15. $6 - 2\sqrt{3} - \sqrt{3} = 6 - 2\sqrt{3} - 1\sqrt{3}$
$= 6 + (-2 - 1)\sqrt{3} = 6 - 3\sqrt{3}$

17. Answers may vary.

19. $\sqrt{12} + \sqrt{27} = \sqrt{4 \cdot 3} + \sqrt{9 \cdot 3}$
$= \sqrt{4} \cdot \sqrt{3} + \sqrt{9} \cdot \sqrt{3} = 2\sqrt{3} + 3\sqrt{3} = 5\sqrt{3}$

21. $\sqrt{45} + 3\sqrt{20} = \sqrt{9 \cdot 5} + 3\sqrt{4 \cdot 5}$
$= \sqrt{9} \cdot \sqrt{5} + 3\sqrt{4} \cdot \sqrt{5} = 3\sqrt{5} + 3 \cdot 2\sqrt{5}$
$= 3\sqrt{5} + 6\sqrt{5} = 9\sqrt{5}$

23. $2\sqrt{54} - \sqrt{20} + \sqrt{45} - \sqrt{24}$
$= 2\sqrt{9 \cdot 6} - \sqrt{4 \cdot 5} + \sqrt{9 \cdot 5} - \sqrt{4 \cdot 6}$
$= 2\sqrt{9} \cdot \sqrt{6} - \sqrt{4} \cdot \sqrt{5} + \sqrt{9} \cdot \sqrt{5} - \sqrt{4} \cdot \sqrt{6}$
$= 2 \cdot 3\sqrt{6} - 2\sqrt{5} + 3\sqrt{5} - 2\sqrt{6}$
$= 6\sqrt{6} - 2\sqrt{5} + 3\sqrt{5} - 2\sqrt{6} = 4\sqrt{6} + \sqrt{5}$

25. $4x - 3\sqrt{x^2} + \sqrt{x} = 4x - 3x + \sqrt{x}$
$= x + \sqrt{x}$

27. $\sqrt{25x} + \sqrt{36x} - 11\sqrt{x}$
$= \sqrt{25 \cdot x} + \sqrt{36 \cdot x} - 11\sqrt{x}$
$= \sqrt{25} \cdot \sqrt{x} + \sqrt{36} \cdot \sqrt{x} - 11\sqrt{x}$
$= 5\sqrt{x} + 6\sqrt{x} - 11\sqrt{x} = 0$

29. $3\sqrt{x^3} - x\sqrt{4x} = 3\sqrt{x^2 \cdot x} - x\sqrt{4 \cdot x}$
$= 3\sqrt{x^2} \cdot \sqrt{x} - x\sqrt{4} \cdot \sqrt{x}$
$= 3x\sqrt{x} - 2x\sqrt{x} = x\sqrt{x}$

31. $\sqrt{75} + \sqrt{48} = \sqrt{25 \cdot 3} + \sqrt{16 \cdot 3}$
$= \sqrt{25} \cdot \sqrt{3} + \sqrt{16} \cdot \sqrt{3} = 5\sqrt{3} + 4\sqrt{3} = 9\sqrt{3}$

33. $\sqrt{8} + \sqrt{9} + \sqrt{18} + \sqrt{81}$
$= \sqrt{4} \cdot \sqrt{2} + 3 + \sqrt{9} \cdot \sqrt{2} + 9$
$= 2\sqrt{2} + 3 + 3\sqrt{2} + 9 = 5\sqrt{2} + 12$

35. $\sqrt{\dfrac{5}{9}} + \sqrt{\dfrac{5}{81}} = \dfrac{\sqrt{5}}{\sqrt{9}} + \dfrac{\sqrt{5}}{\sqrt{81}} = \dfrac{\sqrt{5}}{3} + \dfrac{\sqrt{5}}{9}$
$= \dfrac{3\sqrt{5}}{9} + \dfrac{\sqrt{5}}{9} = \dfrac{3\sqrt{5} + \sqrt{5}}{9} = \dfrac{4\sqrt{5}}{9}$

37. $\sqrt{\dfrac{3}{4}} - \sqrt{\dfrac{3}{64}} = \dfrac{\sqrt{3}}{\sqrt{4}} - \dfrac{\sqrt{3}}{\sqrt{64}} = \dfrac{\sqrt{3}}{2} - \dfrac{\sqrt{3}}{8}$
$= \dfrac{4\sqrt{3}}{8} - \dfrac{\sqrt{3}}{8} = \dfrac{4\sqrt{3} - \sqrt{3}}{8} = \dfrac{3\sqrt{3}}{8}$

39. $2\sqrt{45} - 2\sqrt{20} = 2\sqrt{9 \cdot 5} - 2\sqrt{4 \cdot 5}$
$= 2\sqrt{9} \cdot \sqrt{5} - 2\sqrt{4} \cdot \sqrt{5}$
$= 2 \cdot 3\sqrt{5} - 2 \cdot 2\sqrt{5}$
$= 6\sqrt{5} - 4\sqrt{5} = 2\sqrt{5}$

41. $\sqrt{35} - \sqrt{140} = \sqrt{35} - \sqrt{4 \cdot 35}$
$= \sqrt{35} - \sqrt{4} \cdot \sqrt{35} = \sqrt{35} - 2\sqrt{35}$
$= -\sqrt{35}$

43. $3\sqrt{9x} + 2\sqrt{x} = 3\sqrt{9} \cdot \sqrt{x} + 2\sqrt{x}$
$= 3 \cdot 3\sqrt{x} + 2\sqrt{x} = 9\sqrt{x} + 2\sqrt{x}$
$= 11\sqrt{x}$

45. $\sqrt{9x^2} + \sqrt{81x^2} - 11\sqrt{x}$
$= 3x + 9x - 11\sqrt{x} = 12x - 11\sqrt{x}$

47. $\sqrt{3x^3} + 3x\sqrt{x} = \sqrt{x^2} \cdot \sqrt{3x} + 3x\sqrt{x}$
$= x\sqrt{3x} + 3x\sqrt{x}$

49. $\sqrt{32x^2} + \sqrt{32x^2} + \sqrt{4x^2}$
$= \sqrt{16x^2} \cdot \sqrt{2} + \sqrt{16x^2} \cdot \sqrt{2} + 2x$
$= 4x\sqrt{2} + 4x\sqrt{2} + 2x = 8x\sqrt{2} + 2x$

51. $\sqrt{40x} + \sqrt{40x^4} - 2\sqrt{10x} - \sqrt{5x^4}$

$= \sqrt{4} \cdot \sqrt{10x} + \sqrt{4x^4} \cdot \sqrt{10} - 2\sqrt{10x} - \sqrt{x^4} \cdot \sqrt{5}$

$= 2\sqrt{10x} + 2x^2\sqrt{10} - 2\sqrt{10x} - x^2\sqrt{5}$

$= 2x^2\sqrt{10} - x^2\sqrt{5}$

53. $(x+6)^2 = x^2 + 2(x)(6) + 6^2$

$= x^2 + 12x + 36$

55. $(2x-1)^2 = (2x)^2 - 2(2x)(1) + 1^2$

$4x^2 - 4x + 1$

57. $\begin{cases} x = 2y \\ x + 5y = 14 \end{cases}$

Substitute $2y$ for x in the second equation.

$2y + 5y = 14$

$7y = 14$

$y = 2$

Let $y = 2$ in the first equation

$x = 2(2)$

$x = 4$

The solution is (4, 2).

59. Perimeter = 2(width) + 2(length)

$= 2(\sqrt{5}) + 2(3\sqrt{5})$

$= 2\sqrt{5} + 6\sqrt{5}$

$= (2+6)\sqrt{5}$

$= 8\sqrt{5}$

The perimeter is $8\sqrt{5}$ in.

61. Area = area of two triangles

　　　　　 + area of 2 rectangles

$= 2\left(\dfrac{3\sqrt{27}}{4}\right) + 2(8 \cdot 3)$

$= \dfrac{3\sqrt{9} \cdot \sqrt{3}}{2} + 48$

$= \dfrac{9\sqrt{3}}{2} + 48$

The total area is $\left(\dfrac{9\sqrt{3}}{2} + 48\right)$ sq. ft.

Section 8.4

Practice Problems

1. $\sqrt{5} \cdot \sqrt{2} = \sqrt{5 \cdot 2} = \sqrt{10}$

2. $\sqrt{6} \cdot \sqrt{3} = \sqrt{6 \cdot 3} = \sqrt{18} = \sqrt{9 \cdot 2} = \sqrt{9} \cdot \sqrt{2}$

$= 3\sqrt{2}$

3. $\sqrt{10x} \cdot \sqrt{2x} = \sqrt{10x \cdot 2x} = \sqrt{20x^2}$

$= \sqrt{4x^2 \cdot 5} = \sqrt{4x^2} \cdot \sqrt{5} = 2x\sqrt{5}$

4. a. $\sqrt{7}(\sqrt{7} - \sqrt{3}) = \sqrt{7} \cdot \sqrt{7} - \sqrt{7} \cdot \sqrt{3}$

$= \sqrt{49} - \sqrt{21} = 7 - \sqrt{21}$

b. $(\sqrt{x} + \sqrt{5})(\sqrt{x} - \sqrt{3})$

$= \sqrt{x} \cdot \sqrt{x} - \sqrt{x} \cdot \sqrt{3} + \sqrt{5} \cdot \sqrt{x} - \sqrt{5} \cdot \sqrt{3}$

$= \sqrt{x^2} - \sqrt{3x} + \sqrt{5x} - \sqrt{15}$

$= x - \sqrt{3x} + \sqrt{5x} - \sqrt{15}$

5. a. $(\sqrt{3} + 6)(\sqrt{3} - 6) = (\sqrt{3})^2 - 6^2$

$= 3 - 36 = -33$

b. $(\sqrt{5x} + 4)^2 = (\sqrt{5x})^2 + 2\sqrt{5x}(4) + 4^2$

$= 5x + 8\sqrt{5x} + 16$

6. $\dfrac{\sqrt{15}}{\sqrt{3}} = \sqrt{\dfrac{15}{3}} = \sqrt{5}$

7. $\dfrac{\sqrt{90}}{\sqrt{2}} = \sqrt{\dfrac{90}{2}} = \sqrt{45} = \sqrt{9 \cdot 5} = \sqrt{9} \cdot \sqrt{5}$

$= 3\sqrt{5}$

8. $\dfrac{\sqrt{75x^3}}{\sqrt{5x}} = \sqrt{\dfrac{75x^3}{5x}} = \sqrt{15x^2} = \sqrt{x^2} \cdot \sqrt{15}$

$= x\sqrt{15}$

9. $\dfrac{5}{\sqrt{3}} = \dfrac{5 \cdot \sqrt{3}}{\sqrt{3} \cdot \sqrt{3}} = \dfrac{5\sqrt{3}}{3}$

10. $\dfrac{\sqrt{7}}{\sqrt{20}} = \dfrac{\sqrt{7}}{\sqrt{4}\cdot\sqrt{5}} = \dfrac{\sqrt{7}}{2\cdot\sqrt{5}} = \dfrac{\sqrt{7}\cdot\sqrt{5}}{2\sqrt{5}\cdot\sqrt{5}}$

$= \dfrac{\sqrt{35}}{2\cdot 5} = \dfrac{\sqrt{35}}{10}$

11. $\sqrt{\dfrac{2}{45x}} = \dfrac{\sqrt{2}}{\sqrt{45x}} = \dfrac{\sqrt{2}}{\sqrt{9}\cdot\sqrt{5x}} = \dfrac{\sqrt{2}\cdot\sqrt{5x}}{3\sqrt{5x}\cdot\sqrt{5x}}$

$= \dfrac{\sqrt{2\cdot 5x}}{3(5x)} = \dfrac{\sqrt{10x}}{15x}$

12. $\dfrac{3}{1+\sqrt{7}} = \dfrac{3(1-\sqrt{7})}{(1+\sqrt{7})(1-\sqrt{7})} = \dfrac{3(1-\sqrt{7})}{1^2 - (\sqrt{7})^2}$

$= \dfrac{3(1-\sqrt{7})}{1-7} = \dfrac{3(1-\sqrt{7})}{-6} = \dfrac{1-\sqrt{7}}{-2} = \dfrac{-1+\sqrt{7}}{2}$

13. $\dfrac{\sqrt{2}+5}{\sqrt{2}-1} = \dfrac{(\sqrt{2}+5)(\sqrt{2}+1)}{(\sqrt{2}-1)(\sqrt{2}+1)}$

$= \dfrac{2+\sqrt{2}+5\sqrt{2}+5}{2-1} = 7+6\sqrt{2}$

14. $\dfrac{7}{2-\sqrt{x}} = \dfrac{7(2+\sqrt{x})}{(2-\sqrt{x})(2+\sqrt{x})} = \dfrac{7(2+\sqrt{x})}{4-x}$

Mental Math

1. $\sqrt{2}\cdot\sqrt{3} = \sqrt{6}$

2. $\sqrt{5}\cdot\sqrt{7} = \sqrt{35}$

3. $\sqrt{1}\cdot\sqrt{6} = \sqrt{6}$

4. $\sqrt{7}\cdot\sqrt{x} = \sqrt{7x}$

5. $\sqrt{10}\cdot\sqrt{y} = \sqrt{10y}$

6. $\sqrt{x}\cdot\sqrt{y} = \sqrt{xy}$

Exercise Set 8.4

1. $\sqrt{8}\cdot\sqrt{2} = \sqrt{8\cdot 2} = \sqrt{16} = 4$

3. $\sqrt{10}\cdot\sqrt{5} = \sqrt{10\cdot 5} = \sqrt{50} = \sqrt{25\cdot 2}$
$= \sqrt{25}\cdot\sqrt{2} = 5\sqrt{2}$

5. $\sqrt{6}\cdot\sqrt{6} = \sqrt{6\cdot 6} = \sqrt{36} = 6$

7. $\sqrt{2x}\cdot\sqrt{2x} = \sqrt{2x\cdot 2x} = \sqrt{4x^2} = 2x$

9. $\left(2\sqrt{5}\right)^2 = 2^2\left(\sqrt{5}\right)^2 = 4\cdot 5 = 20$

11. $\left(6\sqrt{x}\right)^2 = 6^2\cdot\left(\sqrt{x}\right)^2 = 36x$

13. $\sqrt{3y}\cdot\sqrt{6x} = \sqrt{3y\cdot 6x} = \sqrt{18xy}$
$= \sqrt{9}\cdot\sqrt{2xy} = 3\sqrt{2xy}$

15. $\sqrt{2xy^2}\cdot\sqrt{8xy} = \sqrt{2xy^2\cdot 8xy}$
$= \sqrt{16x^2y^3} = \sqrt{16x^2y^2}\cdot\sqrt{y} = 4xy\sqrt{y}$

17. $\sqrt{2}\left(\sqrt{5}+1\right) = \sqrt{2}\cdot\sqrt{5} + \sqrt{2}\cdot 1$
$= \sqrt{2\cdot 5} + \sqrt{2} = \sqrt{10} + \sqrt{2}$

19. $\sqrt{10}\left(\sqrt{2}+\sqrt{5}\right) = \sqrt{10}\cdot\sqrt{2} + \sqrt{10}\cdot\sqrt{5}$
$= \sqrt{10\cdot 2} + \sqrt{10\cdot 5} = \sqrt{20} + \sqrt{50}$
$= \sqrt{4}\cdot\sqrt{5} + \sqrt{25}\cdot\sqrt{2} = 2\sqrt{5} + 5\sqrt{2}$

21. $\sqrt{6}\left(\sqrt{5}+\sqrt{7}\right) = \sqrt{6}\cdot\sqrt{5} + \sqrt{6}\cdot\sqrt{7}$
$= \sqrt{6\cdot 5} + \sqrt{6\cdot 7} = \sqrt{30} + \sqrt{42}$

23. $\left(\sqrt{3}+6\right)\left(\sqrt{3}-6\right) = \left(\sqrt{3}\right)^2 - 6^2$
$= 3 - 36 = -33$

25. $\left(\sqrt{3}+\sqrt{5}\right)\left(\sqrt{2}-\sqrt{5}\right)$
$= \sqrt{3}\cdot\sqrt{2} - \sqrt{3}\cdot\sqrt{5} + \sqrt{5}\cdot\sqrt{2} - \sqrt{5}\cdot\sqrt{5}$
$= \sqrt{3\cdot 2} - \sqrt{3\cdot 5} + \sqrt{5\cdot 2} - \sqrt{5\cdot 5}$
$= \sqrt{6} - \sqrt{15} + \sqrt{10} - \sqrt{25}$
$= \sqrt{6} - \sqrt{15} + \sqrt{10} - 5$

27. $\left(2\sqrt{11}+1\right)\left(\sqrt{11}-6\right)$

$= 2\sqrt{11}\cdot\sqrt{11}-2\sqrt{11}\cdot 6+1\cdot\sqrt{11}-1\cdot 6$

$= 2\cdot 11-12\sqrt{11}+\sqrt{11}-6$

$= 22-11\sqrt{11}-6 = 16-11\sqrt{11}$

29. $\left(\sqrt{x}+6\right)\left(\sqrt{x}-6\right)=\left(\sqrt{x}\right)^2-6^2 = x-36$

31. $\left(\sqrt{x}-7\right)^2 =\left(\sqrt{x}\right)^2-2\left(\sqrt{x}\right)(7)+7^2$

$= x-14\sqrt{x}+49$

33. $\left(\sqrt{6y}+1\right)^2 =\left(\sqrt{6y}\right)^2+2\left(\sqrt{6y}\right)(1)+1^2$

$= 6y+2\sqrt{6y}+1$

35. $\dfrac{\sqrt{32}}{\sqrt{2}} = \sqrt{\dfrac{32}{2}} = \sqrt{16} = 4$

37. $\dfrac{\sqrt{21}}{\sqrt{3}} = \sqrt{\dfrac{21}{3}} = \sqrt{7}$

39. $\dfrac{\sqrt{90}}{\sqrt{5}} = \sqrt{\dfrac{90}{5}} = \sqrt{18} = \sqrt{9}\cdot\sqrt{2} = 3\sqrt{2}$

41. $\dfrac{\sqrt{75y^5}}{\sqrt{3y}} = \sqrt{\dfrac{75y^5}{3y}} = \sqrt{25y^4} = 5y^2$

43. $\dfrac{\sqrt{150}}{\sqrt{2}} = \sqrt{\dfrac{150}{2}} = \sqrt{75} = \sqrt{25}\cdot\sqrt{3} = 5\sqrt{3}$

45. $\dfrac{\sqrt{72y^5}}{\sqrt{3y^3}} = \sqrt{\dfrac{72y^5}{3y^3}} = \sqrt{24y^2} = \sqrt{4y^2}\cdot\sqrt{6}$

$= 2y\sqrt{6}$

47. $\dfrac{\sqrt{24x^3y^4}}{\sqrt{2xy}} = \sqrt{\dfrac{24x^3y^4}{2xy}} = \sqrt{12x^2y^3}$

$= \sqrt{4x^2y^2}\cdot\sqrt{3y} = 2xy\sqrt{3y}$

49. $\dfrac{\sqrt{3}}{\sqrt{5}} = \dfrac{\sqrt{3}\cdot\sqrt{5}}{\sqrt{5}\cdot\sqrt{5}} = \dfrac{\sqrt{15}}{5}$

51. $\dfrac{7}{\sqrt{2}} = \dfrac{7\cdot\sqrt{2}}{\sqrt{2}\cdot\sqrt{2}} = \dfrac{7\sqrt{2}}{2}$

53. $\dfrac{1}{\sqrt{6y}} = \dfrac{1\cdot\sqrt{6y}}{\sqrt{6y}\cdot\sqrt{6y}} = \dfrac{\sqrt{6y}}{6y}$

55. $\sqrt{\dfrac{5}{18}} = \dfrac{\sqrt{5}}{\sqrt{18}} = \dfrac{\sqrt{5}}{\sqrt{9}\cdot\sqrt{2}} = \dfrac{\sqrt{5}}{3\sqrt{2}}$

$= \dfrac{\sqrt{5}\cdot\sqrt{2}}{3\sqrt{2}\cdot\sqrt{2}} = \dfrac{\sqrt{10}}{3\cdot 2} = \dfrac{\sqrt{10}}{6}$

57. $\sqrt{\dfrac{3}{x}} = \dfrac{\sqrt{3}}{\sqrt{x}} = \dfrac{\sqrt{3}\cdot\sqrt{x}}{\sqrt{x}\cdot\sqrt{x}} = \dfrac{\sqrt{3x}}{x}$

59. $\sqrt{\dfrac{1}{8}} = \dfrac{\sqrt{1}}{\sqrt{8}} = \dfrac{\sqrt{1}}{\sqrt{4}\cdot\sqrt{2}} = \dfrac{1}{2\sqrt{2}}$

$= \dfrac{1\cdot\sqrt{2}}{2\sqrt{2}\cdot\sqrt{2}} = \dfrac{\sqrt{2}}{2\cdot 2} = \dfrac{\sqrt{2}}{4}$

61. $\sqrt{\dfrac{2}{15}} = \dfrac{\sqrt{2}}{\sqrt{15}} = \dfrac{\sqrt{2}\cdot\sqrt{15}}{\sqrt{15}\cdot\sqrt{15}} = \dfrac{\sqrt{30}}{15}$

63. $\sqrt{\dfrac{3}{20}} = \dfrac{\sqrt{3}}{\sqrt{20}} = \dfrac{\sqrt{3}}{\sqrt{4}\cdot\sqrt{5}} = \dfrac{\sqrt{3}}{2\sqrt{5}}$

$= \dfrac{\sqrt{3}\cdot\sqrt{5}}{2\sqrt{5}\cdot\sqrt{5}} = \dfrac{\sqrt{15}}{2\cdot 5} = \dfrac{\sqrt{15}}{10}$

65. $\dfrac{3x}{\sqrt{2x}} = \dfrac{3x\cdot\sqrt{2x}}{\sqrt{2x}\cdot\sqrt{2x}} = \dfrac{3x\sqrt{2x}}{2x} = \dfrac{3\sqrt{2x}}{2}$

67. $\dfrac{8y}{\sqrt{5}} = \dfrac{8y\cdot\sqrt{5}}{\sqrt{5}\cdot\sqrt{5}} = \dfrac{8y\sqrt{5}}{5}$

69. $\sqrt{\dfrac{y}{12x}} = \dfrac{\sqrt{y}}{\sqrt{12x}} = \dfrac{\sqrt{y}}{\sqrt{4}\cdot\sqrt{3x}} = \dfrac{\sqrt{y}}{2\sqrt{3x}}$

$= \dfrac{\sqrt{y}\cdot\sqrt{3x}}{2\sqrt{3x}\cdot\sqrt{3x}} = \dfrac{\sqrt{3xy}}{2\cdot 3x} = \dfrac{\sqrt{3xy}}{6x}$

71. $\dfrac{3}{\sqrt{2}+1} = \dfrac{3(\sqrt{2}-1)}{(\sqrt{2}+1)(\sqrt{2}-1)} = \dfrac{3(\sqrt{2}-1)}{(\sqrt{2})^2-1^2}$

$= \dfrac{3(\sqrt{2}-1)}{2-1} = \dfrac{3(\sqrt{2}-1)}{1} = 3\sqrt{2}-3$

73. $\dfrac{4}{2-\sqrt{5}} = \dfrac{4(2+\sqrt{5})}{(2-\sqrt{5})(2+\sqrt{5})} = \dfrac{4(2+\sqrt{5})}{2^2-(\sqrt{5})^2}$

$= \dfrac{4(2+\sqrt{5})}{4-5} = \dfrac{4(2+\sqrt{5})}{-1}$

$= -4(2+\sqrt{5}) = -8-4\sqrt{5}$

75. $\dfrac{\sqrt{5}+1}{\sqrt{6}-\sqrt{5}} = \dfrac{(\sqrt{5}+1)(\sqrt{6}+\sqrt{5})}{(\sqrt{6}-\sqrt{5})(\sqrt{6}+\sqrt{5})}$

$= \dfrac{\sqrt{30}+5+\sqrt{6}+\sqrt{5}}{6-5}$

$= \sqrt{30}+5+\sqrt{6}+\sqrt{5}$

77. $\dfrac{\sqrt{3}+1}{\sqrt{2}-1} = \dfrac{(\sqrt{3}+1)(\sqrt{2}+1)}{(\sqrt{2}-1)(\sqrt{2}+1)}$

$= \dfrac{\sqrt{6}+\sqrt{3}+\sqrt{2}+1}{2-1}$

$= \sqrt{6}+\sqrt{3}+\sqrt{2}+1$

79. $\dfrac{5}{2+\sqrt{x}} = \dfrac{5(2-\sqrt{x})}{(2+\sqrt{x})(2-\sqrt{x})}$

$= \dfrac{10-5\sqrt{x}}{4-x}$

81. $\dfrac{3}{\sqrt{x}-4} = \dfrac{3(\sqrt{x}+4)}{(\sqrt{x}-4)(\sqrt{x}+4)}$

$= \dfrac{3\sqrt{x}+12}{x-16}$

83. $x+5 = 7^2$
$x+5 = 49$
$x+5-5 = 49-5$
$x = 44$

85. $4z^2+6z-12 = (2z)^2$
$4z^2+6z-12 = 4z^2$
$4z^2+6z-12-4z^2 = 4z^2-4z^2$
$6z-12 = 0$
$6z-12+12 = 0+12$
$6z = 12$
$\dfrac{6z}{6} = \dfrac{12}{6}$
$z = 2$

87. $9x^2+5x+4 = (3x+1)^2$
$9x^2+5x+4 = 9x^2+6x+1$
$9x^2+5x+4-9x^2 = 9x^2+6x+1-9x^2$
$5x+4 = 6x+1$
$5x+4-5x = 6x+1-5x$
$4 = x+1$
$4-1 = x+1-1$
$3 = x$
$x = 3$

89. Area = length · width
$= 13\sqrt{2} \cdot 5\sqrt{6} = (13 \cdot 5)\sqrt{2 \cdot 6}$
$= 65\sqrt{12} = 65\sqrt{4} \cdot \sqrt{3} = 65 \cdot 2\sqrt{3}$
$= 130\sqrt{3}$
The area is $130\sqrt{3}$ sq. m.

91. Answers may vary.

93. $\dfrac{\sqrt{3}+1}{\sqrt{2}-1} = \dfrac{(\sqrt{3}+1)(\sqrt{3}-1)}{(\sqrt{2}-1)(\sqrt{3}-1)}$

$= \dfrac{3-1}{\sqrt{6}-\sqrt{2}-\sqrt{3}+1}$

$= \dfrac{2}{\sqrt{6}-\sqrt{2}-\sqrt{3}+1}$

Integrated Review

1. $\sqrt{36} = 6$ because $6^2 = 36$ and 6 is positive.

2. $\sqrt{48} = \sqrt{16 \cdot 3} = \sqrt{16} \cdot \sqrt{3} = 4\sqrt{3}$

3. $\sqrt{x^4} = x^2$ because $(x^2)^2 = x^4$.

4. $\sqrt{y^7} = \sqrt{y^6 \cdot y} = \sqrt{y^6} \cdot \sqrt{y} = y^3\sqrt{y}$

5. $\sqrt{16x^2} = 4x$ because $(4x)^2 = 16x^2$.

6. $\sqrt{18x^{11}} = \sqrt{9x^{10} \cdot 2x} = \sqrt{9x^{10}} \cdot \sqrt{2x}$
$= 3x^5\sqrt{2x}$

7. $\sqrt[3]{8} = 2$ because $2^3 = 8$.

8. $\sqrt[4]{81} = 3$ because $3^4 = 81$.

9. $\sqrt[3]{-27} = -3$ because $(-3)^3 = -27$.

10. $\sqrt{-4}$ is not a real number.

11. $\sqrt{\dfrac{11}{9}} = \dfrac{\sqrt{11}}{\sqrt{9}} = \dfrac{\sqrt{11}}{3}$

12. $\sqrt[3]{\dfrac{7}{64}} = \dfrac{\sqrt[3]{7}}{\sqrt[3]{64}} = \dfrac{\sqrt[3]{7}}{4}$

13. $5\sqrt{7} + \sqrt{7} = 5\sqrt{7} + 1\sqrt{7} = (5+1)\sqrt{7} = 6\sqrt{7}$

14. $\sqrt{50} - \sqrt{8}$
$\sqrt{25} \cdot \sqrt{2} - \sqrt{4} \cdot \sqrt{2} = 5\sqrt{2} - 2\sqrt{2}$
$= (5-2)\sqrt{2} = 3\sqrt{2}$

15. $2\sqrt{x} + \sqrt{25x} - \sqrt{36x} + 3x$
$= 2\sqrt{x} + \sqrt{25} \cdot \sqrt{x} - \sqrt{36} \cdot \sqrt{x} + 3x$
$= 2\sqrt{x} + 5\sqrt{x} - 6\sqrt{x} + 3x$
$= \sqrt{x} + 3x$

16. $\sqrt{2} \cdot \sqrt{15} = \sqrt{2 \cdot 15} = \sqrt{30}$

17. $\sqrt{3} \cdot \sqrt{3} = \sqrt{3 \cdot 3} = \sqrt{9} = 3$

18. $\sqrt{3}(\sqrt{11} + 1) = \sqrt{3} \cdot \sqrt{11} + \sqrt{3} \cdot 1$
$= \sqrt{3 \cdot 11} + \sqrt{3}$
$= \sqrt{33} + \sqrt{3}$

19. $(\sqrt{x} - 5)(\sqrt{x} + 2)$
$= \sqrt{x} \cdot \sqrt{x} + 2\sqrt{x} - 5\sqrt{x} - 5 \cdot 2$
$= x - 3\sqrt{x} - 10$

20. $(3 + \sqrt{2})^2 = 3^2 + 2(3)(\sqrt{2}) + (\sqrt{2})^2$
$= 9 + 6\sqrt{2} + 2 = 11 + 6\sqrt{2}$

21. $\dfrac{\sqrt{8}}{\sqrt{2}} = \sqrt{\dfrac{8}{2}} = \sqrt{4} = 2$

22. $\dfrac{\sqrt{45}}{\sqrt{15}} = \sqrt{\dfrac{45}{15}} = \sqrt{3}$

23. $\dfrac{\sqrt{24x^5}}{\sqrt{2x}} = \sqrt{\dfrac{24x^5}{2x}} = \sqrt{12x^4} = \sqrt{4x^4} \cdot \sqrt{3}$
$= 2x^2\sqrt{3}$

24. $\sqrt{\dfrac{1}{6}} = \dfrac{\sqrt{1}}{\sqrt{6}} = \dfrac{1 \cdot \sqrt{6}}{\sqrt{6} \cdot \sqrt{6}} = \dfrac{\sqrt{6}}{6}$

25. $\dfrac{x}{\sqrt{20}} = \dfrac{x}{\sqrt{4} \cdot \sqrt{5}} = \dfrac{x \cdot \sqrt{5}}{2\sqrt{5} \cdot \sqrt{5}} = \dfrac{x\sqrt{5}}{2 \cdot 5} = \dfrac{x\sqrt{5}}{10}$

26. $\dfrac{4}{\sqrt{6} + 1} = \dfrac{4(\sqrt{6} - 1)}{(\sqrt{6} + 1)(\sqrt{6} - 1)} = \dfrac{4(\sqrt{6} - 1)}{6 - 1}$
$= \dfrac{4\sqrt{6} - 4}{5}$

27. $\dfrac{\sqrt{2} + 1}{\sqrt{x} - 5} = \dfrac{(\sqrt{2} + 1)(\sqrt{x} + 5)}{(\sqrt{x} - 5)(\sqrt{x} + 5)}$
$= \dfrac{\sqrt{2} \cdot \sqrt{x} + \sqrt{2} \cdot 5 + 1 \cdot \sqrt{x} + 1 \cdot 5}{x - 5^2}$
$= \dfrac{\sqrt{2x} + 5\sqrt{2} + \sqrt{x} + 5}{x - 25}$

Section 8.5

Practice Problems

1. $\sqrt{x-2} = 7$

$\left(\sqrt{x-2}\right)^2 = 7^2$

$x - 2 = 49$

$x = 51$

2. $\sqrt{x} + 9 = 2$

$\sqrt{x} = -7$

$\left(\sqrt{x}\right)^2 = (-7)^2$

$x = 49$

Check:

$\sqrt{49} + 9 = 2$

$7 + 9 = 2$

$16 = 2$

This is false. It is an extraneous solution.

3. $\sqrt{6x-1} = \sqrt{x}$

$\left(\sqrt{6x-1}\right)^2 = \left(\sqrt{x}\right)^2$

$6x - 1 = x$

$5x = 1$

$x = \dfrac{1}{5}$

4. $\sqrt{9y^2 + 2y - 10} = 3y$

$\left(\sqrt{9y^2 + 2y - 10}\right)^2 = (3y)^2$

$9y^2 + 2y - 10 = 9y^2$

$2y - 10 = 0$

$2y = 10$

$y = 5$

5. $\sqrt{x+1} - x = -5$

$\sqrt{x+1} = x - 5$

$\left(\sqrt{x+1}\right)^2 = (x-5)^2$

$x + 1 = x^2 - 10x + 25$

$0 = x^2 - 11x + 24$

$0 = (x-8)(x-3)$

$x - 8 = 0 \qquad$ or $\quad x - 3 = 0$

$x = 8 \qquad\qquad$ or $\quad x = 3$

Check:

$\sqrt{8+1} - 8 = -5 \qquad\qquad \sqrt{3+1} - 3 = -5$

$\sqrt{9} - 8 = -5 \qquad\qquad\quad \sqrt{4} - 3 = -5$

$3 - 8 = -5 \qquad\qquad\quad\ 2 - 3 = -5$

$-5 = -5 \qquad\qquad\qquad\ -1 = -5$

True $\qquad\qquad\qquad\qquad$ False

The solution is 8.

6. $\sqrt{x} + 3 = \sqrt{x+15}$

$\left(\sqrt{x} + 3\right)^2 = \left(\sqrt{x+15}\right)^2$

$x + 6\sqrt{x} + 9 = x + 15$

$6\sqrt{x} = 6$

$\sqrt{x} = 1$

$\left(\sqrt{x}\right)^2 = 1^2$

$x = 1$

Exercise Set 8.5

1. $\sqrt{x} = 9$

$\left(\sqrt{x}\right)^2 = 9^2$

$x = 81$

3. $\sqrt{x+5} = 2$

$\left(\sqrt{x+5}\right)^2 = 2^2$

$x + 5 = 4$

$x = -1$

5. $\sqrt{2x+6} = 4$

$\left(\sqrt{2x+6}\right)^2 = 4^2$

$2x + 6 = 16$

$2x = 10$

$x = 5$

7. $\sqrt{x} - 2 = 5$

$\sqrt{x} = 7$

$\left(\sqrt{x}\right)^2 = 7^2$

$x = 49$

9. $3\sqrt{x} + 5 = 2$

$3\sqrt{x} = -3$

$\sqrt{x} = -1$

There is no solution since $\sqrt{x}$ cannot equal a negative number.

11. $\sqrt{x+6} + 1 = 3$

$\sqrt{x+6} = 2$

$\left(\sqrt{x+6}\right)^2 = 2^2$

$x + 6 = 4$

$x = -2$

13. $\sqrt{2x+1} + 3 = 5$

$\sqrt{2x+1} = 2$

$\left(\sqrt{2x+1}\right)^2 = 2^2$

$2x + 1 = 4$

$2x = 3$

$x = \dfrac{3}{2}$

15. $\sqrt{x} + 3 = 7$

$\sqrt{x} = 4$

$\left(\sqrt{x}\right)^2 = 4^2$

$x = 16$

17. $\sqrt{x+6} + 5 = 3$

$\sqrt{x+6} = -2$

There is no solution since the result of a square root cannot be negative.

19. $\sqrt{4x-3} = \sqrt{x+3}$

$\left(\sqrt{4x-3}\right)^2 = \left(\sqrt{x+3}\right)^2$

$4x - 3 = x + 3$

$3x = 6$

$x = 2$

21. $\sqrt{x} = \sqrt{3x-8}$

$\left(\sqrt{x}\right)^2 = \left(\sqrt{3x-8}\right)^2$

$x = 3x - 8$

$-2x = -8$

$x = 4$

23. $\sqrt{4x} = \sqrt{2x+6}$

$\left(\sqrt{4x}\right)^2 = \left(\sqrt{2x+6}\right)^2$

$4x = 2x + 6$

$2x = 6$

$x = 3$

25. $\sqrt{9x^2 + 2x - 4} = 3x$

$\left(\sqrt{9x^2 + 2x - 4}\right)^2 = (3x)^2$

$9x^2 + 2x - 4 = 9x^2$

$2x = 4$

$x = 2$

27. $\sqrt{16x^2 - 3x + 6} = 4x$

$\left(\sqrt{16x^2 - 3x + 6}\right)^2 = (4x)^2$

$16x^2 - 3x + 6 = 16x^2$

$3x = 6$

$x = 2$

29. $\sqrt{16x^2 + 2x + 2} = 4x$

$\left(\sqrt{16x^2 + 2x + 2}\right)^2 = (4x)^2$

$16x^2 + 2x + 2 = 16x^2$

$2x = -2$

$x = -1$

A check shows that $x = -1$ is an extraneous solution. Therefore, there is no solution.

31. $\sqrt{2x^2 + 6x + 9} = 3$

$\left(\sqrt{2x^2 + 6x + 9}\right)^2 = 3^2$

$2x^2 + 6x + 9 = 9$

$2x(x + 3) = 0$

$2x = 0 \quad$ or $\quad x + 3 = 0$

$\quad x = 0 \quad$ or $\qquad x = -3$

33. $\sqrt{x+7} = x+5$

$\left(\sqrt{x+7}\right)^2 = (x+5)^2$

$x+7 = x^2 + 10x + 25$

$x^2 + 9x + 18 = 0$

$(x+6)(x+3) = 0$

$x+6 = 0$ or $x+3 = 0$

     $x = -6$ (extraneous) $x = -3$

35. $\sqrt{x} = x-6$

$\left(\sqrt{x}\right)^2 = (x-6)^2$

$x = x^2 - 12x + 36$

$0 = x^2 - 13x + 36$

$0 = (x-9)(x-4)$

$x-9 = 0$ or $x-4 = 0$

   $x = 9$ $x = 4$ (extraneous)

37. $\sqrt{2x+1} = x-7$

$\left(\sqrt{2x+1}\right)^2 = (x-7)^2$

$2x+1 = x^2 - 14x + 49$

$0 = x^2 - 16x + 48$

$0 = (x-12)(x-4)$

$x-12 = 0$ or $x-4 = 0$

    $x = 12$ $x = 4$ (extraneous)

39. $x = \sqrt{2x-2} + 1$

$x-1 = \sqrt{2x-2}$

$(x-1)^2 = \left(\sqrt{2x-2}\right)^2$

$x^2 - 2x + 1 = 2x - 2$

$x^2 - 4x + 3 = 0$

$(x-3)(x-1) = 0$

$x-3 = 0$ or $x-1 = 0$

   $x = 3$ or $x = 1$

41. $\sqrt{1-8x} - x = 4$

$\sqrt{1-8x} = x+4$

$\left(\sqrt{1-8x}\right)^2 = (x+4)^2$

$1 - 8x = x^2 + 8x + 16$

$0 = x^2 + 16x + 15$

$0 = (x+15)(x+1)$

$x + 15 = 0$ or $x+1 = 0$

   $x = -15$ (extraneous) $x = -1$

43. $\sqrt{2x+5} - 1 = x$

$\sqrt{2x+5} = 1+x$

$\left(\sqrt{2x+5}\right)^2 = (1+x)^2$

$2x+5 = 1 + 2x + x^2$

$x^2 = 4$

$x = 2$ or $x = -2$ (extraneous)

45. $\sqrt{x-7} = \sqrt{x} - 1$

$\left(\sqrt{x-7}\right)^2 = \left(\sqrt{x}-1\right)^2$

$x-7 = x - 2\sqrt{x} + 1$

$2\sqrt{x} = 8$

$\sqrt{x} = 4$

$\left(\sqrt{x}\right)^2 = 4^2$

$x = 16$

47. $\sqrt{x} + 3 = \sqrt{x+15}$

$\left(\sqrt{x}+3\right)^2 = \left(\sqrt{x+15}\right)^2$

$x + 6\sqrt{x} + 9 = x + 15$

$6\sqrt{x} = 6$

$\sqrt{x} = 1$

$\left(\sqrt{x}\right)^2 = 1^2$

$x = 1$

49. $\sqrt{x+8} = \sqrt{x} + 2$

$\left(\sqrt{x+8}\right)^2 = \left(\sqrt{x}+2\right)^2$

$x+8 = x + 4\sqrt{x} + 4$

$4 = 4\sqrt{x}$

$1 = \sqrt{x}$

$1^2 = \left(\sqrt{x}\right)^2$

$1 = x$

$x = 1$

51. $3x - 8 = 19$
$3x - 8 + 8 = 19 + 8$
$3x = 27$
$\dfrac{3x}{3} = \dfrac{27}{3}$
$x = 9$

53. Let x = width
$2x$ = length
$2(2x + x) = 24$
$4x + 2x = 24$
$6x = 24$
$\dfrac{6x}{6} = \dfrac{24}{6}$
$x = 4$
$2x = 8$
The length is 8 in.

55. a. $b = \sqrt{\dfrac{V}{2}}$

$b = \sqrt{\dfrac{20}{2}} \approx 3.2$

$b = \sqrt{\dfrac{200}{2}} = 10$

$b = \sqrt{\dfrac{2000}{2}} \approx 31.6$

V	20	200	2000
b	3.2	10	31.6

b. No; it increases by a factor of $\sqrt{10}$.

57. Answers may vary.

59. $\sqrt{x+1} = 2x - 3$
$y_1 = \sqrt{x+1}$
$y_2 = 2x - 3$

The solution is the x-value of the intersection, 2.43.

61. $-\sqrt{x+5} = -7x + 1$
$y_1 = -\sqrt{x+5}$
$y_2 = -7x + 1$

The solution is the x-value of the intersection, 0.48.

Section 8.6

Practice Problems

1. $a^2 + b^2 = c^2$
$3^2 + 4^2 = c^2$
$9 + 16 = c^2$
$25 = c^2$
$5 = c$
The length is 5 centimeters.

2. $a^2 + b^2 = c^2$
$3^2 + b^2 = 6^2$
$9 + b^2 = 36$
$b^2 = 27$
$b = \sqrt{27}$
$b = 3\sqrt{3}$
The length is $3\sqrt{3}$ miles ≈ 5.20 mi.

3. $a^2 + b^2 = c^2$
$(40)^2 + b^2 = (65)^2$
$1600 + b^2 = 4225$
$b^2 = 2625$
$b = \sqrt{2625}$
$b \approx 51.2$
The distance is 51.2 feet.

4. $v = \sqrt{2gh}$

$v = \sqrt{2 \cdot 32 \cdot 20}$

$v = \sqrt{1280}$

$v = \sqrt{256 \cdot 5}$

$v = 16\sqrt{5} \approx 35.8$

The velocity is exactly $16\sqrt{5}$ feet per second or approximately 35.8 feet per second.

Exercise Set 8.6

1. $a^2 + b^2 = c^2$

$2^2 + 3^2 = c^2$

$4 + 9 = c^2$

$13 = c^2$

$\sqrt{13} = c$

The hypotenuse has a length of $\sqrt{13} \approx 3.61$.

3. $a^2 + b^2 = c^2$

$3^2 + b^2 = 6^2$

$9 + b^2 = 36$

$b^2 = 27$

$b = \sqrt{27}$

$b = 3\sqrt{3}$

The unknown side has a length of $3\sqrt{3} \approx 5.20$.

5. $a^2 + b^2 = c^2$

$7^2 + 24^2 = c^2$

$49 + 576 = c^2$

$625 = c^2$

$\sqrt{625} = c$

$25 = c$

The hypotenuse has a length of 25.

7. $a^2 + b^2 = c^2$

$a^2 + \left(\sqrt{3}\right)^2 = 5^2$

$a^2 + 3 = 25$

$a^2 = 22$

$a = \sqrt{22}$

The unknown side has a length of $\sqrt{22} \approx 4.69$.

9. $a^2 + b^2 = c^2$

$4^2 + b^2 = 13^2$

$16 + b^2 = 169$

$b^2 = 153$

$b = \sqrt{153}$

$b = 3\sqrt{17}$

The unknown side has a length of $3\sqrt{17} \approx 12.37$.

11. $a^2 + b^2 = c^2$

$4^2 + 5^2 = c^2$

$16 + 25 = c^2$

$41 = c^2$

$\sqrt{41} = c$

$c = \sqrt{41} \approx 6.40$

13. $a^2 + b^2 = c^2$

$a^2 + 2^2 = 6^2$

$a^2 + 4 = 36$

$a^2 = 32$

$a = \sqrt{32}$

$a = 4\sqrt{2} \approx 5.66$

15. $a^2 + b^2 = c^2$

$\left(\sqrt{10}\right)^2 + b^2 = 10^2$

$10 + b^2 = 100$

$b^2 = 90$

$b = \sqrt{90}$

$b = 3\sqrt{10} \approx 9.49$

17. $a^2 + b^2 = c^2$

$5^2 + 20^2 = c^2$

$25 + 400 = c^2$

$425 = c^2$

$\sqrt{425} = c$

$c \approx 20.6$

The wire is approximately 20.6 feet long.

19. $a^2 + b^2 = c^2$

$6^2 + 10^2 = c^2$

$36 + 100 = c^2$

$136 = c^2$

$\sqrt{136} = c$

$c \approx 11.7$

The brace is approximately 11.7 feet long.

21. $b = \sqrt{\dfrac{3V}{h}}$

$6 = \sqrt{\dfrac{3V}{2}}$

$6^2 = \dfrac{3V}{2}$

$2(36) = 3V$

$\dfrac{2(36)}{3} = V$

$V = 24$

The volume is 24 cubic feet.

23. $s = \sqrt{30\,fd}$

$s = \sqrt{30(0.35)(280)}$

$s = \sqrt{2940}$

$s \approx 54$

It was traveling approximately 54 mph.

25. $r = \sqrt{2.5r}$

$r = \sqrt{2.5(300)}$

$r = \sqrt{750}$

$r \approx 27$

The car can travel at approximately 27 mph.

27. $d = 3.5\sqrt{h}$

$d = 3.5\sqrt{305.4}$

$d \approx 61.2$

You can see approximately 61.2 km.

29. $9 = 3^2$

The number is 3.

31. $100 = 10^2$

The number is 10.

33. $64 = 8^2$

The number is 8.

35. First find the length of the whole base, and label it y.

$y^2 + 3^2 = 7^2$

$y^2 + 9 = 49$

$y^2 = 40$

$y = \sqrt{40}$

$y = 2\sqrt{10}$

Then find the length of the short unknown side, label it z.

$z^2 + 3^2 = 5^2$

$z^2 + 9 = 25$

$z^2 = 16$

$z = 4$

Then find x.

$x = y - z$

$x = 2\sqrt{10} - 4$

37. $a^2 + b^2 = c^2$

$[60(3)]^2 + [30(3)]^2 = c^2$

$(180)^2 + (90)^2 = c^2$

$32,400 + 8100 = c^2$

$40,500 = c^2$

$201 \approx c$

They are approximately 201 miles apart.

39. Answers may vary.

41. Answers may vary.

Chapter 8 Review

1. $\sqrt{81} = 9$ because $9^2 = 81$ and 9 is even.

2. $-\sqrt{49} = -7$ because $\sqrt{49} = 7$.

3. $\sqrt[3]{27} = 3$ because $3^3 = 27$.

4. $\sqrt[4]{16} = 2$ because $2^4 = 16$ and 2 is even.

5. $-\sqrt{\dfrac{9}{64}} = -\dfrac{3}{8}$ because $\sqrt{\dfrac{9}{64}} = \dfrac{3}{8}$.

6. $\sqrt{\dfrac{36}{81}} = \dfrac{6}{9} = \dfrac{2}{3}$ because $\left(\dfrac{6}{9}\right)^2 = \dfrac{36}{81}$.

7. $\sqrt[4]{16} = 2$ because $2^4 = 16$ and 2 is even.

8. $\sqrt[3]{-8} = -2$ because $(-2)^3 = -8$.

9. c

$\sqrt{-4}$ is not a real number because the index is even and the radicand is negative.

10. a and c

$\sqrt{-5}$ and $\sqrt[4]{-5}$ are not real numbers because the index is even and the radicand is negative.

11. $\sqrt{x^{12}} = x^6$ because $\left(x^6\right)^2 = x^{12}$.

12. $\sqrt{x^8} = x^4$ because $\left(x^4\right)^2 = x^8$.

13. $\sqrt{9y^2} = 3y$ because $(3y)^2 = 9y^2$.

14. $\sqrt{25x^4} = 5x^2$ because $\left(5x^2\right)^2 = 25x^4$.

15. $\sqrt{40} = \sqrt{4 \cdot 10} = \sqrt{4} \cdot \sqrt{10} = 2\sqrt{10}$

16. $\sqrt{24} = \sqrt{4 \cdot 6} = \sqrt{4} \cdot \sqrt{6} = 2\sqrt{6}$

17. $\sqrt{54} = \sqrt{9 \cdot 6} = \sqrt{9} \cdot \sqrt{6} = 3\sqrt{6}$

18. $\sqrt{88} = \sqrt{4 \cdot 22} = \sqrt{4} \cdot \sqrt{22} = 2\sqrt{22}$

19. $\sqrt{x^5} = \sqrt{x^4 \cdot x} = \sqrt{x^4} \cdot \sqrt{x} = x^2\sqrt{x}$

20. $\sqrt{y^7} = \sqrt{y^6 \cdot y} = \sqrt{y^6} \cdot \sqrt{y} = y^3\sqrt{y}$

21. $\sqrt{20x^2} = \sqrt{4x^2 \cdot 5} = \sqrt{4x^2} \cdot \sqrt{5} = 2x\sqrt{5}$

22. $\sqrt{50y^4} = \sqrt{25y^4 \cdot 2} = \sqrt{25y^4} \cdot \sqrt{2} = 5y^2\sqrt{2}$

23. $\sqrt[3]{54} = \sqrt[3]{27 \cdot 2} = \sqrt[3]{27} \cdot \sqrt[3]{2} = 3\sqrt[3]{2}$

24. $\sqrt[3]{88} = \sqrt[3]{8 \cdot 11} = \sqrt[3]{8} \cdot \sqrt[3]{11} = 2\sqrt[3]{11}$

25. $\sqrt{\dfrac{18}{25}} = \dfrac{\sqrt{18}}{\sqrt{25}} = \dfrac{\sqrt{9} \cdot \sqrt{2}}{5} = \dfrac{3\sqrt{2}}{5}$

26. $\sqrt{\dfrac{75}{64}} = \dfrac{\sqrt{75}}{\sqrt{64}} = \dfrac{\sqrt{25} \cdot \sqrt{3}}{8} = \dfrac{5\sqrt{3}}{8}$

27. $-\sqrt{\dfrac{50}{9}} = -\dfrac{\sqrt{50}}{\sqrt{9}} = -\dfrac{\sqrt{25} \cdot \sqrt{2}}{3} = -\dfrac{5\sqrt{2}}{3}$

28. $-\sqrt{\dfrac{12}{49}} = -\dfrac{\sqrt{12}}{\sqrt{49}} = -\dfrac{\sqrt{4} \cdot \sqrt{3}}{7} = -\dfrac{2\sqrt{3}}{7}$

29. $\sqrt{\dfrac{11}{x^2}} = \dfrac{\sqrt{11}}{\sqrt{x^2}} = \dfrac{\sqrt{11}}{x}$

30. $\sqrt{\dfrac{7}{y^4}} = \dfrac{\sqrt{7}}{\sqrt{y^4}} = \dfrac{\sqrt{7}}{y^2}$

31. $\sqrt{\dfrac{y^5}{100}} = \dfrac{\sqrt{y^5}}{\sqrt{100}} = \dfrac{\sqrt{y^4} \cdot \sqrt{y}}{10} = \dfrac{y^2\sqrt{y}}{10}$

32. $\sqrt{\dfrac{x^3}{81}} = \dfrac{\sqrt{x^3}}{\sqrt{81}} = \dfrac{\sqrt{x^2} \cdot \sqrt{x}}{9} = \dfrac{x\sqrt{x}}{9}$

33. $5\sqrt{2} - 8\sqrt{2} = (5-8)\sqrt{2} = -3\sqrt{2}$

34.
$\sqrt{3} - 6\sqrt{3} = 1\sqrt{3} - 6\sqrt{3} = (1-6)\sqrt{3}$
$= -5\sqrt{3}$

35.
$6\sqrt{5} + 3\sqrt{6} - 2\sqrt{5} + \sqrt{6}$
$= (6-2)\sqrt{5} + (3+1)\sqrt{6}$
$= 4\sqrt{5} + 4\sqrt{6}$

36. $-\sqrt{7} + 8\sqrt{2} - \sqrt{7} - 6\sqrt{2}$
$= (-1-1)\sqrt{7} + (8-6)\sqrt{2}$
$= -2\sqrt{7} + 2\sqrt{2}$

37. $\sqrt{28} + \sqrt{63} + \sqrt{56}$
$= \sqrt{4} \cdot \sqrt{7} + \sqrt{9} \cdot \sqrt{7} + \sqrt{4} \cdot \sqrt{14}$
$= 2\sqrt{7} + 3\sqrt{7} + 2\sqrt{14} = 5\sqrt{7} + 2\sqrt{14}$

38. $\sqrt{75} + \sqrt{48} - \sqrt{16}$
$= \sqrt{25} \cdot \sqrt{3} + \sqrt{16} \cdot \sqrt{3} - 4$
$= 5\sqrt{3} + 4\sqrt{3} - 4 = 9\sqrt{3} - 4$

39. $\sqrt{\dfrac{5}{9}} - \sqrt{\dfrac{5}{36}} = \dfrac{\sqrt{5}}{\sqrt{9}} - \dfrac{\sqrt{5}}{\sqrt{36}} = \dfrac{\sqrt{5}}{3} - \dfrac{\sqrt{5}}{6}$
$= \dfrac{2\sqrt{5}}{6} - \dfrac{\sqrt{5}}{6} = \dfrac{2\sqrt{5} - \sqrt{5}}{6} = \dfrac{\sqrt{5}}{6}$

40. $\sqrt{\dfrac{11}{25}} + \sqrt{\dfrac{11}{16}} = \dfrac{\sqrt{11}}{\sqrt{25}} + \dfrac{\sqrt{11}}{\sqrt{16}} = \dfrac{\sqrt{11}}{5} + \dfrac{\sqrt{11}}{4}$
$= \dfrac{4\sqrt{11}}{20} + \dfrac{5\sqrt{11}}{20} = \dfrac{9\sqrt{11}}{20}$

41. $\sqrt{45x^2} + 3\sqrt{5x^2} - 7x\sqrt{5} + 10$
$= \sqrt{9x^2} \cdot \sqrt{5} + 3\sqrt{x^2} \cdot \sqrt{5} - 7x\sqrt{5} + 10$
$= 3x\sqrt{5} + 3x\sqrt{5} - 7x\sqrt{5} + 10$
$= -x\sqrt{5} + 10 = 10 - x\sqrt{5}$

42. $\sqrt{50x} - 9\sqrt{2x} + \sqrt{72x} - \sqrt{3x}$
$= \sqrt{25} \cdot \sqrt{2x} - 9\sqrt{2x} + \sqrt{36} \cdot \sqrt{2x} - \sqrt{3x}$
$= 5\sqrt{2x} - 9\sqrt{2x} + 6\sqrt{2x} - \sqrt{3x}$
$= 2\sqrt{2x} - \sqrt{3x}$

43. $\sqrt{3} \cdot \sqrt{6} = \sqrt{3 \cdot 6} = \sqrt{18} = \sqrt{9} \cdot \sqrt{2} = 3\sqrt{2}$

44. $\sqrt{5} \cdot \sqrt{15} = \sqrt{5 \cdot 15} = \sqrt{75} = \sqrt{25} \cdot \sqrt{3}$
$= 5\sqrt{3}$

45. $\sqrt{2}\left(\sqrt{5} - \sqrt{7}\right) = \sqrt{2} \cdot \sqrt{5} - \sqrt{2} \cdot \sqrt{7}$
$= \sqrt{2 \cdot 5} - \sqrt{2 \cdot 7} = \sqrt{10} - \sqrt{14}$

46. $\sqrt{5}\left(\sqrt{11} + \sqrt{3}\right) = \sqrt{5} \cdot \sqrt{11} + \sqrt{5} \cdot \sqrt{3}$
$= \sqrt{5 \cdot 11} + \sqrt{5 \cdot 3} = \sqrt{55} + \sqrt{15}$

47. $\left(\sqrt{3} + 2\right)\left(\sqrt{6} - 5\right)$
$= \sqrt{3} \cdot \sqrt{6} - 5\sqrt{3} + 2\sqrt{6} + 2(-5)$
$= \sqrt{3 \cdot 6} - 5\sqrt{3} + 2\sqrt{6} - 10$
$= \sqrt{18} - 5\sqrt{3} + 2\sqrt{6} - 10$
$= 3\sqrt{2} - 5\sqrt{3} + 2\sqrt{6} - 10$

48. $\left(\sqrt{5} + 1\right)\left(\sqrt{5} - 3\right) = \sqrt{5} \cdot \sqrt{5} - 3\sqrt{5} + \sqrt{5} - 3$
$= 5 - 2\sqrt{5} - 3 = 2 - 2\sqrt{5}$

49. $\left(\sqrt{x} - 2\right)^2 = \left(\sqrt{x}\right)^2 - 2\left(\sqrt{x}\right)(2) + 2^2$
$= x - 4\sqrt{x} + 4$

50. $\left(\sqrt{y} + 4\right)^2 = \left(\sqrt{y}\right)^2 + 2\left(\sqrt{y}\right)(4) + 4^2$
$= y + 8\sqrt{y} + 16$

51. $\dfrac{\sqrt{27}}{\sqrt{3}} = \sqrt{\dfrac{27}{3}} = \sqrt{9} = 3$

52. $\dfrac{\sqrt{20}}{\sqrt{5}} = \sqrt{\dfrac{20}{5}} = \sqrt{4} = 2$

53. $\dfrac{\sqrt{160}}{\sqrt{8}} = \sqrt{\dfrac{160}{8}} = \sqrt{20} = \sqrt{4} \cdot \sqrt{5} = 2\sqrt{5}$

54. $\dfrac{\sqrt{96}}{\sqrt{3}} = \sqrt{\dfrac{96}{3}} = \sqrt{32} = \sqrt{16} \cdot \sqrt{2} = 4\sqrt{2}$

55. $\dfrac{\sqrt{30x^6}}{\sqrt{2x^3}} = \sqrt{\dfrac{30x^6}{2x^3}} = \sqrt{15x^3} = \sqrt{x^2} \cdot \sqrt{15x}$
$= x\sqrt{15x}$

56. $\dfrac{\sqrt{54x^5y^2}}{\sqrt{3xy^2}} = \sqrt{\dfrac{54x^5y^2}{3xy^2}} = \sqrt{18x^4}$
$= \sqrt{9x^4} \cdot \sqrt{2} = 3x^2\sqrt{2}$

57. $\dfrac{\sqrt{2}}{\sqrt{11}} = \dfrac{\sqrt{2} \cdot \sqrt{11}}{\sqrt{11} \cdot \sqrt{11}} = \dfrac{\sqrt{22}}{11}$

58. $\dfrac{\sqrt{3}}{\sqrt{13}} = \dfrac{\sqrt{3} \cdot \sqrt{13}}{\sqrt{13} \cdot \sqrt{13}} = \dfrac{\sqrt{39}}{13}$

59. $\sqrt{\dfrac{5}{6}} = \dfrac{\sqrt{5}}{\sqrt{6}} = \dfrac{\sqrt{5} \cdot \sqrt{6}}{\sqrt{6} \cdot \sqrt{6}} = \dfrac{\sqrt{30}}{6}$

60. $\sqrt{\dfrac{7}{10}} = \dfrac{\sqrt{7}}{\sqrt{10}} = \dfrac{\sqrt{7} \cdot \sqrt{10}}{\sqrt{10} \cdot \sqrt{10}} = \dfrac{\sqrt{70}}{10}$

61. $\dfrac{1}{\sqrt{5x}} = \dfrac{1 \cdot \sqrt{5x}}{\sqrt{5x} \cdot \sqrt{5x}} = \dfrac{\sqrt{5x}}{5x}$

62. $\dfrac{5}{\sqrt{3y}} = \dfrac{5 \cdot \sqrt{3y}}{\sqrt{3y} \cdot \sqrt{3y}} = \dfrac{5\sqrt{3y}}{3y}$

63. $\sqrt{\dfrac{3}{x}} = \dfrac{\sqrt{3}}{\sqrt{x}} = \dfrac{\sqrt{3} \cdot \sqrt{x}}{\sqrt{x} \cdot \sqrt{x}} = \dfrac{\sqrt{3x}}{x}$

64. $\sqrt{\dfrac{6}{y}} = \dfrac{\sqrt{6}}{\sqrt{y}} = \dfrac{\sqrt{6} \cdot \sqrt{y}}{\sqrt{y} \cdot \sqrt{y}} = \dfrac{\sqrt{6y}}{y}$

65. $\dfrac{3}{\sqrt{5}-2} = \dfrac{3(\sqrt{5}+2)}{(\sqrt{5}-2)(\sqrt{5}+2)} = \dfrac{3(\sqrt{5}+2)}{5-4}$

$= 3\sqrt{5} + 6$

66. $\dfrac{8}{\sqrt{10}-3} = \dfrac{8(\sqrt{10}+3)}{(\sqrt{10}-3)(\sqrt{10}+3)}$

$= \dfrac{8(\sqrt{10}+3)}{10-9} = 8\sqrt{10} + 24$

67. $\dfrac{\sqrt{2}+1}{\sqrt{3}-1} = \dfrac{(\sqrt{2}+1)(\sqrt{3}+1)}{(\sqrt{3}-1)(\sqrt{3}+1)}$

$= \dfrac{\sqrt{6}+\sqrt{2}+\sqrt{3}+1}{3-1} = \dfrac{\sqrt{6}+\sqrt{2}+\sqrt{3}+1}{2}$

68. $\dfrac{\sqrt{3}-2}{\sqrt{5}+2} = \dfrac{(\sqrt{3}-2)(\sqrt{5}-2)}{(\sqrt{5}+2)(\sqrt{5}-2)}$

$= \dfrac{\sqrt{15}-2\sqrt{3}-2\sqrt{5}+4}{5-4}$

$= \sqrt{15} - 2\sqrt{3} - 2\sqrt{5} + 4$

69. $\dfrac{10}{\sqrt{x}+5} = \dfrac{10(\sqrt{x}-5)}{(\sqrt{x}+5)(\sqrt{x}-5)} = \dfrac{10\sqrt{x}-50}{x-25}$

70. $\dfrac{8}{\sqrt{x}-1} = \dfrac{8(\sqrt{x}+1)}{(\sqrt{x}-1)(\sqrt{x}+1)} = \dfrac{8\sqrt{x}+8}{x-1}$

71. $\sqrt{2x} = 6$

$\left(\sqrt{2x}\right)^2 = 6^2$

$2x = 36$

$x = 18$

72. $\sqrt{x+3} = 4$

$\left(\sqrt{x+3}\right)^2 = 4^2$

$x + 3 = 16$

$x = 13$

73. $\sqrt{x} + 3 = 8$

$\sqrt{x} = 5$

$\left(\sqrt{x}\right)^2 = 5^2$

$x = 25$

74. $\sqrt{x} + 8 = 3$

$\sqrt{x} = -5$

There is no solution because $\sqrt{8}$ cannot equal a negative number.

75. $\sqrt{2x+1} = x - 7$

$\left(\sqrt{2x+1}\right)^2 = (x-7)^2$

$2x + 1 = x^2 - 14x + 49$

$0 = x^2 - 16x + 48$

$$0 = (x - 12)(x - 4)$$
$$x - 12 = 0 \quad \text{or} \quad x - 4 = 0$$
$$x = 12 \qquad\qquad x = 4 \text{ (extraneous)}$$

76. $\sqrt{3x+1} = x - 1$

$$\left(\sqrt{3x+1}\right)^2 = (x-1)^2$$
$$3x + 1 = x^2 - 2x + 1$$
$$0 = x^2 - 5x$$
$$0 = x(x - 5)$$
$$x = 0 \text{ (extraneous)} \quad \text{or} \quad x - 5 = 0$$
$$x = 5$$

77. $\sqrt{x} + 3 = \sqrt{x+15}$

$$\left(\sqrt{x}+3\right)^2 = \left(\sqrt{x+15}\right)^2$$
$$x + 6\sqrt{x} + 9 = x + 15$$
$$6\sqrt{x} = 6$$
$$\sqrt{x} = 1$$
$$\left(\sqrt{x}\right)^2 = 1^2$$
$$x = 1$$

78. $\sqrt{x-5} = \sqrt{x} - 1$

$$\left(\sqrt{x-5}\right)^2 = \left(\sqrt{x}-1\right)^2$$
$$x - 5 = x - 2\sqrt{x} + 1$$
$$2\sqrt{x} = 6$$
$$\sqrt{x} = 3$$
$$\left(\sqrt{x}\right)^2 = 3^2$$
$$x = 9$$

79. $a^2 + b^2 = c^2$

$$5^2 + b^2 = 9^2$$
$$25 + b^2 = 81$$
$$b^2 = 56$$
$$b = \sqrt{56}$$
$$b = 2\sqrt{14} \approx 7.48$$

80. $a^2 + b^2 = c^2$

$$6^2 + 9^2 = c^2$$
$$36 + 81 = c^2$$
$$117 = c^2$$
$$\sqrt{117} = c$$
$$c = \sqrt{117} \approx 10.82$$

81. $a^2 + b^2 = c^2$

$$(20)^2 + (12)^2 = c^2$$
$$400 + 144 = c^2$$
$$544 = c^2$$
$$\sqrt{544} = c$$
$$c = 4\sqrt{34}$$

They are $4\sqrt{34}$ feet apart.

82. $a^2 + b^2 = c^2$

$$a^2 + 5^2 = 10^2$$
$$a^2 + 25 = 100$$
$$a^2 = 75$$
$$a = \sqrt{75}$$
$$a = 5\sqrt{3}$$

The length is $5\sqrt{3}$ inches.

83. $r = \sqrt{\dfrac{S}{4\pi}}$

$$S = 72$$
$$r = \sqrt{\dfrac{72}{4\pi}}$$
$$r \approx 2.4 \text{ in.}$$

The radius is approximately 2.4 in.

84. $r = \sqrt{\dfrac{S}{4\pi}}$

$$r = 6$$
$$6 = \sqrt{\dfrac{S}{4\pi}}$$
$$(6)^2 = \left(\sqrt{\dfrac{S}{4\pi}}\right)^2$$
$$36 = \dfrac{S}{4\pi}$$
$$36(4\pi) = S$$
$$144\pi = S$$

The surface area is 144π sq. in.

Chapter 8 Test

1. $\sqrt{16} = 4$ because $4^2 = 16$ and 4 is positive.

2. $\sqrt[3]{125} = 5$ because $5^3 = 125$.

3. $\sqrt[4]{81} = 3$ because $3^4 = 81$ and 3 is positive.

4. $\sqrt{\dfrac{9}{16}} = \dfrac{3}{4}$ because $\left(\dfrac{3}{4}\right)^2 = \dfrac{9}{16}$ and $\dfrac{3}{4}$ is positive.

5. $\sqrt[4]{-81}$ is not a real number because the index is even and the radicand is negative.

6. $\sqrt{x^{10}} = x^5$ because $\left(x^5\right)^2 = x^{10}$.

7. $\sqrt{54} = \sqrt{9 \cdot 6} = \sqrt{9} \cdot \sqrt{6} = 3\sqrt{6}$

8. $\sqrt{92} = \sqrt{4 \cdot 23} = \sqrt{4} \cdot \sqrt{23} = 2\sqrt{23}$

9. $\sqrt{y^7} = \sqrt{y^6 \cdot y} = \sqrt{y^6} \cdot \sqrt{y} = y^3\sqrt{y}$

10. $\sqrt{24x^8} = \sqrt{4x^8 \cdot 6} = \sqrt{4x^8} \cdot \sqrt{6} = 2x^4\sqrt{6}$

11. $\sqrt[3]{27} = 3$

12. $\sqrt[3]{16} = \sqrt[3]{8 \cdot 2} = \sqrt[3]{8} \cdot \sqrt[3]{2} = 2\sqrt[3]{2}$

13. $\sqrt{\dfrac{5}{16}} = \dfrac{\sqrt{5}}{\sqrt{16}} = \dfrac{\sqrt{5}}{4}$

14. $\sqrt{\dfrac{y^3}{25}} = \dfrac{\sqrt{y^3}}{\sqrt{25}} = \dfrac{\sqrt{y^2} \cdot \sqrt{y}}{5} = \dfrac{y\sqrt{y}}{5}$

15. $\sqrt{13} + \sqrt{13} - 4\sqrt{13} = 1\sqrt{13} + 1\sqrt{13} - 4\sqrt{13}$
$= (1 + 1 - 4)\sqrt{13} = -2\sqrt{13}$

16. $\sqrt{18} - \sqrt{75} + 7\sqrt{3} - \sqrt{8}$
$= \sqrt{9} \cdot \sqrt{2} - \sqrt{25} \cdot \sqrt{3} + 7\sqrt{3} - \sqrt{4} \cdot \sqrt{2}$
$= 3\sqrt{2} - 5\sqrt{3} + 7\sqrt{3} - 2\sqrt{2}$
$= (3 - 2)\sqrt{2} + (-5 + 7)\sqrt{3}$
$= \sqrt{2} + 2\sqrt{3}$

17. $\sqrt{\dfrac{3}{4}} + \sqrt{\dfrac{3}{25}} = \dfrac{\sqrt{3}}{\sqrt{4}} + \dfrac{\sqrt{3}}{\sqrt{25}} = \dfrac{\sqrt{3}}{2} + \dfrac{\sqrt{3}}{5}$
$= \dfrac{5\sqrt{3}}{10} + \dfrac{2\sqrt{3}}{10} = \dfrac{5\sqrt{3} + 2\sqrt{3}}{10} = \dfrac{7\sqrt{3}}{10}$

18.
$\sqrt{7} \cdot \sqrt{14} = \sqrt{7 \cdot 14} = \sqrt{98} = \sqrt{49} \cdot \sqrt{2}$
$= 7\sqrt{2}$

19. $\sqrt{2}\left(\sqrt{6} - \sqrt{5}\right) = \sqrt{2} \cdot \sqrt{6} - \sqrt{2} \cdot \sqrt{5}$
$= \sqrt{2 \cdot 6} - \sqrt{2 \cdot 5} = \sqrt{12} - \sqrt{10}$
$= \sqrt{4} \cdot \sqrt{3} - \sqrt{10}$
$= 2\sqrt{3} - \sqrt{10}$

20. $\left(\sqrt{x} + 2\right)\left(\sqrt{x} - 3\right)$
$= \left(\sqrt{x}\right)^2 - 3\sqrt{x} + 2\sqrt{x} - 6$
$= x - \sqrt{x} - 6$

21. $\dfrac{\sqrt{50}}{\sqrt{10}} = \sqrt{\dfrac{50}{10}} = \sqrt{5}$

22. $\dfrac{\sqrt{40x^4}}{\sqrt{2x}} = \sqrt{\dfrac{40x^4}{2x}} = \sqrt{20x^3} = \sqrt{4x^2} \cdot \sqrt{5x}$
$= 2x\sqrt{5x}$

23. $\sqrt{\dfrac{2}{3}} = \dfrac{\sqrt{2}}{\sqrt{3}} = \dfrac{\sqrt{2} \cdot \sqrt{3}}{\sqrt{3} \cdot \sqrt{3}} = \dfrac{\sqrt{6}}{3}$

24. $\dfrac{8}{\sqrt{5y}} = \dfrac{8 \cdot \sqrt{5y}}{\sqrt{5y} \cdot \sqrt{5y}} = \dfrac{8\sqrt{5y}}{5y}$

25. $\dfrac{8}{\sqrt{6}+2} = \dfrac{8\left(\sqrt{6}-2\right)}{\left(\sqrt{6}+2\right)\left(\sqrt{6}-2\right)} = \dfrac{8\left(\sqrt{6}-2\right)}{6-4}$

$= \dfrac{8\left(\sqrt{6}-2\right)}{2} = 4\left(\sqrt{6}-2\right) = 4\sqrt{6}-8$

26. $\dfrac{1}{3-\sqrt{x}} = \dfrac{1\left(3+\sqrt{x}\right)}{\left(3-\sqrt{x}\right)\left(3+\sqrt{x}\right)} = \dfrac{3+\sqrt{x}}{9-x}$

27. $\sqrt{x}+8=11$

$\sqrt{x}=3$

$\left(\sqrt{x}\right)^2 = 3^2$

$x=9$

28. $\sqrt{3x-6} = \sqrt{x+4}$

$\left(\sqrt{3x-6}\right)^2 = \left(\sqrt{x+4}\right)^2$

$3x-6 = x+4$

$2x=10$

$x=5$

29. $\sqrt{2x-2} = x-5$

$\left(\sqrt{2x-2}\right)^2 = (x-5)^2$

$2x-2 = x^2 -10x+25$

$0 = x^2 -12x+27$

$0 = (x-9)(x-3)$

$x-9=0$ or $x-3=0$

$\quad x=9 \qquad\qquad x=3$ (extraneous)

30. $a^2 + b^2 = c^2$

$8^2 + b^2 = 12^2$

$64 + b^2 = 144$

$b^2 = 80$

$b = \sqrt{80}$

$b = 4\sqrt{5}$

The length is $4\sqrt{5}$ inches.

31. $r = \sqrt{\dfrac{A}{\pi}}$

$r = \sqrt{\dfrac{15}{\pi}}$

$x \approx 2.19$

The radius is approximately 2.19 meters.

Chapter 8 Cumulative Review

1. $-5(-10) = 50$

2. $-\dfrac{2}{3} \cdot \dfrac{4}{7} = \dfrac{-2 \cdot 4}{3 \cdot 7} = \dfrac{-8}{21} = -\dfrac{8}{21}$

3. $4(2x-3)+7 = 3x+5$

$4(2x)-4(3)+7 = 3x+5$

$8x-12+7 = 3x+5$

$8x-5 = 3x+5$

$8x-3x-5 = 3x+5-3x$

$5x-5 = 5$

$5x-5+5 = 5+5$

$5x = 10$

$\dfrac{5x}{5} = \dfrac{10}{5}$

$x = 2$

4. a. From the circle graph, we see that 45% of homeowners spend under $250 per year on home maintenance.

b. From the circle graph, we know that 45% of homeowners spend under $250 per year and 38% of homeowners spend $250–$999 per year, so that the sum 45% + 38% or 83% of homeowners spend less than $1000 per year.

c. Since 45% of homeowners spend under $250 per year on maintenance, we find 45% of 22,000.
45% of 22,000 = 0.45(22,000) = 9900
We might then expect that 9900 homeowners in Fairview spend under $250 per year on home maintenance.

5. a. $1.02 \times 10^5 = 102,000$

b. $7.358 \times 10^{-3} = 0.007358$

c. $\quad 8.4 \times 10^7 = 84,000,000$

d. $\quad 3.007 \times 10^{-5} = 0.00003007$

6. $(3x + 2)(2x - 5)$
$= 3x(2x) + 3x(-5) + 2(2x) + 2(-5)$
$= 6x^2 - 15x + 4x - 10$
$= 6x^2 - 11x - 10$

7. $xy + 2x + 3y + 6 = x(y + 2) + 3(y + 2)$
$= (y + 2)(x + 3)$

8. $3x^2 + 11x + 6 = (3x + 2)(x + 3)$

9. a. The denominator of $\dfrac{x}{x - 3}$ is 0 when
$x - 3 = 0$ or when $x = 3$. Thus, when $x = 3$, the expression $\dfrac{x}{x - 3}$ is undefined.

b. Set the denominator equal to zero.
$x^2 - 3x + 2 = 0$
$(x - 2)(x - 1) = 0$
$x - 2 = 0 \qquad$ or $\qquad x - 1 = 0$
$x = 2 \qquad$ or $\qquad x = 1$

Thus, when $x = 2$ or $x = 1$, the denominator $x^2 - 3x + 2$ is 0. So the rational expression $\dfrac{x^2 + 2}{x^2 - 3x + 2}$ is undefined when $x = 2$ or when $x = 1$.

c. The denominator of $\dfrac{x^3 - 6x^2 - 10x}{3}$ is never zero, so there are no values of x for which this expression is undefined.

10. $\dfrac{x^2 + 4x + 4}{x^2 + 2x} = \dfrac{(x + 2)(x + 2)}{x(x + 2)} = \dfrac{x + 2}{x}$

11. a. $\dfrac{a}{4} - \dfrac{2a}{8} = \dfrac{a(2)}{4(2)} - \dfrac{2a}{8} = \dfrac{2a}{8} - \dfrac{2a}{8}$
$= \dfrac{2a - 2a}{8} = \dfrac{0}{8} = 0$

b. $\dfrac{3}{10x^2} + \dfrac{7}{25x} = \dfrac{3(5)}{10x^2(5)} + \dfrac{7(2x)}{25x(2x)}$
$= \dfrac{15}{50x^2} + \dfrac{14x}{50x^2} = \dfrac{15 + 14x}{50x^2}$

12. $\dfrac{4x}{x^2 - 25} + \dfrac{2}{x - 5} = \dfrac{1}{x + 5}$
If we factor $x^2 - 25$, we see that the LCD is $(x + 5)(x - 5)$. Multiply both sides of the equation by this LCD, then simplify.

$(x + 5)(x - 5)\left(\dfrac{4x}{x^2 - 25} + \dfrac{2}{x - 5} \right) = (x + 5)(x - 5)\left(\dfrac{1}{x + 5} \right)$

$4x + 2(x + 5) = x - 5$
$4x + 2x + 10 = x - 5$
$6x + 10 = x - 5$
$5x = -15$
$x = -3$

13.

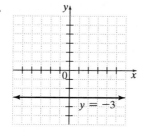

14. Using the slope-intercept form of an equation of a line with $m = \dfrac{1}{4}$ and $b = -3$ we have,

$y = mx + b$

$y = \dfrac{1}{4}x - 3$

15. $\begin{cases} 3x + 4y = 13 \\ 5x - 9y = 6 \end{cases}$

Multiply the first equation by 9 and the second equation by 4, then add the equations.

$\begin{aligned} 27x + 36y &= 117 \\ 20x - 36y &= 24 \\ \hline 47x &= 141 \\ x &= 3 \end{aligned}$

To find the y-value, let $x = 3$ in the first equation.

$3x + 4y = 13$

$3\,3) + 4y = 13$

$9 + 4y = 13$

$4y = 4$

$y = 1$

The solution is (3, 1).

16. The relationship between distance, speed, and time is given by the formula $d = rt$, where d is the distance traveled, r is the rate, and t is the time.

Let x = Albert's rate in miles per hour

y = Louis's rate in miles per hour.

	r $\cdot$	$t =$	d
Albert	x	2	$2x$
Louis	y	2	$2y$

Since Louis walks one mile per hour faster than Albert, we write

$y = x + 1$

Since the total distance traveled is 15 miles, we write

$2x + 2y = 15$

$\begin{cases} y = x + 1 \\ 2x + 2y = 15 \end{cases}$

Substitute $x + 1$ for y in the second equation.

$2x + 2(x + 1) = 15$

$2x + 2x + 2 = 15$

$4x + 2 = 15$

$4x = 13$

$x = \dfrac{13}{4}$

$x = 3.25$

To find the y-value, let $x = 3.25$ in the first equation.

$y = x + 1$

$y = 3.25 + 1$

$y = 4.25$

Albert walks at a rate of 3.25 mph and Louis walks at a rate of 4.25 mph.

17. $\sqrt[3]{1} = 1$

18. $\sqrt[3]{-27} = -3$

19. $\sqrt[3]{\dfrac{1}{125}} = \dfrac{\sqrt[3]{1}}{\sqrt[3]{125}} = \dfrac{1}{5}$

20. $\sqrt{54} = \sqrt{9 \cdot 6} = \sqrt{9} \cdot \sqrt{6} = 3\sqrt{6}$

21. $\sqrt{200} = \sqrt{100 \cdot 2} = \sqrt{100} \cdot \sqrt{2} = 10\sqrt{2}$

22. $7\sqrt{12} - \sqrt{75} = 7\sqrt{4} \cdot \sqrt{3} - \sqrt{25} \cdot \sqrt{3}$
$= 7 \cdot 2\sqrt{3} - 5\sqrt{3} = 14\sqrt{3} - 5\sqrt{3} = 9\sqrt{3}$

23. $2\sqrt{x^2} - \sqrt{25x} + \sqrt{x}$
$= 2\sqrt{x^2} - \sqrt{25} \cdot \sqrt{x} + \sqrt{x}$
$= 2x - 5\sqrt{x} + \sqrt{x} = 2x - 4\sqrt{x}$

24. $\dfrac{2}{\sqrt{7}} = \dfrac{2 \cdot \sqrt{7}}{\sqrt{7} \cdot \sqrt{7}} = \dfrac{2\sqrt{7}}{7}$

25. $\sqrt{x} = \sqrt{5x - 2}$

$\left(\sqrt{x}\right)^2 = \left(\sqrt{5x - 2}\right)^2$

$x = 5x - 2$

$-4x = -2$

$x = \dfrac{1}{2}$

Chapter 9

1. $a^2 - 6a = 0$
   $a(a - 6) = 0$
   $a = 0$ or $a - 6 = 0$
                       $a = 6$
   The solutions are 0 and 6.

2. $2x^2 - 11x = 6$
   $2x^2 - 11x - 6 = 0$
   $(2x + 1)(x - 6) = 0$
   $2x + 1 = 0$ or $x - 6 = 0$
   $2x = -1$ $x = 6$
   $x = -\dfrac{1}{2}$

   The solutions are $-\dfrac{1}{2}$ and 6.

3. $b^2 = 144$
   $b = \sqrt{144}$ or $b = -\sqrt{144}$
   $b = 12$ $b = -12$
   The solutions are 12 and -12.

4. $(2y - 7)^2 = 24$
   $2y - 7 = \sqrt{24}$ or $2y - 7 = -\sqrt{24}$
   $2y = 7 + \sqrt{24}$ $2y = 7 - \sqrt{24}$
   $y = \dfrac{7 + \sqrt{24}}{2}$ $y = \dfrac{7 - \sqrt{24}}{2}$
   $y = \dfrac{7 + 2\sqrt{6}}{2}$ $y = \dfrac{7 - 2\sqrt{6}}{2}$

   The solutions are $\dfrac{7 + 2\sqrt{6}}{2}$ and $\dfrac{7 - 2\sqrt{6}}{2}$.

5. $x^2 - 14x + 48 = 0$
   $x^2 - 14x = -48$
   $x^2 - 14x + 49 = -48 + 49$
   $(x - 7)^2 = 1$
   $x - 7 = \sqrt{1}$ or $x - 7 = -\sqrt{1}$
   $x = 7 + \sqrt{1}$ $x = 7 - \sqrt{1}$
   $x = 7 + 1$ $x = 7 - 1$
   $x = 8$ $x = 6$
   The solutions are 8 and 6.

6. $3x^2 - 5x = 2$
   $x^2 - \dfrac{5}{3}x = \dfrac{2}{3}$
   $x^2 - \dfrac{5}{3}x + \dfrac{25}{36} = \dfrac{2}{3} + \dfrac{25}{36}$
   $\left(x - \dfrac{5}{6}\right)^2 = \dfrac{49}{36}$
   $x - \dfrac{5}{6} = \sqrt{\dfrac{49}{36}}$ or $x - \dfrac{5}{6} = -\sqrt{\dfrac{49}{36}}$
   $x - \dfrac{5}{6} = \dfrac{7}{6}$ $x - \dfrac{5}{6} = -\dfrac{7}{6}$
   $x = \dfrac{12}{6}$ $x = -\dfrac{2}{6}$
   $x = 2$ $x = -\dfrac{1}{3}$

   The solutions are 2 and $-\dfrac{1}{3}$.

7. The equation is in standard form with $a = 1$, $b = -6$, and $c = -27$.
   $$x = \frac{-b \pm \sqrt{b^2 - 4ac}}{2a}$$
   $$x = \frac{-(-6) \pm \sqrt{(-6)^2 - 4(1)(-27)}}{2(1)}$$
   $$= \frac{6 \pm \sqrt{36 + 108}}{2} = \frac{6 \pm \sqrt{144}}{2} = \frac{6 \pm 12}{2}$$
   $$x = \frac{6 + 12}{2} = 9 \text{ or } x = \frac{6 - 12}{2} = -3$$
   The solutions are 9 and -3.

8. The equation is in standard form with $a = 1$, $b = -\dfrac{7}{4}$, and $c = -\dfrac{3}{2}$.
   $$m = \frac{-b \pm \sqrt{b^2 - 4ac}}{2a}$$

$$m = \frac{-\left(-\frac{7}{4}\right) \pm \sqrt{\left(-\frac{7}{4}\right)^2 - 4(1)\left(-\frac{3}{2}\right)}}{2(1)}$$

$$= \frac{\frac{7}{4} \pm \sqrt{\frac{49}{16} + \frac{12}{2}}}{2} = \frac{\frac{7}{4} \pm \sqrt{\frac{145}{16}}}{2} = \frac{\frac{7}{4} \pm \frac{\sqrt{145}}{4}}{2}$$

$$= \frac{7 \pm \sqrt{145}}{8}$$

$$x = \frac{7 + \sqrt{145}}{8} \quad \text{or} \quad x = \frac{7 - \sqrt{145}}{8}$$

The solutions are $\dfrac{7 + \sqrt{145}}{8}$ and $\dfrac{7 - \sqrt{145}}{8}$.

9. $(2x + 3)(x - 1) = 6$

$2x^2 + x - 3 = 6$

$2x^2 + x - 9 = 0$

Use the quadratic formula with $a = 2$, $b = 1$, and $c = -9$.

$$x = \frac{-b \pm \sqrt{1^2 - 4(2)(-9)}}{2(2)}$$

$$x = \frac{-1 \pm \sqrt{1 + 72}}{4} = \frac{-1 \pm \sqrt{73}}{4}$$

$$x = \frac{-1 + \sqrt{73}}{4} \quad \text{or} \quad x = \frac{-1 - \sqrt{73}}{4}$$

The solutions are $\dfrac{-1 + \sqrt{73}}{4}$ and $\dfrac{-1 - \sqrt{73}}{4}$.

10. $(5x + 3)^2 = 18$

Use the square root property.

$5x + 3 = \sqrt{18}$ or $5x + 3 = -\sqrt{18}$

$5x = -3 + \sqrt{18}$ $5x = -3 - \sqrt{18}$

$x = \dfrac{-3 + \sqrt{18}}{5}$ $x = \dfrac{-3 - \sqrt{18}}{5}$

$x = \dfrac{-3 + 3\sqrt{2}}{5}$ $x = \dfrac{-3 - 3\sqrt{2}}{5}$

The solutions are $\dfrac{-3 + 3\sqrt{2}}{5}$ and $\dfrac{-3 - 3\sqrt{2}}{5}$.

11. $8x^2 + 18x + 9 = 0$

Use the quadratic formula with $a = 8$, $b = 18$, and $c = 9$.

$$x = \frac{-b \pm \sqrt{b^2 - 4ac}}{2a}$$

$$x = \frac{-18 \pm \sqrt{18^2 - 4(8)(9)}}{2(8)}$$

$$= \frac{-18 \pm \sqrt{324 - 288}}{16}$$

$$= \frac{-18 \pm \sqrt{36}}{16} = \frac{-18 \pm 6}{16}$$

$$x = \frac{-18 + 6}{16} = -\frac{3}{4} \quad \text{or} \quad x = \frac{-18 - 6}{16} = -\frac{3}{2}$$

The solutions are $-\dfrac{3}{4}$ and $-\dfrac{3}{2}$.

12. $m^2 - 6m = -3$

Complete the square.

$m^2 - 6m + 9 = -3 + 9$

$(m - 3)^2 = 6$

$m - 3 = \sqrt{6}$ or $m - 3 = -\sqrt{6}$

$m = 3 + \sqrt{6}$ or $m = 3 - \sqrt{6}$

The solutions are $3 + \sqrt{6}$ and $3 - \sqrt{6}$.

13. $\dfrac{1}{4}x^2 + x - \dfrac{1}{8} = 0$

Multiply both sides of the equation by 4.

$x^2 + 4x - \dfrac{1}{2} = 0$

$x^2 + 4x = \dfrac{1}{2}$

Complete the square.

$x^2 + 4x + 4 = \dfrac{1}{2} + 4$

$(x + 2)^2 = \dfrac{9}{2}$

$$x + 2 = \sqrt{\frac{9}{2}} \qquad \text{or} \qquad x + 2 = -\sqrt{\frac{9}{2}}$$

$$x = -2 + \sqrt{\frac{9}{2}} \qquad\qquad x = -2 - \sqrt{\frac{9}{2}}$$

$$x = -2 + \frac{3\sqrt{2}}{2} \qquad\qquad x = -2 - \frac{3\sqrt{2}}{2}$$

$$x = \frac{-4 + 3\sqrt{2}}{2} \qquad\qquad x = \frac{-4 - 3\sqrt{2}}{2}$$

The solutions are $\dfrac{-4 + 3\sqrt{2}}{2}$ and $\dfrac{-4 - 3\sqrt{2}}{2}$.

14. $(y + 7)^2 - 5 = 0$

$(y + 7)^2 = 5$

Use the square root property.

$$y + 7 = \sqrt{5} \qquad \text{or} \qquad y + 7 = -\sqrt{5}$$

$$y = -7 + \sqrt{5} \qquad \text{or} \qquad y = -7 - \sqrt{5}$$

The solutions are $-7 + \sqrt{5}$ and $-7 - \sqrt{5}$.

15. $y = -3x^2$

x	y
0	0
1	−3
2	−12
−1	−3
−2	−12

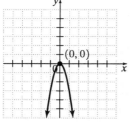

16. $y = x^2 + 3$

x	y
−2	7
−1	4
0	3
1	4
2	7

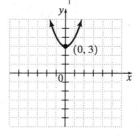

17. $y = x^2 + 4x$

x	y
0	0
−1	−3
−2	−4
−3	−3
−4	0

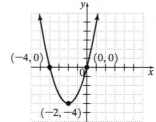

18. $y = x^2 + 2x - 3$

x	y
-3	0
-2	-3
-1	-4
0	-3
1	0

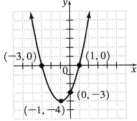

Section 9.1

Practice Problems

1. $x^2 - 25 = 0$
$(x + 5)(x - 5) = 0$
$x + 5 = 0$ or $x - 5 = 0$
$x = -5$ or $x = 5$
The solutions are -5 and 5.

2. $2x^2 - 3x = 9$
$2x^2 - 3x - 9 = 0$
$(2x + 3)(x - 3) = 0$
$2x + 3 = 0$ or $x - 3 = 0$
$x = -\dfrac{3}{2}$ or $x = 3$

The solutions are $-\dfrac{3}{2}$ and 3.

3. $x^2 - 16 = 0$
$x^2 = 16$
$x = \sqrt{16}$ or $x = -\sqrt{16}$
$x = 4$ or $x = -4$
The solutions are 4 and -4.

4. $3x^2 = 11$
$x^2 = \dfrac{11}{3}$

$x = \sqrt{\dfrac{11}{3}}$ or $x = -\sqrt{\dfrac{11}{3}}$

$x = \dfrac{\sqrt{11} \cdot \sqrt{3}}{\sqrt{3} \cdot \sqrt{3}}$ or $x = -\dfrac{\sqrt{11} \cdot \sqrt{3}}{\sqrt{3} \cdot \sqrt{3}}$

$x = \dfrac{\sqrt{33}}{3}$ or $x = -\dfrac{\sqrt{33}}{3}$

The solutions are $\dfrac{\sqrt{33}}{3}$ and $-\dfrac{\sqrt{33}}{3}$.

5. $(x - 4)^2 = 49$
$x - 4 = \sqrt{49}$ or $x - 4 = -\sqrt{49}$
$x - 4 = 7$ or $x - 4 = -7$
$x = 4 + 7$ or $x = 4 - 7$
$x = 11$ or $x = -3$
The solutions are 11 and -3.

6. $(x - 5)^2 = 18$
$x - 5 = \sqrt{18}$ or $x - 5 = -\sqrt{18}$
$x - 5 = 3\sqrt{2}$ or $x - 5 = -3\sqrt{2}$
$x = 5 + 3\sqrt{2}$ or $x = 5 - 3\sqrt{2}$
The solutions are $5 \pm 3\sqrt{2}$.

7. $(x + 3)^2 = -5$
This equation has no real solution because the square root of -5 is not a real number.

8. $(4x + 1)^2 = 15$
$4x + 1 = \sqrt{15}$ or $4x + 1 = -\sqrt{15}$
$4x = -1 + \sqrt{15}$ or $4x = -1 - \sqrt{15}$
$x = \dfrac{-1 + \sqrt{15}}{4}$ or $x = \dfrac{-1 - \sqrt{15}}{4}$

The solutions are $\dfrac{-1 \pm \sqrt{15}}{4}$.

Exercise Set 9.1

1. $k^2 - 9 = 0$
$(k+3)(k-3) = 0$
$k+3 = 0 \quad \text{or} \quad k-3 = 0$
$k = -3 \quad \text{or} \quad k = 3$
The solutions are −3 and 3.

3. $m^2 + 2m = 15$
$m^2 + 2m - 15 = 0$
$(m+5)(m-3) = 0$
$m+5 = 0 \quad \text{or} \quad m-3 = 0$
$m = -5 \quad \text{or} \quad m = 3$
The solutions are −5 and 3.

5. $2x^2 - 32 = 0$
$2\left(x^2 - 16\right) = 0$
$2(x+4)(x-4) = 0$
$x+4 = 0 \quad \text{or} \quad x-4 = 0$
$x = -4 \quad \text{or} \quad x = 4$
The solutions are −4 and 4.

7. $4a^2 - 36 = 0$
$4\left(a^2 - 9\right) = 0$
$4(a+3)(a-3) = 0$
$a+3 = 0 \quad \text{or} \quad a-3 = 0$
$a = -3 \quad \text{or} \quad a = 3$
The solutions are −3 and 3.

9. $x^2 + 7x = -10$
$x^2 + 7x + 10 = 0$
$(x+5)(x+2) = 0$
$x+5 = 0 \quad \text{or} \quad x+2 = 0$
$x = -5 \quad \text{or} \quad x = -2$
The solutions are −5 and −2.

11. $x^2 = 64$
$x = \pm\sqrt{64}$
$x = \pm 8$
The solutions are ±8.

13. $x^2 = 21$
$x = \pm\sqrt{21}$
The solutions are $\pm\sqrt{21}$.

15. $x^2 = \dfrac{1}{25}$
$x = \pm\sqrt{\dfrac{1}{25}}$
$x = \pm\dfrac{\sqrt{1}}{\sqrt{25}}$
$x = \pm\dfrac{1}{5}$
The solutions are $\pm\dfrac{1}{5}$.

17. $x^2 = -4$
This equation has no real solution because the square root of −4 is not a real number.

19. $3x^2 = 13$
$x^2 = \dfrac{13}{3}$
$x = \pm\sqrt{\dfrac{13}{3}}$
$x = \pm\dfrac{\sqrt{13}}{\sqrt{3}}$
$x = \pm\dfrac{\sqrt{13}\cdot\sqrt{3}}{\sqrt{3}\cdot\sqrt{3}}$
$x = \pm\dfrac{\sqrt{39}}{3}$
The solutions are $\pm\dfrac{\sqrt{39}}{3}$

21. $7x^2 = 4$
$x^2 = \dfrac{4}{7}$
$x = \pm\sqrt{\dfrac{4}{7}}$
$x = \pm\dfrac{\sqrt{4}}{\sqrt{7}}$
$x = \pm\dfrac{2\cdot\sqrt{7}}{\sqrt{7}\cdot\sqrt{7}}$
$x = \pm\dfrac{2\sqrt{7}}{7}$
The solutions are $\pm\dfrac{2\sqrt{7}}{7}$.

23. $x^2 - 2 = 0$

$x^2 = 2$

$x = \pm\sqrt{2}$

The solutions are $\pm\sqrt{2}$.

25. Answers may vary.

27. $(x - 5)^2 = 49$

$x - 5 = \pm\sqrt{49}$

$x - 5 = \pm 7$

$x = 5 \pm 7$

$x = 5 - 7$ or $x = 5 + 7$

$x = -2$ or $x = 12$

The solutions are –2 and 12.

29. $(x + 2)^2 = 7$

$x + 2 = \pm\sqrt{7}$

$x = -2 \pm \sqrt{7}$

The solutions are $-2 \pm \sqrt{7}$.

31. $\left(m - \dfrac{1}{2}\right)^2 = \dfrac{1}{4}$

$m - \dfrac{1}{2} = \pm\sqrt{\dfrac{1}{4}}$

$m - \dfrac{1}{2} = \pm\dfrac{1}{2}$

$m = \dfrac{1}{2} \pm \dfrac{1}{2}$

$m = \dfrac{1}{2} - \dfrac{1}{2}$ or $m = \dfrac{1}{2} + \dfrac{1}{2}$

$m = 0$ or $m = 1$

The solutions are 0 and 1.

33. $(p + 2)^2 = 10$

$p + 2 = \pm\sqrt{10}$

$p = -2 \pm \sqrt{10}$

The solutions are $-2 \pm \sqrt{10}$

35. $(3y + 2)^2 = 100$

$3y + 2 = \pm\sqrt{100}$

$3y + 2 = \pm 10$

$3y = -2 \pm 10$

$y = \dfrac{-2 \pm 10}{3}$

$y = \dfrac{-2 - 10}{3}$ or $y = \dfrac{-2 + 10}{3}$

$y = -4$ or $y = \dfrac{8}{3}$

The solutions are –4 and $\dfrac{8}{3}$.

37. $(z - 4)^2 = -9$

This equation has no real solution because the square root of –9 is not a real number.

39. $(2x - 11)^2 = 50$

$2x - 11 = \pm\sqrt{50}$

$2x - 11 = \pm 5\sqrt{2}$

$2x = 11 \pm 5\sqrt{2}$

$x = \dfrac{11 \pm 5\sqrt{2}}{2}$

The solutions are $\dfrac{11 \pm 5\sqrt{2}}{2}$.

41. $(3x - 7)^2 = 32$

$3x - 7 = \pm\sqrt{32}$

$3x - 7 = \pm 4\sqrt{2}$

$3x = 7 \pm 4\sqrt{2}$

$x = \dfrac{7 \pm 4\sqrt{2}}{3}$

The solutions are $x = \dfrac{7 \pm 4\sqrt{2}}{3}$.

43. $x^2 + 6x + 9 = x^2 + 2(3)(x) + 3^2 = (x + 3)^2$

45. $x^2 - 4x + 4 = x^2 - 2(2)(x) + 2^2 = (x - 2)^2$

47. $x^2 + 4x + 4 = 16$

$(x+2)^2 = 16$

$x + 2 = \pm\sqrt{16}$

$x + 2 = \pm 4$

$x = -2 \pm 4$

$x = -2 - 4$ or $x = -2 + 4$

$x = -6$ or $x = 2$

The solutions are –6 and 2.

49. $A = \pi r^2$

$36\pi = \pi r^2$

$36 = r^2$

$\sqrt{36} = r$

$6 = r$

The radius is 6 in.

51. $d = 16t^2$

$400 = 16t^2$

$\dfrac{400}{16} = t^2$

$\sqrt{\dfrac{400}{16}} = t$

$\dfrac{\sqrt{400}}{\sqrt{16}} = t$

$\dfrac{20}{4} = t$

$5 = t$

It will take 5 seconds.

53. $(x - 1.37)^2 = 5.71$

$x - 1.37 = \pm\sqrt{5.71}$

$x = 1.37 \pm \sqrt{5.71}$

$x = 1.37 - \sqrt{5.71}$ $x = 1.37 + \sqrt{5.71}$

$x \approx -1.02$ $x \approx 3.76$

The solutions are –1.02 and 3.76.

55. $y = 5x^2 + 33$

$213 = 5x^2 + 33$

$180 = 5x^2$

$36 = x^2$

$\sqrt{36} = x$

$6 = x$

$1993 + 6 = 1999$

The year will be 1999.

Section 9.2

Practice Problems

1. $x^2 + 8x + 1 = 0$

$x^2 + 8x = -1$

$\left(\dfrac{8}{2}\right)^2 = 4^2 = 16$

$x^2 + 8x + 16 = -1 + 16$

$(x + 4)^2 = 15$

$x + 4 = \sqrt{15}$ or $x + 4 = -\sqrt{15}$

$x = -4 + \sqrt{15}$ or $x = -4 - \sqrt{15}$

The solutions are $-4 \pm \sqrt{15}$.

2. $x^2 - 14x = -32$

$\left(\dfrac{-14}{2}\right)^2 = (-7)^2 = 49$

$(x - 7)^2 = 17$

$x - 7 = \sqrt{17}$ or $x - 7 = -\sqrt{17}$

$x = 7 + \sqrt{17}$ or $x = 7 - \sqrt{17}$

The solutions are $7 \pm \sqrt{17}$.

3. $4x^2 - 16x - 9 = 0$

$x^2 - 4x - \dfrac{9}{4} = 0$

$x^2 - 4x = \dfrac{9}{4}$

$x^2 - 4x + 4 = \dfrac{9}{4} + 4$

$(x - 2)^2 = \dfrac{25}{4}$

$x - 2 = \sqrt{\dfrac{25}{4}}$ or $x - 2 = -\sqrt{\dfrac{25}{4}}$

$x = 2 + \dfrac{5}{2}$ or $x = 2 - \dfrac{5}{2}$

$x = \dfrac{9}{2}$ or $x = -\dfrac{1}{2}$

The solutions are $\dfrac{9}{2}$ and $-\dfrac{1}{2}$.

4. $2x^2 + 10x = -13$

$x^2 + 5x = -\dfrac{13}{2}$

$\left(\dfrac{5}{2}\right)^2 = \dfrac{25}{4}$

$x^2 + 5x + \dfrac{25}{4} = -\dfrac{13}{2} + \dfrac{25}{4}$

$\left(x + \dfrac{5}{2}\right)^2 = -\dfrac{1}{4}$

There is no real solution.

5. $2x^2 = -3x + 2$

$x^2 = -\dfrac{3}{2}x + 1$

$x^2 + \dfrac{3}{2}x = 1$

$\left(\dfrac{3}{4}\right)^2 = \dfrac{9}{16}$

$x^2 + \dfrac{3}{2}x + \dfrac{9}{16} = 1 + \dfrac{9}{16}$

$\left(x + \dfrac{3}{4}\right)^2 = \dfrac{25}{16}$

$x + \dfrac{3}{4} = \sqrt{\dfrac{25}{16}}$ or $x + \dfrac{3}{4} = -\sqrt{\dfrac{25}{16}}$

$x = -\dfrac{3}{4} + \dfrac{5}{4}$ or $x = -\dfrac{3}{4} - \dfrac{5}{4}$

$x = \dfrac{1}{2}$ or $x = -2$

the solutions are $\dfrac{1}{2}$ and -2.

Mental Math

1. $p^2 + 8p$

$\left(\dfrac{8}{2}\right)^2 = 4^2 = 16$

2. $p^2 + 6p$

$\left(\dfrac{6}{2}\right)^2 = 3^2 = 9$

3. $x^2 + 20x$

$\left(\dfrac{20}{2}\right)^2 = 10^2 = 100$

4. $x^2 + 18x$

$\left(\dfrac{18}{2}\right)^2 = 9^2 = 81$

5. $y^2 + 14y$

$\left(\dfrac{14}{2}\right)^2 = 7^2 = 49$

6. $y^2 + 2y$

$\left(\dfrac{2}{2}\right)^2 = 1^2 = 1$

Exercise Set 9.2

1. $x^2 + 8x = -12$

$x^2 + 8x + 16 = -12 + 16$

$(x + 4)^2 = 4$

$x + 4 = \pm\sqrt{4}$

$x = -4 \pm 2$

$x = -6$ or $x = -2$

The solutions are -6 and -2.

3. $x^2 + 2x - 5 = 0$

$x^2 + 2x = 5$

$x^2 + 2x + 1 = 5 + 1$

$(x + 1)^2 = 6$

$x + 1 = \pm\sqrt{6}$

$x = -1 \pm \sqrt{6}$

The solutions are $-1 \pm \sqrt{6}$.

5. $x^2 - 6x = 0$

$x^2 - 6x + 9 = 0 + 9$

$(x - 3)^2 = 9$

$x - 3 = \pm\sqrt{9}$

$x = 3 \pm 3$

$x = 0$ or $x = 6$

The solutions are 0 and 6.

7. $z^2 + 5z = 7$

$z^2 + 5z + \frac{25}{4} = 7 + \frac{25}{4}$

$\left(z + \frac{5}{2}\right)^2 = \frac{53}{4}$

$z = \frac{5}{2} = \pm\sqrt{\frac{53}{4}}$

$z = -\frac{5}{2} \pm \frac{\sqrt{53}}{2}$

$z = \frac{-5 \pm \sqrt{53}}{2}$

The solutions are $\frac{-5 \pm \sqrt{53}}{2}$.

9. $x^2 - 2x - 1 = 0$

$x^2 - 2x = 1$

$x^2 - 2x + 1 = 1 + 1$

$(x - 1)^2 = 2$

$x - 1 = \pm\sqrt{2}$

$x = 1 \pm \sqrt{2}$

The solutions are $1 \pm \sqrt{2}$.

11. $y^2 + 5y + 4 = 0$

$y^2 + 5y = -4$

$y^2 + 5y + \frac{25}{4} = -4 + \frac{25}{4}$

$\left(y + \frac{5}{2}\right)^2 = \frac{9}{4}$

$y + \frac{5}{2} = \pm\sqrt{\frac{9}{4}}$

$y = -\frac{5}{2} \pm \frac{3}{2}$

$y = \frac{-5 - 3}{2}$ or $y = \frac{-5 + 3}{2}$

$y = -4$ or $y = -1$

The solutions are -4 and -1.

13. $x^2 + 6x - 25 = 0$

$x^2 + 6x = 25$

$x^2 + 6x + 9 = 25 + 9$

$(x + 3)^2 = 34$

$x + 3 = \pm\sqrt{34}$

$x = -3 \pm \sqrt{34}$

The solutions are $-3 \pm \sqrt{34}$.

15. $x(x + 3) = 18$

$x^2 + 3x = 18$

$x^2 + 3x + \frac{9}{4} = 18 + \frac{9}{4}$

$\left(x + \frac{3}{2}\right)^2 = \frac{81}{4}$

$x + \frac{3}{2} = \pm\sqrt{\frac{81}{4}}$

$x = -\frac{3}{2} \pm \frac{9}{2}$

$x = \frac{-3 - 9}{2}$ or $x = \frac{-3 + 9}{2}$

$x = -6$ or $x = 3$

The solutions are -6 and 3.

17. $4x^2 - 24x = 13$

$x^2 - 6x = \frac{13}{4}$

$x^2 - 6x + 9 = \frac{13}{4} + 9$

$(x - 3)^2 = \frac{49}{4}$

$x - 3 = \pm\sqrt{\frac{49}{4}}$

$x = 3 \pm \frac{7}{2}$

$x = \frac{6 - 7}{2}$ or $x = \frac{6 + 7}{2}$

$x = -\frac{1}{2}$ or $x = \frac{13}{2}$

The solutions are $-\frac{1}{2}$ and $\frac{13}{2}$.

19. $5x^2 + 10x + 6 = 0$

$5x^2 + 10x = -6$

$x^2 + 2x = -\frac{6}{5}$

$x^2 + 2x + 1 = -\frac{6}{5} + 1$

$(x + 1)^2 = -\frac{1}{5}$

There is no real solution because the square root of a negative number is not a real number.

21. $2x^2 = 6x + 5$

$2x^2 - 6x = 5$

$x^2 - 3x = \frac{5}{2}$

$x^2 - 3x + \frac{9}{4} = \frac{5}{2} + \frac{9}{4}$

$\left(x - \frac{3}{2}\right)^2 = \frac{19}{4}$

$x - \frac{3}{2} = \pm\sqrt{\frac{19}{4}}$

$x = \frac{3}{2} \pm \frac{\sqrt{19}}{2}$

The solutions are $\dfrac{3 \pm \sqrt{19}}{2}$.

23. $3x^2 - 6x = 24$

$x^2 - 2x = 8$

$x^2 - 2x + 1 = 8 + 1$

$(x - 1)^2 = 9$

$x - 1 = \pm\sqrt{9}$

$x = 1 \pm 3$

$x = -2 \quad \text{or} \quad x = 4$

The solutions are -2 and 4.

25. $2y^2 + 8y + 5 = 0$

$2y^2 + 8y = -5$

$y^2 + 4y = -\frac{5}{2}$

$y^2 + 4y + 4 = -\frac{5}{2} + 4$

$(y + 2)^2 = \frac{3}{2}$

$y + 2 = \pm\sqrt{\frac{3}{2}}$

$y = -2 \pm \frac{\sqrt{6}}{2}$

The solutions are $-2 \pm \dfrac{\sqrt{6}}{2}$.

27. $2y^2 - 3y + 1 = 0$

$2y^2 - 3y = -1$

$y^2 - \frac{3}{2}y = -\frac{1}{2}$

$y^2 - \frac{3}{2}y + \frac{9}{16} = -\frac{1}{2} + \frac{9}{16}$

$\left(y - \frac{3}{4}\right)^2 = \frac{1}{16}$

$y - \frac{3}{4} = \pm\sqrt{\frac{1}{16}}$

$y = \frac{3}{4} \pm \frac{1}{4}$

$y = \frac{1}{2} \quad \text{or} \quad y = 1$

The solutions are $\frac{1}{2}$ and 1.

29. $\dfrac{3}{4} - \sqrt{\dfrac{25}{16}} = \dfrac{3}{4} - \dfrac{\sqrt{25}}{\sqrt{16}} = \dfrac{3}{4} - \dfrac{5}{4} = \dfrac{3 - 5}{4}$

$\qquad = -\dfrac{2}{4} = -\dfrac{1}{2}$

31. $\dfrac{1}{2} - \sqrt{\dfrac{9}{4}} = \dfrac{1}{2} - \dfrac{\sqrt{9}}{\sqrt{4}} = \dfrac{1}{2} - \dfrac{3}{2} = \dfrac{1 - 3}{2}$

$\qquad = -\dfrac{2}{2} = -1$

33. $\dfrac{6 + 4\sqrt{5}}{2} = \dfrac{6}{2} + \dfrac{4\sqrt{5}}{2} = 3 + 2\sqrt{5}$

35. $\dfrac{3 - 9\sqrt{2}}{6} = \dfrac{3\left(1 - 3\sqrt{2}\right)}{3(2)} = \dfrac{1 - 3\sqrt{2}}{2}$

37. $x^2 + kx + 16$

$\left(\dfrac{k}{2}\right)^2 = 16$

$\dfrac{k^2}{4} = 16$

$k^2 = 64$

$k = \pm\sqrt{64}$

$k = \pm 8$

39. $y = 3x^2 + 80x + 340$

$815 = 3x^2 + 80x + 340$

$475 = 3x^2 + 80x$

$\dfrac{475}{3} = x^2 + \dfrac{80}{3}x$

$\dfrac{475}{3} + \dfrac{6400}{36} = x^2 + \dfrac{80}{3} + \dfrac{6400}{36}$

$\dfrac{12,100}{36} = \left(x + \dfrac{80}{6}\right)^2$

$\dfrac{110}{6} = x + \dfrac{80}{6}$

$\dfrac{30}{6} = x$

$5 = x$

$1994 + 5 = 1999$

The year will be 1999.

41. $x^2 + 8x = -12$

$y_1 = x^2 + 8x$

$y_2 = -12$

The solutions are the x-coordinates of the
intersections, –6 and –2.

43. $2x^2 = 6x + 5$

$y_1 = 2x^2$

$y_2 = 6x + 5$

The solutions are the x-coordinates of the
intersections, $x \approx -0.68$, 3.68.

Section 9.3

Practice Problems

1. $2x^2 - x - 5 = 0$

$a = 2,\ b = -1,\ c = -5$

$x = \dfrac{-b \pm \sqrt{b^2 - 4ac}}{2a}$

$x = \dfrac{-(-1) \pm \sqrt{(-1)^2 - 4(2)(-5)}}{2(2)}$

$x = \dfrac{1 \pm \sqrt{1 + 40}}{4}$

$x = \dfrac{1 \pm \sqrt{41}}{4}$

The solutions are $\dfrac{1 - \sqrt{41}}{4}$ and $\dfrac{1 + \sqrt{41}}{4}$.

2. $3x^2 + 8x = 3$

$3x^2 + 8x - 3 = 0$

$a = 3,\ b = 8,\ c = -3$

$x = \dfrac{-b \pm \sqrt{b^2 - 4ac}}{2a}$

$x = \dfrac{-8 \pm \sqrt{8^2 - 4(3)(-3)}}{2(3)}$

$x = \dfrac{-8 \pm \sqrt{64 + 36}}{6}$

$x = \dfrac{-8 \pm \sqrt{100}}{6}$

$x = \dfrac{-8 \pm 10}{6}$

$x = \dfrac{-8 - 10}{6}$ or $x = \dfrac{-8 + 10}{6}$

$x = -3$ or $x = \dfrac{1}{3}$

The solutions are –3 and $\dfrac{1}{3}$.

3. $5x^2 = 2$

$5x^2 - 2 = 0$

$a = 5, \; b = 0, \; c = -2$

$x = \dfrac{-b \pm \sqrt{b^2 - 4ac}}{2a}$

$x = \dfrac{0 \pm \sqrt{0^2 - 4(5)(-2)}}{2(5)}$

$x = \dfrac{\pm\sqrt{40}}{10}$

$x = \dfrac{\pm 2\sqrt{10}}{10}$

$x = \pm \dfrac{\sqrt{10}}{5}$

The solutions are $-\dfrac{\sqrt{10}}{5}$ and $\dfrac{\sqrt{10}}{5}$.

4. $x^2 = -2x - 3$

$x^2 + 2x + 3 = 0$

$a = 1, \; b = 2, \; c = 3$

$x = \dfrac{-b \pm \sqrt{b^2 - 4ac}}{2a}$

$x = \dfrac{-2 \pm \sqrt{2^2 - 4(1)(3)}}{2(1)}$

$x = \dfrac{-2 \pm \sqrt{4 - 12}}{2}$

$x = \dfrac{-2 \pm \sqrt{-8}}{2}$

There is no real number solution because $\sqrt{-8}$ is not a real number.

5. $\dfrac{1}{3}x^2 - x = 1$

$\dfrac{1}{3}x^2 - x - 1 = 0$

$x^2 - 3x - 3 = 0$

$a = 1, \; b = -3, \; c = -3$

$x = \dfrac{-b \pm \sqrt{b^2 - 4ac}}{2a}$

$x = \dfrac{-(-3) \pm \sqrt{(-3)^2 - 4(1)(-3)}}{2(1)}$

$x = \dfrac{3 \pm \sqrt{9 + 12}}{2}$

$x = \dfrac{3 \pm \sqrt{21}}{2}$

The solutions are $\dfrac{3 - \sqrt{21}}{2}$ and $\dfrac{3 + \sqrt{21}}{2}$.

Mental Math

1. $2x^2 + 5x + 3 = 0$

$a = 2, \; b = 5, \; c = 3$

2. $5x^2 - 7x + 1 = 0$

$a = 5, \; b = -7, \; c = 1$

3. $10x^2 - 13x - 2 = 0$

$a = 10, \; b = -13, \; c = -2$

4. $x^2 + 3x - 7 = 0$

$a = 1, \; b = 3, \; c = -7$

5. $x^2 - 6 = 0$

$a = 1, \; b = 0, \; c = -6$

6. $9x^2 - 4 = 0$

$a = 9, \; b = 0, \; c = -4$

Exercise Set 9.3

1. $x^2 - 3x + 2 = 0$

$a = 1, \; b = -3, \; c = 2$

$x = \dfrac{-b \pm \sqrt{b^2 - 4ac}}{2a}$

$x = \dfrac{-(-3) \pm \sqrt{(-3)^2 - 4(1)(2)}}{2(1)}$

$x = \dfrac{3 \pm \sqrt{9 - 8}}{2}$

$x = \dfrac{3 \pm 1}{2}$

$x = 1 \quad \text{or} \quad x = 2$

The solutions are 1 and 2.

3. $3k^2 + 7k + 1 = 0$

$a = 3,\ b = 7,\ c = 1$

$k = \dfrac{-7 \pm \sqrt{7^2 - 4(3)(1)}}{2(3)}$

$k = \dfrac{-7 \pm \sqrt{49 - 12}}{6}$

$k = \dfrac{-7 \pm \sqrt{37}}{6}$

The solutions are $\dfrac{-7 \pm \sqrt{37}}{6}$.

5. $49x^2 - 4 = 0$

$a = 49,\ b = 0,\ c = -4$

$x = \dfrac{-0 \pm \sqrt{0^2 - 4(49)(-4)}}{2(49)}$

$x = \dfrac{\pm\sqrt{784}}{98}$

$x = \pm\dfrac{28}{98}$

$x = \pm\dfrac{2}{7}$

The solutions are $\pm\dfrac{2}{7}$.

7. $5z^2 - 4z + 3 = 0$

$a = 5,\ b = -4,\ c = 3$

$z = \dfrac{-(-4) \pm \sqrt{(-4)^2 - 4(5)(3)}}{2(5)}$

$z = \dfrac{4 \pm \sqrt{16 - 60}}{10}$

$z = \dfrac{4 \pm \sqrt{-44}}{10}$

There is no real solution because the square root of a negative number is not a real number.

9. $y^2 = 7y + 30$

$y^2 - 7y - 30 = 0$

$a = 1,\ b = -7,\ c = -30$

$y = \dfrac{-(-7) \pm \sqrt{(-7)^2 - 4(1)(-30)}}{2(1)}$

$y = \dfrac{7 \pm \sqrt{49 + 120}}{2}$

$y = \dfrac{7 \pm \sqrt{169}}{2}$

$y = \dfrac{7 \pm 13}{2}$

$y = -3$ or $y = 10$

The solutions are -3 and 10.

11. $2x^2 = 10$

$2x^2 - 10 = 0$

$a = 2,\ b = 0,\ c = -10$

$x = \dfrac{-0 \pm \sqrt{0^2 - 4(2)(-10)}}{2(2)}$

$x = \dfrac{\pm\sqrt{80}}{4}$

$x = \dfrac{\pm 4\sqrt{5}}{4}$

$x = \pm\sqrt{5}$

The solutions are $\pm\sqrt{5}$.

13. $m^2 - 12 = m$

$m^2 - m - 12 = 0$

$a = 1,\ b = -1,\ c = -12$

$m = \dfrac{-(-1) \pm \sqrt{(-1)^2 - 4(1)(-12)}}{2(1)}$

$m = \dfrac{1 \pm \sqrt{49}}{2}$

$m = \dfrac{1 \pm 7}{2}$

$m = -3$ or $m = 4$

The solutions are -3 and 4.

15. $3 - x^2 = 4x$

$-x^2 - 4x + 3 = 0$

$a = -1,\ b = -4,\ c = 3$

$x = \dfrac{-(-4) \pm \sqrt{(-4)^2 - 4(-1)(3)}}{2(-1)}$

$x = \dfrac{4 \pm \sqrt{28}}{-2}$

$x = \dfrac{4 \pm 2\sqrt{7}}{-2}$

$x = -2 \pm \sqrt{7}$

The solutions are $-2 \pm \sqrt{7}$.

17. $6x^2 + 9x = 2$

$6x^2 + 9x - 2 = 0$

$a = 6,\ b = 9,\ c = -2$

$x = \dfrac{-9 \pm \sqrt{9^2 - 4(6)(-2)}}{2(6)}$

$x = \dfrac{-9 \pm \sqrt{129}}{12}$

The solutions are $\dfrac{-9 \pm \sqrt{129}}{12}$.

19. $7p^2 + 2 = 8p$

$7p^2 - 8p + 2 = 0$

$a = 7,\ b = -8,\ c = 2$

$p = \dfrac{-(-8) \pm \sqrt{(-8)^2 - 4(7)(2)}}{2(7)}$

$p = \dfrac{8 \pm \sqrt{8}}{14}$

$p = \dfrac{8 \pm 2\sqrt{2}}{14}$

$p = \dfrac{4 \pm \sqrt{2}}{7}$

The solutions are $\dfrac{4 \pm \sqrt{2}}{7}$.

21. $a^2 - 6a + 2 = 0$

$a = 1,\ b = -6,\ c = 2$

$a = \dfrac{-(-6) \pm \sqrt{(-6)^2 - 4(1)(2)}}{2(1)}$

$a = \dfrac{6 \pm \sqrt{28}}{2}$

$a = \dfrac{6 \pm 2\sqrt{7}}{2}$

$a = 3 \pm \sqrt{7}$

The solutions are $3 \pm \sqrt{7}$.

23. $2x^2 - 6x + 3 = 0$

$a = 2,\ b = -6,\ c = 3$

$x = \dfrac{-(-6) \pm \sqrt{(-6)^2 - 4(2)(3)}}{2(2)}$

$x = \dfrac{6 \pm \sqrt{12}}{4}$

$x = \dfrac{6 \pm 2\sqrt{3}}{4}$

$x = \dfrac{3 \pm \sqrt{3}}{2}$

The solutions are $\dfrac{3 \pm \sqrt{3}}{2}$.

25. $3x^2 = 1 - 2x$

$3x^2 + 2x - 1 = 0$

$a = 3,\ b = 2,\ c = -1$

$x = \dfrac{-2 \pm \sqrt{2^2 - 4(3)(-1)}}{2(3)}$

$x = \dfrac{-2 \pm \sqrt{16}}{6}$

$x = \dfrac{-2 \pm 4}{6}$

$x = -1$ or $x = \dfrac{1}{3}$

The solutions are -1 and $\dfrac{1}{3}$.

27. $4y^2 = 6y + 1$

$4y^2 - 6y - 1 = 0$

$a = 4,\ b = -6,\ c = -1$

$y = \dfrac{-(-6) \pm \sqrt{(-6)^2 - 4(4)(-1)}}{2(4)}$

$y = \dfrac{6 \pm \sqrt{52}}{8}$

$y = \dfrac{6 \pm 2\sqrt{13}}{8}$

$y = \dfrac{3 \pm \sqrt{13}}{4}$.

The solutions are $\dfrac{3 \pm \sqrt{13}}{4}$.

29. $20y^2 = 3 - 11y$

$20y^2 + 11y - 3 = 0$

$a = 20,\ b = 11,\ c = -3$

$y = \dfrac{-11 \pm \sqrt{(11)^2 - 4(20)(-3)}}{2(20)}$

$y = \dfrac{-11 \pm \sqrt{361}}{40}$

$y = \dfrac{-11 \pm 19}{40}$

$y = -\dfrac{3}{4}$ or $y = \dfrac{1}{5}$

The solutions are $-\dfrac{3}{4}$ and $\dfrac{1}{5}$.

31. $x^2 + x + 2 = 0$

$a = 1,\ b = 1,\ c = 2$

$$x = \frac{-1 \pm \sqrt{1^2 - 4(1)(2)}}{2(1)}$$

$$x = \frac{-1 \pm \sqrt{-7}}{2}$$

There is no real solution.

33. $3p^2 - \frac{2}{3}p + 1 = 0$

$9p^2 - 2p + 3 = 0$

$a = 9,\ b = -2,\ c = 3$

$$p = \frac{-(-2) \pm \sqrt{(-2)^2 - 4(9)(3)}}{2(9)}$$

$$p = \frac{2 \pm \sqrt{-104}}{18}$$

There is no real solution.

35. $\frac{m^2}{2} = m + \frac{1}{2}$

$m^2 - 2m - 1 = 0$

$a = 1,\ b = -2,\ c = -1$

$$m = \frac{-(-2) \pm \sqrt{(-2)^2 - 4(1)(-1)}}{2(1)}$$

$$m = \frac{2 \pm \sqrt{8}}{2}$$

$$m = \frac{2 \pm 2\sqrt{2}}{2}$$

$$m = 1 \pm \sqrt{2}$$

The solutions are $1 \pm \sqrt{2}$.

37. $4p^2 + \frac{3}{2} = -5p$

$8p^2 + 10p + 3 = 0$

$a = 8,\ b = 10,\ c = 3$

$$p = \frac{-10 \pm \sqrt{10^2 - 4(8)(3)}}{2(8)}$$

$$p = \frac{-10 \pm \sqrt{4}}{16}$$

$$p = \frac{-10 \pm 2}{16}$$

$$p = -\frac{3}{4} \quad \text{or} \quad p = -\frac{1}{2}$$

The solutions are $-\frac{3}{4}$ and $-\frac{1}{2}$.

39. $5x^2 = \frac{7}{2}x + 1$

$10x^2 - 7x - 2 = 0$

$a = 10,\ b = -7,\ c = -2$

$$x = \frac{-(-7) \pm \sqrt{(-7)^2 - 4(10)(-2)}}{2(10)}$$

$$x = \frac{7 \pm \sqrt{129}}{20}$$

The solutions are $\frac{7 \pm \sqrt{129}}{20}$.

41. $28x^2 + 5x + \frac{11}{4} = 0$

$112x^2 + 20x + 11 = 0$

$a = 112,\ b = 20,\ c = 11$

$$x = \frac{-20 \pm \sqrt{20^2 - 4(112)(11)}}{2(112)}$$

$$x = \frac{-20 \pm \sqrt{-4528}}{224}$$

There is no real solution.

43. $5z^2 - 2z = \frac{1}{5}$

$25z^2 - 10z - 1 = 0$

$a = 25,\ b = -10,\ c = -1$

$$z = \frac{-(-10) \pm \sqrt{(-10)^2 - 4(25)(-1)}}{2(25)}$$

$$z = \frac{10 \pm \sqrt{200}}{50}$$

$$z = \frac{10 \pm 10\sqrt{2}}{50}$$

$$z = \frac{1 \pm \sqrt{2}}{5}$$

The solutions are $\frac{1 \pm \sqrt{2}}{5}$.

45. $y = -3$

47. $y = 3x - 2$

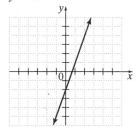

49. $a^2 + b^2 = c^2$

$x^2 + 7^2 = 10^2$

$x^2 + 49 = 100$

$x^2 = 51$

$x = \sqrt{51}$

The unknown side is $\sqrt{51}$ meters.

51. $x^2 + 3\sqrt{2}x - 5 = 0$

$a = 1, \ b = 3\sqrt{2}, \ c = -5$

$$x = \frac{-3\sqrt{2} \pm \sqrt{\left(3\sqrt{2}\right)^2 - 4(1)(-5)}}{2(1)}$$

$$x = \frac{-3\sqrt{2} \pm \sqrt{38}}{2}$$

The solutions are $\dfrac{-3\sqrt{2} \pm \sqrt{38}}{2}$.

53. Answers may vary.

55. $y^2 - y = 11$

$y^2 - y - 11 = 0$

$a = 1, b = -1, c = -11$

$$y = \frac{-(-1) \pm \sqrt{(-1)^2 - 4(1)(-11)}}{2(1)}$$

$$y = \frac{1 \pm \sqrt{45}}{2}$$

$y \approx -2.9$ or $y \approx 3.9$

The approximate solutions are -2.9 and 3.9.

57. $7.3z^2 + 5.4z - 1.1 = 0$

$a = 7.3, b = 5.4, c = -1.1$

$$z = \frac{-5.4 \pm \sqrt{(5.4)^2 - 4(7.3)(-1.1)}}{2(7.3)}$$

$$z = \frac{-5.4 \pm \sqrt{61.28}}{14.6}$$

$z \approx -0.9$ or $z \approx 0.2$

The approximate solutions are -0.9 and 0.2.

59. $h = -16t^2 + 120t + 80$

$0 = -16t^2 + 120t + 80$

$a = -16, \ b = 120, \ c = 80$

$$t = \frac{-120 \pm \sqrt{(120)^2 - 4(-16)(80)}}{2(-16)}$$

$$t = \frac{-120 \pm \sqrt{19,520}}{-32}$$

$t \approx -0.6$ or $t \approx 8.1$

Disregard the negative time. It will take approximately 8.1 seconds.

61. $y = -0.5x^2 + 12.5x + 82$

$150 = -0.5x^2 + 12.5x + 82$

$0 = -0.5x^2 + 12.5x - 68$

$a = -0.5, b = 12.5, c = -68$

$$x = \frac{-12.5 \pm \sqrt{(12.5)^2 - 4(-0.5)(-68)}}{2(-0.5)}$$

$$x = \frac{-12.5 \pm \sqrt{20.25}}{-1}$$

$x = 12.5 \pm 4.5$

$x = 8$ or $x = 17$

$1994 + 8 = 2002$

The first year that the sales will reach \$150 billion is in 2002.

Integrated Review

1. $5x^2 - 11x + 2 = 0$

$a = 5, b = -11, c = 2$

$$x = \frac{-(-11) \pm \sqrt{(-11)^2 - 4(5)(2)}}{2(5)}$$

$$x = \frac{11 \pm \sqrt{81}}{10}$$
$$x = \frac{11 \pm 9}{10}$$
$$x = \frac{1}{5} \quad \text{or} \quad x = 2$$

The solutions are $\frac{1}{5}$ and 2.

2. $5x^2 + 13x - 6 = 0$
   $a = 5,\ b = 13,\ c = -6$
$$x = \frac{-13 \pm \sqrt{(13)^2 - 4(5)(-6)}}{2(5)}$$
$$x = \frac{-13 \pm \sqrt{289}}{10}$$
$$x = \frac{-13 \pm 17}{10}$$
$$x = -3 \quad \text{or} \quad x = \frac{2}{5}$$

The solutions are -3 and $\frac{2}{5}$.

3. $x^2 - 1 = 2x$
   $x^2 - 2x = -1$
   $x^2 - 2x + 1 = 1 + 1$
   $(x - 1)^2 = 2$
   $x - 1 = \pm\sqrt{2}$
   $x = 1 \pm \sqrt{2}$
   The solutions are $1 \pm \sqrt{2}$.

4. $x^2 + 7 = 6x$
   $x^2 - 6x = -7$
   $x^2 - 6x + 9 = -7 + 9$
   $(x - 3)^2 = 2$
   $x - 3 = \pm\sqrt{2}$
   $x = 3 \pm \sqrt{2}$
   The solutions are $3 \pm \sqrt{2}$.

5. $a^2 = 20$
   $a = \pm\sqrt{20}$
   $a = \pm 2\sqrt{5}$
   The solutions are $\pm 2\sqrt{5}$.

6. $a^2 = 72$
   $a = \pm\sqrt{72}$
   $a = \pm 6\sqrt{2}$
   The solutions are $\pm 6\sqrt{2}$.

7. $x^2 - x + 4 = 0$
   $x^2 - x = -4$
   $x^2 - x + \frac{1}{4} = -4 + \frac{1}{4}$
   $\left(x - \frac{1}{2}\right)^2 = -\frac{15}{4}$
   There is no real solution.

8. $x^2 - 2x + 7 = 0$
   $x^2 - 2x = -7$
   $x^2 - 2x + 1 = -7 + 1$
   $(x - 1)^2 = -6$
   There is no real solution.

9. $3x^2 - 12x + 12 = 0$
   $x^2 - 4x + 4 = 0$
   $(x - 2)^2 = 0$
   $x - 2 = 0$
   $x = 2$
   The solution is 2.

10. $5x^2 - 30x + 45 = 0$
   $x^2 - 6x + 9 = 0$
   $(x - 3)^2 = 0$
   $x - 3 = 0$
   $x = 3$
   The solution is 3.

11. $9 - 6p + p^2 = 0$
   $(p - 3)^2 = 0$
   $p - 3 = 0$
   $p = 3$
   The solution is 3.

12. $49 - 28p + 4p^2 = 0$

$(2p - 7)^2 = 0$

$2p - 7 = 0$

$p = \dfrac{7}{2}$

The solution is $\dfrac{7}{2}$.

13. $4y^2 - 16 = 0$

$4y^2 = 16$

$y^2 = 4$

$y = \pm\sqrt{4}$

$y = \pm 2$

The solution is ± 2.

14. $3y^2 - 27 = 0$

$3y^2 = 27$

$y^2 = 9$

$y = \pm\sqrt{9}$

$y = \pm 3$

The solutions are ± 3.

15. $x^4 - 3x^3 + 2x^2 = 0$

$x^2\left(x^2 - 3x + 2\right) = 0$

$x^2(x - 2)(x - 1) = 0$

$x = 0$ or $x = 2$ or $x = 1$

The solutions are 0, 1, and 2.

16. $x^3 + 7x^2 + 12x = 0$

$x\left(x^2 + 7x + 12\right) = 0$

$x(x + 4)(x + 3) = 0$

$x = 0$ or $x = -4$ or $x = -3$

The solutions are -4, -3, and 0.

17. $(2z + 5)^2 = 25$

$2z + 5 = \pm\sqrt{25}$

$2z + 5 = \pm 5$

$z = \dfrac{-5 \pm 5}{2}$

$z = -5$ or $z = 0$

The solutions are -5 and 0.

18. $(3z - 4)^2 = 16$

$3z - 4 = \pm\sqrt{16}$

$3z - 4 = \pm 4$

$z = \dfrac{4 \pm 4}{3}$

$z = 0$ or $z = \dfrac{8}{3}$

The solutions are 0 and $\dfrac{8}{3}$.

19. $30x = 25x^2 + 2$

$0 = 25x^2 - 30x + 2$

$a = 25,\ b = -30,\ c = 2$

$x = \dfrac{-(-30) \pm \sqrt{(-30)^2 - 4(25)(2)}}{2(25)}$

$x = \dfrac{30 \pm \sqrt{700}}{50}$

$x = \dfrac{30 \pm 10\sqrt{7}}{50}$

$x = \dfrac{3 \pm \sqrt{7}}{5}$

The solutions are $\dfrac{3 \pm \sqrt{7}}{5}$.

20. $12x = 4x^2 + 4$

$0 = 4x^2 - 12x + 4$

$0 = x^2 - 3x + 1$

$a = 1,\ b = -3,\ c = 1$

$x = \dfrac{-(-3) \pm \sqrt{(-3)^2 - 4(1)(1)}}{2(1)}$

$x = \dfrac{3 \pm \sqrt{5}}{2}$

The solutions are $\dfrac{3 \pm \sqrt{5}}{2}$.

21. $\dfrac{2}{3}m^2 - \dfrac{1}{3}m - 1 = 0$

$2m^2 - m - 3 = 0$

$a = 2,\ b = -1,\ c = -3$

$m = \dfrac{-(-1) \pm \sqrt{(-1)^2 - 4(2)(-3)}}{2(2)}$

$m = \dfrac{1 \pm \sqrt{25}}{4}$

$m = \dfrac{1 \pm 5}{4}$

$m = -1$ or $m = \dfrac{3}{2}$

The solutions are -1 and $\dfrac{3}{2}$.

22. $\dfrac{5}{8}m^2 + m - \dfrac{1}{2} = 0$

$5m^2 + 8m - 4 = 0$

$a = 5,\ b = 8,\ c = -4$

$m = \dfrac{-8 \pm \sqrt{8^2 - 4(5)(-4)}}{2(5)}$

$m = \dfrac{-8 \pm \sqrt{144}}{10}$

$m = \dfrac{-8 \pm 12}{10}$

$m = -2$ or $m = \dfrac{2}{5}$

The solutions are -2 and $\dfrac{2}{5}$.

23. $x^2 - \dfrac{1}{2}x - \dfrac{1}{5} = 0$

$10x^2 - 5x - 2 = 0$

$a = 10,\ b = -5,\ c = -2$

$x = \dfrac{-(-5) \pm \sqrt{(-5)^2 - 4(10)(-2)}}{2(10)}$

$x = \dfrac{5 \pm \sqrt{105}}{20}$

The solutions are $\dfrac{5 \pm \sqrt{105}}{20}$.

24. $x^2 + \dfrac{1}{2}x - \dfrac{1}{8} = 0$

$8x^2 + 4x - 1 = 0$

$a = 8,\ b = 4,\ c = -1$

$x = \dfrac{-4 \pm \sqrt{4^2 - 4(8)(-1)}}{2(8)}$

$x = \dfrac{-4 \pm \sqrt{48}}{16}$

$x = \dfrac{-4 \pm 4\sqrt{3}}{16}$

$x = \dfrac{-1 \pm \sqrt{3}}{4}$

The solutions are $\dfrac{-1 \pm \sqrt{3}}{4}$.

25. $4x^2 - 27x + 35 = 0$

$a = 4,\ b = -27,\ c = 35$

$x = \dfrac{-(-27) \pm \sqrt{(-27)^2 - 4(4)(35)}}{2(4)}$

$x = \dfrac{27 \pm \sqrt{169}}{8}$

$x = \dfrac{27 \pm 13}{8}$

$x = \dfrac{7}{4}$ or $x = 5$

The solutions are $\dfrac{7}{4}$ and 5.

26. $9x^2 - 16x + 7 = 0$

$a = 9,\ b = -16,\ c = 7$

$x = \dfrac{-(-16) \pm \sqrt{(-16)^2 - 4(9)(7)}}{2(9)}$

$x = \dfrac{16 \pm \sqrt{4}}{18}$

$x = \dfrac{16 \pm 2}{18}$

$x = \dfrac{7}{9}$ or $x = 1$

The solutions are $\dfrac{7}{9}$ and 1.

27. $(7 - 5x)^2 = 18$

$7 - 5x = \pm\sqrt{18}$

$7 - 5x = \pm 3\sqrt{2}$

$x = \dfrac{-7 \pm 3\sqrt{2}}{-5}$

$x = \dfrac{7 \pm 3\sqrt{2}}{5}$

The solutions are $\dfrac{7 \pm 3\sqrt{2}}{5}$.

28. $(5 - 4x)^2 = 75$

$5 - 4x = \pm\sqrt{75}$

$5 - 4x = \pm 5\sqrt{3}$

$x = \dfrac{-5 \pm 5\sqrt{3}}{-4}$

$x = \dfrac{5 \pm 5\sqrt{3}}{4}$

The solutions are $\dfrac{5 \pm 5\sqrt{3}}{4}$.

29. $3z^2 - 7z = 12$
$3z^2 - 7z - 12 = 0$
$a = 3,\ b = -7,\ c = -12$
$$z = \frac{-(-7) \pm \sqrt{(-7)^2 - 4(3)(-12)}}{2(3)}$$
$$x = \frac{7 \pm \sqrt{193}}{6}$$
The solutions are $\dfrac{7 \pm \sqrt{193}}{6}$.

30. $6z^2 + 7z = 6$
$6z^2 + 7z - 6 = 0$
$a = 6,\ b = 7,\ c = -6$
$$z = \frac{-7 \pm \sqrt{7^2 - 4(6)(-6)}}{2(6)}$$
$$z = \frac{-7 \pm \sqrt{193}}{12}$$
The solutions are $\dfrac{-7 \pm \sqrt{193}}{12}$.

31. $x = x^2 - 110$
$110 = x^2 - x$
$110 + \dfrac{1}{4} = x^2 - x + \dfrac{1}{4}$
$$\frac{441}{4} = \left(x - \frac{1}{2}\right)^2$$
$$\pm\frac{21}{2} = x - \frac{1}{2}$$
$$\frac{1}{2} \pm \frac{21}{2} = x$$
$x = -10$ or $x = 11$
The solutions are -10 and 11.

32. $x = 56 - x^2$
$x^2 + x = 56$
$x^2 + x + \dfrac{1}{4} = 56 + \dfrac{1}{4}$
$$\left(x + \frac{1}{2}\right)^2 = \frac{225}{4}$$
$$x + \frac{1}{2} = \pm\frac{15}{2}$$
$$x = -\frac{1}{2} \pm \frac{15}{2}$$
$x = -8$ or $x = 7$
The solutions are -8 and 7.

33. $\dfrac{3}{4}x^2 - \dfrac{5}{2}x - 2 = 0$
$3x^2 - 10x - 8 = 0$
$a = 3,\ b = -10,\ c = -8$
$$x = \frac{-(-10) \pm \sqrt{(-10)^2 - 4(3)(-8)}}{2(3)}$$
$$x = \frac{10 \pm \sqrt{196}}{6}$$
$$x = \frac{10 \pm 14}{6}$$
$$x = -\frac{2}{3} \quad \text{or} \quad x = 4$$
The solutions are $-\dfrac{2}{3}$ and 4.

34. $x^2 - \dfrac{6}{5}x - \dfrac{8}{5} = 0$
$5x^2 - 6x - 8 = 0$
$a = 5,\ b = -6,\ c = -8$
$$x = \frac{-(-6) \pm \sqrt{(-6)^2 - 4(5)(-8)}}{2(5)}$$
$$x = \frac{6 \pm \sqrt{196}}{10}$$
$$x = \frac{6 \pm 14}{10}$$
$$x = -\frac{4}{5} \quad \text{or} \quad x = 2$$
The solutions are $-\dfrac{4}{5}$ and 2.

35. $x^2 - 0.6x + 0.05 = 0$

$a = 1,\ b = -0.6,\ c = 0.05$

$$x = \frac{-(-0.6) \pm \sqrt{(-0.6)^2 - 4(1)(0.05)}}{2(1)}$$

$$x = \frac{0.6 \pm \sqrt{0.16}}{2}$$

$$x = \frac{0.6 \pm 0.4}{2}$$

$x = 0.1$ or $x = 0.5$

The solutions are 0.1 and 0.5, or $\frac{1}{2}$ and $\frac{1}{10}$.

36. $x^2 - 0.1x - 0.06 = 0$

$a = 1,\ b = -0.1,\ c = -0.06$

$$x = \frac{-(-0.1) \pm \sqrt{(-0.1)^2 - 4(1)(-0.06)}}{2(1)}$$

$$x = \frac{0.1 \pm \sqrt{0.25}}{2}$$

$$x = \frac{0.1 \pm 0.5}{2}$$

$x = -0.2$ or $x = 0.3$

The solutions are -0.2 and 0.3.

37. $10x^2 - 11x + 2 = 0$

$a = 10,\ b = -11,\ c = 2$

$$x = \frac{-(-11) \pm \sqrt{(-11)^2 - 4(10)(2)}}{2(10)}$$

$$x = \frac{11 \pm \sqrt{41}}{20}$$

The solutions are $\frac{11 \pm \sqrt{41}}{20}$.

38. $20x^2 - 11x + 1 = 0$

$a = 20,\ b = -11,\ c = 1$

$$x = \frac{-(-11) \pm \sqrt{(-11)^2 - 4(20)(1)}}{2(20)}$$

$$x = \frac{11 \pm \sqrt{41}}{40}$$

The solutions are $\frac{11 \pm \sqrt{41}}{40}$.

39. $\frac{1}{2}z^2 - 2z + \frac{3}{4} = 0$

$$z^2 - 4z = -\frac{3}{2}$$

$$z^2 - 4z + 4 = -\frac{3}{2} + 4$$

$$(z - 2)^2 = \frac{5}{2}$$

$$z - 2 = \sqrt{\frac{5}{2}}$$

$$z = 2 \pm \frac{\sqrt{10}}{2}$$

$$z = \frac{4 \pm \sqrt{10}}{2}$$

The solutions are $\frac{4 \pm \sqrt{10}}{2}$.

40. $\frac{1}{5}z^2 - \frac{1}{2}z - 2 = 0$

$2z^2 - 5z - 20 = 0$

$a = 2,\ b = -5,\ c = -20$

$$z = \frac{-(-5) \pm \sqrt{(-5)^2 - 4(2)(-20)}}{2(2)}$$

$$z = \frac{5 \pm \sqrt{185}}{4}$$

The solutions are $\frac{5 \pm \sqrt{185}}{4}$.

41. Answers may vary.

Section 9.4

Practice Problems

1. $y = -3x^2$

$a = -3$

Since a is negative, the parabola opens downward. Connect selected points.

x	$y = -3x^2$
-2	-12
-1	-3
0	0
1	-3
2	-12

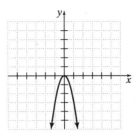

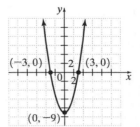

2. $y = x^2 - 9$

$a = 1$

Since a is positive, the parabola opens upward. Find y-intercept by letting $x = 0$.

$y = 0^2 - 9$

$y = -9$

The y-intercept is $(0, -9)$.

Find x-intercept by letting $y = 0$.

$0 = x^2 - 9$

$9 = x^2$

$\pm\sqrt{9} = x$

$\pm 3 = x$

The x-intercepts are $(-3, 0)$ and $(3, 0)$.

Connect selected points.

x	$y = x^2 - 9$
-4	7
-3	0
-2	-5
-1	-8
0	-9
1	-8
2	-5
3	0
4	7

3. $y = x^2 - 2x - 3$

Find the vertex.

$$x = \frac{-b}{2a} = \frac{-(-2)}{2(1)} = 1$$

$y = (1)^2 - 2(1) - 3 = -4$

The vertex is $(1, -4)$.

Find the y-intercept by setting $x = 0$.

$y = 0^2 - 2(0) - 3 = -3$

The y-intercept is $(0, -3)$.

Find the x-intercepts by setting $y = 0$.

$0 = x^2 - 2x - 3$

$0 = (x - 3)(x + 1)$

$x = 3$ or $x = -1$

the x-intercepts are $(-1, 0)$ and $(3, 0)$.

Connect selected points.

x	$y = x^2 - 2x - 3$
-2	5
-1	0
0	-3
1	-4
2	-3
3	0
4	5

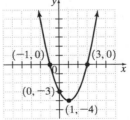

4. $y = x^2 - 3x + 1$

Find the vertex.

$$x = \frac{-b}{2a} = \frac{-(-3)}{2(1)} = \frac{3}{2}$$

$$y = \left(\frac{3}{2}\right)^2 - 3\left(\frac{3}{2}\right) + 1 = -\frac{5}{4}$$

The vertex is $\left(\frac{3}{2}, -\frac{5}{4}\right)$.

Find the y-intercept by setting $x = 0$.

$$y = 0^2 - 3(0) + 1 = 1$$

The y-intercept is $(0, 1)$.

Find the x-intercepts by setting $y = 0$.

$$0 = x^2 - 3x + 1$$

$$x = \frac{-b \pm \sqrt{b^2 - 4ac}}{2a}$$

$$x = \frac{-(-3) \pm \sqrt{(-3)^2 - 4(1)(1)}}{2(1)}$$

$$x = \frac{3 \pm \sqrt{5}}{2}$$

The x-intercepts are $\left(\frac{3 - \sqrt{5}}{2}, 0\right)$ and

$\left(\frac{3 + \sqrt{5}}{2}, 0\right)$.

Connect selected points.

x	$y = x^2 - 3x + 1$
-1	5
0	1
1	-1
2	-1
3	1
4	5

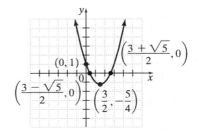

Exercise Set 9.4

1.

x	$y = 2x^2$
-2	8
-1	2
0	0
1	2
2	8

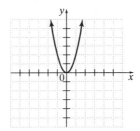

3.

y	$y = -x^2$
-3	-9
-2	-4
-1	-1
0	0
1	-1
2	-4
3	-9

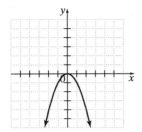

5.

x	$y = \frac{1}{3}x^2$
-5	$\dfrac{25}{3}$
-3	3
0	0
3	3
5	$\dfrac{25}{3}$

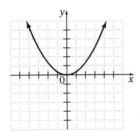

7. $y = x^2 - 1$
Find vertex.
$$x = -\frac{b}{2a} = -\frac{0}{2(1)} = 0$$
$$y = 0^2 - 1 = -1$$
Vertex = $(0, -1)$
y-intercept = $(0, -1)$
Find x-intercept.
$$0 = x^2 - 1$$
$$x = \pm 1$$
x-intercepts = $(-1, 0)$, $(1, 0)$

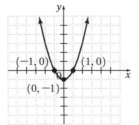

9. $y = x^2 + 4$
Find vertex.
$$x = -\frac{b}{2a} = -\frac{0}{2(1)} = 0$$
$$y = 0 + 4 = 4$$
vertex = $(0, 4)$
y-intercept = $(0, 4)$
Find x-intercepts.
$$0 = x^2 + 4$$
$$x^2 = -4$$
There are no x-intercepts because there is no solution to this equation.

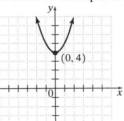

11. $y = x^2 + 6x$
Find vertex.
$$x = \frac{-b}{2a} = -\frac{6}{2(1)} = -3$$
$$y = (-3)^2 + 6(-3) = -9$$
vertex = $(-3, -9)$
Find x-intercepts.
$$0 = x^2 + 6x$$
$$0 = x(x + 6)$$
$$x = 0 \quad \text{or} \quad x = -6$$
x-intercepts = $(0, 0)$, $(-6, 0)$
y-intercepts = $(0, 0)$

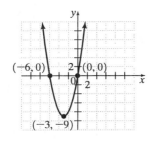

13. $y = x^2 + 2x - 8$

Find vertex.

$x = -\dfrac{b}{2a} = -\dfrac{2}{2(1)} = -1$

$y = (-1)^2 + 2(-1) - 8 = -9$

vertex $= (-1, -9)$

Find x-intercepts:

$0 = x^2 + 2x - 8$

$0 = (x + 4)(x - 2)$

$x = -4$ or $x = 2$

x-intercepts: $(-4, 0)$, $(2, 0)$

Find y-intercept.

$y = 0^2 + 2(0) - 8 = -8$

y-intercept $= (0, -8)$

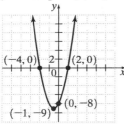

15. $y = -x^2 + x + 2$

Find vertex.

$x = -\dfrac{b}{2a} = -\dfrac{1}{2(-1)} = \dfrac{1}{2}$

$y = -\left(\dfrac{1}{2}\right)^2 + \dfrac{1}{2} + 2 = \dfrac{9}{4}$

vertex $= \left(\dfrac{1}{2}, \dfrac{9}{4}\right)$

Find x-intercepts,

$0 = -x^2 + x + 2$

$0 = x^2 - x - 2$

$0 = (x + 1)(x - 2)$

$x = -1$ or $x = 2$

x-intercepts $= (-1, 0)$, $(2, 0)$

Find y-intercept.

$y = 0 + 0 + 2 = 2$

y-intercept $= (0, 2)$

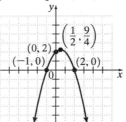

17. $y = x^2 + 5x + 4$

Find vertex.

$x = -\dfrac{b}{2a} = \dfrac{-5}{2(1)} = -\dfrac{5}{2}$

$y = \left(-\dfrac{5}{2}\right)^2 + 5\left(-\dfrac{5}{2}\right) + 4 = -\dfrac{9}{4}$

vertex $= \left(-\dfrac{5}{2}, -\dfrac{9}{4}\right)$

Find x-intercepts.

$0 = x^2 + 5x + 4$

$0 = (x + 4)(x + 1)$

$x = -4$ or $x = -1$

x-intercepts $= (-4, 0)$, $(-1, 0)$

Find y-intercept.

$y = 0 + 5(0) + 4 = 4$

y-intercept $= (0, 4)$

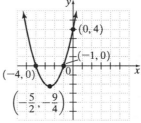

19. $y = -x^2 + 4x - 3$

Find vertex.

$x = -\dfrac{b}{2a} = -\dfrac{4}{2(-1)} = 2$

$y = -(2)^2 + 4(2) - 3 = 1$

vertex $= (2, 1)$

Find x-intercepts.
$$0 = -x^2 + 4x - 3$$
$$0 = x^2 - 4x - 3$$
$$0 = (x - 1)(x - 3)$$
$$x = 1 \quad \text{or} \quad x = 3$$
x-intercepts $= (1, 0), (3, 0)$
Find y-intercept.
$$y = 0 + 4(0) - 3 = -3$$
y-intercept $= (0, -3)$

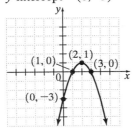

21. $\dfrac{\frac{1}{7}}{\frac{2}{5}} = \dfrac{1}{7} \cdot \dfrac{5}{2} = \dfrac{5}{14}$

23. $\dfrac{\frac{1}{x}}{\frac{2}{x^2}} = \dfrac{1}{x} \cdot \dfrac{x^2}{2} = \dfrac{x}{2}$

25. $\dfrac{2x}{1 - \frac{1}{x}} = \dfrac{2x}{\frac{x-1}{x}} = \dfrac{2x}{1} \cdot \dfrac{x}{x-1} = \dfrac{2x^2}{x-1}$

27. $\dfrac{\frac{a-b}{2b}}{\frac{b-a}{8b^2}} = \dfrac{a-b}{2b} \cdot \dfrac{8b^2}{b-a} = \dfrac{a-b}{2b} \cdot \dfrac{8b^2}{-1(a-b)}$
$$= -4b$$

29. $h = -16t^2 + 128t$

    a. Find the value of h at vertex.
$$t = \dfrac{-b}{2a} = -\dfrac{128}{2(-16)} = 4$$
$$h = -16(4)^2 + 128(4) = 256$$
The fireball will reach a height of 256 feet.

    b. Find the value of t at vertex.
$$t = 4$$
The fireball will reach a maximum height at 4 seconds.

    c. Let $h = 0$. Solve for t.
$$0 = -16t^2 + 128t$$
$$0 = -16t(t - 8)$$
$$t = 0 \quad \text{or} \quad t = 8$$
The fireball will return to the ground after 8 seconds.

31. The graph opens down, $a < 0$, and crosses the x-axis in one place. Graph E matches the description.

33. The graph opens up, $a > 0$, and it does not cross the x-axis. Graph C matches the description.

35. The graph opens down, $a < 0$, and it crosses the x-axis in two places. Graph B matches the description.

Chapter 9 Review

1. $(x - 4)(5x + 3) = 0$
$$x - 4 = 0 \quad \text{or} \quad 5x + 3 = 0$$
$$x = 4 \qquad \text{or} \quad x = -\dfrac{3}{5}$$
The solutions are $-\dfrac{3}{5}$ and 4.

2. $(x + 7)(3x + 4) = 0$
$$x + 7 = 0 \quad \text{or} \quad 3x + 4 = 0$$
$$x = -7 \qquad \text{or} \quad x = -\dfrac{4}{3}$$
The solutions are -7 and $-\dfrac{4}{3}$.

3. $3m^2 - 5m = 2$
$$3m^2 - 5m - 2 = 0$$
$$(3m + 1)(m - 2) = 0$$
$$3m + 1 = 0 \quad \text{or} \quad m - 2 = 0$$
$$m = -\dfrac{1}{3} \quad \text{or} \quad m = 2$$
The solutions are $-\dfrac{1}{3}$ and 2.

4. $7m^2 + 2m = 5$
$$7m^2 + 2m - 5 = 0$$
$$(7m - 5)(m + 1) = 0$$
$$7m - 5 = 0 \quad \text{or} \quad m + 1 = 0$$
$$m = \dfrac{5}{7} \qquad \text{or} \quad m = -1$$
The solutions are -1 and $\dfrac{5}{7}$.

5. $6x^3 - 54x = 0$

$6x\left(x^2 - 9\right) = 0$

$6x(x + 3)(x - 3) = 0$

$x = 0 \quad$ or $\quad x = -3 \quad$ or $\quad x = 3$

The solutions are –3, 0, and 3.

6. $2x^2 - 8 = 0$

$2\left(x^2 - 4\right) = 0$

$2(x + 2)(x - 2) = 0$

$x + 2 = 0 \quad$ or $\quad x - 2 = 0$

$x = -2 \qquad$ or $\quad x = 2$

The solutions are –2 and 2.

7. $x^2 = 36$

$x = \pm\sqrt{36}$

$x = \pm 6$

The solutions are ± 6.

8. $x^2 = 81$

$x = \pm\sqrt{81}$

$x = \pm 9$

The solutions are ± 9.

9. $k^2 = 50$

$k = \pm\sqrt{50}$

$k = \pm 5\sqrt{2}$

The solutions are $\pm 5\sqrt{2}$.

10. $k^2 = 45$

$k = \pm\sqrt{45}$

$k = \pm 3\sqrt{5}$

The solutions are $\pm 3\sqrt{5}$.

11. $(x - 11)^2 = 49$

$x - 11 = \pm\sqrt{49}$

$x - 11 = \pm 7$

$x = 11 \pm 7$

$x = 4 \quad$ or $\quad x = 18$

The solutions are 4 and 18.

12. $(x + 3)^2 = 100$

$x + 3 = \pm\sqrt{100}$

$x + 3 = \pm 10$

$x = -3 \pm 10$

$x = -13 \quad$ or $\quad x = 7$

The solutions are –13 and 7.

13. $(4p + 2)^2 = 100$

$4p + 2 = \pm\sqrt{100}$

$4p + 2 = \pm 10$

$p = \dfrac{-2 \pm 10}{4}$

$p = -3 \quad$ or $\quad p = 2$

The solutions are –3 and 2.

14. $(3p + 6)^2 = 81$

$3p + 6 = \pm\sqrt{81}$

$3p + 6 = \pm 9$

$p = \dfrac{-6 \pm 9}{3}$

$p = -5 \quad$ or $\quad p = 1$

The solutions are –5 and 1.

15. $A = P(1 + r)^2$

$1690 = 1000(1 + r)^2$

$1.69 = (1 + r)^2$

$\pm\sqrt{1.69} = 1 + r$

$-1 \pm 1.3 = r$ (Disregard negative interest.)

$0.3 = r$

The interest rate is 30%.

16. $A = P(1 + r)^2$

$1210 = 1000(1 + r)^2$

$1.21 = (1 + r)^2$

$\pm\sqrt{1.21} = 1 + r$

$-1 \pm 1.1 = r$ (Disregard negative interest.)

$0.1 = r$

The interest rate is 10%.

17. $x^2 + 4x = 1$

$x^2 + 4x + 4 = 1 + 4$

$(x + 2)^2 = 5$

$x + 2 = \pm\sqrt{5}$

$x = -2 \pm\sqrt{5}$

The solutions are $-2 \pm\sqrt{5}$

18. $x^2 - 8x = 3$
$x^2 - 8x + 16 = 3 + 16$
$(x - 4)^2 = 19$
$x - 4 = \pm\sqrt{19}$
$x = 4 \pm \sqrt{19}$
The solutions are $4 \pm \sqrt{19}$.

19. $x^2 - 6x + 7 = 0$
$x^2 - 6x = -7$
$x^2 - 6x + 9 = -7 + 9$
$(x - 3)^2 = 2$
$x - 3 = \pm\sqrt{2}$
$x = 3 \pm \sqrt{2}$
The solutions are $3 \pm \sqrt{2}$.

20. $x^2 + 6x + 7 = 0$
$x^2 + 6x = -7$
$x^2 + 6x + 9 = -7 + 9$
$(x + 3)^2 = 2$
$x + 3 = \pm\sqrt{2}$
$x = -3 \pm \sqrt{2}$
The solutions are $-3 \pm \sqrt{2}$.

21. $2y^2 + y - 1 = 0$
$y^2 + \frac{1}{2}y = \frac{1}{2}$
$y^2 + \frac{1}{2}y + \frac{1}{16} = \frac{1}{2} + \frac{1}{16}$
$\left(y + \frac{1}{4}\right)^2 = \frac{9}{16}$
$y + \frac{1}{4} = \pm\sqrt{\frac{9}{16}}$
$y = -\frac{1}{4} \pm \frac{3}{4}$
$y = -1$ or $y = \frac{1}{2}$
The solutions are -1 and $\frac{1}{2}$.

22. $y^2 + 3y - 1 = 0$
$y^2 + 3y = 1$
$y^2 + 3y + \frac{9}{4} = 1 + \frac{9}{4}$

$\left(y + \frac{3}{2}\right)^2 = \frac{13}{4}$
$y + \frac{3}{2} = \pm\sqrt{\frac{13}{4}}$
$y = -\frac{3}{2} \pm \frac{\sqrt{13}}{2}$
The solutions are $\dfrac{-3 \pm \sqrt{13}}{2}$.

23. $x^2 - 10x + 7 = 0$
$a = 1,\ b = -10,\ c = 7$
$x = \dfrac{-(-10) \pm \sqrt{(-10)^2 - 4(1)(7)}}{2(1)}$
$x = \dfrac{10 \pm \sqrt{72}}{2}$
$x = \dfrac{10 \pm 6\sqrt{2}}{2}$
$x = 5 \pm 3\sqrt{2}$
The solutions are $5 \pm 3\sqrt{2}$.

24. $x^2 + 4x - 7 = 0$
$a = 1,\ b = 4,\ c = -7$
$x = \dfrac{-4 \pm \sqrt{4^2 - 4(1)(-7)}}{2(1)}$
$x = \dfrac{-4 \pm \sqrt{44}}{2}$
$x = \dfrac{-4 \pm 2\sqrt{11}}{2}$
$x = -2 \pm \sqrt{11}$
The solutions are $-2 \pm \sqrt{11}$.

25. $2x^2 + x - 1 = 0$
$a = 2,\ b = 1,\ c = -1$
$x = \dfrac{-1 \pm \sqrt{1^2 - 4(2)(-1)}}{2(2)}$
$x = \dfrac{-1 \pm \sqrt{9}}{4}$
$x = \dfrac{-1 \pm 3}{4}$
$x = -1$ or $x = \frac{1}{2}$
The solutions are -1 and $\frac{1}{2}$.

26. $x^2 + 3x - 1 = 0$

$a = 1, b = 3, c = -1$

$$x = \frac{-3 \pm \sqrt{3^2 - 4(1)(-1)}}{2(1)}$$

$$x = \frac{-3 \pm \sqrt{13}}{2}$$

The solutions are $\frac{-3 \pm \sqrt{13}}{2}$.

27. $9x^2 + 30x + 25 = 0$

$a = 9, b = 30, c = 25$

$$x = \frac{-30 \pm \sqrt{30^2 - 4(9)(25)}}{2(9)}$$

$$x = \frac{-30 \pm \sqrt{0}}{18}$$

$$x = -\frac{5}{3}$$

The solution is $-\frac{5}{3}$.

28. $16x^2 - 72x + 81 = 0$

$a = 16, b = -72, c = 81$

$$x = \frac{-(-72) \pm \sqrt{(-72)^2 - 4(16)(81)}}{2(16)}$$

$$x = \frac{72 \pm \sqrt{0}}{32}$$

$$x = \frac{9}{4}$$

The solution is $\frac{9}{4}$.

29. $15x^2 + 2 = 11x$

$15x^2 - 11x + 2 = 0$

$a = 15, b = -11, c = 2$

$$x = \frac{-(-11) \pm \sqrt{(-11)^2 - 4(15)(2)}}{2(15)}$$

$$x = \frac{11 \pm \sqrt{1}}{30}$$

$$x = \frac{11 \pm 1}{30}$$

$$x = \frac{1}{3} \quad \text{or} \quad x = \frac{2}{5}$$

The solutions are $\frac{1}{3}$ and $\frac{2}{5}$.

30. $15x^2 + 2 = 13x$

$15x^2 - 13x + 2 = 0$

$a = 15, b = -13, c = 2$

$$x = \frac{-(-13) \pm \sqrt{(-13)^2 - 4(15)(2)}}{2(15)}$$

$$x = \frac{13 \pm \sqrt{49}}{30}$$

$$x = \frac{13 \pm 7}{30}$$

$$x = \frac{1}{5} \quad \text{or} \quad x = \frac{2}{3}$$

The solutions are $\frac{1}{5}$ and $\frac{2}{3}$.

31. $2x^2 + x + 5 = 0$

$a = 2, b = 1, c = 5$

$$x = \frac{-1 \pm \sqrt{1^2 - 4(2)(5)}}{2(2)}$$

$$x = \frac{-1 \pm \sqrt{-39}}{4}$$

There is no real solution.

32. $7x^2 - 3x + 1 = 0$

$a = 7, b = -3, c = 1$

$$x = \frac{-(-3) \pm \sqrt{(-3)^2 - 4(7)(1)}}{2(7)}$$

$$x = \frac{3 \pm \sqrt{-19}}{14}$$

There is no real solution.

33. $\dfrac{AB}{AC} = \dfrac{AC}{CB}$

$\dfrac{x}{1} = \dfrac{1}{x-1}$

$x(x - 1) = 1$

$x^2 - x = 1$

$x^2 - x + \frac{1}{4} = 1 + \frac{1}{4}$

$\left(x - \frac{1}{2}\right)^2 = \frac{5}{4}$

$x - \frac{1}{2} = \pm\sqrt{\frac{5}{4}}$

$x = \frac{1}{2} \pm \frac{\sqrt{5}}{2}$

Disregard a negative length.

$$AB = \frac{1 + \sqrt{5}}{2}$$

34. $y = 3x^2$

Find the x-intercept.

$0 = 3x^2$

$x^2 = 0$

$x = 0$

The x-intercept and y-intercept is $(0, 0)$.

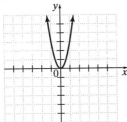

35. $y = -\frac{1}{2}x^2$

Find x-intercept.

$0 = -\frac{1}{2}x^2$

$x^2 = 0$

$x = 0$

The x-intercept and y-intercept is $(0, 0)$.

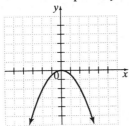

36. $y = x^2 - 25$

Find vertex.

$x = -\frac{b}{2a} = -\frac{0}{2(1)} = 0$

$y = 0^2 - 25 = -25$

vertex $= (0, -25)$

y-intercept $= (0, -25)$

Find x-intercepts.

$0 = x^2 - 25$

$x^2 = 25$

$x = \pm 5$

x-intercepts $= (-5, 0), (5, 0)$

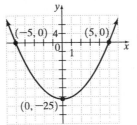

37. $y = x^2 - 36$

Find vertex.

$x = -\frac{b}{2a} = -\frac{0}{2(1)} = 0$

$y = 0 - 36 = -36$

vertex $= (0, -36)$

y-intercept $= (0, -36)$

Find x-intercepts.

$0 = x^2 - 36$

$x^2 = 36$

$x = \pm 6$

x-intercepts $= (-6, 0), (6, 0)$

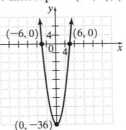

38. $y = x^2 + 3$

Find vertex.

$x = -\frac{b}{2a} = -\frac{0}{2(1)} = 0$

$y = 0 + 3 = 3$

vertex $= (0, 3)$

y-intercept $= (0, 3)$

Find x-intercepts.

$0 = x^2 + 3$

$x^2 = -3$

There are no x-intercepts because there is no solution to this equation.

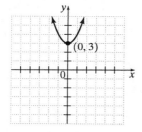

x-intercepts $= \left(-\sqrt{2}, 0\right), \left(\sqrt{2}, 0\right)$

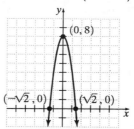

39. $y = x^2 + 8$

Find vertex.

$x = -\dfrac{b}{2a} = -\dfrac{0}{2(1)} = 0$

$y = 0 + 8 = 8$

vertex $= (0, 8)$

y-intercept $= (0, 8)$

Find x-intercepts.

$0 = x^2 + 8$

$x^2 = -8$

There is no x-intercepts because there is no solution to the equation.

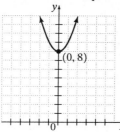

40. $y = -4x^2 + 8$

Find vertex.

$x = -\dfrac{b}{2a} = -\dfrac{0}{2(-4)} = 0$

$y = -4(0) + 8 = 8$

vertex $= (0, 8)$

y-intercept $= (0, 8)$

Find x-intercepts.

$0 = -4x^2 + 8$

$x^2 = 2$

$x = \pm\sqrt{2}$

41. $y = -3x^2 + 9$

Find vertex.

$x = \dfrac{-b}{2a} = -\dfrac{0}{2(-3)} = 0$

$y = -3(0) + 9 = 9$

vertex $= (0, 9)$

y-intercepts $= (0, 9)$

Find x-intercepts.

$0 = -3x^2 + 9$

$x^2 = 3$

$x = \pm\sqrt{3}$

x-intercepts $= \left(-\sqrt{3}, 0\right), \left(\sqrt{3}, 0\right)$

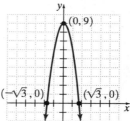

42. $y = x^2 + 3x - 10$

Find vertex.

$x = -\dfrac{b}{2a} = -\dfrac{3}{2(1)} = -\dfrac{3}{2}$

$y = \left(-\dfrac{3}{2}\right)^2 + 3\left(-\dfrac{3}{2}\right) - 10 = -\dfrac{49}{4}$

vertex $= \left(-\dfrac{3}{2}, -\dfrac{49}{4}\right)$

Find x-intercepts.

$0 = x^2 + 3x - 10$

$0 = (x + 5)(x - 2)$

$x = -5$ or $x = 2$

x-intercepts $= (-5, 0), (2, 0)$

Find y-intercept.

$$y = 0^2 + 3(0) - 10 = -10$$
$$y\text{-intercept} = (0, -10)$$

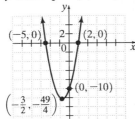

43. $y = x^2 + 3x - 4$
Find vertex.
$$x = -\frac{b}{2a} = -\frac{3}{2(1)} = -\frac{3}{2}$$
$$y = \left(-\frac{3}{2}\right)^2 + 3\left(-\frac{3}{2}\right) - 4 = -\frac{25}{4}$$
$$\text{vertex} = \left(-\frac{3}{2}, -\frac{25}{4}\right)$$
Find x-intercepts.
$$0 = x^2 + 3x - 4$$
$$0 = (x + 4)(x - 1)$$
$$x = -4 \quad \text{or} \quad x = 1$$
$$x\text{-intercepts} = (-4, 0), (1, 0)$$
Find y-intercept.
$$y = 0^2 + 3(0) - 4 = -4$$
$$y\text{-intercept} = (0, -4)$$

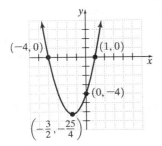

44. $y = -x^2 - 5x - 6$
Find vertex.
$$x = -\frac{b}{2a} = -\frac{(-5)}{2(-1)} = -\frac{5}{2}$$
$$y = -\left(-\frac{5}{2}\right)^2 - 5\left(-\frac{5}{2}\right) - 6 = \frac{1}{4}$$
$$\text{vertex} = \left(-\frac{5}{2}, \frac{1}{4}\right)$$
Find x-intercepts.
$$0 = -x^2 - 5x - 6$$
$$0 = x^2 + 5x + 6$$
$$0 = (x + 2)(x + 3)$$
$$x = -2 \quad \text{or} \quad x = -3$$
$$x\text{-intercepts} = (-3, 0), (-2, 0)$$
Find y-intercept.
$$y = 0 - 5(0) - 6 = -6$$
$$y\text{-intercept} = (0, -6)$$

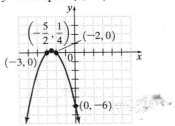

45. $y = -x^2 + 4x + 8$
Find vertex.
$$x = -\frac{b}{2a} = -\frac{4}{2(-1)} = 2$$
$$y = -(2)^2 + 4(2) + 8 = 12$$
$$\text{vertex} = (2, 12)$$
Find x-intercepts.
$$0 = -x^2 + 4x + 8$$
$$a = -1, \ b = 4, \ c = 8$$
$$x = \frac{-4 \pm \sqrt{4^2 - 4(-1)(8)}}{2(-1)}$$
$$x = \frac{-4 \pm \sqrt{48}}{-2}$$
$$= \frac{-4 \pm 4\sqrt{3}}{-2}$$
$$x = 2 \pm 2\sqrt{3}$$
$$x\text{-intercepts} = \left(2 - 2\sqrt{3}, \ 0\right), \left(2 + 2\sqrt{3}, \ 0\right)$$
Find y-intercept.

$$y = -0^2 + 4(0) + 8 = 8$$
y-intercept = (0, 8)

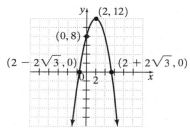

46. $y = 2x^2 - 11x - 6$
Find vertex.
$$x = -\frac{b}{2a} = -\frac{(-11)}{2(2)} = \frac{11}{4}$$
$$y = 2\left(\frac{11}{4}\right)^2 - 11\left(\frac{11}{4}\right) - 6 = -\frac{169}{8}$$
vertex $= \left(\frac{11}{4}, -\frac{169}{8}\right)$
Find x-intercepts.
$$0 = 2x^2 - 11x - 6$$
$$0 = (2x + 1)(x - 6)$$
$2x + 1 = 0$ or $x - 6 = 0$
$x = -\frac{1}{2}$ or $x = 6$

x-intercepts $= \left(-\frac{1}{2}, 0\right)$, (6, 0)
Find y-intercept.
$$y = 2(0)^2 - 11(0) - 6 = -6$$
y-intercept = (0, –6)

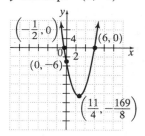

47. $y = 3x^2 - x - 2$
Find vertex.
$$x = -\frac{6}{2a} = -\frac{(-1)}{2(3)} = \frac{1}{6}$$
$$y = 3\left(\frac{1}{6}\right)^2 - \frac{1}{6} - 2 = -\frac{25}{12}$$
vertex $= \left(\frac{1}{6}, -\frac{25}{12}\right)$
Find x-intercepts.
$$0 = 3x^2 - x - 2$$
$$0 = (3x + 2)(x - 1)$$
$3x + 2 = 0$ or $x - 1 = 0$
$x = -\frac{2}{3}$ or $x = 1$

x-intercepts $= \left(-\frac{2}{3}, 0\right)$, (1, 0)
Find y-intercept.
$$y = 3(0)^2 - 0 - 2 = -2$$
y-intercept: (0, –2)

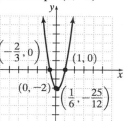

48. The equation has one solution because the graph crosses the x-axis in one place, $x = -2$.

49. The equation has two solutions because the graph crosses the x-axis in two places, $x = -\frac{3}{2}$, 3.

50. The equation has no real solutions because the graph does not cross the x-axis.

51. The equation has two solutions because the graph crosses the x-axis in two places, $x = -2, 2$.

52. $y = 2x^2$
Graph A is the graph from this equation. It opens up, $a > 0$, and the x- and y-intercepts are (0, 0).

53. $y = -x^2$

Graph D is the graph from this equation. It opens down, $a < 0$, and the x- and y-intercepts are $(0, 0)$.

54. $y = x^2 + 4x + 4$

Graph B is the graph from this equation. It's x-intercept is $(-2, 0)$.

$0 = (x + 2)^2$

$x + 2 = 0$

$x = -2$

55. $y = x^2 + 5x + 4$

Graph C is the graph from this equation. The vertex is $(-2.5, -2.25)$.

$x = -\dfrac{b}{2a} = -\dfrac{5}{2(1)} = -2.5$

$y = (-2.5)^2 + 5(-2.5) + 4 = -2.25$

Chapter 9 Test

1. $2x^2 - 11x = 21$

$2x^2 - 11x - 21 = 0$

$(2x + 3)(x - 7) = 0$

$2x + 3 = 0$ or $x - 7 = 0$

$x = -\dfrac{3}{2}$ or $x = 7$

The solutions are $-\dfrac{3}{2}$ and 7.

2. $x^4 + x^3 - 2x^2 = 0$

$x^2\left(x^2 + x - 2\right) = 0$

$x^2(x + 2)(x - 1) = 0$

$x = 0$ or $x = -2$ or $x = 1$

The solutions are -2, 0, and 1.

3. $5k^2 = 80$

$k^2 = 16$

$k = \pm\sqrt{16}$

$k = \pm 4$

The solutions are ± 4.

4. $(3m - 5)^2 = 8$

$3m - 5 = \pm\sqrt{8}$

$3m - 5 = \pm 2\sqrt{2}$

$3m = 5 \pm 2\sqrt{2}$

$m = \dfrac{5 \pm 2\sqrt{2}}{3}$

The solutions are $\dfrac{5 \pm 2\sqrt{2}}{3}$.

5. $x^2 - 26x + 160 = 0$

$x^2 - 26x = -160$

$x^2 - 26x + 169 = -160 + 169$

$(x - 13)^2 = 9$

$x - 13 = \pm\sqrt{9}$

$x = 13 \pm 3$

$x = 10$ or $x = 16$

The solutions are 10 and 16.

6. $5x^2 + 9x = 2$

$x^2 + \dfrac{9}{5}x = \dfrac{2}{5}$

$x^2 + \dfrac{9}{5}x + \dfrac{81}{100} = \dfrac{2}{5} + \dfrac{81}{100}$

$\left(x + \dfrac{9}{10}\right)^2 = \dfrac{121}{100}$

$x + \dfrac{9}{10} = \pm\sqrt{\dfrac{121}{100}}$

$x = -\dfrac{9}{10} \pm \dfrac{11}{10}$

$x = -2$ or $x = \dfrac{1}{5}$

The solutions are -2 and $\dfrac{1}{5}$.

7. $x^2 - 3x - 10 = 0$

$a = 1, b = -3, c = -10$

$x = \dfrac{-(-3) \pm \sqrt{(-3)^2 - 4(1)(-10)}}{2(1)}$

$x = \dfrac{3 \pm \sqrt{49}}{2}$

$x = \dfrac{3 \pm 7}{2}$

$x = -2$ or $x = 5$

The solutions are -2 and 5.

8. $p^2 - \frac{5}{3}p - \frac{1}{3} = 0$

$3p^2 - 5p - 1 = 0$

$a = 3, \ b = -5, \ c = -1$

$p = \dfrac{-(-5) \pm \sqrt{(-5)^2 - 4(3)(-1)}}{2(3)}$

$p = \dfrac{5 \pm \sqrt{37}}{6}$

The solutions are $\dfrac{5 \pm \sqrt{37}}{6}$.

9. $(3x - 5)(x + 2) = -6$

$3x^2 + 6x - 5x - 10 = -6$

$3x^2 + x - 4 = 0$

$(3x + 4)(x - 1) = 0$

$3x + 4 = 0 \quad \text{or} \quad x - 1 = 0$

$x = -\dfrac{4}{3} \quad \text{or} \quad x = 1$

The solutions are $-\dfrac{4}{3}$ and 1.

10. $(3x - 1)^2 = 16$

$3x - 1 = \pm\sqrt{16}$

$3x - 1 = \pm 4$

$3x = 1 \pm 4$

$x = \dfrac{1 \pm 4}{3}$

$x = -1 \quad \text{or} \quad x = \dfrac{5}{3}$

The solutions are -1 and $\dfrac{5}{3}$.

11. $3x^2 - 7x - 2 = 0$

$a = 3, \ b = -7, \ c = -2$

$x = \dfrac{-(-7) \pm \sqrt{(-7)^2 - 4(3)(-2)}}{2(3)}$

$x = \dfrac{7 \pm \sqrt{73}}{6}$

The solutions are $\dfrac{7 \pm \sqrt{73}}{6}$.

12. $x^2 - 4x - 5 = 0$

$(x - 5)(x + 1) = 0$

$x - 5 = 0 \quad \text{or} \quad x + 1 = 0$

$x = 5 \quad \text{or} \quad x = -1$

The solutions are -1 and 5.

13. $3x^2 - 7x + 2 = 0$

$a = 3, \ b = -7, \ c = 2$

$x = \dfrac{-(-7) \pm \sqrt{(-7)^2 - 4(3)(2)}}{2(3)}$

$x = \dfrac{7 \pm \sqrt{25}}{6}$

$x = \dfrac{7 \pm 5}{6}$

$x = \dfrac{1}{3} \quad \text{or} \quad x = 2$

The solutions are $\dfrac{1}{3}$ and 2.

14. $2x^2 - 6x + 1 = 0$

$a = 2, \ b = -6, \ c = 1$

$x = \dfrac{-(-6) \pm \sqrt{(-6)^2 - 4(2)(1)}}{2(2)}$

$x = \dfrac{6 \pm \sqrt{28}}{4}$

$x = \dfrac{6 \pm 2\sqrt{7}}{4}$

$x = \dfrac{3 \pm \sqrt{7}}{2}$

The solutions are $\dfrac{3 \pm \sqrt{7}}{2}$.

15. $A = \frac{1}{2}bh$

$18 = \frac{1}{2}x(4x)$

$18 = 2x^2$

$9 = x^2$

$\sqrt{9} = x^2$

$3 = x$

$12 = 4x$

The base is 3 feet and the height is 12 feet.

16. $y = -5x^2$

Find the vertex.

$x = -\dfrac{b}{2a} = -\dfrac{0}{2(-5)} = 0$

$y = -5(0)^2 = 0$

vertex $= (0, 0)$

x-intercept $= (0, 0)$

y-intercept $= (0, 0)$

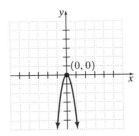

17. $y = x^2 - 4$
Find the vertex.
$$x = -\frac{b}{2a} = -\frac{0}{2(1)} = 0$$
$$y = 0^2 - 4 = -4$$
vertex = $(0, -4)$
y-intercept = $(0, -4)$
Find the x-intercepts.
$$0 = x^2 - 4$$
$$x^2 = 4$$
$$x = \pm\sqrt{4}$$
$$x = \pm 2$$
x-intercepts = $(-2, 0)$, $(2, 0)$

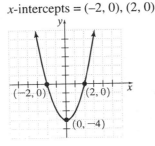

18. $y = x^2 - 7x + 10$
Find the vertex.
$$x = -\frac{b}{2a} = -\frac{(-7)}{2(1)} = \frac{7}{2}$$
$$y = \left(\frac{7}{2}\right)^2 - 7\left(\frac{7}{2}\right) + 10 = -\frac{9}{4}$$
vertex = $\left(\frac{7}{2}, -\frac{9}{4}\right)$
Find the x-intercepts.
$$0 = x^2 - 7x + 10$$
$$0 = (x - 5)(x - 2)$$
$$x = 5 \quad \text{or} \quad x = 2$$
x-intercepts = $(2, 0)$, $(5, 0)$
Find the y-intercept.

$$y = 0^2 - 7(0) + 10 = 10$$
y-intercept = $(0, 10)$

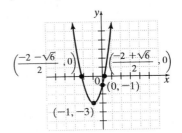

19. $y = 2x^2 + 4x - 1$
Find the vertex.
$$x = -\frac{b}{2a} = -\frac{4}{2(2)} = -1$$
$$y = 2(-1)^2 + 4(-1) - 1 = -3$$
vertex = $(-1, -3)$
Find the x-intercepts.
$$0 = 2x^2 + 4x - 1$$
$$a = 2, \ b = 4, \ c = -1$$
$$x = \frac{-4 \pm \sqrt{4^2 - 4(2)(-1)}}{2(2)}$$
$$x = \frac{-4 \pm \sqrt{24}}{4}$$
$$x = \frac{-4 \pm 2\sqrt{6}}{4}$$
$$x = \frac{-2 \pm \sqrt{6}}{2}$$
x-intercepts $= \left(\frac{-2 - \sqrt{6}}{2}, 0\right), \left(\frac{-2 + \sqrt{6}}{2}, 0\right)$

Find the y-intercept.
$$y = 2(0)^2 + 4(0) - 1 = -1$$
y-intercept = $(0, -1)$

20. $d = \dfrac{n^2 - 3n}{2}$

$9 = \dfrac{n^2 - 3n}{2}$

$18 = n^2 - 3n$

$0 = n^2 - 3n - 18$

$0 = (n - 6)(n + 3)$

$n - 6 = 0$ or $n + 3 = 0$

$n = 6$ or $n = -3$

Disregard a negative number of sides. The polygon has 6 sides.

Chapter 9 Cumulative Review

1. $y + 0.6 = -1.0$

$y + 0.6 - 0.6 = -1.0 - 0.6$

$y = -1.6$

2. $8(2 - t) = -5t$

$16 - 8t = -5t$

$16 - 8t + 8t = -5t + 8t$

$16 = 3t$

$\dfrac{16}{3} = \dfrac{3t}{3}$

$\dfrac{16}{3} = t$

3. Let x = the number of minutes. Then $0.36x$ is the charge per minute of phone use.

$50 + 0.36x = 99.68$

$50 + 0.36x - 50 = 99.68 - 50$

$0.36x = 49.68$

$\dfrac{0.36x}{0.36} = \dfrac{49.68}{0.36}$

$x = 138$

Elaine spent 138 minutes on her cellular phone this month.

4. $3^0 = 1$

5. $(5x^3 y^2)^0 = 1$

6. $-4^0 = -1$

7. $(3y + 2)^2 = (3y)^2 + 2(3y)(2) + (2)^2$

$= 9y^2 + 12y + 4$

8. $x + 3 \overline{\smash{)}\begin{array}{l} x + 4 \\ \hline x^2 + 7x + 12 \end{array}}$

$\underline{x^2 - 3x}$

$4x + 12$

$\underline{4x + 12}$

0

Thus, $\dfrac{x^2 + 7x + 12}{x + 3} = x + 4.$

9. $r^2 - r - 42 = (r + 6)(r - 7)$

10. $10x^2 - 13xy - 3y^2 = (2x - 3y)(5x + y)$

11. $8x^2 - 14x + 5 = 8x^2 - 10x - 4x + 5$

$= 2x(4x - 5) - 1(4x - 5) = (4x - 5)(2x - 1)$

12. a.

$4x^3 - 49x = x(4x^2 - 49) = x((2x)^2 - 7^2)$

$= x(2x + 7)(2x - 7)$

b.

$162x^4 - 2 = 2(81x^4 - 1) = 2((9x^2)^2 - 1^2)$

$= 2(9x^2 + 1)(9x^2 - 1)$

$= 2(9x^2 + 1)((3x)^2 - 1^2)$

$= 2(9x^2 + 1)(3x + 1)(3x - 1)$

13. $(5x - 1)(2x^2 + 15x + 18) = 0$

$(5x - 1)(2x + 3)(x + 6) = 0$

$5x - 1 = 0$ or $2x + 3 = 0$ or $x + 6 = 0$

$5x = 1$ $2x = -3$ $x = -6$

$x = \dfrac{1}{5}$ $x = -\dfrac{3}{2}$

The solutions are $\dfrac{1}{5}$, $-\dfrac{3}{2}$, and -6.

14. $\dfrac{x^2 + 8x + 7}{x^2 - 4x - 5} = \dfrac{(x + 7)(x + 1)}{(x - 5)(x + 1)} = \dfrac{x + 7}{x - 5}$

15. Let x = the unknown number.
$$\frac{x}{6} - \frac{5}{3} = \frac{x}{2}$$
$$6\left(\frac{x}{6} - \frac{5}{3}\right) = 6\left(\frac{x}{2}\right)$$
$$x - 10 = 3x$$
$$-10 = 2x$$
$$-5 = x$$
The number is –5.

16.

	x	y
a.	–1	–3
b.	0	0
c.	–3	–9

17. a.   The slope of the line $y = -\frac{1}{5}x + 1$ is

$-\frac{1}{5}$. Find the slope of the second line
by solving it for y.
$$2x + 10y = 3$$
$$10y = -2x + 3$$
$$y = -\frac{1}{5}x + \frac{3}{10}$$

The slope of this line is also $-\frac{1}{5}$.

Since the lines have the same slope,
they are parallel.

b.   To find each slope, solve each equation
for y.
$$x + y = 3$$
$$y = -x + 3$$
The slope is –1.
$$-x + y = 4$$
$$y = x + 4$$
The slope is 1.
Since the product of the slopes is –1, the
lines are perpendicular.

c.   Solve each equation for y to find each
slope.
$$3x + y = 5$$
$$y = -3x + 5$$
The slope of this line is –3.

$$2x + 3y = 6$$
$$3y = -2x + 6$$
$$y = -\frac{2}{3}x + 2$$

The slope of this line is $-\frac{2}{3}$.

Since the slopes are not the same, and
since their product is not –1, they are
neither parallel nor perpendicular.

18. a.   Since each x-value is assigned to only
one y-value, the relation is a function.

b.   Since the x-value 0 is paired with two
y-values, –2 and 3, this relation is not a
function.

19. $\begin{cases} 2x + y = 10 \\ x = y + 2 \end{cases}$

Substitute $y + 2$ for x in the first equation.
$$2x + y = 10$$
$$2(y + 2) + y = 10$$
$$2y + 4 + y = 10$$
$$3y + 4 = 10$$
$$3y = 6$$
$$y = 2$$
To find the x-value, let $y = 2$ in the second
equation.
$$x = y + 2$$
$$x = 2 + 2$$
$$x = 4$$
The solution is (4, 2).

20. $\begin{cases} 2x - y = 7 \\ 8x - 4y = 1 \end{cases}$

Multiply both sides of the first equation by
–4, then add the equations.
$$-8x + 4y = -28$$
$$\underline{8x - 4y = 1}$$
$$0 = -27$$
Since $0 = -27$ is a false statement, the
system has no solution.

21. $\sqrt{36} = 6$

22. $\sqrt{\dfrac{9}{100}} = \dfrac{3}{10}$

23. $\dfrac{2}{1+\sqrt{3}} = \dfrac{2\left(1-\sqrt{3}\right)}{\left(1+\sqrt{3}\right)\left(1-\sqrt{3}\right)} = \dfrac{2\left(1-\sqrt{3}\right)}{1^2 - \left(\sqrt{3}\right)^2}$

$= \dfrac{2\left(1-\sqrt{3}\right)}{1-3} = \dfrac{2\left(1-\sqrt{3}\right)}{-2} = \dfrac{2\left(1-\sqrt{3}\right)}{-2}$

$= -1\left(1-\sqrt{3}\right) = -1 + \sqrt{3}$

24. $(x-3)^2 = 16$

$\quad x - 3 = \sqrt{16} \qquad \text{or} \qquad x - 3 = -\sqrt{16}$

$\quad x - 3 = 4 \qquad\qquad\qquad x - 3 = -4$

$\quad x = 7 \qquad\qquad\qquad\quad x = -1$

The solutions are 7 and -1.

25. Write the equation in standard form, then clear the equation of fractions by multiplying both sides by the LCD 2.

$\dfrac{1}{2}x^2 - x = 2$

$\dfrac{1}{2}x^2 - x - 2 = 0$

$2\left(\dfrac{1}{2}x^2 - x - 2\right) = 2(0)$

$x^2 - 2x - 4 = 0$

Here $a = 1$, $b = -2$, and $c = -4$.

$x = \dfrac{-b \pm \sqrt{b^2 - 4ac}}{2a}$

$x = \dfrac{-(-2) \pm \sqrt{(-2)^2 - 4(1)(-4)}}{2(1)}$

$= \dfrac{2 \pm \sqrt{4+16}}{2} = \dfrac{2 \pm \sqrt{20}}{2} = \dfrac{2 \pm 2\sqrt{5}}{2}$

$= \dfrac{2\left(1 \pm \sqrt{5}\right)}{2} = 1 \pm \sqrt{5}$

$x = 1 + \sqrt{5}$ or $x = 1 - \sqrt{5}$

The solutions are $1 + \sqrt{5}$ and $1 - \sqrt{5}$.